ELECTROMAGNETIC FIELDS AND WAVES

ELECTROMAGNETIC FIELDS AND WAVES

SECOND EDITION

V V SARWATE

Former Professor
Government Engineering College, Jabalpur
and
Director of Technical Education
Madhya Pradesh

NEW AGE INTERNATIONAL (P) LIMITED, PUBLISHERS
LONDON • NEW DELHI • NAIROBI
Bangalore • Chennai • Cochin • Guwahati • Hyderabad • Kolkata • Lucknow • Mumbai
Visit us at **www.newagepublishers.com**

Published by New Age International (P) Ltd., Publishers
First Edition: 1993
Second Edition: 2018

GLOBAL OFFICES

- **New Delhi** **NEW AGE INTERNATIONAL (P) LIMITED, PUBLISHERS**
7/30 A, Daryaganj, New Delhi-110002, (INDIA)
Tel.: (011) 23253771, 23253472, **Telefax:** 23267437, 43551305
E-mail: contactus@newagepublishers.com • Visit us at www.newagepublishers.com
- **London** **NEW AGE INTERNATIONAL (UK) LTD.**
27 Old Gloucester Street, London, WC1N 3AX, UK
E-mail: info@newacademicscience.co.uk • Visit us at www.newacademicscience.co.uk
- **Nairobi** **NEW AGE GOLDEN (EAST AFRICA) LTD.**
Ground Floor, Westlands Arcade, Chiromo Road (Next to Naivas Supermarket)
Westlands, Nairobi, KENYA, **Tel.:** 00-254-713848772, 00-254-725700286
E-mail: kenya@newagepublishers.com

BRANCHES

- **Bangalore** 37/10, 8th Cross (Near Hanuman Temple), Azad Nagar, Chamarajpet, Bangalore- 560 018
Tel.: (080) 26756823, **Telefax:** 26756820, **E-mail: bangalore@newagepublishers.com**
- **Chennai** 26, Damodaran Street, T. Nagar, Chennai-600 017, **Tel.:** (044) 24353401
Telefax: 24351463, **E-mail: chennai@newagepublishers.com**
- **Cochin** CC-39/1016, Carrier Station Road, Ernakulam South, Cochin-682 016
Tel.: (0484) 2377303, **Telefax:** 4051304, **E-mail: cochin@newagepublishers.com**
- **Guwahati** Hemsen Complex, Mohd. Shah Road, Paltan Bazar, Near Starline Hotel
Guwahati-781 008, **Tel.:** (0361) 2513881, **Telefax:** 2543669
E-mail: guwahati@newagepublishers.com
- **Hyderabad** 105, 1st Floor, Madhiray Kaveri Tower, 3-2-19, Azam Jahi Road, Near Kumar Theater
Nimboliadda Kachiguda, Hyderabad-500 027, **Tel.:** (040) 24652456, **Telefax:** 24652457
E-mail: hyderabad@newagepublishers.com
- **Kolkata** RDB Chambers (Formerly Lotus Cinema) 106A, 1st Floor, S N Banerjee Road
Kolkata-700 014, **Tel.:** (033) 22273773, **Telefax:** 22275247
E-mail: kolkata@newagepublishers.com
- **Lucknow** 16-A, Jopling Road, Lucknow-226 001, **Tel.:** (0522) 2209578, 4045297, **Telefax:** 2204098
E-mail: lucknow@newagepublishers.com
- **Mumbai** 142C, Victor House, Ground Floor, N.M. Joshi Marg, Lower Parel, Mumbai-400 013
Tel.: (022) 24927869, **Telefax:** 24915415, **E-mail: mumbai@newagepublishers.com**
- **New Delhi** 22, Golden House, Daryaganj, New Delhi-110 002, **Tel.:** (011) 23262368, 23262370
Telefax: 43551305, **E-mail: sales@newagepublishers.com**

ISBN: 978-81-224-3780-5

C-17-06-10535

Printed in India at Ajit Printing Press, Delhi.
Typeset at Excellent Graphics, Delhi.

NEW AGE INTERNATIONAL (P) LIMITED, PUBLISHERS
7/30 A, Daryaganj, New Delhi-110002
Visit us at **www.newagepublishers.com**
(CIN: U74899DL1966PTC004618)

Dedicated

To

My Students

Preface to the Second Edition

After the good response of the earlier edition it is a pleasure for us to present the second edition. We would like to acknowledge the valuable advice and suggestions of many instructors and students who used the earlier edition. We would be grateful to our esteem readers for suggestions in the interest of the subject.

Except for the correction of a few printers's errors which, in spite of the best care and attention, had crept in the final print, the present edition is a reprint of the first edition, necessitated by the fact that we had run short of copies. Besides it also given effect to the suggestions made by many teachers and students.

Publisher

Preface to the First Edition

This book has been specially written keeping in mind the difficulties faced by the majority of students these days. Their problem is best illustrated by the Ekalavya episode in the Indian epic "Mahabharat". Dronacharya, the renowned teacher and expert in the use of various weapons of warfare, declined to accept Ekalavya as his student. Though feeling discouraged, Ekalavya decided to use a stratagem. He prepared a statue of Dronacharya, and in the presence of the statue, he embarked on a course of self-study, deriving his inspiration from the statue. In this, he succeeded so well that one day, Dronacharya, was amazed to discover the Ekalavya had acquired so much skill in archery that he was able to silence a barking dog by shooting a number of arrows into his mouth without hurting him. The moral of the story is that no wisdom can be acquired without the assistance of a teacher, even if he is in the form of a statue.

The almost ten-fold increase in this decade in the number of students taking a course in electromagnetic fields has created a very serious educational problem. A distressingly large percentage of these students is unable to keep pace with the teacher in the class; these students are forced to depend on self-study to prepare for the examination. There is another group which also faces the same need for self-study. These are the persons who are compelled by economic considerations to abandon their studies to seek employment in the industrial sector. But these persons have an urge to better their prospects by further study. This "Earn-and-Learn" group does self-study and appears in the examinations conducted by various universities and engineering institutions.

Though there is no dearth of excellent books on the subject, all of them have been written specifically to be used as textbooks by a teacher giving this course. When learning any new subject, many doubts arise in the mind of the pupil because his thinking process does not exactly coincide with that of the teacher. Normally these doubts are resolved by discussion with the teacher in a tutorial session, but this facility is denied to the person engaged in self-study. For him the book itself must raise these doubts and answer them, or it must present the subject, step by step, in such a manner that the doubts may not arise. Both these methods have been employed in this book.

As the sub-title shows, this book deliberately presents the subject in a manner suitable for self-study. It does not stop at merely giving information about the subject or of the mathematical tools available for dealing with it. On every page an attempt is made to guide the student to learn how to think about the subject and why certain tools are used for attacking different situations. Just as a person climbing a mountain stops to look back to get an overview of what has been attained and to plan the path to be taken for climbing further up, similarly in this book, at the end of every chapter, there is a summary of what has been learnt, and a plan for further study. This will help to build up the student's self-confidence and also sustain his interest in further study. The language style used in the book has also been deliberately chosen to help build up this self-confidence, by making the student feel that at every stage, the teacher is at his elbow to help him out, if he tends to go off at a tangent. The practical applications of the principles learnt will help the student to feel that the knowledge is not merely of theoretical interest.

Many guide-books popular amongst students give what are called "solved examples". Most often these involve mere substitution of numerical values in a formula derived in the text and sometimes they also require special mathematical manipulative skill. Since the object of this book is to mould the thinking process of the student, in everyone of the over one hundred examples solved here, the method of attack is discussed at length. This is to ensure that it results in "permanent" education of the student. The problems at the end of each chapter are intended to do exactly what their title indicates: exercise the mind to stimulate correct and logical thinking. That is why cumbersome calculations have been avoided in them. Mathematics is the bugbear of most students. It is presented here in a manner that has been found by three decades of teaching experience, to eliminate this fear. The process of converting a statement in the form of sentences into mathematical expressions is clearly brought out, and the objective to be attained by various mathematical procedures are clearly enunciated to help in building self-confidence. The author wishes to express his grateful thanks to Prof. S.V. Palshikar of Bharati Vidyapeeth's College of Engineering, Pune for his suggestions.

V.V. Sarwate

CONTENTS

13. The Poynting Vector 285–297

14. Guided Waves and Wave-Guides 298–326

15. Antenna and Antenna Arrrays 327–362

CHAPTER 1

NEEDED MATHEMATICS

1.1 BINOMIAL THEOREM

It was while studying for his B.A. degree (1661–64) at Trinity College, Cambridge that Isaac Newton (1642–1727) discovered the binomial theorem which may be expressed as follows:

If $|x| < 1$, then

$$(1+x)^n = 1 + nx + \frac{n(n-1)}{1.2}x^2 + \frac{n(n-1)(n-2)}{1.2.3}x^3 + \cdots \qquad ...(1.1)$$

where n is a positive or negative integer or fraction.

If $|x| \ll 1$, then x^2 and higher terms may be regarded as begin negligibly small, and hence one can write

$$(1+x)^{-1} = \frac{1}{(1+x)} \approx 1 - x \qquad ...(1.2)$$

$$(1-x)^{-1} = \frac{1}{(1-x)} \approx 1 + x \qquad ...(1.3)$$

1.2 EXPONENTIAL SERIES

By writing the binomial expansions of $\left(1+\frac{1}{n}\right)^{nx}$ and $\left(1+\frac{1}{n}\right)^{n}$ and then letting n increase indefinitely to infinity, we get by noting that

$$\left(1+\frac{1}{n}\right)^{nx} = \left\{\left(1+\frac{1}{n}\right)^{n}\right\}^{x},$$

$$\lim_{n\to\infty}\left[\left(1+\frac{1}{n}\right)^{n}\right] = e = 1 + \frac{1}{1!} + \frac{1}{2!} + \frac{1}{3!} + \cdots$$

$$= 2.718\,281\,828\,4\,\ldots \qquad ...(1.4)$$

and

$$e^x = 1 + x + \frac{x^2}{2!} + \frac{x^3}{3!} + \frac{x^4}{4!} + \cdots \qquad ...(1.5)$$

1.2.1 Exponential Decay

A function of the form

$$y = Ae^{-t/T} \quad ...(1.6)$$

is shown in Fig. 1.1. At time $t = 0$, its magnitude is equal to A, and the rate at which it is changing at that moment is

$$\left[\frac{dy}{dt}\right]_{t=0} = \left[-\frac{A}{T}e^{-t/T}\right]_{t=0} = -\frac{A}{T} \quad ...(1.7)$$

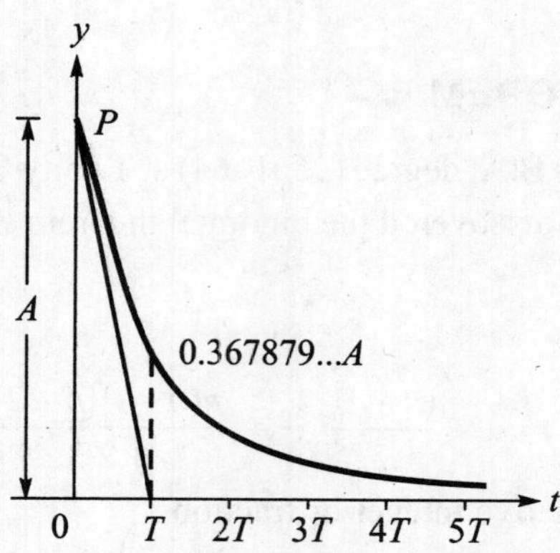

Fig. 1.1. Exponential decay

The tangent to the curve at that point $P(O, A)$ is the straight line PT whose equation is $y = -\frac{A}{T}t + A$. It intersects the time axis at $t = T$, which indicates that if the initial rate of decay had remained constant, the function would decay to zero value in time T, which is called its decay time constant.

However, since the value of $\frac{dy}{dt}$ in the brackets in Eqn. (1.7) involves t, it changes with time and the rate of decay decreases with time. Hence in reality, in time T, the function is only reduced to 0.367 879 ... of its value at $t = 0$, as shown in Table 1.1.

Table 1.1. Exponential Decay Function

t	Values of $\frac{y}{A} = e^{-t/T}$
0	$e^{0} = 1.000 \quad = 100.00\ \%$
T	$e^{-1} = 0.367\ 879... \quad \approx 36.79\ \%$
$2T$	$e^{-2} = 0.135\ 335... \quad \approx 13.53\ \%$
$3T$	$e^{-3} = 0.049\ 787... \quad \approx 4.98\ \%$
$4T$	$e^{-4} = 0.018\ 316... \quad \approx 1.83\ \%$
$5T$	$e^{-5} = 0.006\ 733... \quad \approx 0.67\ \%$

The figures to be remembered are that the value decays to 36.79% in one time-constant and becomes negligible in five time-constants.

1.3 LOGARITHMS

If $e^x = N$, then x is called the logarithm of N to the base e. Logarithms were invented by John Napier (1550–1617) and hence x is also called the Napierian (or Naperian) or natural logarithm of N and is written as

$$x = \log_e N \quad \text{or} \quad \ln N$$

Logarithms to base other than e were also soon developed. In particular Henry Briggs (1561–1631) used base 10. If $10^y = N$, then $y = \log_{10} N$. Here y, which is the logarithm of N to base 10, is called the common logarithm. Briggs computed the first table of common logarithms.

Since $a^0 = 1$ for all values of a, therefore $\log 1 = 0$ whatever the base may be. Other useful relations are:

$$\log_a N = (\log_b N)\,(\log_a b) \qquad \text{...(1.8)}$$

$$(\log_b a)\,(\log_a b) = 1 \qquad \text{...(1.9)}$$

$$\log_{10} e = 0.434\,294\,..., \ \log_e 10 = 2.302\,585\,... \qquad \text{...(1.10)}$$

1.4 EXPONENTIAL THEOREM AND LOGARITHMIC SERIES

Using Eqn. (1.5), we get

$$(e^c)^x = e^{cx} = 1 + cx + \frac{c^2x^2}{2!} + \frac{c^3x^3}{3!} + \cdots$$

In this equation, let $e^c = a$ so that $c = \log_e a$. Substituting this value of c, we get the Exponential Theorem as

$$a^x = 1 + x \log_e a + \frac{x^2(\log_e a)^2}{2!} + \frac{x^3(\log_e a)^3}{3!} + \cdots \qquad \text{...(1.11)}$$

Using this theorem we can write

$$(1+x)^y = 1 + y\log_e(1+x) + \frac{y^2\log_e(1+x)^2}{2!} - \frac{y^3\log_e(1+x)^3}{3!} + \cdots \qquad \text{...(1.12)}$$

But if $|x| < 1$, we can also use Binomial theorem and get

$$(1+x)^y = 1 + yx + \frac{y(y-1)}{2!}x^2 + \frac{y(y-1)(y-2)}{3!}x^3 + \cdots \qquad \text{...(1.13)}$$

Equating the coefficients of y in Eqns. (1.12) and (1.13), we have the Logarithmic Series

$$\log_e(1+x) = x + \frac{(-1)}{1.2}x^2 + \frac{(-1)(-2)}{1.2.3}x^3 + \cdots$$

or $$\log_e (1 + x) = x - \frac{x^2}{2} + \frac{x^3}{3} - \frac{x^4}{4} + \cdots, \text{ where } |x| < 1 \qquad ...(1.14)$$

1.5 NUMBER SYSTEM USING ORDERED PAIRS

The system of natural numbers which mankind began using thousands of years ago was subsequently extended to include other types of numbers. Thus we have

N = The set of natural numbers $N = \{1, 2, 3, 4, ...\}$

W = The set of whole numbers $W = \{0, 1, 2, 3, 4, ...\}$

I = The set of integers $I = \{..., -3, -2, -1, 0, 1, 2, 3, ...\}$

Q = The set of rational numbers which can be expressed as p/q, where both p and q belong to the set of integers and q is not equal to zero.

$$Q = \left\{\frac{p}{q} : p, q \in I \text{ and } q \neq 0\right\}.$$

The numbers which cannot be expressed as p/q as above are called irrational numbers or surds. When an attempt is made to express an irrational number in the decimal form, the decimal never terminates. Examples of irrational numbers are $\sqrt{2}$, $\sqrt{5}$, $3^{1/5}$, e, π, etc. The set of irrational numbers is denoted by S.

Greek geometricians began denoting numbers by lengths measured along a straight line from an origin O, the lengths measured to the right being positive and those measured towards the left of O being negative. They found that even irrational numbers can be expressed as an *exact* length on this number line. For instance, in a right angled triangle, where the sides forming the right angle are of lengths 1 and 2 respectively, the length of the hypotenuse is exactly $\sqrt{5}$.

The set consisting of the union of the rational numbers and the irrational numbers is called the set of real numbers and is denoted by R. Thus

$$R = Q \cup S = \{x : x \in Q, x \in S\}$$

The horizontal line extending to infinity in both directions from the origin O is such that the distance from O to any point on the line represents a real number. The line is called the axis of real numbers. All possible real numbers are included as points on this line.

The final extension of the system of numbers was made by the French philosopher René Descartes (1596–1650) when he provided answers to the following two questions which he asked himself:

(*i*) If every point on the real number line represents a real number uniquely, can we have a number system in which every point on a plane will represent a number in a unique manner?

(*ii*) Can we have a system in which the geometry developed by Euclid (circa 300 B.C.) and the Greeks could be combined with the algebra developed over the centuries by the Indian mathematicians like Aryabhatta (b. 376 A.D.), Brahmagupta (b. 598 A.D.) and Bhaskaracharya II (b. 1114 A.D.)?

Descartes developed this new extension of the number system by taking two *real* numbers a and b and arranging them to form an ordered pair written as (a, b). The number a is to be measured in the usual manner from the origin O along a real number axis, while the real number b is to be measured along another axis passing through the origin O but at right angles to the first axis. The number (a, b) is called an "ordered pair" because the order determines which number is to be measured along a real number axis and which at right angles to this axis. Thus $(a, b) \neq (b\ .\ a)$.

The algebraic and geometric forms of writing this ordered pair is discussed in the next section. The ordered pair of numbers like (x, y) is called the co-ordinates of a point on the $x - y$ plane in the Cartesian (= relating to Descartes) system of co-ordinates, but in the algebraic form, it is called a complex number and the set of complex numbers is denoted by C.

Therefore, we see that each extension of the number system includes all the previous systems as subsets. Thus we have

$$N \subset W \subset I \subset Q \subset R \subset C$$

The rules for the addition and multiplication of ordered pairs of numbers are as follows:

$$(a, b) + (c, d) = (a + c, b + d) \qquad ...(1.15)$$

$$(a, b) + (c, d) = (ac - bd, ad + bc) \qquad ...(1.16)$$

1.6 COMPLEX NUMBERS

Calculations involving ordered pairs become very cumbersome to take advantage of the simplicity of algebraic calculations, the ordered pair (a, b) is written in the form $a + jb$ and ordinary algebraic rules are applied to it. We may think of j as a red flag, warning us that the number associated with it is the second number of the ordered pair. However, it is better to think of j as an operator, requiring us to perform an operation on b while at the same time informing us that b is the second term of the ordered pair. Thus a and b are real numbers measured along different axes, but what significance is to be attached to the operator j is yet to be decided in a way that is consistent with rules of algebra.

The addition and multiplication of two complex numbers $a + jb$ and $c + jd$ using rules of algebra gives

$$(a + jb) + (c + jd) = (a + c) + j(b + d) \qquad ...(1.17)$$

$$(a + jb) + (c + jd) = (ac + j^2bd) + j(ad + bc) \qquad ...(1.18)$$

All the terms in Eqns. (1.17) and (1.18) are seen to be the algebraic forms of the ordered pairs in Eqns. (1.15) and (1.16), provided we all agree that $j^2 = -1$. Applying the rules of algebra to this result, we can write

$$j = +\sqrt{-1}, \qquad j^2 = -1,$$

$$j^3 = -j, \qquad j^4 = +1 \qquad \text{...(1.19)}$$

This then is the algebraic significance of the operator j. It must be specially noted that even though in algebra we have two square roots for any number N^2 as $+N$ and $-N$, when taking the square root of $j^2 = -1$, only the positive value $j = +\sqrt{-1}$ is taken (j is *never* taken to be $-\sqrt{-1}$).

The geometric significance of the operator j is shown in Fig. 1.2. If a, b, c and d are real numbers, the ordered pairs (a, b) and $(-c, -d)$ are obtained by measuring these numbers along the real number line and then rotating b and $-d$ counterclockwise some 90° to get jb and $j(-d)$. Since in trigonometry, a counterclockwise rotation of the radius vector generates a positive angle, this geometric interpretation of the operator j is consistent with the algebraic rule of taking only the positive square root of -1.

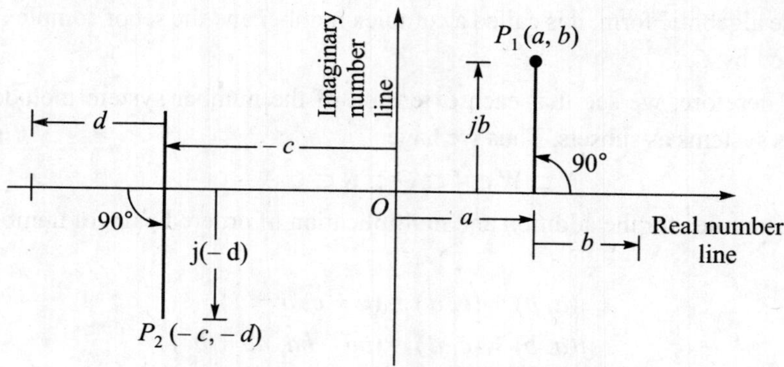

Fig. 1.2. Geometric significance of operator j

The part a is called the real part of the complex number $(a + jb)$ and jb is called the imaginary part. However, it must be clearly understood that the word imaginary, is not used here as a derogatory term calling in question the genuineness of that part. In fact as seen from Fig. 1.2, the line of length jb is as truly existing as the line of length b. The word imaginary, therefore is only to be understood in the sense "not along the real number line but at right angles to it". The orthogonal axis is therefore called the imaginary number line.

In Fig. 1.3 the length r of the line OP and the angle θ ($= \angle POX$) could be used to give another—a geometrical—form of expressing, the ordered pair $(x \,.\, y)$. This is called a polar form and is written as $r \angle\theta$. Thus we have ordered pair

$$(x, y) = (x + jy) = r\angle\theta. \qquad \text{...(1.20)}$$

where,

$$x = r \cos \theta$$

$$y = r \sin \theta \qquad \text{...(1.21)}$$

$$r = +\sqrt{x^2 + y^2},$$

$$\theta = \tan^{-1}\left(\frac{y}{x}\right) \qquad \text{...(1.22)}$$

Notice the +sign as all lengths measured away from O are treated as positive.

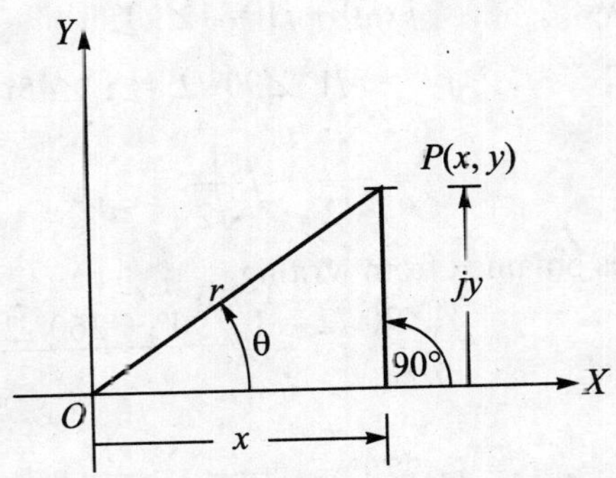

Fig. 1.3. Polar form of complex number

In practice the algebraic form is the best for addition and subtraction of complex numbers, but the geometrical form is simpler for their multiplication and division. Thus if

$$r_1 \angle\theta_1 = a + jb \text{ and } r_2 \angle\theta_2 = c + jd,$$

then

$$r_1 = \sqrt{a^2 + b^2}, \tan\theta_1 = \frac{b}{a},$$

$$r_2 = \sqrt{c^2 + d^2}, \tan\theta_2 = \frac{d}{c}$$

and $\quad (r_1 \angle\theta_1)(r_2 \angle\theta_1) = (a + jb)(c + jd)$

$$= (ac - bd) - j(ad + bc)$$

$$= \sqrt{(ac - bd)^2 + (ad + bc)^2} \,\Big/ \tan^{-1}\left(\frac{ad + bc}{ac - bd}\right) \quad \text{...by Eqn. (1.22)}$$

$$= \sqrt{a^2c^2 + b^2d^2 + a^2d^2 + b^2c^2} \,\Big/ \tan^{-1}\left(\frac{b/a + d/c}{1 - \frac{b}{a}\frac{d}{c}}\right)$$

$$= \sqrt{(a^2 + b^2)(c^2 + d^2)} \,\Big/ \tan^{-1}\left(\frac{\tan\theta_1 + \tan\theta_2}{1 - \tan\theta_1 \tan\theta_2}\right)$$

or $\quad (r_1 \angle\theta_1)(r_2 \angle\theta_2) = r_1 r_2 \,\underline{/\theta_1 + \theta_2}$ ...(1.23)

Similarly

$$\frac{(r_1 \angle\theta_1)}{(r_2 \angle\theta_2)} = \frac{r_1}{r_2} \,\underline{/\theta_1 + \theta_2} \quad \text{...(1.24)}$$

From Eqn. (1.23), it is obvious that

$$\sqrt{r\angle\theta} = \sqrt{r} \,\Big/ \frac{\theta}{2}. \quad \text{...(1.25)}$$

Example: *If* $j = +\sqrt{-1}$, what is $\sqrt{j}$?

Solution: Now $j = 0 + j1 = 1\angle 90°$ by Eqn. (1.22)

Hence, $\sqrt{j} = \sqrt{1}\ \angle 90°/2 = 1\ \angle 45°$ by Eqn. (1.25)

$$= \frac{1}{\sqrt{2}} + j\frac{1}{\sqrt{2}}$$

Another value is obtained from writing

$$j = 1\ \angle 90°/2 = 1\ \angle 90° + 360°/2$$

and getting

$$\sqrt{j} = 1\angle 225° = -\frac{1}{\sqrt{2}} - j\frac{1}{\sqrt{2}}$$

as shown in Fig. 1.4.

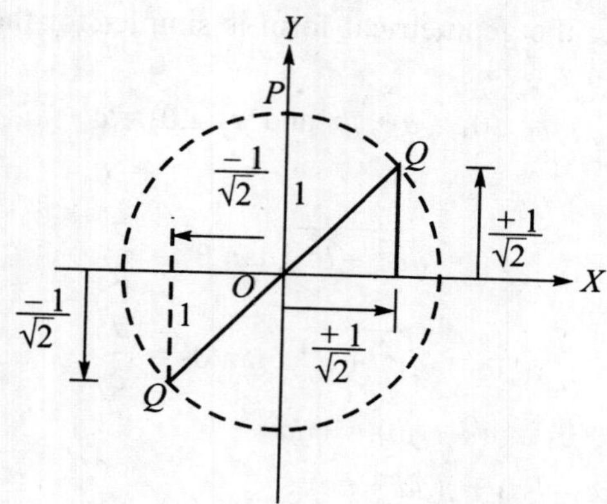

Fig. 1.4. Square root of *j*

The algebraic form could be used as follows:

Let $\sqrt{j} = a + jb,$

Hence squaring,

$$j = 0 + j\,1 = (a + jb)^2 = (a^2 - b^2) + j2ab$$

Equating real and imaginary parts,

$$a^2 - b^2 = 0 \quad \text{and} \quad 2ab = 1$$

These show that $a = b = \pm\,\frac{1}{\sqrt{2}}$

So that

$$\sqrt{j} = \frac{1}{\sqrt{2}} + j\frac{1}{\sqrt{2}} \quad \text{or} \quad -\frac{1}{\sqrt{2}} - j\frac{1}{\sqrt{2}} \qquad ...(1.26)$$

When two complex numbers differ only in the sign of their imaginary parts, they are said to be conjugate. The conjugate of a complex number *V* is denoted by *V**. Thus if $V = a + jb$, then $V^* = a - jb$. A real number results when two conjugate numbers are added or are multiplied. Thus $V + V^* = 2a$ and $VV^* = a^2 + b^2$. The conjugate of a more complicated expression is obtained merely by changing the sign of j everywhere in that expression.

1.7 NATURAL AND HYPERBOLIC SINES AND COSINES

The following trigonometric relations are frequently needed:

$$\sin(-x) = -\sin x, \ \cos(-x) = \cos x \qquad \text{...(1.27)}$$

$$\left.\begin{aligned} \sin(x \pm y) &= \sin x \cos y \pm \cos x \sin y \\ \cos(x \pm y) &= \cos x \cos y \quad\ \sin x \sin y \\ \tan(x \pm y) &= \frac{\tan x \pm \tan y}{1 \quad \tan x \tan y} \end{aligned}\right\} \qquad \text{...(1.28)}$$

$$\left.\begin{aligned} \sin 2x &= 2 \sin x \cos x \\ \cos 2x &= \cos^2 x - \sin^2 x = 2\cos^2 x - 1 \\ &= 1 - 2\sin^2 x \end{aligned}\right\} \qquad \text{...(1.29)}$$

In any triangle ABC (see Fig. 1.5)

$$\frac{a}{\sin A} = \frac{b}{\sin B} = \frac{c}{\sin C} \qquad \text{...(1.30)}$$

(Law of sines)

$$c^2 = a^2 + b^2 - 2ab \cos C \qquad \text{...(1.31)}$$

(Law of cosines)

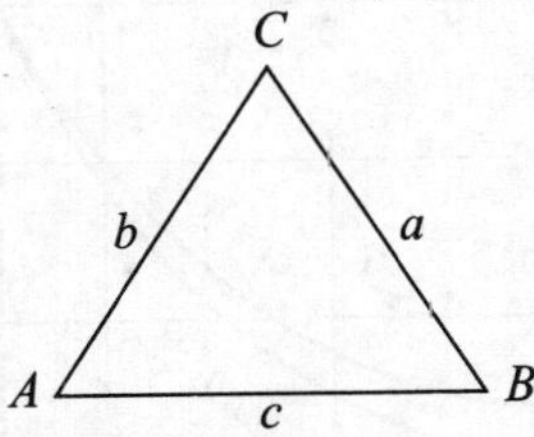

Fig. 1.5. Triangle

From the following Maclaurin expansions

$$\sin x = x - \frac{x^3}{3!} + \frac{x^5}{5!} - \frac{x^7}{7!} + \cdots \qquad \text{...(1.32)}$$

$$\cos x = 1 - \frac{x^2}{2!} + \frac{x^4}{4!} - \frac{x^6}{6!} + \cdots \qquad \text{...(1.33)}$$

we get $\cos x + j \sin x = e^{jx}$ ...(1.34)

and $\cos x - j \sin x = e^{-jx}$ ...(1.35)

Hence we get Euler's equations

$$\cos x = \frac{e^{jx} + e^{-jx}}{2},$$

$$\sin x = \frac{e^{jx} - e^{-jx}}{2j} \qquad \text{...(1.36)}$$

and De Moivre's theorem that

$$(\cos x + j \sin x)^n = [e^{jx}]^n = e^{jnx} = \cos nx + j \sin nx \qquad \text{...(1.37)}$$

The hyperbolic sine and cosine functions are defined as

$$\sin h\, x = \frac{e^x - e^{-x}}{2}$$

and
$$\cos h\, x = \frac{e^x + e^{-x}}{2} \qquad \text{...(1.38)}$$

From Eqns. (1.36) and (1.38) we get

$$\sin h\, jx = j \sin x,\ \cos h\, jx = \cos x \qquad \text{...(1.39)}$$

$$\sin h\, x = -j \sin jx,\ \cos h\, x = \cos jx \qquad \text{...(1.40)}$$

For real values of x varying from zero to infinity, sinh x increases from 0 to ∞ and cosh x increases from 1 to ∞. As shown in Fig. 1.6, for $x > 3$, sin h$x \approx$ cos hx within 0.5%, because

$$\sin h\, 3 = 10.0179 \ldots \text{ and } \cos h\, 3 = 10.0677 \ldots$$

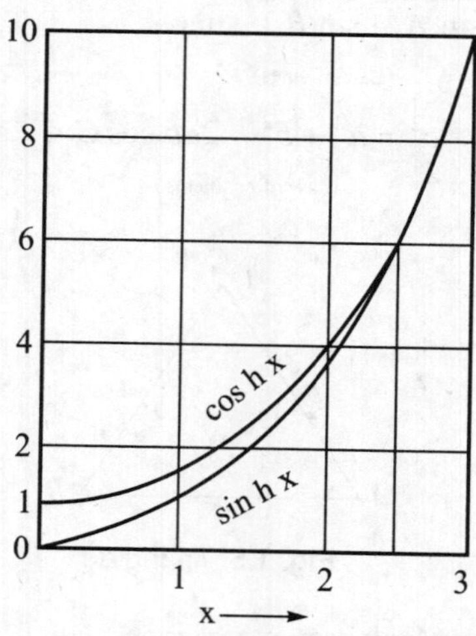

Fig. 1.6. Curves for hyperbolic sine and cosine

From these relations we conclude that hyperbolic functions may be regarded as natural trigonometric functions of complex numbers. Thus

$$\sin^2 jx + \cos^2 jx = 1$$

becomes
$$\cos h^2\, x - \sin h^2\, x = 1 \qquad \text{...(1.41)}$$

Also we have

$$\sin h\,(x + jy) = \sin h\, x \cos y + j \cos h\, x \sin y$$

$$\cos h\,(x + jy) = \cos h\, x \cos y + j \sin h\, x \sin y \qquad \text{...(1.42)}$$

The relationship between natural logarithms and inverse hyperbolic functions is easily established. Let $\sinh^{-1} x = z$. Then

$$x = \sin h\, z = \frac{e^z - e^{-z}}{2} = \frac{e^{2z} - 1}{2e^z}$$

or $$e^{2z} - 2xe^z - 1 = 0.$$

The roots of this quadratic equation are

$$e^z = x \pm \sqrt{x^2 + 1} \quad \text{or} \quad z = \log_e (x \pm \sqrt{x^2 + 1})$$

or $$\sin h^{-1} x = \ln (x \pm \sqrt{x^2 + 1})$$

Similarly

$$\cos h^{-1} x = \ln (x \pm \sqrt{x^2 - 1}) \qquad \text{...(1.43)}$$

1.8 CO-ORDINATE SYSTEMS

Three co-ordinate systems have been used in this book to define the position of a point in three-dimensional space. Through any arbitrarily selected origin O, three mutually perpendicular (= orthogonal) lines OX, OY and OZ are drawn, and the following co-ordinated systems are then defined:

1.8.1 Rectangular or Cartesian Co-ordinate System

The right-handed screw is the standard used in this system to define the positive directions of the three axes OX, OY and OZ (Fig. 1.7). Writing the letters in cyclic order as XYZ (or YZX or ZXY), if any axis such as OX (or OY or OZ) is rotated through 90° to bring it in line with the next axis OY (or OZ or OX), then the direction in which this rotation will cause a right-handed screw to travel is the positive direction of the third axis OZ (or OX or OY).

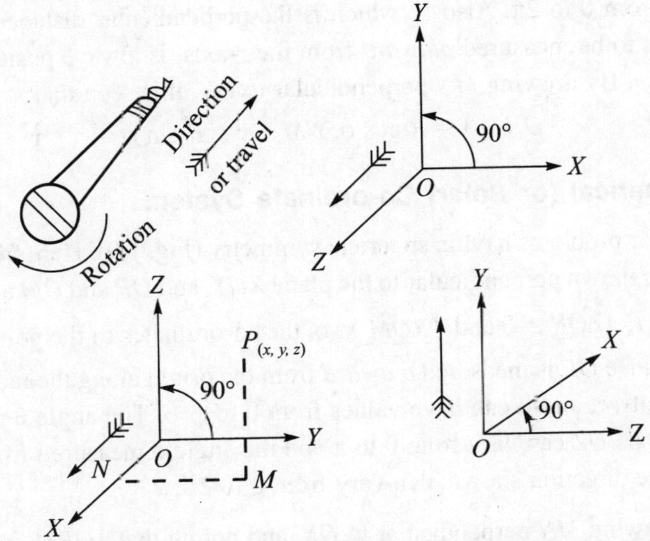

Fig. 1.7. Rectangular or cartesian co-ordinate system based on right-handed screw

If from the point P, a line PM is drawn perpendicular to the plane XOY and then from the point M in the plane XOY, a line MN is drawn perpendicular to the axis OX, then putting $ON = x$, $NM = y$ and $MP = z$, the co-ordinates of P become (x, y, z). Each of these co-ordinates can have values from $-\infty$ to $+\infty$.

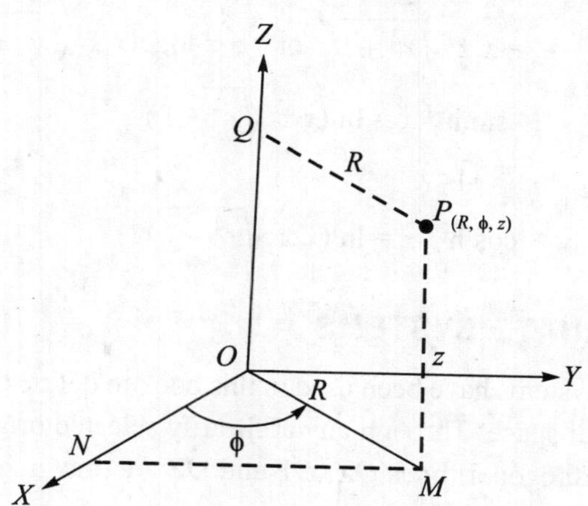

Fig. 1.8. Cylindrical co-ordinate system

1.8.2 Cylindrical Co-ordinate System

This is best for problems involving cylindrical symmetry (Fig. 1.8). Here perpendiculars PQ and PM are drawn from the point P to the z-axis and the plane XOY respectively, and line OM is drawn in plane XOY. Then putting $QP = R$ (= OM), $\underline{/XOM}$ ϕ and $MP = z$, the cylindrical co-ordinates of the point P are written as (R, ϕ, z). Notice that the angle ϕ is positive when it is measured from OX in the direction shown. The angle ϕ can vary from 0 to 2π. Also R, which is the perpendicular distance of P from the z-axis, and is to be measured *outward* from the z-axis, is always positive. It can vary from 0 to $+\infty$. By drawing MN perpendicular to OX, it is seen that

$$ON = x = R \cos \phi, \; NM = y = R \sin \phi \quad \text{...(1.44)}$$

1.8.3 Spherical (or Polar) Co-ordinate System

This is best for problems having spherical symmetry (Fig. 1.9). Here from the point P, the line PM is drawn perpendicular to the plane XOY, and OP and OM are joined. Then putting $OP = r$, $\underline{/ZOP} = \theta$ and $\underline{/XOM} = \phi$, the co-ordinates of the point P are written as (r, θ, ϕ). Since OP is measured *outward* from the origin along the radius vector, OP is always positive, and it can have values from 0 to $+\infty$. The angle θ measured from the positive axis OZ can vary from 0 to π and the angle ϕ measured from the positive axis OX in the direction shown, can vary from 0 to 2π.

By drawing MN perpendicular to OX, and noting that $\underline{/OPM} = \theta$,

we get
$$OM = R = r \sin \theta, \; MP = z = r \cos \theta$$

$$ON = x = r \sin \theta \cos \phi, \; NM = y = r \sin \theta \sin \phi \quad \text{...(1.45)}$$

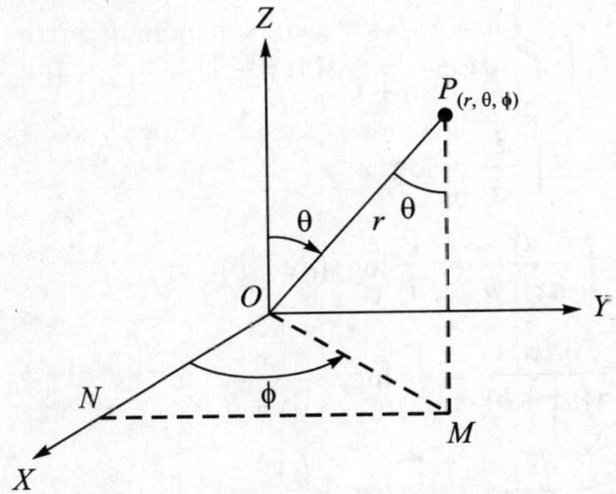

Fig. 1.9. Spherical (Polar) co-ordinate system

1.9 DIFFERENTIAL CALCULUS, INTEGRAL CALCULUS AND DIFFERENTIAL EQUATIONS

If u and v are functions of x, and a, b, c and n are independent of x, then

$$\frac{dv}{dx} = \frac{dv}{du} \cdot \frac{du}{dx},$$

$$\frac{dy}{dx} = \frac{1}{\frac{dx}{dy}} \qquad \text{if} \quad \frac{dx}{dy} \neq 0 \qquad \text{...(1.46)}$$

$$\frac{d}{dx}(uv) = u\frac{dv}{dx} + v\frac{du}{dx}$$

$$\frac{d}{dx}\left(\frac{u}{v}\right) = \frac{v\frac{du}{dx} - u\frac{dv}{dx}}{v^2} \qquad \text{...(1.47)}$$

$$\frac{d}{dx}(u^v) = vu^{v-1}\frac{du}{dx} + [\log_e(u)]u^v\frac{dv}{dx} \qquad \text{...(1.48)}$$

$$\frac{d}{dx}(e^{ax}) = ae^{ax}$$

$$\frac{d}{dx}(x^n) = nx^{n-1} \qquad \text{...(1.49)}$$

$$\frac{d}{dx}(\sin x) = \cos x$$

$$\frac{d}{dx}(\cos x) = -\sin x \qquad \text{...(1.50)}$$

$$\int uv\, dx = u\int v\, dx - \int\left[\frac{du}{dx}\int v\, dx\right] dx \qquad \text{...(1.51)}$$

(Integration by parts)

$$\int x^n \; dx = \frac{x^{n+1}}{n+1} \text{ if } n \neq -1$$

$$\int \frac{dx}{x} = \log_e x \qquad ...(1.52)$$

$$\int \frac{dx}{ax+b} = \frac{1}{a} \log_e (ax+b)$$

$$\int \frac{dx}{x(ax+b)} = \frac{1}{b} \log_e \left(\frac{x}{ax+b}\right) \qquad ...(1.53)$$

$$\int \frac{dx}{x^2+a^2} = \frac{1}{a} \tan^{-1}\left(\frac{x}{a}\right)$$

$$\int \frac{dx}{x^2-a^2} = -\frac{1}{a} \tan^{-1}\left(\frac{a}{x}\right) = \frac{1}{2a} \log_e \frac{x-a}{x+a} \qquad ...(1.54)$$

$$\int \sin ax \; dx = -\frac{1}{a} \cos ax, \int \cos ax \; dx = -\frac{1}{a} \sin ax$$

$$\int \tan x = -\log_e \cos x \qquad ...(1.55)$$

$$\int \frac{dx}{\sqrt{x^2 \pm a^2}} = \log_e (x + \sqrt{x^2 \pm a^2}\,)$$

$$\int \frac{dx}{\sqrt{a^2 - x^2}} = \sin^{-1} \frac{x}{a} \qquad ...(1.56)$$

$$\int \sqrt{a^2 - x^2} \; dx = \frac{1}{2}\left[x\sqrt{a^2 - x^2} + a^2 \sin^{-1}\left(\frac{x}{a}\right)\right] \qquad ...(1.56a)$$

$$\int e^{ax} . \, dx = \frac{1}{a} e^{ax}, \int xe^x \; dx = e^x (x-1) \qquad ...(1.57)$$

In the linear differential equation $\frac{dv}{dx} + uv = 0$, the variables can be separated in the form $\frac{dv}{v} = -u \, dx$. Hence integrating

$$\log_e v = -\int u \; dx + c \qquad ...(1.58)$$

where c is the constant of integration.

The solution of the second order differential equation

$$a\frac{d^2 y}{dx^2} + b\frac{dy}{dx} + cy = 0$$

is

$$y = c_1 \, e^{m_1 x} + c_2 \, e^{m_2 x} \qquad ...(1.59)$$

where m_1 and m_2 are the roots of the auxiliary equation

$$am^2 + bm + c = 0$$

and m_1 and m_2 are the roots of the auxiliary equation

$$am^2 + bm + c = 0$$

and c_1 and c_2 are constants of integration.

In particular, the solution of the equation $\frac{d^2y}{dx^2} + c^2y = 0$ is

$$y = a \cos cx + b \sin cx \quad \text{...(1.60)}$$

The solution of Bessel's equation of order n (= zero or positive integer)

$$x^2 \frac{d^2y}{dx^2} + x\frac{dy}{dx} + (x^2 - n^2)y = 0$$

is the Bessel's function of the first kind of order n denoted by $J_n(x)$, where

$$J_n(x) = \sum_{k=0}^{\infty} (-1)^k \frac{x^{n+2k}}{2^{n+2k}\, k!(n+k)!}, \ (n = 0, 1, 2, ...) \quad \text{...(1.61)}$$

1.10 FOURIER SERIES

If a function of time t denoted by $f(t)$ satisfies the Dirichlet conditions, viz., that $f(t)$ is single-valued, is finite and repeats with a periodicity T, and has a finite number of discontinuities, and also a finite number of maxima and minima in the interval T for one oscillation, then the function $f(t)$ can be represented from $t = -\infty$ to $t = +\infty$ (except at the discontinuities) by a series of simple harmonic functions with frequencies which are integral multiples of the fundamental (radian) frequency ω_0, as

$$f(t) = \frac{a_0}{2} + \sum_{n=1}^{\infty} (a_n \cos n\omega_0 t + b_n \sin n\omega_0 t) \quad \text{...(1.62)}$$

where $\omega_0 = \frac{2\pi}{T}$

The coefficients a_n and b_n ($n = 0, 1, 2, ...$) can be obtained from

$$a_n = \frac{2}{T}\int_0^T f(t) \cos n\, \omega_0 t\, dt$$

and

$$b_n = \frac{2}{T}\int_0^T f(t) \sin n\, \omega_0 t\, dt \quad \text{...(1.63)}$$

Another way of writing the Fourier series is

$$f(x) = \frac{a_0}{2} + \sum_{n=1}^{\infty} (a_n \cos nx + b_n \sin nx) \quad \text{...(1.64)}$$

where

$$a_n = \frac{1}{\pi}\int_c^{2\pi+c} f(x) \cos nx\, dx$$

and

$$b_n = \frac{1}{\pi}\int_c^{2\pi+c} f(x) \sin nx\, dx$$

1.10.1 Half-range Series

In many problems we are only interested in describing a function $f(y)$ over an interval $y = C_1$ to $y = C_2$, and we do not care what values it has from $y = -\infty$ to C_1 and from $y = C_2$ to $+\infty$. In this interval C_1 to C_2, $f(y)$ may vary in any manner (not necessarily repetitive) or it may not vary at all.

To make a repetitive function $F(x)$ from this, we arbitrarily put interval C_1 to C_2 as 0 to π for function $f(x)$. Hence the name "half-range series". The other half of the repetitive function lying between $-\pi$ and 0 can be chosen in two ways, in both of which $f(x)$ and $F(x)$ are identical over the interval $x = 0$ to $x = \pi$, or for $0 < x < \pi$. Over the interval $-\pi$ to 0, if we choose $F(x) = -f(-x)$, then we have odd-function symmetry and the Fourier series has only sine terms as $a_n = 0$. But if we choose $F(x) = +f(-x)$ over the interval $-\pi$ to 0, then we have even-function symmetry and only cosine terms are present as $b_n = 0$.

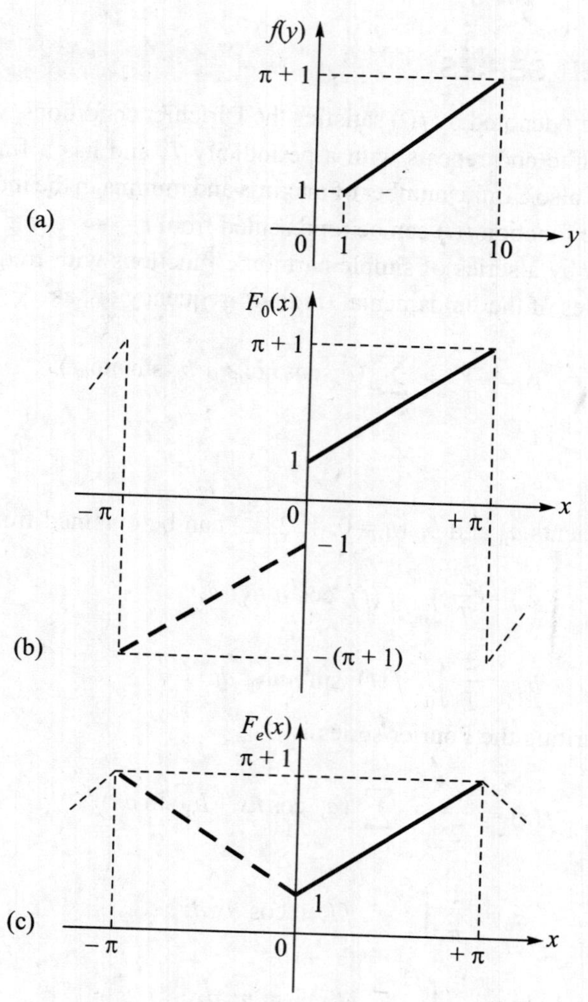

Fig. 1.10. Use of halfrange series

To illustrate the method, let us try to find a Fourier series for $F(y) = \frac{\pi}{9}(y-1) + 1$ over the range $1 < y < 10$, assuming that we do not care what value $f(y)$ is made to have *outside* these limits [Fig. 1.10 (*a*)]. At $y = 1, f(y) = 1$ and at $y = 10, f(y) = \pi + 1$.

First we change the range to 0 to π by putting $x = \frac{\pi}{9}(y-1)$, so that when $y = 1$, $x = 0$ and when $y = 10$, $x = \pi$. Over this half-range, $f(x) = x + 1$.

1.10.2 Odd Function Symmetry

For this we make

$$F_0(x) = f(x), \qquad 0 < x < \pi$$

but

$$F_0(x) = -f(-x), \qquad -\pi < x < 0$$

or

$$F_0(x) = x + 1, \qquad 0 < x < \pi$$

$$F_0(x) = -(-x+1) = x - 1, \qquad -\pi < x < 0 \qquad ...(1.65)$$

$$F_0(x) = F_0(x + 2\pi k),\ k \text{ is any integer.}$$

The function $F_0(x)$ is shown in Fig. 1.10 (*b*). The coefficients of the Fourier series to represent $F_0(x)$ are found by Eqn. (1.64). Then $a_n = 0$ and

$$b_n = \frac{1}{\pi}\int_{-\pi}^{0} (x-1)\sin nx\,dx + \frac{1}{\pi}\int_{0}^{\pi} (x+1)\sin nx\,dx$$

$$= \frac{2}{\pi} \times \frac{1-(\pi+1)\cos n\pi}{n}$$

Thus

$$F_0(x) = \frac{2}{\pi}\left[(\pi+2)\sin x - \pi\frac{\sin 2x}{2} + (\pi+2)\frac{\sin 3x}{3} - \pi\frac{\sin 4x}{4} + ...\right] \qquad ...(1.66)$$

Thus $f(x) = F_0(x)$ over the desired range $0 < x < \pi$, or replacing x by $\frac{\pi}{9}(y-1)$ on right hand side of Eqn. (1.66) we get $f(y)$ over the range $1 < y < 10$.

1.10.3 Even-Function Symmetry

For this we make

$$F_e(x) = f(x), \qquad 0 < x < \pi$$

but

$$F_e(x) = +f(-x), \qquad -\pi < x < 0$$

or

$$\left.\begin{aligned} F_e(x) &= x + 1, && 0 < x < \pi \\ F_e(x) &= -x + 1, && -\pi < x < 0 \\ F_e(x) &= F_e(x + 2\pi k), && k \text{ is any integer} \end{aligned}\right\} \qquad ...(1.67)$$

The function $F_e(x)$ is shown in Fig. 1.10 (*c*). Notice that it is identical with $F_0(x)$ only over the range $0 < x < \pi$. Now for $F_e(x)$.

$$b_n = 0,\ a_0 = \frac{2}{\pi}\int_0^{\pi} (x+1)\ dx = \pi + 2$$

and

$$a_n = \frac{1}{\pi}\int_{-\pi}^{0} (1-x)\ \cos nx\ dx + \frac{1}{\pi}\int_0^{\pi} (x+1)\ \cos nx\ dx$$

$$= \frac{2}{\pi} \times \frac{(\cos n\pi) - 1}{n^2},\ \text{(Therefore } a_n = 0 \text{ if } n \text{ is even)}$$

Then

$$F_e(x) = \left(\frac{\pi}{2} + 1\right) - \frac{4}{\pi}\left[\cos x + \frac{\cos 3x}{3^2} + \frac{\cos 5x}{5^2} + \cdots\right] \quad ...(1.68)$$

from which $f(y)$ can be found as before.

1.11 VECTOR ANALYSIS

For a study of electric and magnetic fields, vector analysis is the most powerful mathematical tool available. It provides a very convenient short-hand notation for writing the mathematical relationships between various physical quantities in a compact form. Moreover, these relations are independent of the co-ordinate system. This results in great economy in writing space and in time needed for writing them. There is also an economy in the thinking process, because ideas and concepts are handled as a whole rather than in their component parts. It is like comprehending the argument developed in a paragraph by understanding the concepts in each successive sentence rather than by studying each word in a sentence. Of course, to interpret the final result, it has to be broken up into its component parts using some co-ordinate system, like a stenographer's short-hand symbols being converted into a typescript. In this way each vector equation yields three scalar equations.

In vector analysis all physical quantities are classified as scalars or vectors. A scalar quantity is completely specified when its magnitude and algebraic sign are given. The magnitude gives the size or the number of multiples of any unit, and the algebraic sign indicates addition or subtraction, gain or loss. Examples of scalar quantities are mass, temperature, altitude above sea-level, work etc. Scalar quantities do not have any specific orientation in space. On the other hand, vector quantities have not only a magnitude but direction or orientation in space, this direction getting reversed when the sign of the vector is changed. Examples of vector quantities are displacement, velocity, acceleration, force, temperature gradient, slope of land, etc.

However, the magnitude of a vector $\bar{A}$ is a scalar $|A|$ or A, and hence in books, the same letter symbol is used for both. The letter in italics represents the magnitude of the vector and the same letter in clarendon or bold-faced type represents the vector. Since a person cannot write bold-faced letters in a notebook, he has to denote a vector by a bar above the letter representing it. Hence this method is deliberately followed in this book.

Geometrically a vector is represented by a line segment, the length being proportional to the magnitude of the vector, its orientation denoting the direction-line in space and an arrow at one end denoting the sense or the positive direction of the vector.

A region in space, at every point of which, a physical quantity can be represented by a scalar or a vector, is called a scalar or a vector field. A relief map of India made in a park, or a map of India in an atlas showing the temperatures in July are examples depicting scalar fields, but a world map showing ocean currents or wind directions, or the region inside a high voltage laboratory where an electric force exists at every point are examples of vector fields.

1.12 VECTOR ALGEBRA

Ordinary rules of arithmetic give the values of scalar sums and differences like $A \pm B$, but a rule similar to the law of parallelogram of forces to find the resultant force, has to be applied to find the sum of vectors $\overline{A}$ and $\overline{B}$ [Fig. 1.11 (*a*)], as a third vector $(\overline{A} + \overline{B})$. It is obvious from the geometric construction that

$$\overline{A} + \overline{B} = \overline{B} + \overline{A} \qquad \text{...(1.69)}$$

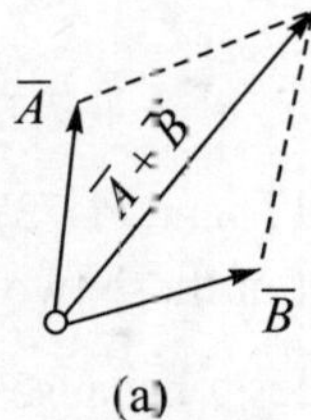

(a)

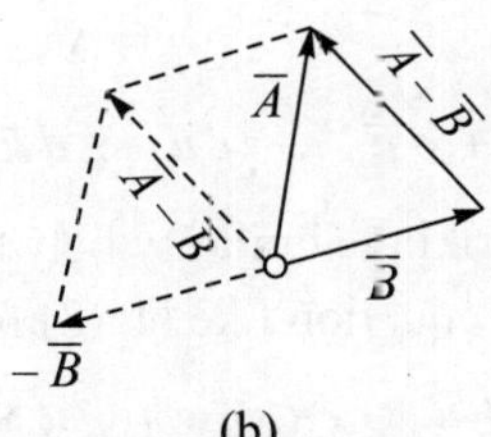

(b)

Fig. 1.11. (*a*) Addition and (*b*) Subtraction of vectors

or vector addition obeys the cummutative law. For finding the difference, we first get the vector $-\overline{B}$ equal in magnitude to $\overline{B}$ but in reverse direction and then find the sum of the vectors $\overline{A}$ and $(-\overline{B})$. Another way to look at it is to write $\overline{A} = \overline{B} + (\overline{A} - \overline{B})$ and then find $(\overline{A} - \overline{B})$ as that vector which when added to $\overline{B}$ gives the vector $\overline{A}$ [Fig. 1.11 (*b*)].

1.12.1 Unit Vector

The above geometrical procedure is very cumbersome. So we search for some suitable procedure involving ordinary rules of algebra. We notice that a vector $\overline{A}$ has a magnitude A and a direction denoted by a line with arrowhead. As a first step, we separate these two aspects of a vector by introducing the concept of a unit vector $\overline{u}_A$ whose magnitude is equal to unity, but whose direction is the same as that of the vector $\overline{A}$. Hence, we write the vector $\overline{A}$ as $A\,\overline{u}_A$, using two separate symbols to denote the magnitude and the direction of $\overline{A}$. It follows that when a vector $\overline{A}$ is multiplied by a scalar c, only the magnitude of the vector becomes c times its previous value, but the direction does not change, or

Scalar c multiplied by a vector $\overline{A} = c(A\,\overline{u}_A) = (cA)\,\overline{u}_A = c\,\overline{A}$ ...(1.70)

From this we conclude that when vectors in the same direction are added, their magnitudes get added but direction remains the same, or $c\,\overline{A} + d\,\overline{A} = (c + d)\,\overline{A}$. To take advantage of this idea, we choose three unit vectors along three mutually perpendicular lines and write any vector as the sum of its resolved parts along these axes. Thus in cartesian co-ordinates, we write

$$\overline{A} = A_x\,\overline{u}_x + A_y\,\overline{u}_y + A_z\,\overline{u}_z \qquad ...(1.71)$$

Addition (or subtraction) of vectors then becomes a simple algebraic process, for

$$\overline{A} \pm \overline{B} = (A_x\overline{u}_x + A_y\overline{u}_y + A_z\overline{u}_z) \pm (B_x\overline{u}_x + B_y\overline{u}_y + B_z\overline{u}_z)$$

or

$$\overline{A} \pm \overline{B} = (A_x \pm B_x)\,\overline{u}_x + (A_y \pm B_y)\,\overline{u}_y + (A_z \pm B_z)\,\overline{u}_z \qquad ...(1.72)$$

A note of warning here is necessary. Let two vectors, both lying entirely in the $x - y$ plane be represented as $(a\,\overline{u}_x + b\,\overline{u}_y)$ and $(c\,\overline{u}_x + d\,\overline{u}_y)$. Then adding them as above, we get

$$(a\,\overline{u}_x + b\,\overline{u}_y) + (c\,\overline{u}_x + d\,\overline{u}_y) = (a + c)\,\overline{u}_x + (b + d)\,\overline{u}_y. \qquad ...(1.73)$$

There is a striking similarity between Eqns. (1.17) and (1.73), and one may be tempted to write an equation like (1.18) for the product of these two vectors, but such a product $(a\,\overline{u}_x + b\,\overline{u}_y) \times (c\,\overline{u}_x + d\,\overline{u}_y)$ is meaningless, because the quantities in (Eqn. (1.18)) are *complex numbers* and two *numbers* can be multiplied but two *vectors* cannot be multiplied according to ordinary algebraic rules.

1.12.2 Scalar (dot) and Vector (cross) Product of Two Vectors

Multiplication in algebra is indicated in three different ways, like AB, $A \cdot B$ and $A \times B$, all three of which are treated as meaning the same thing. Since two vector quantities cannot be multiplied in the ordinary algebraic manner, special meanings are given to product of two vectors, and the dot and the cross are reserved to denote this special significance.

Out of the two quantities, where one is a vector and the other quantity is a scalar, the ordinary algebraic multiplication is possible according to Eqn. (1.70), and the fact that algebraic multiplication is indicated is shown by the *absence* of a dot or a cross between the two quantities. Thus products like $A\,\overline{B}$ or $B\,\overline{A}$ are permissible.

If the product of two vectors $\overline{A}$ and $\overline{B}$ is a scalar quantity whose magnitude is equal to the product of the magnitude of these vectors and the cosine of the angle between them, this product is indicated by a dot between the two vectors. Thus from Fig. 1.12 (a), we write

$$\overline{A} \cdot \overline{B} = AB \cos\theta \quad ...(1.74)$$

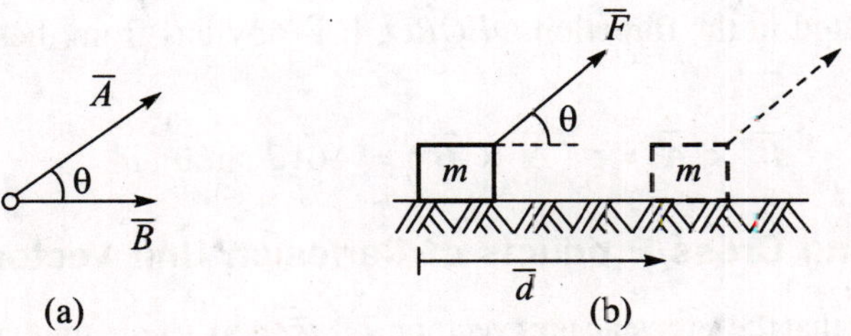

Fig. 1.12. Scalar or dot product of two vectors

Because the result is a scalar and because a dot is placed between the two vectors in the defining Eqn. (1.74), this is called a scalar or dot product of two vectors. Form this definition it follows that

$$\overline{A} \cdot \overline{B} = \overline{B} \quad \overline{A} = A\,(B\cos\theta) = (A\cos\theta)\,B \quad ...(1.75)$$

In Fig. 1.12 (*b*) a problem in mechanics is shown. The force $\overline{F}$ causes the mass to be moved over a distance $\overline{d}$ along the ground and the work done is given by $Fd\cos\theta = \overline{F} \cdot \overline{d}$.

If the product of two vectors $\overline{A}$ and $\overline{B}$ is a vector quantity $\overline{C}$ whose magnitude is equal to the product of the magnitude of these vectors and the sine of the angle between the vectors, and also whose direction is such that $\overline{A}$, $\overline{B}$ and $\overline{C}$ form a right-handed system (see Section 1.8), this fact is indicated by a cross between the vectors. Thus for 1.13, we write

$$\overline{A} \times \overline{B} = \overline{C}\ (AB\sin\theta)\ \overline{u}_c \quad ...(1.76)$$

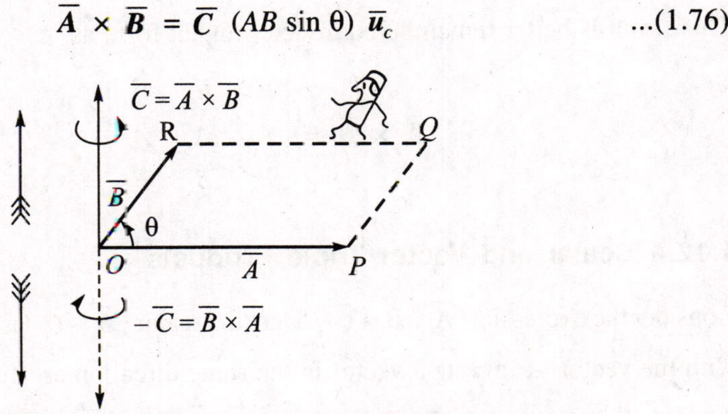

Fig. 1.13. Vector or cross product of two vectors

Because the result is a vector and because a cross is placed between the vectors in the defining Eqn. (1.76), this is called a vector or cross product of two vectors.

Geometrically $AB\sin\theta$ is the area of the parallelogram *OPQR*, of which the vectors $\overline{A}$ and $\overline{B}$ are the two adjacent sides. In mathematics an area is represented by a vector normal to the area, and directed away from the area. Which of the two outward directions is to be used is decided by the right-handed screw convention. If a person goes round the periphery *OPQRO* of an area, so that the area is always on his *left* hand side, then the area is said to be *positive,* and the vector to denote it is drawn normal to the area and pointing in the direction in which a right-handed screw will travel when rotated in the direction *OPQRO.* It is obvious from these considerations that

$$\overline{\boldsymbol{B}} \times \overline{\boldsymbol{A}} = -(\overline{\boldsymbol{A}} \times \overline{\boldsymbol{B}}) = -(AB \sin \theta)\, \overline{\boldsymbol{u}}_c \qquad ...(1.77)$$

1.12.3 Dot and Cross Products of Cartesian Unit Vectors

By remembering that the cartesian unit vectors $\overline{\boldsymbol{u}}_x$, $\overline{\boldsymbol{u}}_y$ and $\overline{\boldsymbol{u}}_z$ are mutually perpendicular to each other ($\theta = 90°$), using Eqns. (1.74) and (1.76), we have

$$\overline{\boldsymbol{u}}_x \cdot \overline{\boldsymbol{u}}_x = \overline{\boldsymbol{u}}_y \cdot \overline{\boldsymbol{u}}_y = \overline{\boldsymbol{u}}_z \cdot \overline{\boldsymbol{u}}_z = 1$$

$$\overline{\boldsymbol{u}}_x \times \overline{\boldsymbol{u}}_x = \overline{\boldsymbol{u}}_y \times \overline{\boldsymbol{u}}_y = \overline{\boldsymbol{u}}_z \times \overline{\boldsymbol{u}}_z = 0 \qquad ...(1.78)$$

$$\overline{\boldsymbol{u}}_x \cdot \overline{\boldsymbol{u}}_y = \overline{\boldsymbol{u}}_y \cdot \overline{\boldsymbol{u}}_z = \overline{\boldsymbol{u}}_z \cdot \overline{\boldsymbol{u}}_x = 0$$

$$\overline{\boldsymbol{u}}_x \times \overline{\boldsymbol{u}}_y = \overline{\boldsymbol{u}}_z,\ \overline{\boldsymbol{u}}_y \times \overline{\boldsymbol{u}}_z = \overline{\boldsymbol{u}}_x,\ \overline{\boldsymbol{u}}_z \times \overline{\boldsymbol{u}}_x = \overline{\boldsymbol{u}}_y \qquad ...(1.79)$$

Using these relations, Eqn. (1.74) can be modified as

$$\begin{aligned}\overline{\boldsymbol{A}} \cdot \overline{\boldsymbol{B}} &= (A_x \overline{\boldsymbol{u}}_x + A_y \overline{\boldsymbol{u}}_y + A_z \overline{\boldsymbol{u}}_z) \,.\, (B_x \overline{\boldsymbol{u}}_x + B_y \overline{\boldsymbol{u}}_y + B_z \overline{\boldsymbol{u}}_z) \\ &= A_x B_x + A_y B_y + A_z B_z \qquad ...(1.80)\end{aligned}$$

Similarly, by a straight forward algebraic multiplication again, we get, using Eqns. (1.78) and (1.79).

$$\begin{aligned}\overline{\boldsymbol{A}} \times \overline{\boldsymbol{B}} &= (A_x \overline{\boldsymbol{u}}_x + A_y \overline{\boldsymbol{u}}_y + A_z \overline{\boldsymbol{u}}_z) \times (B_x \overline{\boldsymbol{u}}_x + B_y \overline{\boldsymbol{u}}_y + B_z \overline{\boldsymbol{u}}_z) \\ &= (A_y B_z - A_z B_y)\overline{\boldsymbol{u}}_x + (A_z B_x - A_x B_z)\overline{\boldsymbol{u}}_y + (A_x B_y - A_y B_x)\overline{\boldsymbol{u}}_z\end{aligned}$$

This result is better remembered in determinant form as

$$\overline{\boldsymbol{A}} \times \overline{\boldsymbol{B}} = \begin{vmatrix} \overline{\boldsymbol{u}}_x & \overline{\boldsymbol{u}}_y & \overline{\boldsymbol{u}}_z \\ A_x & A_y & A_z \\ B_x & B_y & B_z \end{vmatrix} \qquad ...(1.81)$$

1.12.4 Scalar and Vector Triple Products

Consider the expression $\overline{\boldsymbol{A}}\,(\overline{\boldsymbol{B}} \cdot \overline{\boldsymbol{C}})$. Here the scalar $(\overline{\boldsymbol{B}} \cdot \overline{\boldsymbol{C}})$ is algebraically multiplied with the vector $\overline{\boldsymbol{A}}$ giving a vector in the same direction as $\overline{\boldsymbol{A}}$. It is therefore obvious that $(\overline{\boldsymbol{A}} \cdot \overline{\boldsymbol{B}})\,\overline{\boldsymbol{C}}$ is not the same vector as $\overline{\boldsymbol{A}}\,(\overline{\boldsymbol{B}} \cdot \overline{\boldsymbol{C}})$.

By actual multiplication of the cartesian components of the three vectors, it can be shown that

$$\overline{\boldsymbol{A}} \,.\, (\overline{\boldsymbol{B}} \times \overline{\boldsymbol{C}}) = \overline{\boldsymbol{B}} \cdot (\overline{\boldsymbol{C}} \times \overline{\boldsymbol{A}}) = \overline{\boldsymbol{C}} \,.\, (\overline{\boldsymbol{A}} \times \overline{\boldsymbol{B}}) = \begin{vmatrix} A_x & A_y & A_z \\ B_x & B_y & B_z \\ C_x & C_y & C_z \end{vmatrix} \qquad ...(1.82)$$

But a simpler, physical aspect gives a better visualisation. For it is seen that in each of these expressions, the cross-product of the vectors represents an area-vector at right angles to the plane containing these two vectors, and the dot product with the third vector is the multiplication of the magnitude of the area-vector by the projection of the third vector in the direction of the area-vector. Each expression therefore represents the volume of the parallelopiped having the three vectors as its edges.

To find the vector triple product $\bar{A} \times (\bar{B} \times \bar{C})$, we first form a mental picture of it. Thus $(\bar{B} \times \bar{C})$ gives the area-vector at right angles to the plane containing $\bar{B}$ and $\bar{C}$. The cross multiplication of $\bar{A}$ with the area-vector will give a vector at right angles to the area-vector, that is, the resultant vector will lie in the plane containing $\bar{B}$ and $\bar{C}$. By resolving along $\bar{B}$ and $\bar{C}$, the resultant vector can be written in the form $p\bar{B} + q\bar{C}$, where p and q are scalars.

To simplify the calculations involved in determining p and q, choose the x and y axes in the plane of the vectors $\bar{B}$ and $\bar{C}$ with x-axis in the direction of $\bar{B}$. Then we write

$$\bar{B} = B_x\bar{u}_x, \ \bar{C} = C_x\bar{u}_x + C_y\bar{u}_y$$

and

$$\bar{A} = A_x\bar{u}_x + A_y\bar{u}_y + A_z\bar{u}_z$$

Using Eqn. (1.79), we get

$$\begin{aligned}
\bar{A} \times (\bar{B} \times \bar{C}) &= (A_x\bar{u}_x + A_y\bar{u}_y + A_z\bar{u}_z) \times (B_xC_y\bar{u}_z) \\
&= -(A_xB_xC_y)\bar{u}_y + (A_yB_xC_y)\bar{u}_x \\
&= (B_x\bar{u}_x)(A_yC_y) + (B_x\bar{u}_x)(A_xC_x) - (B_x\bar{u}_x)(A_xC_x) - (C_y\bar{u}_y)(A_xB_x) \\
&= (B_x\bar{u}_x)(A_xC_x + A_yC_y) - (C_x\bar{u}_x + C_y\bar{u}_y)(A_xB_x)
\end{aligned}$$

or

$$\bar{A} \times (\bar{B} \times \bar{C}) = \bar{B}(\bar{A} \cdot \bar{C}) - \bar{C}(\bar{A} \cdot \bar{B}) \qquad \text{...(1.83)}$$

showing that $p = \bar{A} \cdot \bar{C}$ and $q = -(\bar{A} \cdot \bar{B})$. Thus by a similar analysis we can show that

$$(\bar{A} \times \bar{B}) \times \bar{C} = \bar{B}(\bar{A} \cdot \bar{C}) - \bar{A}(\bar{B} \cdot \bar{C}) \qquad \text{...(1.84)}$$

This result can however be more readily derived by applying Eqn. (1.77) to Eqn. (1.83) and writing

$$\bar{A} \times (\bar{B} \times \bar{C}) = -(\bar{B} \times \bar{C}) \times \bar{A} = (\bar{C} \times \bar{B}) \times \bar{A}$$

or

$$(\bar{C} \times \bar{B}) \times \bar{A} = \bar{B}(\bar{A} \cdot \bar{C}) - \bar{C}(\bar{A} \cdot \bar{B})$$

Then interchanging the symbols $\bar{A}$ and $\bar{C}$, we get

$$(\bar{A} \times \bar{B}) \times \bar{C} = \bar{B}(\bar{C} \cdot \bar{A}) - \bar{A}(\bar{C} \cdot \bar{B})$$

which is the same as Eqn. (1.84), because of Eqn. (1.75).

1.13 VECTOR DIFFERENTIAL OPERATOR

To apply the ideas of differential calculus to vectors, we first notice the definition of differentiation. If $y = f(x)$, then

$$\frac{dy}{dx} = \lim_{h \to 0} \left[\frac{f(x+h) - f(x)}{h} \right] \qquad ...(1.85)$$

However, when finding the particular integral part of solution of a differential equation, we introduce an operator D to replace $\frac{d}{dx}$, D^2 for $\frac{d^2}{dx^2}$, etc. and then proceed to regard $Dy \left(= \frac{dy}{dx} \right)$ not only as an operation defined by Eqn. (1.85) but simply as an algebraic product of D and y. In a similar way, a vector differential operator, written as inverted delta, and called "del" (or "nabla" or "alted", reverse spelling of delta), is defined as

$$\overline{\nabla} = \overline{u}_x \frac{\partial}{\partial x} + \overline{u}_y \frac{\partial}{\partial y} + \overline{u}_z \frac{\partial}{\partial z} \qquad ...(1.86)$$

Notice the bar above the inverted delta, which is meant to remind us that del is to be treated as a *vector*, through *it has no physical existence in space.* The definition also shows that the del operation involves differentiation with respect to distance in space. Eqns. (1.70), (1.74) and (1.76) show the three different ways in which the vector differential operator del can be multiplied.

1.13.1 Gradient of a Scalar *A*, or Grad *A* or $\overline{\nabla} A$

Let $P(x, y, z)$ be a point in a scalar field at a distance $OP = r$ from the origin, so that

$$\overline{r} = x\overline{u}_x + y\overline{u}_y + z\overline{u}_z \qquad ...(1.87)$$

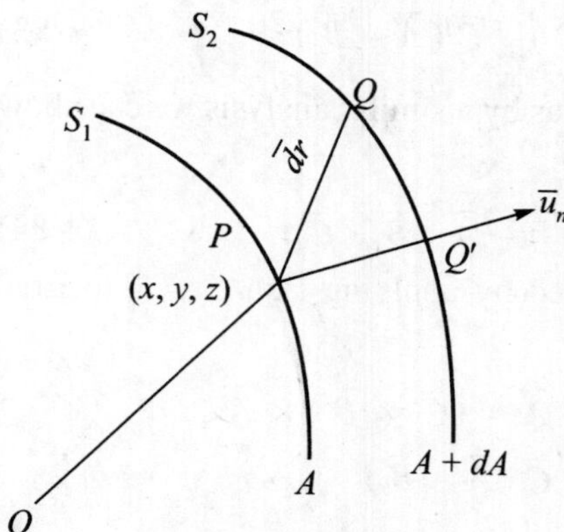

Fig. 1.14. Gradient *A*

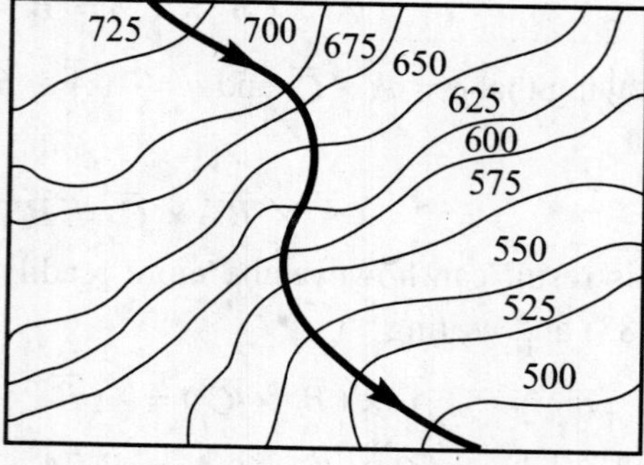

Fig. 1.15. Contour map

If the scalar has a value A at the point P, then the locus of all points where the scalar has the value A will be a surface S_1 passing through P (Fig. 1.14). Similarly on a surface S_2, the scalar has a value $A + dA$, where the complete differential

$$dA = \frac{\partial A}{\partial x}dx + \frac{\partial A}{\partial y}dy + \frac{\partial A}{\partial z}dz \qquad \text{...(1.88)}$$

which may be written as

$$dA = \left(\overline{u}_x \frac{\partial A}{\partial x} + \overline{u}_y \frac{\partial A}{\partial y} + \overline{u}_z \frac{\partial A}{\partial z}\right) \cdot (\overline{u}_x\, dx + \overline{u}_y\, dy + \overline{u}_z\, dz)$$

$$= \overline{\nabla} A \,.\, \overline{dr} \qquad \text{...(1.89)}$$

The quantity $\overline{\nabla} A$ is called the gradient of A. Thus

$$\text{gradient } A = \text{grad } A = \overline{\nabla} A$$

$$= \overline{u}_x \frac{\partial A}{\partial x} + \overline{u}_y \frac{\partial A}{\partial y} + \overline{u}_z \frac{\partial A}{\partial z} \qquad \text{...(1.90)}$$

algebraic product of del with a scalar A.

Let $\overline{u}_n$ be the unit vector normal to the surface S_1 at the point P in the direction of increasing A. If Q is any point on surface S_2, PQ will be denoted by $\overline{dr}$. If Q' is the point where the unit vector $\overline{u}_n$ intersects S_2, then PQ' will be the smallest value of $\overline{dr}$, where

$$PQ' = dn = \overline{u}_n \,.\, \overline{dr}$$

But we can write

$$dA = \frac{dA}{dn}dn = \frac{dA}{dn}\overline{u}_n \,.\, \overline{dr} \qquad \text{...(1.91)}$$

Combining Eqns. (1.89) and (1.91)

$$\overline{\nabla} A \cdot \overline{dr} = \frac{dA}{dn}\overline{u}_n \,.\, \overline{dr}$$

so that

$$\text{gradient } A = \text{grad } A = \overline{\nabla} A = \frac{dA}{dn}\overline{u}_n \qquad \text{...(1.92)}$$

Since dn is the smallest value of $\overline{dr}$, $\dfrac{dA}{dn}$ is the maximum rate of change of A at the point P. This shows that the gradient of a scalar is a vector normal to the surface S_1, for which A has a constant value, the vector being in the direction of increasing A, and its magnitude being equal to the greatest rate of increase of A.

If A is the altitude in metres above mean sea level in a contour map (Fig. 1.15), then $-\ \overline{\nabla} A$ indicates the gradient or slope in the direction in which the given decrease (because of the minus sign) of A occurs in the shortest distance. It is the direction in which water will flow down if poured continuously at the point P.

1.13.2 Divergence of Vector $\overline{A}$ or div $\overline{A}$ or $\overline{\nabla} \cdot \overline{A}$

Divergence of a vector $\overline{A}$ is defined and written as

$$\text{div } \overline{A} = \overline{\nabla} \cdot \overline{A}$$

$$= \left(\overline{u}_x \frac{\partial}{\partial x} + \overline{u}_y \frac{\partial}{\partial y} + \overline{u}_z \frac{\partial}{\partial z} \right) \cdot (A_x \overline{u}_x + A_y \overline{u}_y + A_z \overline{u}_z)$$

or

$$\overline{\nabla} \cdot \overline{A} = \frac{\partial A_x}{\partial x} + \frac{\partial A_y}{\partial x} + \frac{\partial A_z}{\partial x} \qquad ...(1.93)$$

The divergence of a vector $\overline{A}$ is therefore a scalar quantity.

Divergence arises only when there is a flow of something, whether physically real as in the case of a liquid or a gas flowing or a gas flowing in a tube, or fictitious like electric or magnetic flux lines flowing in a vector space, or radius vector lines radiating from the origin in spherical co-ordinates.

Consider an infinitesimally small volume Δv enclosed inside a surface ΔS around the given point P. If we erect normals to the surface at each point of the surface ΔS, then we can find whether the component of the vector $\overline{A}$ along each normal is entering the volume or leaving the volume Δv. The sum total of all the normal inward flows into the volume Δv is denoted by ΣA_{in}, and the sum total of all the normal outward flows from the volume Δv is denoted as ΣA_{out}. Then the divergence of the vector $\overline{A}$ at the given point P is divergence of vector $\overline{A}$,

$$\text{div } \overline{A} = \overline{\nabla} \cdot \overline{A} = \lim_{\Delta v \to 0} \left[\frac{\Sigma A_{\text{out}} - \Sigma A_{\text{in}}}{\Delta v} \right] \qquad ...(1.94)$$

In other words, *the divergence in the net outward flow of* $\overline{A}$ *per unit volume at the point P.* It is obvious that if ΣA_{in} is greater than ΣA_{out}, then the divergence will be negative or it will be convergence.

If we prick an inflated rubber balloon with a pin, then at each point throughout the volume inside the balloon, a positive value of divergence exists as the air leaks away. On the other hand, when the alarm chain is pulled in a running railway train, then at each point inside the vacuum drum under the carriage, a negative divergence is produced as the air rushes in.

1.13.3 Curl of Vector $\overline{A}$ or Curl $\overline{A}$ or $\overline{\nabla} \times \overline{A}$

The curl of a vector $\overline{A}$ is defined and written as

$$\text{curl } \overline{A} = \overline{\nabla} \times \overline{A} = \left(\frac{\partial A_z}{\partial y} - \frac{\partial A_y}{\partial z} \right) \overline{u}_x + \left(\frac{\partial A_x}{\partial z} - \frac{\partial A_z}{\partial x} \right) \overline{u}_y + \left(\frac{\partial A_y}{\partial x} - \frac{\partial A_x}{\partial y} \right) \overline{u}_z$$

or

$$\text{curl } \overline{A} = \overline{\nabla} \times \overline{A} = \begin{vmatrix} \overline{u}_x & \overline{u}_y & \overline{u}_z \\ \frac{\partial}{\partial x} & \frac{\partial}{\partial y} & \frac{\partial}{\partial z} \\ A_x & A_y & A_z \end{vmatrix} \qquad ...(1.95)$$

Hence the curl of vector $\overline{A}$ is a vector. Physically the curl is a measure of the tendency of the vector to curl or to cause rotation or twisting. It can be illustrated by again taking the analogy of flow of the vector $\overline{A}$. However it must be noted that this rotation is not like that produced by the circular motion of water in eddies in a flooded river during the monsoon. The rotational tendency is caused by the fact that the values of A on opposite sides of the point P are unequal (Fig. 1.16) even in a streamline flow. The figure shows a tiny paper boat which a child has launched at P in a flowing stream close to the bank.

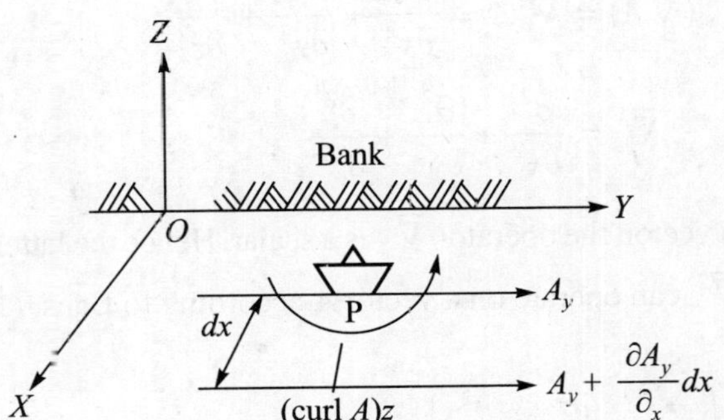

Fig. 1.16. Curl of a vector

Assume that the velocity of water is everywhere the same on a vertical plane x = constant. At the bank $x = 0$ and $A_y = 0$, but A_y increases as x increases. At the point P the velocity is A_y and this will cause the boat to drift along the straight line parallel to the y-axis. However, because of the changing value of A_y in the x-direction, Eqn. (1.95) shows that all terms, except one, being zero, there is only a z-component of the curl $\overline{A}$ which is written as

$$(\text{curl } \overline{A})_z = \text{curl}_z \overline{A} = (\overline{\nabla} \times \overline{A})_z = \overline{\nabla}_z \times \overline{A} = \frac{\partial A_y}{\partial x} \overline{u}_z \qquad \text{...(1.96)}$$

This will cause the boat to rotate about the point P in the direction shown. This is the positive direction for (curl $\overline{A}$)$_z$, because a right-handed screw rotated in this direction will move forward in the positive direction of the z-axis.

It must be noted that this curl exists only because of the variation of $\overline{A}$ in the x-direction. Especially one should not associate this rotation with a change of direction of $\overline{A}$ in a vector field; in the above example, the direction of $\overline{A}$ does not change anywhere.

1.13.4 Laplacian Operator or Divergence of (Gradient *A*)

Since the gradient of a scalar is a vector, and we can find the divergence of any vector, it follows that we can find the divergence of (gradient A) as

$$\text{div (grad } A) = \overline{\nabla} \,.\, (\overline{\nabla} A) = (\overline{\nabla} \,.\, \overline{\nabla}) A = \overline{\nabla}^2 A \qquad \text{...(1.97)}$$

The operator $\overline{\nabla}^2$ is called the Laplacian in honour of the French mathematician Pierre Simon Laplace (1749–1825). It should be clearly understood that del-square is only an abbreviation for the dot product (?) ($\overline{\nabla} \cdot \overline{\nabla}$) and it should not be interpreted as being like an algebraic quantity such as a^2, because two vectors cannot be multiplied without a dot or a cross between them. (We can similarly write $\overline{A}^2$ to mean $\overline{A} \cdot \overline{A}$).

By writing their cartesian expanded forms, we get using Eqns. (1.86) and (1.90).

$$\text{div (grad } A) = \left(\overline{u}_x \frac{\partial}{\partial x} + \overline{u}_y \frac{\partial}{\partial y} + \overline{u}_z \frac{\partial}{\partial z}\right) \cdot \left(\overline{u}_x \frac{\partial A}{\partial x} + \overline{u}_y \frac{\partial A}{\partial y} + \overline{u}_z \frac{\partial A}{\partial z}\right)$$

or
$$\overline{\nabla} \cdot (\overline{\nabla} A) = \overline{\nabla}^2 A = \frac{\partial^2 A}{\partial x^2} + \frac{\partial^2 A}{\partial y^2} + \frac{\partial^2 A}{\partial z^2} \qquad \text{...(1.98)}$$

so that
$$\overline{\nabla}^2 = \frac{\partial^2}{\partial x^2} + \frac{\partial^2}{\partial y^2} + \frac{\partial^2}{\partial z^2} \qquad \text{...(1.99)}$$

Thus while $\overline{\nabla}$ is a vector, the operator $\overline{\nabla}^2$ is a scalar. Hence the latter can be multiplied with a vector or $\overline{\nabla}^2$ can operate on a vector $\boldsymbol{A}$ according to Eqns. (1.70) and (1.71) to give

$$\overline{\nabla}^2 \overline{A} = \left(\frac{\partial^2}{\partial x^2} + \frac{\partial^2}{\partial y^2} + \frac{\partial^2}{\partial z^2}\right) (\overline{u}_x A_x + \overline{u}_y A_y + \overline{u}_z A_z)$$

$$= \overline{u}_x \left(\frac{\partial^2 A_x}{\partial x^2} + \frac{\partial^2 A_x}{\partial y^2} + \frac{\partial^2 A_x}{\partial z^2}\right) + \overline{u}_y \left(\frac{\partial^2 A_y}{\partial x^2} + \frac{\partial^2 A_y}{\partial y^2} + \frac{\partial^2 A_y}{\partial z^2}\right)$$

$$+ \overline{u}_z \left(\frac{\partial^2 A_z}{\partial x^2} + \frac{\partial^2 A_z}{\partial y^2} + \frac{\partial^2 A_z}{\partial z^2}\right)$$

$$= \overline{u}_x \overline{\nabla}^2 A_x + \overline{u}_y \overline{\nabla}^2 A_y + \overline{u}_z \overline{\nabla}^2 A_z \qquad \text{...(1.100)}$$

1.14 UNIT VECTORS AND DIFFERENTIAL ELEMENTS OF VOLUME

1.14.1 Rectangular or Cartesian Co-ordinates

For the point P (x, y, z) in Fig. 1.17, the differential element of volume $dx\, dy\, dz$ is a rectangular parallelopiped with sides $PA = dx$, $PB = dy$ and $PC = dz$. The unit vectors are $\overline{u}_x$, $\overline{u}_y$ and $\overline{u}_z$, so that

$$\overline{ds} = \overline{u}_x\, dx + \overline{u}_y\, dy + \overline{u}_z\, dz \qquad \text{...(1.101)}$$

The desired volume can be found from

$$\text{volume} = \iiint dx\, dy\, dz \qquad \text{...(1.101}a\text{)}$$

the triple integral being evaluated between appropriate limits.

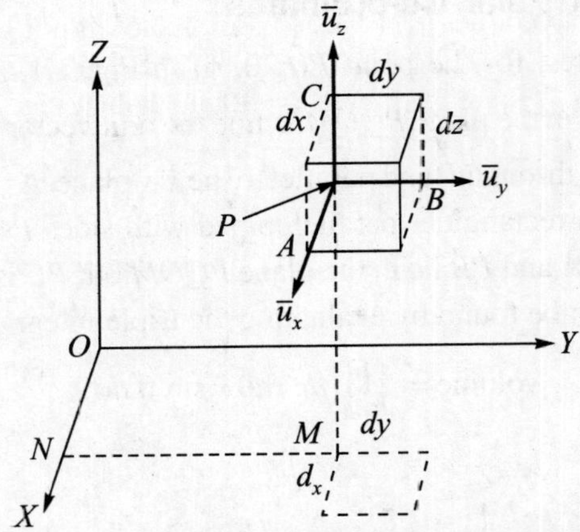

Fig. 1.17. Differential element of volume in cartesian co-ordinates

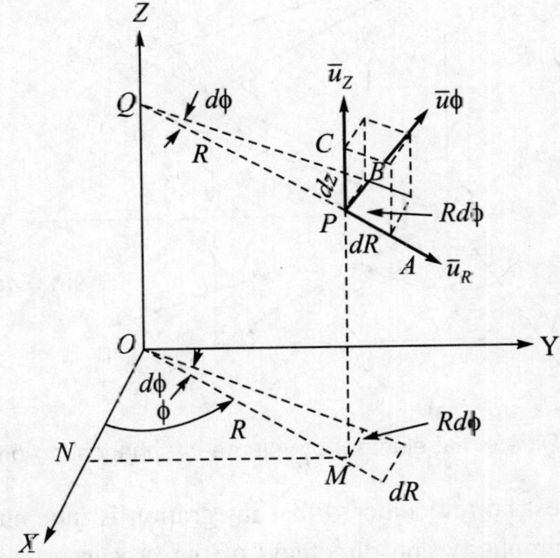

Fig. 1.18. Differential element of volume in cylindrical co-ordinates

1.14.2 Cylindrical Co-ordinates

In cylindrical co-ordinates for the point $P(R, \phi, z)$ in Fig. 1.18, the unit vector $\overline{u}_\phi$ is tangential to the arc PB at the point P. But in the limit when dR, $R\,d\phi$, and dz tend to become infinitesimally small, $R\,d\phi$ is regarded as a straight line coincident with $\overline{u}_\phi$. Hence the differential element of volume is a rectangular parallelopiped with sides $PA = dR$, $PB = R\,d\phi$ and $PC = dz$. The desired volume can be found by evaluating the triple integral

$$\text{Volume} = \iiint dR\ R d\phi\ dz \qquad \ldots(1.102)$$

between appropriate limits.

1.14.3 Spherical or Polar Co-ordinates

In spherical co-ordinates, for the point $P(r, \theta, \phi)$ in Fig. 1.19, the unit vector $\boldsymbol{u}_\theta$ is tangential to the arc PB in the plane $PQOM$, while the unit vector $\bar{u}_\theta$ is tangential to the arc PC in plane passing through P and parallel to the x-y plane. As before the differential element of volume is a rectangular parallelopiped with sides $PA = dr$, $PB = r\, d\theta$ and $PC = r \sin\theta\, d\phi$. Both PA and PB lie in the plane $PQOM$ but PC is normal to this plane. The desired volume can be found by evaluating the triple integral

$$\text{Volume} = \iiint dr\, r d\theta\, r \sin\theta\, d\phi \qquad ...(1.103)$$

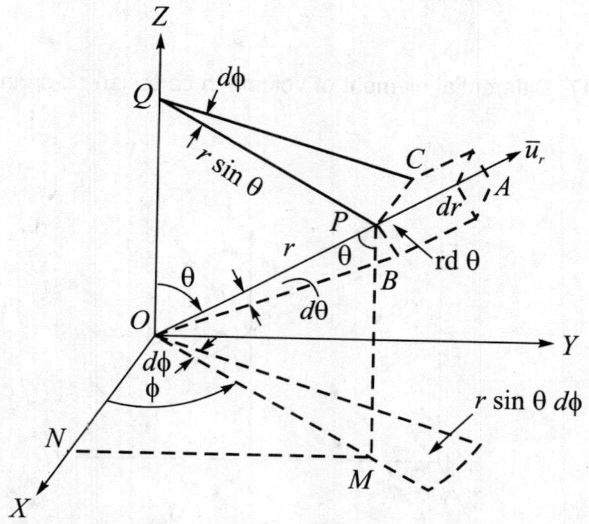

Fig. 1.19. Differential element of volume in spherical co-ordinates

However it must be remembered that integration is inherently a complicated process lacking the simplicity and directness of the process of differentiation, and triple integrals are often too cumbersome for easy handling. Hence in this book, wherever possible, the use of triple integrals is avoided by using physical concepts to reduce them to integration with a single variable.

1.15 DEL OPERATOR IN THE THREE CO-ORDINATE SYSTEMS

Mathematics books on vector analysis show how with the help of orthogonal curvilinear co-ordinates, we can transform the equations from one system of co-ordinates to another. The results established in Section 1.13 are in the cartesian system of co-ordinates and they are in their simplest form. They are grouped together below and this is followed by their transformation into other co-ordinates systems.

1.15.1 Cartesian Co-ordinates (*x, y, z*)

$$\text{grad } V = \overline{\nabla}\, V = \bar{\boldsymbol{u}}_x \frac{\partial V}{\partial x} + \bar{\boldsymbol{u}}_y \frac{\partial V}{\partial y} + \bar{\boldsymbol{u}}_z \frac{\partial V}{\partial z} \qquad \text{from Eqn. (1.90)}$$

$$\text{div } \overline{\boldsymbol{D}} = \overline{\nabla} \times \overline{\boldsymbol{D}} = \frac{\partial D_x}{\partial x} + \frac{\partial D_y}{\partial y} + \frac{\partial D_z}{\partial z} \quad \text{from Eqn. (1.93)}$$

$$\text{curl } \overline{H} = \overline{\nabla} \times \overline{H} = \begin{vmatrix} \overline{\boldsymbol{u}}_x & \overline{\boldsymbol{u}}_y & \overline{\boldsymbol{u}}_z \\ \dfrac{\partial}{\partial x} & \dfrac{\partial}{\partial y} & \dfrac{\partial}{\partial z} \\ H_x & H_y & H_z \end{vmatrix} \quad \text{from Eqn. (1.95)}$$

$$\text{div grad } V = \overline{\nabla} \,.\, (\overline{\nabla} V) = \overline{\nabla}^2 V$$

$$= \frac{\partial^2 V}{\partial x^2} + \frac{\partial^2 V}{\partial y^2} + \frac{\partial^2 V}{\partial z^2} \quad \text{from Eqn. (1.98)}$$

$$(\overline{\nabla} \,.\, \overline{\nabla})\, \overline{\boldsymbol{A}} = \overline{\nabla}^2\, \overline{\boldsymbol{A}}$$

$$= \overline{\boldsymbol{u}}_x\, \overline{\nabla}^2 A_x + \overline{\boldsymbol{u}}_y\, \overline{\nabla}^2 A_y + \overline{\boldsymbol{u}}_z\, \overline{\nabla}^2 A_z \qquad \text{from Eqn. (1.100)}$$

1.15.2 Cylindrical Co-ordinates (*R*, ϕ, *z*)

$$\text{grad } V = \overline{\nabla}\, V = \overline{\boldsymbol{u}}_R \frac{\partial V}{\partial R} + \overline{\boldsymbol{u}}_\phi \frac{1}{R}\frac{\partial V}{\partial \phi} + \overline{\boldsymbol{u}}_z \frac{\partial V}{\partial z} \qquad \text{...(1.104)}$$

$$\text{div } \overline{\boldsymbol{D}} = \overline{\nabla} \,.\, \overline{\boldsymbol{D}} = \frac{1}{R}\frac{\partial}{\partial R}(RD_R) + \frac{1}{R}\frac{\partial D_\phi}{\partial \phi} + \frac{\partial D_z}{\partial z} \qquad \text{...(1.105)}$$

$$\text{curl } \overline{H} = \overline{\nabla} \times \overline{H} = \overline{\boldsymbol{u}}_R \left[\frac{1}{R}\frac{\partial H_z}{\partial \phi} - \frac{\partial H_\phi}{\partial z}\right] + \overline{\boldsymbol{u}}_\phi \left[\frac{\partial H_R}{\partial z} - \frac{\partial H_z}{\partial R}\right] +$$

$$\overline{\boldsymbol{u}} \left[\frac{1}{R}\frac{\partial}{\partial R}(RH_\phi) - \frac{1}{R}\frac{\partial H_R}{\partial \phi}\right] \qquad \text{...(1.106)}$$

$$\text{div grad } V = \overline{\nabla}^2 V = \frac{1}{R}\frac{\partial}{\partial R}\left(R\frac{\partial V}{\partial R}\right) + \frac{1}{R^2}\frac{\partial^2 V}{\partial \phi^2} + \frac{\partial^2 V}{\partial z^2} \qquad \text{...(1.107)}$$

$$(\overline{\nabla} \,.\, \overline{\nabla})\, \overline{\boldsymbol{A}} = \overline{\nabla}^2\, \overline{\boldsymbol{A}} = \overline{\boldsymbol{u}}_R \left(\overline{\nabla}^2 A_R - \frac{2}{R^2}\frac{\partial A_\phi}{\partial \phi} - \frac{A_R}{R^2}\right)$$

$$+ \overline{\boldsymbol{u}}_\phi \left(\overline{\nabla}^2 A_\phi + \frac{2}{R^2}\frac{\partial A_R}{\partial \phi} - \frac{A_\phi}{R^2}\right) + \overline{\boldsymbol{u}}_z\, (\overline{\nabla}^2 A_z) \qquad \text{...(1.108)}$$

1.15.3 Spherical Co-ordinates (*r*, θ, ϕ)

$$\text{grad } V = \overline{\nabla}\, V = \overline{\boldsymbol{u}}_r \frac{\partial V}{\partial r} + \overline{\boldsymbol{u}}_\theta \frac{1}{r}\frac{\partial V}{\partial \theta} + \overline{\boldsymbol{u}}_\phi \frac{1}{r \sin\theta}\frac{\partial V}{\partial \phi} \qquad \text{...(1.109)}$$

$$\text{div } \overline{\boldsymbol{D}} = \overline{\nabla} \,.\, \overline{\boldsymbol{D}} =$$

$$\frac{1}{r^2}\frac{\partial}{\partial r}(r^2 D_r) + \frac{1}{r \sin\theta}\frac{\partial}{\partial \theta}(\sin\theta\, D_\theta) + \frac{1}{r \sin\theta}\frac{\partial D_\phi}{\partial \phi} \qquad \text{...(1.110)}$$

$$\text{curl } \bar{H} = \bar{\nabla} \times \bar{H} = \bar{u}_r \frac{1}{r \sin\theta}\left[\frac{\partial}{\partial\theta}(H_\phi \sin\theta) - \frac{\partial H_\phi}{\partial\phi}\right]$$

$$+ \bar{u}_\theta \frac{1}{r}\left[\frac{1}{\sin\theta}\frac{\partial H_r}{\partial\phi} - \frac{\partial}{\partial r}(rH_\phi)\right]$$

$$+ \bar{u}_\phi \frac{1}{r}\left[\frac{\partial}{\partial r}(rH_\theta) - \frac{\partial H_r}{\partial\theta}\right] \quad ...(1.111)$$

$$\text{div grad } V = \bar{\nabla}^2 V = \frac{1}{r^2}\frac{\partial}{\partial r}\left(r^2 \frac{\partial V}{\partial r}\right) + \frac{1}{r^2 \sin\theta}\frac{\partial}{\partial\theta}\left(\sin\theta \frac{\partial V}{\partial\theta}\right)$$

$$+ \frac{1}{r^2 \sin^2\theta}\frac{\partial^2 V}{\partial\phi^2} \quad ...(1.112)$$

$$(\bar{\nabla}\cdot\bar{\nabla})\bar{A} = \bar{\nabla}^2 \bar{A}$$

$$= \bar{u}_r\left[\bar{\nabla}^2 A_r - \frac{2}{r^2}\left(A_r + \cot\theta\, A_\theta + \frac{\partial A_\theta}{\partial\theta} + \text{cosec}\,\theta \frac{\partial A_\phi}{\partial\phi}\right)\right]$$

$$+ \bar{u}_\theta\left[\bar{\nabla}^2 A_\theta - \frac{1}{r^2}\left(\text{cosec}^2\theta\, A_\theta - 2\frac{\partial A_r}{\partial\theta} + 2\cot\theta\,\text{cosec}\,\theta \frac{\partial A_\phi}{\partial\phi}\right)\right]$$

$$+ \bar{u}_\phi\left[\bar{\nabla}^2 A_\phi - \frac{1}{2}\left(\text{cosec}^2\theta\, A_\phi - 2\,\text{cosec}\,\theta \frac{\partial A_r}{\partial\phi} - 2\cot\theta\,\text{cosec}\,\theta \frac{\partial A_\theta}{\partial\phi}\right)\right] \quad ...(1.113)$$

1.15.4 Other Equations Involving $\bar{\nabla}$ Operator

By expanding $\bar{\nabla}$ and the vectors in terms of their components using Eqns. (1.86) and (1.71), the following useful results can be easily established.

$$\text{grad}(uv) = \bar{\nabla}(uv) = u\bar{\nabla} v + v\bar{\nabla} u \quad ...(1.114)$$

$$\text{div}(u\bar{A}) = \bar{\nabla}\,.\,(u\bar{A}) = u(\bar{\nabla}\,.\,\bar{A}) + \bar{A}\cdot(\bar{\nabla} u) \quad ...(1.115)$$

$$\text{curl}(u\bar{A}) = \bar{\nabla}\times(u\bar{A}) = u(\bar{\nabla}\times\bar{A}) + (\bar{\nabla} u)\times\bar{A} \quad ...(1.116)$$

$$\text{div}(\text{curl } \bar{A}) = \bar{\nabla}\cdot(\bar{\nabla}\times\bar{A}) = 0 \quad ...(1.117)$$

$$\text{curl}(\text{grad } u) = \bar{\nabla}\times(\bar{\nabla} u) = 0 \quad ...(1.118)$$

$$\text{curl}(\text{curl } \bar{E}) = \text{grad}(\text{div } \bar{E}) - \bar{\nabla}^2\bar{E}$$

or

$$\bar{\nabla}\times(\bar{\nabla}\times\bar{E}) = \bar{\nabla}(\bar{\nabla}\,.\,\bar{E}) - \bar{\nabla}^2\bar{E} \quad ...(1.119)$$

1.16 STANDARD CURVES

All standard equations pertaining to these curves assume the origin to be at the point (0.0). If the origin is shifted to the point $(+h, +k)$, in all equations put $(x - h)$ in place of x and $(y - k)$ in place of y.

1.16.1 Circle

Circle of radius a with centre at origin [Fig. 1.20 (a)] is

$$x^2 + y^2 = a^2 \quad ...(1.120)$$

Angles in the same segment are equal [Fig. 1.20 (b)]. Common chord = BC.

$$m\angle BAC = m\ \angle BA'X \quad ...(1.121)$$

Angle in a semicircle is a right angle.

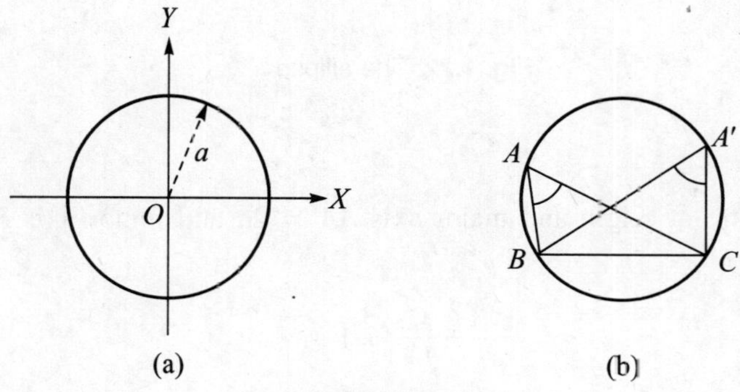

Fig. 1.20. The circle

1.16.2 Parabola

With origin at vertex, the equation to the x-parabola open to the right is (Fig. 1.21)

$$y^2 = 4ax \quad ...(1.122a)$$

Focus at $(a, 0)$

x-parabola open to the left (dotted in Fig. 1.21) is

$$y^2 = -\ 4ax \quad ...(1.122b)$$

Focus at $(-\ a, 0)$

Similarly y-parabolas are

$$x^2 = \pm\ 4ay \quad ...(1.122c)$$

with plus sign referring to parabola open upwards.

Eccentricity of parabola = 1

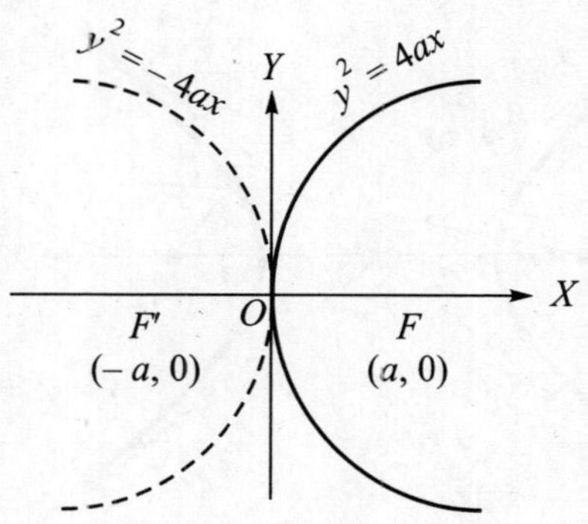

Fig. 1.21. The parabola

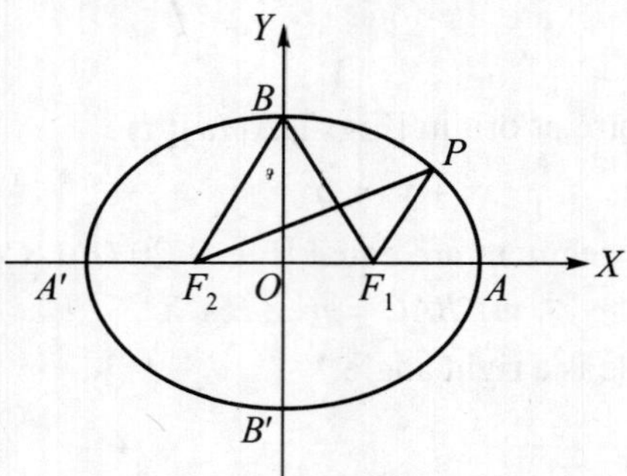

Fig. 1.22. The ellipse

1.16.3 Ellipse

Ellipse with centre at origin and major axis $AA' = 2a$ and minor axis $BB' = 2b$ is (Fig. 1.22)

$$\frac{x^2}{a^2} + \frac{y^2}{b^2} = 1 \qquad ...(1.123)$$

$F_1(ae, 0)$ and $F_2(-ae, 0)$ are the foci

$$BF_1 = BF_2 = a \qquad ...(1.124)$$

For any point P on the ellipse

$$PF_1 + PF_2 = 2a = \text{Constant} \qquad ...(1.125)$$

Eccentricity e is given by

$$b^2 = a^2 (1 - e^2),\ e < 1 \qquad ...(1.126)$$

1.16.4 Hyperbola

The x-hyperbola with centre at origin and transverse axis $AA' = 2a$ is (Fig. 1.23).

$$\frac{x^2}{a^2} - \frac{y^2}{b^2} = 1 \qquad ...(1.127)$$

where eccentricity $e(> 1)$ is given by

$$b^2 = a^2(e^2 - 1),\ e > 1 \qquad ...(1.128)$$

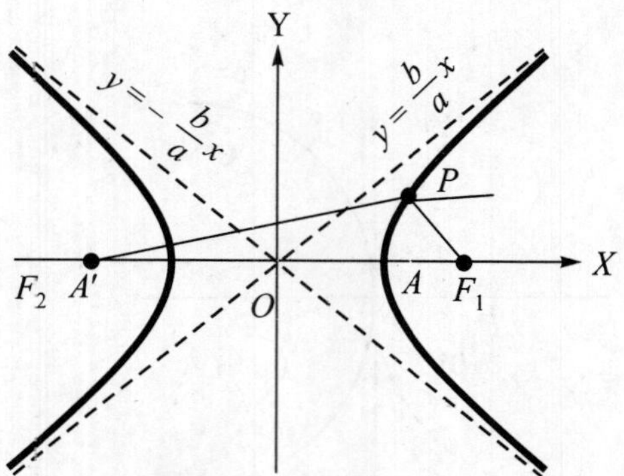

Fig. 1.23. The hyperbola

$F_1(ae, 0)$ and $F_2(-ae, 0)$ are the foci.

For any point P on the hyperbola

$$|PF_2 - PF_1| = 2a \qquad ...(1.129)$$

Asymptotes to the hyperbola are

$$y = \pm \frac{b}{a} x \qquad ...(1.130)$$

The y-hyperbola is

$$\frac{y^2}{b^2} - \frac{x^2}{a^2} = 1 \qquad ...(1.131)$$

It is also called the conjugate hyperbola.

Rectangular hyperbola is

$$xy = \text{Constant} \qquad ...(1.132)$$

in which the axis of x and the axis of y are the asymptotes.

1.16.5 Solid Angle

At a point P, distance $OP = r$ from the origin O, is a small area dA on a surface passing through P (Fig. 1.24). As per Fig. 1.12, it can be represented by a vector $\overline{dA} = dA\, \overline{u}_n$ where $\overline{u}_n$ is the vector normal to the area at P.

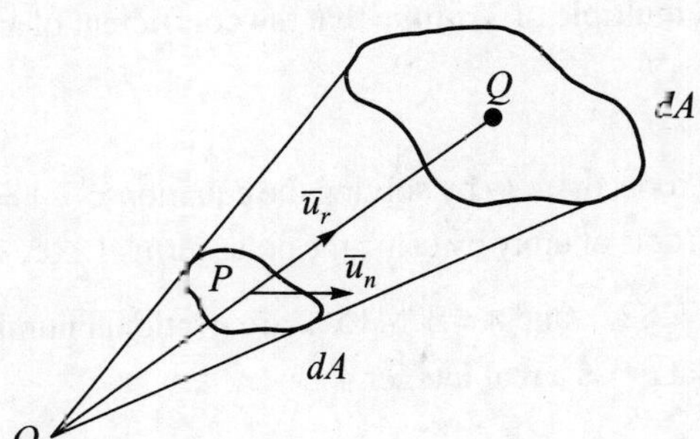

Fig. 1.24. Solid angle

If O is joined to every point on the periphery of area dA by straight lines, a cone will be formed with vertex at O. The projection of area dA on plane normal to the line OP (or vector $\overline{u}_r$) is $\overline{u}_r \,.\, (\overline{u}_n\ dA)$. It is also the area intercepted by the cone from the sphere of radius r with centre O. Then by definition, solid angle $d\Omega$ subtended by area dA at O is

$$d\Omega = \frac{\overline{u}_r \,.\, (\overline{u}_n dA)}{r^2} \qquad ...(1.133)$$

The solid angle can also be thought of as the area dA' intercepted by the cone from a sphere of *unit* radius with centre at origin O. If a point source of light is placed at O, then dA' is the shadow of dA on the unit sphere (r is assumed to be less than unity for this simile). The solid angle subtended by any angle surface which completely encloses the point 0 is

$$\frac{4\pi r^2}{r^2} = 4\pi \text{ steradians} \qquad ...(1.134)$$

EXERCISE 1

1. Expand $(1-x)^{-1}$ by Binomial Theorem.
2. Sum to n terms the geometric series
$$1 + x + x^2 + x^3 + \ldots,$$
and then let n become infinitely large.
3. Show that for *any* point on the curve $y = Ae^{-t/T}$, the rate of decay is such that, if this rate remains constant for the next interval of time $= T$, then the value of y would be reduced to zero.
4. If $\log(m+n) = \log m + \log n$, find m as a function of n.
5. If n and r are positive integers, by putting $n = 0$ in the expression $\frac{n!}{(n-r)!}$, find the value of $\frac{l}{(-r)!}$.
6. If the integer n is a multiple of 3, prove that the coefficient of x^n in the expansion of $\log_e(1+x+x^2)$ is $-\frac{2}{n}$.
7. Find the three cube roots of unity by solving the equation $x^3 - 1 = 0$. [Hint: Factorise].
8. Find the three cube roots of unity by using the polar forms $1\angle 0°$, $1\angle 360°$ and $1\angle 720°$.
9. Even though $e = 2.7182$ ... and $\pi = 3.1415$ ... are irrational numbers and $j = \sqrt{-1}$ is imaginary, show that $e^{j\pi}$ is a real integer.
10. Prove that
$$\int \sqrt{a^2 - x^2}\, dx = \frac{1}{2}\left[x\sqrt{a^2 - x^2} + a^2 \sin^{-1}\left(\frac{x}{a}\right)\right]$$
11. By choosing (*i*) $x = \frac{\pi}{4}$ and (*ii*) $x = \frac{\pi}{2}$, verify that the value of $F_0(x)$ given by Eqn. (1.66) and of $F_e(x)$ given by Eqn. (1.68) are the same for $0 < x < \pi$.
12. A function of y has a constant value V_0 over the interval $0 < y < C$. If we do not care what value $f(y)$ has outside this interval, show that V_0 can be represented by the Fourier series
$$V_0 = \sum_{n=1}^{\infty} b_n \sin \frac{n\pi y}{C}$$

$$\text{where } \begin{cases} b_n = \dfrac{4V_0}{n\pi}, & n \text{ odd} \\ b_n = 0, & n \text{ even} \end{cases}$$

or $$V_0 = \frac{4V_0}{\pi}\left[\sin\left(\frac{\pi y}{C}\right) + \frac{\sin\left(\frac{3\pi y}{C}\right)}{3} + \frac{\sin\left(\frac{5\pi y}{C}\right)}{5} + \cdots\right]$$

where $0 < y < C$.

13. Using vector methods, show that:

(*i*) The diagonals of a parallelogram bisect each other.

(*ii*) The lines joining the mid-points of the opposite sides of a quadrilateral bisect each other.

(*iii*) The bisectors of the angles of a triangle intersect each other at one point.

14. Using vector methods derive Eqn. (1.31), the Law of Cosines for a triangle.

15. In cartesian co-ordinates if the three unit vectors from origin O form the three edges of a cube, find the scalar product of any two diagonals of the cube, and also find the angle between these diagonals.

16. Show that

$$\overline{A} \times (\overline{B} \times \overline{C}) + \overline{B} \times (\overline{C} \times \overline{A}) + \overline{C} \times (\overline{A} \times \overline{B}) = 0$$

17. Show that

$$\overline{\nabla} \cdot (\overline{A} \times \overline{B}) = \overline{B} \,.\, (\overline{\nabla} \times \overline{A}) - \overline{A} \,.\, (\overline{\nabla} \times \overline{B})$$

18. Show that

$$(\overline{A} \times \overline{B}) \cdot (\overline{C} \times \overline{d}) = \begin{vmatrix} \overline{A} \cdot \overline{C} & \overline{B} \cdot \overline{C} \\ \overline{A} \cdot \overline{D} & \overline{B} \cdot \overline{D} \end{vmatrix}$$

19. Show that

$$(\overline{A} \times \overline{B}) \times (\overline{C} \times \overline{D}) = [(\overline{A} \times \overline{B}) \cdot \overline{D}]\,\overline{C} - [(\overline{A} \times \overline{B}) \cdot \overline{C}]\,\overline{D}$$

20. Show that

$$\overline{\nabla} \times (\overline{A} \times \overline{B}) = [\overline{A}(\overline{\nabla} \,.\, \overline{B}) - \overline{B}(\overline{\nabla} \,.\, \overline{A})] + [(\overline{B} \,.\, \overline{\nabla})\overline{A} - (\overline{A} \,.\, \overline{\nabla})\overline{B}]$$

21. If $r = \sqrt{x^2 + y^2 + z^2}$ and n is any number, show that

$$\overline{\nabla}^2 (r^n) = n(n+1) r^{n-2}$$

22. If $\overline{r} \,.\, \overline{dr} = 0$, show that r = Constant.

23. Show that div $\overline{r} = \overline{\nabla} \,.\, \overline{r} = 3$.

24. If $r = \sqrt{x^2 + y^2 + z^2}$ and the vector $\boldsymbol{A}$ is constant, show that

$$\overline{A} \,.\, \overline{\nabla}\left(\frac{1}{r}\right) = -\frac{1}{r^3}(\overline{A} \,.\, \overline{r})$$

25. If $\overline{V}$ is a vector function of x, y and z, show that

$$\overline{\nabla}(\overline{V} \,.\, \overline{V}) = \overline{\nabla V}^2 = 2(\overline{V} \,.\, \overline{\nabla})\overline{V} + 2\,\overline{V} \times (\overline{\nabla} \times \overline{V})$$

26. In cylindrical co-ordinates, for the plane $z = 0$, plot the points $P\left(R, \frac{\pi}{3}\right)$ and $Q\left(R, \frac{3\pi}{4}\right)$. At P and Q draw the unit vectors $\bar{u}_R$ and $\bar{u}_\phi$. Indicate the angle ϕ for each point. Hence derive the relations given by Eqn. (1.44) at each point.
27. Purely from physical considerations (without finding any areas), find the solid angle subtended by one face of a cube at the centre of the cube.
28. Using the result of Problem 27 above, find the solid angle subtended by one face of a cube at one corner of the opposite face.
29. Find the solid angle subtended by an infinite plane at a point not lying on the plane.
30. If a cone with a semi-vertical angle θ, cuts out a sectorial cap from the surface of a sphere whose centre is at the vertex of the cone, show that the solid angle subtended by this cap at the centre of the sphere is $2\pi (1 - \cos \theta)$.

CHAPTER

2 COULOMB'S LAW AND STATIC ELECTRIC FIELDS

2.1 COULOMB'S LAW

When the English astronomer Edmund Halley (1656–1742) published his friend Newton's book *Principia Mathematica* in 1687, the world became familiar with the inverse square law of gravitation which Newton had used along with calculus to explain Kepler's laws of planetary motion. Later the French physicist Charles Augustine de Coulomb (1736–1806) established in 1785 the inverse square laws of electricity and magnetism. Light illumination was also found to obey an inverse square law. It has Albert Einstein (1879–1955) who showed that the inverse square law was really a property associated with space, and that was the reason why all these different phenomena occurring in space had this common characteristic.

From his experience with tiny, charged pith balls representing electric point charges, Coulomb found that each electric point charge exerts a mechanical force on the other, and this force exhibits five characteristics as follows:

(*i*) The force is proportional to the product of the magnitudes Q_1 and Q_2 of the two electric charges.

(*ii*) Following the suggestion made in 1746 by the American statesman Benjamin Franklin (1706–1790) to classify the electric charges as + and –, Coulomb found that like charges (both + or both –) repelled each other, but unlike charges (one + and the other –) attracted each other.

(*iii*) The force is inversely proportional to the square of the distance between the two charges.

(*iv*) The force depends upon the medium in which the two charges are located.

(*v*) The force always acts along the straight line joining the two point charges Q_1 and Q_2.

These conclusions were combined and expressed mathematically in the form (Fig. 2.1).

Force exerted by Q_1 on Q_2

$$= \overline{\boldsymbol{F}}_{12} = k\frac{(+Q_1)(+Q_2)}{\varepsilon r^2}\overline{\boldsymbol{u}}_{12} \qquad ...(2.1)$$

Force exerted by Q_2 on Q_1

$$= \overline{\boldsymbol{F}}_{21} = k\frac{(+Q_1)(+Q_2)}{\varepsilon r^2}\overline{\boldsymbol{u}}_{21} \qquad ...(2.2)$$

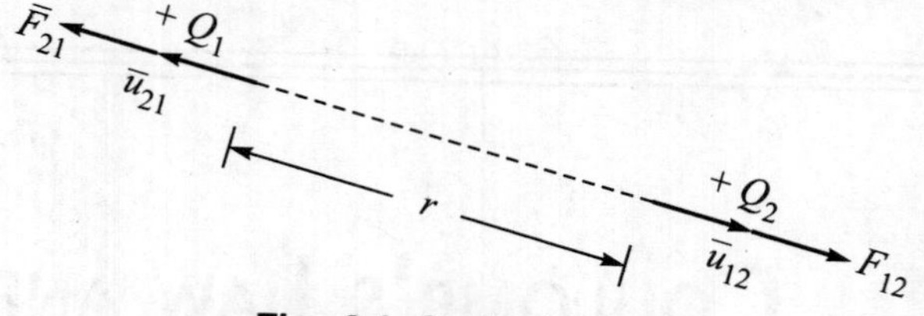

Fig. 2.1. Coulomb's law

where Q_1 and Q_2 are the electric point charges separated by a distance r,

k is a constant of proportionality,

ε is a constant depending on the characteristics of the medium in which Q_1 and Q_2 are situated,

$\overline{u}_{12}$ is the unit vector directed along the straight line from point charge Q_1 towards point charge Q_2, and

$\overline{u}_{21}$ is the unit vector directed along the straight line from point charge Q_2 towards point charge Q_1.

Obviously $\overline{u}_{21} = -\overline{u}_{12}$.

Equation (2.1) [or Eqn. (2.2)] is the mathematical expression for Coulomb's Law for electrical point charges.

In the CGS electrostatic system of units, the unit of electric charge – called a statcoulomb—was defined on the basis of Coulomb's Law by arbitrarily putting $k = 1$ and $\varepsilon = 1$ for vacuum, and then stating that if two equal point charges separated in vacuum by a distance of one centimeter experience a mechanical force of one dyne each, under the influence of the other, then each charge has a magnitude of one statcoulomb. Thus in vacuum

$$(\text{One dyne}) = \frac{(\text{One statcoulomb})(\text{One statcoulomb})}{(\text{One cm})^2} \qquad \text{...(2.3)}$$

The present form of Coulomb's law is very much different from Eqns. (2.1) to (2.3) because of the value chosen for k and the system of units used. The first change was to use, as suggested in 1901 by Italian Professor Giorgi, the Metre-Kilogram-Second (MKS) in place of the Centimeter Gram-Second (CGS) system of units. This with $k = 1$ and $\varepsilon = \varepsilon_0$ for vacuum gave the Coulomb's law in the unrationalized MKS system of units.

It was the British physicist Oliver Heaviside (1850–1925) who suggested that, since the factor 2π frequently comes up in calculations for problems involving a cylindrical geometry, and the factor 4π appears similarly in case of problems with spherical symmetry, a factor 4π may be introduced in the basic Coulomb's law itself. This was done by arbitrarily putting $k = \frac{1}{4\pi}$ giving the rationalized MKS system of units.

Most of the countries of the world have now adopted the International System (SI) of units (see Appendix I). The SI units are used in this book and in these units, Coulomb's law becomes, for vacuum,

$$\overline{F}_{12} = \frac{(+Q_1)(+Q_2)}{4\pi\varepsilon_0 r^2}\overline{u}_{12} \qquad ...(2.4)$$

where $\overline{F}_{12}$ is in newton, Q_1 and Q_2 in coulomb and r in metres. The units for ε_0 are discussed in the next section.

2.2 PERMITTIVITY

The quantity ε_0 in Coulomb's law Eqn. (2.4) is called the permittivity of free space in vacuum. If the charges Q_1 and Q_2 are located in some material, then the symbol ε is used instead of ε_0 in Eqn. (2.4), where

$$\varepsilon = \varepsilon_r \varepsilon_0,$$

or

$$\varepsilon_r = \frac{\varepsilon}{\varepsilon_0} \qquad ...(2.5)$$

It is obvious that ε_r is a dimensionless constant, which may be a real number or a complex number. Therefore ε_r is called the relative permittivity or dielectric constant, because the insulating material separating the charges is called a dielectric. Under the influence of the Coulomb forces or electric field, the molecules in the dielectric get polarized (see Sec. 7.4). If the electric field varies sinusoidally, so will the polarisation in the dielectric vary sinusoidally, but if these variations are not in phase with each other, then this phase difference gives rise to the imaginary part of the relative permittivity.

In SI units, the ampere and the second are two of the fundamental units. Hence the magnitude of the derived unit coulomb is fixed, because a current of one ampere means that a charge of one coulomb is carried per second. Knowing the relation between the units in the CGS and the SI systems, it is possible to find out the magnitude of ε_0.

Thus 1 newton = 10^5 dynes and 1 coulomb $\approx 3 \times 10^9$ statcoulomb and 1 metre = 100 cm. The relation between the coulomb and the statcoulomb is approximate, because it involves the value of the velocity of propagation of electromagnetic waves in free space, whose approximate value 3×10^8 metres per second (m/s) is assumed here, though the accurate value currently accepted is 299 792 457.4 ± 1.1 m/s.

We express Eqn. (2.4) in the form.

$$\varepsilon_0 = \frac{Q_1 Q_2}{4\pi r^2 \overline{F}_{12}}$$

If we have

$Q_1 = Q_2 = 1$ statcoulomb $\approx \dfrac{1}{3\times 10^9}$ coulomb and $r = 1$ cm $= 10^{-2}$ m, we know

from Eqn. (2.3) that $\overline{F}_{12} = 1$ dyne $= 10^{-5}$ N.

Inserting these values in the above equation, we have

$$\varepsilon_0 \approx \frac{\left(\dfrac{1}{3\times 10^9}\right)^2}{4\pi(10^{-2})^2 10^{-5}} \frac{(\text{Coulomb})^2}{\text{Newton (metre)}^2}$$

or
$$\varepsilon_0 \approx \frac{10^{-9}}{36\pi} \frac{(\text{Coulomb})^2}{\text{Newton (metre)}^2} \quad ...(2.6)$$

To put the units for ε_0 in a more familiar form, we note the following:

Stored energy in a capacitor

$$= \frac{1}{2} CV^2 = \frac{1}{2} C\left(\frac{Q}{C}\right)^2 = \frac{1}{2}\frac{Q^2}{C}$$

The units on the two sides of this equation are

$$(\text{Joule}) = \frac{(\text{Coulomb})^2}{(\text{Farad})}$$

But since work = Force × Distance, (Joule) = (Newton) × (metre).

Hence
$$(\text{Farad}) = \frac{(\text{Coulomb})^2}{(\text{Joule})}$$

$$= \frac{(\text{Coulomb})^2}{(\text{Newton})(\text{metre})} \quad ...(2.7)$$

Using Eqn. (2.7) in (2.6), we get

$$\varepsilon_0 \approx \frac{10^{-9}}{36\pi} \text{ Farad/Metre} \quad ...(2.8)$$

If we had used the more accurate value of the velocity of propagation of electromagnetic waves for converting statcoulomb into coulomb, then we would have got the accurate value of ε_0 as

$$\varepsilon_0 = 8.854 \times 10^{-12} \text{ Farad/Metre} \quad ...(2.9)$$

which is the value we shall always use in calculations.

2.3 COMMENTS ON COULOMB'S LAW

Coulomb's Law, Eqn. (2.1) or (2.4) enables us to determine the magnitude and the direction of the mechanical force experienced by a point charge Q_2 under the influence of a point charge Q_1. It gives a unique (that is, a single) answer to each of the questions:

(*i*) What is the magnitude of the force on Q_2?

(*ii*) What is the direction in which the force on Q_2 acts?

However Coulomb's law fails to give unique answers to the inverse problem: Given the magnitude and the direction of the mechanical force on + Q_2.

(*i*) What is the magnitude of the charge Q_1?

(*ii*) What is its polarity?

(*iii*) What is its location and distance from charge Q_2?

All that one can say definitely is that Q_1 must be located on the straight line passing through Q_2 along which the force acts, but an infinite number of combinations of values of Q_1 and r is possible to give the same constant value for $\frac{Q_1}{r^2}$. In the direction of the arrowhead of $\overline{F}$, the polarity of Q_1 will be negative (since Q_2 is positive) and on the opposite side of Q_2, it will be positive.

Next we note that Coulomb's law can only be applied between two *point* charges. If electric charges are distributed over an area or over a volume, then Coulomb's law cannot be applied directly to such total charges. Fortunately, a method of attack on such problems is made possible by the work of Newton, who showed that in mechanics, the effect of a force is not influenced in any way by the presence or absence of any other forces. Hence the total effect of a number of forces acting at a point can be calculated by the concept of superposition and adding the effects of all individual forces. It follows that instead of adding the effects, we may add the forces themselves to get the resultant force which will produce the same total effect.

Thus the resultant force on a test point charge $+ Q_t$ due to the various point charges $Q_1, Q_2, Q_3, ..., Q_n$ is obtained by finding the forces $\overline{F}_{1t}, \overline{F}_{2t}, \overline{F}_{3t}, ... \overline{F}_{nt}$ for the pairs of point charges $(Q_1, Q_t), (Q_2, Q_t), (Q_3, Q_t), ..., (Q_n, Q_t)$ and adding them as follows:

$$\overline{F}_r = \overline{F}_{1t} + \overline{F}_{2t} + \overline{F}_{3t} + ... + \overline{F}_{nt}$$

$$= \frac{Q_1 Q_t}{4\pi\varepsilon_0 r_{1t}^2}\overline{u}_{1t} + \frac{Q_2 Q_t}{4\pi\varepsilon_0 r_{2t}^2}\overline{u}_{2t} + ... + \frac{Q_n Q_t}{4\pi\varepsilon_0 r_{nt}^2}\overline{u}_{nt}$$

or

$$\overline{F}_r = \sum_{n=1}^{n} \frac{Q_n Q_t}{4\pi\varepsilon_0 r_{nt}^2}\overline{u}_{nt} \qquad ...(2.10)$$

By regarding electric charges distributed over an area or over a volume as being the sum of infinitesimally small point charges, the resultant force can be found by applying Eqn. (2.10) and replacing the summation by integrating over suitable limits.

The summation indicated on the right-hand side of Eqn. (2.10) is a vector summation, which can be done in three different ways. The first is a geometrical method of applying the law of parallelogram of forces repeatedly to sum each pair of forces. The second is a trigonometric application of the law of cosines, Eqn. (1.31) repeatedly. The third, and the best method is to resolve each force $\overline{F}_{nt}$ along three mutually orthogonal unit vectors at point where $\overline{Q}_t$ is located (see Sec. 1.14). This permits *scalar* addition of force components along each unit vector separately.

Finally, we consider the characteristics of the medium in which the point charges are located, the characteristic being accounted for by replacing ε_0 by ε (= $\varepsilon_r\varepsilon_0$) in Coulomb's law, given by Eqn. (2.4). The medium is called homogeneous (or inhomogeneous) if the characteristics of the medium do not (or do) depend upon the position of the point charges. Thus air in the laboratory is regarded as being homogeneous, but between a ground station and a satellite, it is inhomogeneous because of variations caused by changes of pressure, temperature, humidity etc.

If the characteristics of the medium do not change with change in the direction of the Coulomb's law force in space, the medium is said to be isotropic. If the characteristics varies with direction, the medium is anisotropic. Most materials are isotropic, but many crystals, and ferrites biased by steady magnetic field are anisotropic.

A medium is said to be linear when the effect of a force is directly proportional to the magnitude of the force, no matter how much the force is increased. Fortunately most media are linear for the normal ranges of forces considered, and hence the principle of superposition on which Eqn. (2.10) is based is valid.

Normally the characteristics of a medium do not change with time. Such a medium is said to be time-invariant. But by passing an ultrasonic wave through air or water, the compressions and rarefactions produced by it at any point will cause the characteristics to vary with time.

Throughout this elementary textbook, the medium will be assumed to be homogeneous, isotropic, linear and time-invariant, unless otherwise stated.

Finally, a question may arise as to what happens to Coulomb's law given by Eqns. (2.1) and (2.2) if the permittivity of the medium is ε_1 at the location of Q_1 and ε_2 at the location of Q_2. In each case, the value to be used in the formula is for permittivity *at the point where the force is acting*. Thus we have the modified forms of these equations as

$$\overline{F}_{12} = \frac{(+Q_1)(+Q_2)}{4\pi\varepsilon_2 r^2}\overline{u}_{12},$$

$$\overline{F}_{21} = \frac{(+Q_1)(+Q_2)}{4\pi\varepsilon_1 r^2}\overline{u}_{21}, \qquad \text{...(2.11)}$$

as seen in Chap 4.

2.4 ELECTRIC FIELD INTENSITY

A region in which a test charge Q_t experiences Coulomb force as a result of the presence of one or more electric point charges [as per Eqn. 2.10] is said to be an electric field of these point charges. Since no restriction is placed on the magnitude of r in Coulomb's law, it follows that the electric field extends to infinity in all directions from the point charges.

To describe the field, we make use of Eqn. (2.10) and adopt a standard procedure by dividing both sides of Eqn. (2.10) by $+Q_t$, the magnitude of the test charge, and then calling this quantity the electric field intensity at that point where Q_t is located. Thus

$$\overline{E}_t = \frac{\overline{F}_r}{Q_t}$$

$$= \sum_{n=1}^{\infty} \frac{Q_n}{4\pi\varepsilon_2 r^2}\overline{u}_{nt} \text{ newton/coulomb} \qquad \text{...(2.12)}$$

Notice that this same result would also have been obtained if we had made Q_t equal to one coulomb in Eqn. (2.10). It appears that newton per coulomb is the unit of electric field intensity. The commonly used unit is volts per metre as shown in Sec. 3.3, or derived as follows:

In any electrical circuit, power = (Volts) × (Amperes) or the units are

$$\frac{\text{(Joule)}}{\text{(Second)}} = \text{(Volt)} \times \frac{\text{(Coulomb)}}{\text{(Second)}}$$

or $$\text{(Newton)} \times \text{(Metre)} = \text{(Volt)} \times \text{(Coulomb)}$$

so that $$\frac{\text{(Newton)}}{\text{(Coulomb)}} = \frac{\text{(Volt)}}{\text{(Metre)}} \qquad ...(2.13)$$

One may well wonder whether this procedure of determining the electric field intensity by placing a test charge $+Q_t$ at the point, determining the value of the resultant force $\overline{F}_r$ by Eqn. (2.10) and then obtaining $\overline{E}_t$ by dividing $\overline{F}_r$ by Q_t is really satisfactory. After all every electric charge produces an electric field. Therefore, this test charge Q_t will produce its own electric field and this will alter the electric field due to the other point charges which we wish to measure. Even though we can obtain Eqn. (2.12) from Eqn. (2.10) by making Q_t equal to one coulomb, this would be most unsatisfactory because in electrostatics, one coulomb is a tremendously large amount of charge (see Problem 4.16) and such a huge test charge will materially alter the field intensity we set out to measure. To reduce its disturbing effect, we must make the test charge very small.

From Eqn. (2.10) it is seen that $\overline{F}_r$ is directly proportional to the test charge Q_t. Ideally we would like to make $Q_t = 0$, so that it may not disturb the electric field being measured, but with $Q_t = 0$, $\overline{F}_r$ also becomes zero. But from Eqn. (2.12) we see that their ratio $\dfrac{\overline{F}_r}{Q_t}$ does not become zero! This is the justification for introducing the concept of the electric field intensity in the above form, since $\overline{E}_t$ retains the same value at any point, even when we do not place any test charge there!! Thus we ensure that no distortion or deviation of the electric field is caused by the introduction of a test charge to measure the field intensity.

Throughout this text as well as in the exercises, unless otherwise stated, free space conditions are to be assumed with $\varepsilon = \varepsilon_0$.

2.5 ELECTRICAL FIELD MAP WITH DIRECTION LINES

Since the electric field intensity is a vector quantity, at every point in space it has both magnitude and direction. As it is not possible to depict a three dimensional space on a two-dimensional paper, a plane is selected, and the electric field is mapped on it. On the field map it is easy to show direction of the field intensity by a short segment of a line with an arrow head. If we also try to show the magnitude of the electric field intensity by varying either the length or the width of the arrow, to scale, the map soon becomes exceedingly clumsy and confusing. Hence all attempts to show magnitude on the field map in this manner are abandoned, and we confine ourselves to drawing curved lines on the map in such a way that the tangent to the curve at any point will give the direction of the electric field intensity at that point. Field maps therefore show direction lines only. Arrow heads placed on the direction lines show the positive direction of the field intensity, or the direction in which a positive test charge will experience a force. (By standard convention, the test charge is always positive).

Now electric field intensity exists at every point of the field. If curves are drawn through every point on the map, the whole map will become black and will convey no information at all. Hence only some selected lines are shown on the field map. This selection is used to display any symmetry in the field. Thus for a point charge, we always select lines radiating from the point at equal angular distances from each other. In other cases, this angular symmetry may exist at very great distances away. With this method of selection of direction lines, a rough picture of the magnitude of the electric field is conveyed to the viewer. The number of lines passing through equal area will give an idea of the magnitude of the field intensity; intensity is greater where the lines are crowded together.

Other terms used by different writers to describe direction lines are lines of force, field lines, flux line and stream lines.

2.6 EQUATION TO DIRECTION LINE ON FIELD MAP

A simple step-by-step procedure enables one to derive the equation to a direction line on a field map where the electric field is produced by a known distribution of charge.

Step 1: From physical considerations, one should try to form a rough idea of the shapes of the direction lines near, and also at very great distances away from the given charge distribution. This helps in taking the second step.

Step 2: From considerations of symmetry, a suitable plane is selected for which the field map is to be drawn. This will also indicate which coordinate system–cartesian, cylindrical or spherical—should be selected so as to minimize complexity in calculations.

Step 3: Choosing an infinitesimally small charge in the charge distribution, and choosing an arbitrary point P in space (but lying on the plane chosen in Step 2), the electric field intensity $\overline{dE}$ produced at P by the point charge is found. It acts along the straight line joining the charge to P. Then choosing three mutually orthogonal unit vectors as per Sec. 1.14, the components of $\overline{dE}$ along these vectors are found out.

Step 4: As the infinitesimally small charge is progressively shifted to cover the entire charge distribution, an integration, over appropriate limits, gives the resultant components of the total field intensity $\overline{E}$ at P along the three units vectors. For this integration, the co-ordinates of P are treated as constants.

Step 5: Since the electric field intensity at P is in the direction of the tangent to the direction line passing through P, an infinitesimally small segment $\overline{ds}$ of the direction line at P can be regarded as representing the electric field intensity, both in magnitude (to some scale) and direction. Hence by resolving $\overline{ds}$ along the three mutually orthogonal unit vectors at P as shown in Sec. 1.14, we can obtain any one of the following equations:

Cartesian co-ordinates : $\dfrac{dx}{E_x} = \dfrac{dy}{E_y} = \dfrac{dz}{E_z}$ ...(2.14)

Cylindrical co-ordinates : $\dfrac{dR}{E_R} = \dfrac{Rd\phi}{E_\phi} = \dfrac{dz}{E_z}$ ...(2.15)

Spherical co-ordinates : $\dfrac{dr}{E_r} = \dfrac{r\,d\theta}{E_\theta} = \dfrac{r \sin \theta \, d\phi}{E_\phi}$ ...(2.16)

Step 6: The above equations incorporate (and are based on) the data about the charge distribution, the co-ordinates of the point P, and the differential elements at P along the three unit vectors. By solving these differential equations, we shall get relationships involving the co-ordinates of any point on the direction line and some arbitrary constants. These will be the equations of the direction lines. By giving different values to the arbitrary constants, we can get the different direction lines.

2.7 ELECTRIC FIELD MAP FOR A POINT CHARGE

From Coulomb's law it is obvious that everywhere, the electric field intensity will be directed along the radius vector radiating from a positive charge + Q coulomb. Also the field will have spherical symmetry. In Fig. 2.2 the field map is shown for any plane passing through the point where the point charge + Q is located.

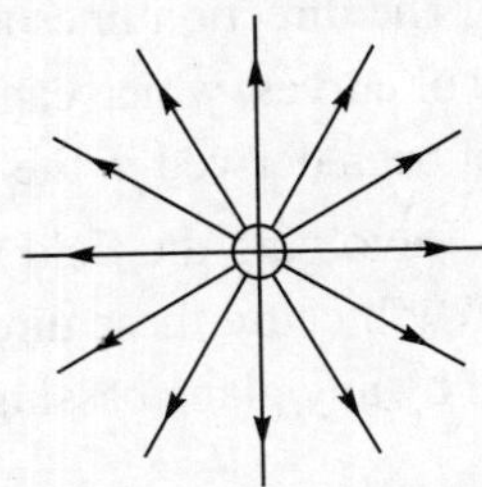

Fig. 2.2. Field map of point charge + Q

If the point charge was – Q coulomb, the same field map will be obtained, but all its arrow heads will be reversed. In both cases the field lines are supposed to extend up to infinity in all directions. The three-dimensional field can be visualised by remembering that the same map is obtained for every plane passing through the point charge, but it cannot be produced by rotating the figure about any axis.

2.8 ELECTRIC FIELD MAP OF TWO LIKE AND EQUAL POINT CHARGES

Let the charges be $+Q$ coulomb each, separated by a distance $2a$. As the field will have cylindrical symmetry about the line joining the charges, the z-axis is chosen as this line with the origin midway between the point charges. The field map is drawn for any plane passing through the z-axis (that is for ϕ = constant) as in Fig. 2.3.

By symmetry, for $R = 0$, $E_R = 0$ and for $z = 0$, $E_z = 0$, and at the origin, there is no field (neutral point).

Along the z-axis, from $-\infty < z < -a$ and from $+a < z < +\infty$ the electric field intensities due to both the charges act in the same direction away from O, but for $-a < z < +a$, they act in opposite directions, but with unequal magnitudes because of unequal distances from the charges. Hence the fields are as shown in Fig. 2.3.

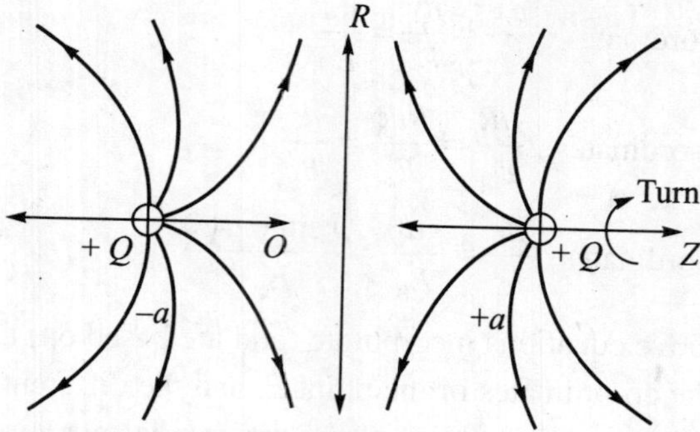

Fig. 2.3. Field map of two like and equal point charges

At distances very far away from the origin O, the two charges appear as if they are both located at the origin. Hence all direction lines at very great distances will tend to become radial lines passing through O. Such equally spaced radial lines are therefore selected for drawing the field map of Fig. 2.3.

The direction lines are curves of many different types, and they do not form a family of curves, which can be described by a single equation. Hence such an equation has not been derived in this case.

Revolving the field map about the z-axis in the direction of the arrow marked "Turn" will cause these direction lines to describe three-dimensional surfaces, whose section by any plane passing through the z-axis is given in Fig. 2.3.

2.9 ELECTRIC FIELD MAP OF TWO UNLIKE AND EQUAL CHARGES (ELECTRIC DIPOLE)

Two equal but opposite electric charges $-Q$ and $+Q$ coulomb separated by a distance $2a$ form what is called an electric dipole. The product $(2aQ)$ is called the dipole moment $\overline{P}$, which is a vector quantity of magnitudes $2aQ$ and which is directed from $-Q$ to $+Q$.

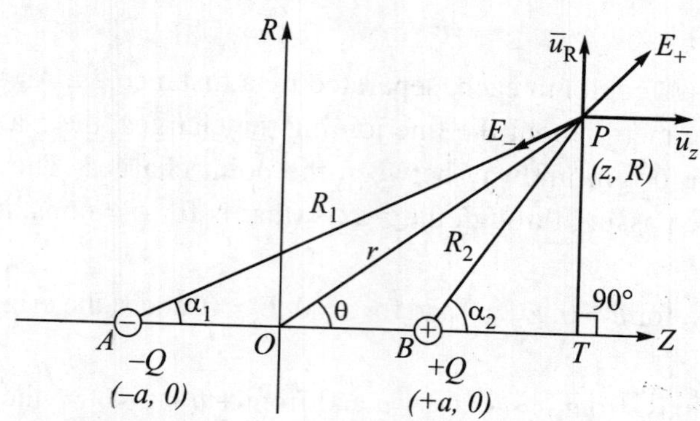

Fig. 2.4. To find electric field intensity produced by an electric dipole

It is obvious that the field map will have cylindrical symmetry. As in the previous section, the z-axis is chosen to pass through the two charges and the origin O is midway between the charges. The field map is to be made for any plane passing through the z-axis (for ϕ = constant). The line through O and at right angle to the z-axis is here called the R-axis (Fig. 2.4), with co-ordinates of any point given as (z, R). Thus charge $-Q$ is located at $A(-a, 0)$ and $+Q$ at $B(+a, 0)$.

At the point $P(z, R)$ where $OT = z$, $TP = R$, $AP = R_1$ and $BP = R_2$, the electric field intensity due to the positive and the negative charges is given by

$$E_+ = \frac{+Q}{4\pi\varepsilon_0 R_2^2}$$

$$E_- = \frac{-Q}{4\pi\varepsilon_0 R_1^2}$$

Taking their resolved parts along the unit vectors $\bar{\boldsymbol{u}}_z$ and $\bar{\boldsymbol{u}}_R$, we get

$$E_Z = \frac{1}{4\pi\varepsilon_0}\left[\frac{Q(z-a)}{R_2^2 R_2} + \frac{-Q(z+a)}{R_1^2 R_1}\right] \quad ...(2.17a)$$

$$E_R = \frac{1}{4\pi\varepsilon_0}\left[\frac{QR}{R_2^2 R_2} + \frac{-QR}{R_1^2 R_1}\right] \quad ...(2.17b)$$

From these we get, using Eqn. (2.15)

$$\frac{dR}{dz} = \frac{E_R}{E_Z} = \frac{\dfrac{R}{R_2^3} - \dfrac{R}{R_1^3}}{\dfrac{(z-a)}{R_2^3} - \dfrac{(z+a)}{R_1^3}}$$

$$= \frac{R(R_1^3 - R_2^3)}{z(R_1^3 - R_2^3) - a(R_1^3 + R_2^3)} \quad ...(2.17)$$

This differential equation cannot be solved easily, because both R_1 and R_2 are functions of z and R. Hence the simplifying assumption is made that z and R are large as compared to a. Then using Binomial Theorem Eqn. (1.1) and neglecting a^2 as compared to az, we get

$$R_1^3 = \left[\sqrt{(z+a)^2 + R^2}\right]^3$$

$$= \left[\sqrt{(R^2 + z^2) + (2az + a^2)}\right]^3$$

$$= \left(R^2 + z^2\right)^{\frac{3}{2}}\left[1 + \frac{2az + a^2}{R^2 + z^2}\right]^{\frac{3}{2}}$$

$$= \left(R^2 + z^2\right)^{\frac{3}{2}}\left[1 + \frac{3}{2}\cdot\frac{2az}{R^2 + z^2}\right]$$

$$R_2^3 = \left[\sqrt{(z-a)^2 + R^2}\right]^3$$

$$= \left(R^2 + z^2\right)^{\frac{3}{2}}\left[1 + \frac{3}{2}\cdot\frac{(-2az)}{R^2 + z^2}\right]$$

Substituting these values in Eqn. (2.17) and cancelling out the common factor $\left(R^2 + z^2\right)^{\frac{3}{2}}$, we get

$$\frac{dR}{dz} = \frac{6azR}{6az^2 - 2a(R^2 + z^2)} = \frac{3zR}{2z^2 - R^2}$$

or
$$2z^2\, dR = R^2\, dR + 3\, R\, zdz$$
$$= (R^2\, dR + R\, zdz) + 2R\, zdz$$

or
$$2z^2\, dR = R(R\, dR + zdz) + 2R\, zdz$$

Adding $2R^2\, dR$ to both the sides

$$2(R^2 + z^2)\, dR = R(R\, dR + zdz) + 2R(zdz + R\, dR)$$

or
$$2\frac{dR}{R} = \frac{3}{2}\,\frac{2R\,dR + 2\,zdz}{R^2 + z^2} = \frac{3}{2}\,\frac{d(R^2 + z^2)}{R^2 + z^2}$$

Integrating, we get the equation to the direction line as

$$2 \log_e R = \frac{3}{2} \log_e (R^2 + z^2) + \text{Constant}$$

Writing the constant as $-\log_e A$ where A is a constant, the equation to the direction line becomes

$$\left(R^2 + z^2\right)^{\frac{3}{2}} = AR^2 \tag{2.18}$$

By giving different values to A, we can plot the system of direction lines as in Fig. 2.5, where $\overline{\boldsymbol{P}}$ shows the direction of the dipole moment. Though each direction line must start at the charge $+\,Q$ and end on the charge $-\,Q$, they are not so shown on the field map because Eqn. (2.18) is not valid for small values of z and R near the dipole.

From a point very far away from the origin, the field is as if the two charges are located at the origin. But the two charges being equal and opposite, this makes the total charge as zero at the origin. This shows that the dipole has very little effect at very great distances from it. This is indicated in Fig. 2.5 by direction lines leaving $+\,Q$ curving round to converge towards $-\,Q$.

A very special feature of this field map is the direction of the direction lines at all points on the plane $z = 0$ shown dotted in Fig. 2.5. For every point on this plane, $R_1 = R_2$. Therefore $\overline{\boldsymbol{E}}_+$ and $\overline{\boldsymbol{E}}_-$ are equal in magnitude and they make equal and opposite angles with the unit vector $(-\,\overline{\boldsymbol{u}}_z)$. The resolved components of $\overline{\boldsymbol{E}}_+$ and $\overline{\boldsymbol{E}}_-$ parallel to the R-axis cancel out [Eqn. (2.17b)], and the components parallel to the z-axis add to give a resultant $\overline{\boldsymbol{E}}$ that is always normal to the plane $z = 0$ and directed from right to left. A very interesting use of this special feature is made later in Sec. 5.2.

The three dimensional surface is obtained by revolving the field map about the z-axis as indicated by arrow marked "TURN".

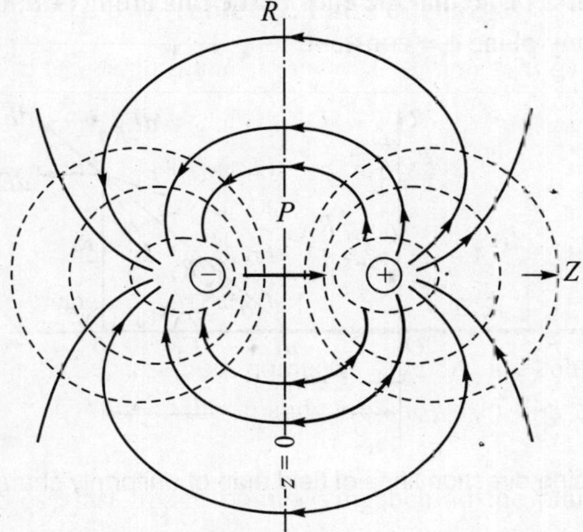

Fig. 2.5. Electric field map of electric dipole showing equipotentials as dotted lines

2.10 ELECTRIC FIELD MAP OF UNIFORMLY CHARGED SHORT LINE

A uniform linear charge density of + q_l coulomb per metre exists along a short straight line of length $2a$ metres. At very short distances, the Coulomb's force is very large, being inversely proportional to r^2. Hence for any point close to the charged line, the charge opposite to the point will contribute the major part of the field intensity [see Fig. 2.6(a)].

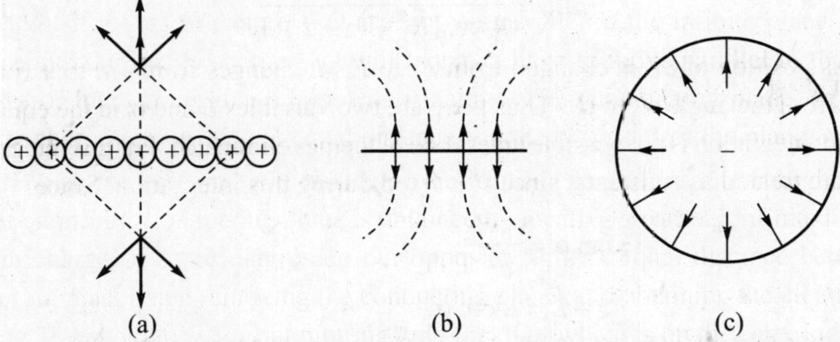

Fig. 2.6. Field at short (a, b) and at great (c) distances from charged line

Moreover, symmetrically situated charges will produce equal and symmetrical forces, whose components parallel to the line will cancel (except near the ends). Hence the direction lines will be normal to the line at all points [Fig. 2.6(b)]. For very large distances, however, the short line will appear like a point charge of ($+2aq_1$) coulomb located at the centre of the line and direction lines will appear to be tending to become straight lines radiating from the centre of the short line [Fig. 2.6(c)].

As there is cylindrical symmetry in the system, the field map is found using cylindrical co-ordinates with the z-axis along the charged line and the origin at the centre of the line (Fig. 2.7), so that the ends of the line are $F_1(+a, 0)$ and $F_2(-a, 0)$ in z-R co-ordinates for any plane ϕ = constant.

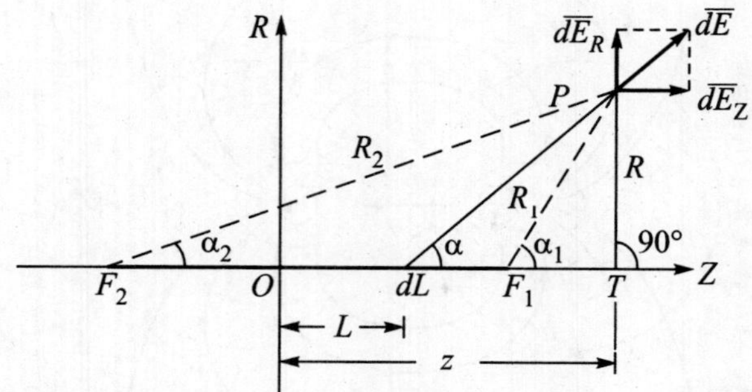

Fig. 2.7. For finding direction lines of field map of uniformly charged short line

Choose an infinitesimally small length dL of the line at a distance L from the origin O. The magnitude of the "point" charge is then $(+\,q_l\,dL)$. For any point $P(z, R)$, $OT = z$, $TP = R$, and the distance from dL to P is $\sqrt{R^2 + (z-L)^2}$. The electric field intensity produced by $(+\,q_l\,dL)$ at P is

$$dE = \frac{q_l\,dL}{4\pi\varepsilon_0[R^2 + (z-L)^2]}$$

in the direction shown. Its components along the z and R axes are

$$dE_z = \frac{+\,q_l\,dL\cos\alpha}{4\pi\varepsilon_0[R^2 + (z-L)^2]}$$

and

$$dE_R = \frac{+\,q_l\,dL\sin\alpha}{4\pi\varepsilon_0[R^2 + (z-L)^2]} \qquad ...(2.19)$$

As position of dL is changed from F_2 to F_1, L changes from $-\,a$ to $+\,a$ and α changes from the angle α_2 to α_1. Thus there are two variables L and α in the equations (2.10) and integration is not possible unless one of them is eliminated. For this integration, z and R are treated as constants, since P is fixed during this integration. Since

$$\tan\alpha = \frac{R}{z-L},$$

$$z - L = R\cot\alpha.$$

Differentiating

$$-\,dL = -\,R\,\text{cosec}^2\,\alpha\,d\alpha$$

$$= -\,R\,\frac{[R^2 + (z-L)^2]}{R^2}\,d\alpha$$

Substituting this value of dL in Eqn. (2.19) and integrating

$$E_z = \int_{\alpha_2}^{\alpha_1} \frac{q_l\cos\alpha}{4\pi\varepsilon_0[R^2 + (z-L)^2]}\,\frac{[R^2 + (z-L)^2]\,d\alpha}{R}$$

$$= \frac{q_l}{4\pi\varepsilon_0 R}\int_{\alpha_2}^{\alpha_1}\cos\alpha\,d\alpha$$

so that
$$E_z = \frac{q_l}{4\pi\varepsilon_0 R}(\sin\alpha_1 - \sin\alpha_2) \quad ...(2.20)$$

and
$$E_R = \int_{\alpha_2}^{\alpha_1} \frac{q_l \sin\alpha}{4\pi\varepsilon_0[R^2 + (z-L)^2]} \frac{[R^2 + (z-L)^2]\, d\alpha}{R}$$
$$= \frac{q_l}{4\pi\varepsilon_0 R}\int_{\alpha_2}^{\alpha_1} \sin\alpha \, d\alpha$$

or
$$E_R = \frac{q_l}{4\pi\varepsilon_0 R}(\cos\alpha_2 - \cos\alpha_1) \quad ...(2.21)$$

To get the equation to the direction line, we first use Eqn. (2.15) and write
$$\frac{dR}{dz} = \frac{E_R}{E_z} = \frac{\cos\alpha_2 - \cos\alpha_1}{\sin\alpha_1 - \sin\alpha_2}$$
$$= \frac{\dfrac{z+a}{R_2} - \dfrac{z-a}{R_1}}{\dfrac{R}{R_1} - \dfrac{R}{R_2}}$$

where $F_1P = R_1$ and $F_2P = R_2$.

Hence $$\frac{(z+a)\,dz}{R_2} + \frac{R\,dR}{R_2} = \frac{(z-a)\,dz}{R_1} + \frac{R\,dR}{R_1} \quad ...(2.22)$$

But $$R_1^2 = (z-a)^2 + R^2$$

and $$R_2^2 = (z+a)^2 + R^2 \quad ...(2.23)$$

which give on differentiation
$$2\,R_1\,dR_1 = 2(z-a)\,dz + 2R\,dR$$
$$2\,R_2\,dR_2 = 2(z+a)\,dz + 2R\,dR$$

Using these values Eqn. (2.22) reduces to
$$dR_2 = dR_1$$

the solution for which is
$$|\,R_2 - R_1\,| = \text{Constant} \quad ...(2.24)$$

Comparing Eqns. (2.24) and (1.129) we find that the direction lines are a family of hyperbola with F_1 and F_2 as the two foci.

Caution: In Eqn. (1.129), $2a = AA'$ but in this problem, the distance F_1F_2 is labelled as $2a$.

The equation to the direction line is then easily obtained. Since $|\,R_2 - R_1\,| = c$, where c is a constant, we get on squaring and transposing
$$2\,R_2\,R_1 = R_2^2 + R_1^2 - c^2$$

Substituting the values of R_1 and R_2 from Eqn. (2.23)
$$2\sqrt{(z^2 + a^2 + R^2 + 2az)(z^2 + a^2 + R^2 - 2az)} =$$
$$(z+a)^2 + R^2 + (z-a)^2 + R^2 - c^2$$

or $$2\sqrt{(z^2+a^2+R^2)^2-4a^2z^2}=2(z^2+a^2+R^2)-c^2$$

Squaring again

$$4(z^2+a^2+R^2)^2-16a^2z^2=4(z^2+a^2+R^2)^2+c^4-4c^2(z^2+a^2+R^2)$$

Removing the common first term, and transposing

$$4c^2(z^2+R^2)-16a^2z^2=c^4-4c^2a^2$$

or $$4(c^2-4a^2)z^2+4c^2R^2=c^2(c^2-4a^2)$$

or $$\frac{z^2}{\left(\frac{c}{2}\right)^2}-\frac{R^2}{a^2-\left(\frac{c}{2}\right)^2}=1 \qquad \text{...(2.25)}$$

the equation to the hyperbola in the standard form of Eqn. (1.127). Since the distance $F_1F_2 = 2a$, it follows from Eqn. (2.24) that c can have values only between 0 and $2a$. The asymptotes of (2.25) are obtained using Eqn. (1.130) as

$$R=\pm\frac{\sqrt{a^2-\left(\frac{c}{2}\right)^2}}{\frac{c}{2}}z \qquad \text{...(2.26)}$$

In Fig. 2.8 direction lines are selected to provide equally spaced asymptotes at 90°, 60° and 30° for values of $c = 0$, a, and $\sqrt{3}a$ respectively, as suggested in Fig. 2.6(*c*).

The three dimensional surface is obtained as hyperboloids of revolution by revolving the field map about the z-axis as indicated by the arrow marked "TURN".

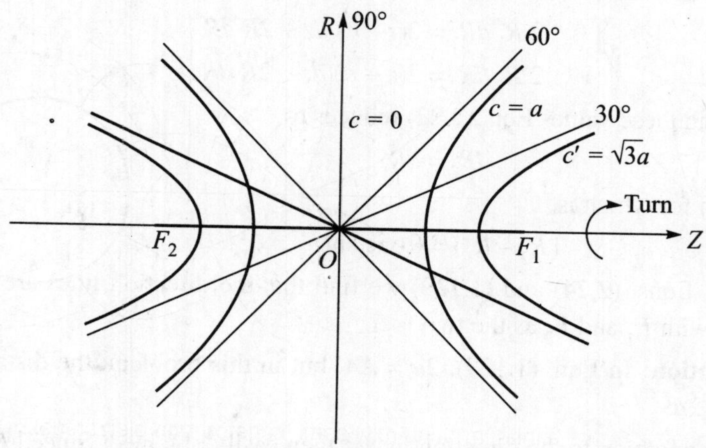

Fig. 2.8. Electric field map of uniformly charged short line

2.11 UNIFORMLY CHARGED INFINITE LINE

Uniform charge density of $+q_l$ coulomb per metre distributed along an infinitely long straight line can be thought of as an extension of the line of Fig. 2.7, in which a is made infinite. This makes $\alpha_1 = \pi$ and $\alpha_2 = 0$ and reduces Eqn. (2.20) to $E_z = 0$ and Eqn. (2.21) to

$$E_R = \frac{q_l}{4\pi\varepsilon_0 R}(\cos 0 - \cos \pi)$$

or

$$\bar{E}_R = \frac{q_l}{2\pi\varepsilon_0 R}\bar{u}_R \qquad ...(2.27)$$

Since the line is infinitely long extending from $z = -\infty$ to $z = +\infty$, for every point P in space, there are two symmetrically situated points on the line equidistant from P which produce electric field intensities at P, such that their z-components cancel each other. Therefore the electric field intensity is everywhere directed away from the charged line, in a perpendicular direction to the line. On every plane at right angles to the infinite line, the field map will look like the field map of a point charge (Fig. 2.2). The three-dimensional picture of the field map of the infinite line can be visualized as being obtained by moving Fig. 2.2 parallel to itself, keeping its centre always on the infinite line.

Equation (2.27) is so important that, like the point charge, the infinitely long uniformly charged line can be used as a fundamental building block for studying more complicated charge distributions.

2.12 FIELD MAP OF TWO INFINITE PARALLEL LINES WITH EQUAL AND OPPOSITE UNIFORM LINEAR CHARGES

In cartesian co-ordinates, we assume that these lines are parallel to the z-axis (Fig. 2.9) and intersecting the x-axis at $B(x = -a)$ and $C(x = +a)$. The linear electric charge density is $(-q_l)$ coulomb/metre on the B-line and $(+q_l)$ C/m on the C-line.

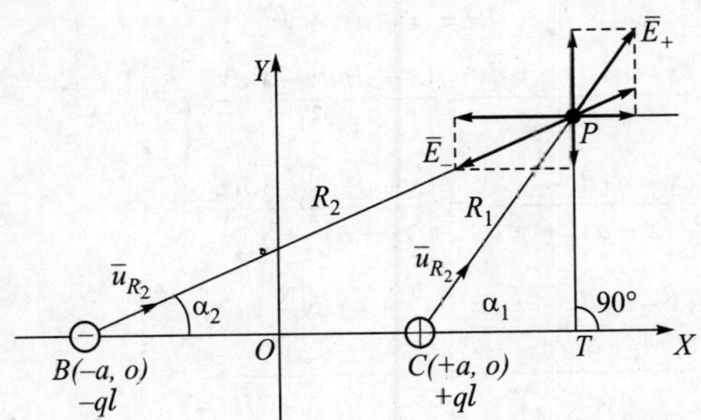

Fig. 2.9. Electric field intensity produced by parallel charged lines of opposite polarity

Let P be any point with PT perpendicular to x-axis, so that $OT = x$ and $TP = y$. Then denoting $CP = R_1$, $BP = R_2$, $m\angle PCT = \alpha_1$ and $m\angle PBT = \alpha_2$, we get, using Eqn. (2.27), the electric field intensities at P produced by these two infinite lines as

$$\overline{E}_+ = \frac{(+q_l)}{2\pi\varepsilon_0 R_1}\overline{u}_{R_1}$$

and $$\overline{E}_- = \frac{(-q_l)}{2\pi\varepsilon_0 R_2}\overline{u}_{R_2}$$

Taking their components along the x and y axes, we get

$$E_x = \frac{(q_l)}{2\pi\varepsilon_0}\left(\frac{\cos\alpha_1}{R_1} - \frac{\cos\alpha_2}{R_2}\right)$$

$$= \frac{(q_l)}{2\pi\varepsilon_0}\left(\frac{x-a}{R_1^2} - \frac{x+a}{R_2^2}\right) \qquad ...(2.28)$$

and $$E_y = \frac{(q_l)}{2\pi\varepsilon_0}\left(\frac{\sin\alpha_1}{R_1} - \frac{\sin\alpha_2}{R_2}\right)$$

$$= \frac{(q_l)}{2\pi\varepsilon_0}\left(\frac{y}{R_1^2} - \frac{y}{R_2^2}\right) \qquad ...(2.29)$$

Using Eqn. (2.14)

$$\frac{dy}{dx} = \frac{E_y}{E_x}$$

$$= \frac{\dfrac{y}{R_1^2} - \dfrac{y}{R_2^2}}{\dfrac{x-a}{R_1^2} - \dfrac{x+a}{R_2^2}} \qquad ...(2.30)$$

Cross multiplying and putting

$$R_1^2 = (x-a)^2 + y^2$$

and $$R_2^2 = (x+a)^2 + y^2$$

we get $$\frac{(x-a)\,dy - y\,dx}{R_1^2} = \frac{(x+a)\,dy - y\,dx}{R_2^2}$$

$$\frac{(x-a)\,dy - y\,dx}{(x-a)^2 + y^2} = \frac{(x+a)\,dy - y\,dx}{(x+a)^2 + y^2}$$

$$\frac{\dfrac{(x-a)\,dy - y\,dx}{(x-a)^2}}{1+\left(\dfrac{y}{x-a}\right)^2} = \frac{\dfrac{(x+a)\,dy - y\,dx}{(x+a)^2}}{1+\left(\dfrac{y}{x+a}\right)^2} \qquad ...(2.31)$$

The purpose of these manipulations which have only yielded these complicated looking expressions will now become clear when we substitute

$$\frac{y}{x-a} = u$$

and $$\frac{y}{x+a} = v \qquad ...(2.32)$$

and compare the numerators with Eqn. (1.47) and note that Eqn. (2.31) reduces to

$$\frac{du}{1+u^2} = \frac{dv}{1+v^2}$$

Integrating this using Eqn. (1.54) we get the solution

$$\tan^{-1} u = \tan^{-1} v + \text{constant} \qquad \text{...(2.33}a\text{)}$$

But $$u = \frac{y}{x-a} = \tan \alpha_1$$

and $$v = \frac{y}{x+a} = \tan \alpha_2$$

This changes Eqn. (2.33*a*) to

$$\alpha_1 - \alpha_2 = \text{Constant} \qquad \text{...(2.33}b\text{)}$$

or $$m\angle PCT - m\angle PBT = \text{Constant}$$

$$= A(\text{say}) \qquad \text{...(2.33)}$$

But in triangle *PBC*, $\angle PCT$ is the exterior angle and hence

$$m\angle PCT = m\angle PBT - m\angle BPC$$

so that Eqn. (2.33) means that

$$m\angle BPC = \text{Constant} = A \qquad \text{...(2.34)}$$

Since *BC* is fixed, this equation implies [Fig. 1.20(*b*)] that *P* lies on the segment of a circle of which *BC* is a chord. If the circle intersects the *y*-axis at the point *M*, then (Fig. 2.10).

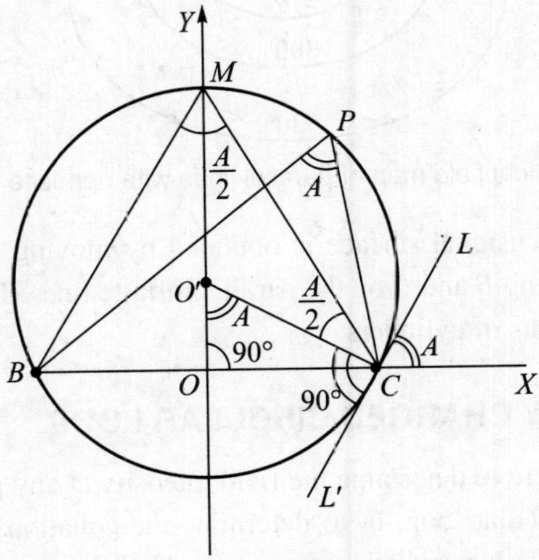

Fig. 2.10. Geometrical construction for direction line for Fig. 2.9

$$m\angle PBT - m\angle BMC = A$$

If O' is the centre of the circle and $O'M$ and $O'C$ are the radii, then the geometrical construction for drawing the circle becomes apparent. At C, a line LCL' is drawn making an angle

$$\angle LCX = \angle L'CO = A.$$

Then from the angles indicated in Fig. 2.10, LCL' is a tangent to the circle at C. By drawing the perpendicular CO' to LCL' and finding the point O' where it intersects the y-axis, the centre O' is located, and $O'C$ becomes the radius.

The electric field map will then be obtained as a system of circles, when different tangent lines are selected by giving different values to the angle A. The family of segments of circle (see Fig. 2.11) always pass through the points B and C. The circles are marked with the values of the angle A that the tangent to the circle at C makes with the axis of x.

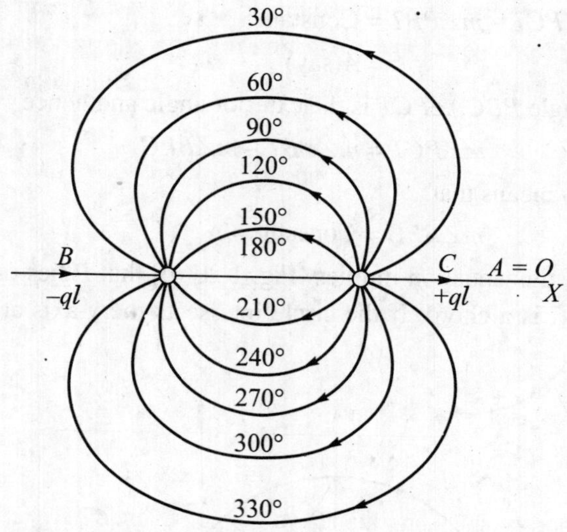

Fig. 2.11. Electrical field map of parallel lines with opposite charge densities

The three dimensional surface is obtained by moving the plane of Fig. 2.11 parallel to itself, keeping B and C on the parallel infinite lines. The segments of circles will then form segments of cylinders.

2.13 UNIFORMLY CHARGED CIRCULAR LOOP

No attempt is made here to determine the field intensity at any point in space due to a uniformly charged circular loop, or to determine the equation to the direction lines, because the process involves a difficulty explained below.

If in the case of an ellipse, we attempt to find (Fig. 1.22) the length of the arc BP by putting $\angle BOP = \phi$, and writing $x = a \sin \phi$ and $y = b \cos \phi$ and eccentricity of ellipse as

$$k = \sqrt{1 - \frac{b^2}{a^2}}, \text{ we get}$$

$$BP = \int dS = \int \sqrt{dx^2 + dy^2}$$

$$= \int_0^{\phi} \sqrt{a^2 \cos^2 \phi + b^2 \sin^2 \phi}\, d\phi$$

or

$$E(k, \phi) = a \int_0^{\phi} \sqrt{1 - k^2 \sin^2 \phi}\, d\phi, \; (0 < k < 1) \qquad ...(2.35)$$

Because of its origin, the expression $E(k, \phi)$ defined by Eqn. (2.35) is called the elliptic integral of the second kind. A corresponding integral $F(k, \phi)$ called the elliptic integral of the first kind has the radical term in the denominator. It occurs in the analysis of a simple pendulum, if the amplitude of oscillations is large and we cannot use the simplification $\sin \phi \approx \phi$. Unfortunately mathematicians have not been able to find a solution to these elliptic integrals in finite form using trigonometric or inverse trigonometric, exponential or logarithmic functions. Problems involving elliptic integrals can only be solved using tables giving values of the integral for different values of ϕ and $\sin^{-1} k$.

For our problem of the uniformly charged circular loop, the elliptic integral appears in the analysis for points not lying on the axis of the loop. That is why we confine our analysis to a point P located on the z-axis, when the loop is located in the plane $z = 0$ (Fig. 2.12) and the loop can be described by the equation $x^2 + y^2 = a^2$ in cartesian co-ordinates or by $R = a$ in cylindrical co-ordinates. The charge density is $+ q_l$ coulomb per metre along the circumference of the loop.

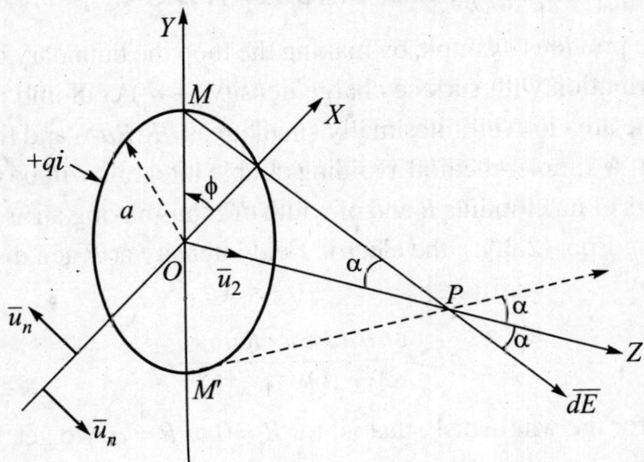

Fig. 2.12. Electric field intensity at a point on the axis of uniformly charged circular loop

A small arc of length $a\, d\phi$ located at M will produce an electric field intensity $\overline{dE}$ at P as shown, where

$$dE = \frac{q_l\, a\, d\phi}{4 \pi \varepsilon_0 (a^2 + z^2)}$$

since $OM = a$ and $OP = z$. The field intensity dE can be resolved along OZ and at right angles to OZ. For a symmetrical point M' diametrically opposite to M, the field intensity (shown dotted) will be equal in magnitude to dE but with an equal inclination on the other side of OZ. The resolved parts of these two intensities will add along the z-axis but the resolved parts at right angles to OZ will cancel each other. Hence we shall have to integrate only the component of dE along OZ. Then with ϕ varying from 0 to 2π, $m\angle MPO = \alpha$.

$$E_z = \int_0^{2\pi} \frac{q_l\, a\, d\phi}{4\pi\varepsilon_0(a^2+z^2)} \cos\alpha$$

$$= \int_0^{2\pi} \frac{q_l\, a\, d\phi\, z}{4\pi\varepsilon_0(a^2+z^2)^{3/2}} = \frac{q_l(2\pi a)z}{4\pi\varepsilon_0(a^2+z^2)^{3/2}}$$

or
$$\overline{E}_z = \frac{(q_l\, 2\pi a)z}{4\pi\varepsilon(a^2+z^2)^{3/2}}\overline{u}_z,\ \overline{E}_R = 0 \qquad \text{...(2.36)}$$

On the negative side of the z-axis, the field will be in the reverse direction but of the same magnitude as given in Eqn. (2.36), because only the sign of z will become negative. On both sides therefore, the electric field intensity is directed along the *outward* normal to the plane, denoted by $\overline{u}_n$. Then putting $(+\, q_l\, 2\pi\, a) = +\, Q$, the total charge on the circular loop, Eqn. (2.36) can also be written as

$$\overline{E}_z = \frac{Q|z|}{4\pi\varepsilon_0(a^2+z^2)^{3/2}}\overline{u}_n, \qquad \text{...(2.36a)}$$

2.14 UNIFORMLY CHARGED CIRCULAR DISK

If we modify the previous example by making the loop the boundary of a circular disk-like charge distribution with surface charge density $(+\, q_s)$ coulomb per square metre, we can divide the area into infinitesimally small areas $dR(Rd\phi)$ and then proceed as in previous section. A simpler method avoiding double integrals will be to divide the area into circular rings of inner radius R and of width dR, thus having an area $(2\pi R)\, dR$ (Fig. 2.13). Then using Eqn. (2.36*a*), the electric field intensity at P at a distance z from the origin is given by

$$\overline{dE} = \frac{q_s(2\pi R\, dR)|z|}{4\pi\varepsilon_0(R^2+z^2)^{3/2}}\overline{u}_n$$

Integrating this for the whole disk, that is, for $R = 0$ to $R = a$, we get

$$\overline{E} = \frac{2\pi q_s|z|\overline{u}_n}{4\pi\varepsilon_0}\int_0^a \frac{R\, dR}{(R^2+z^2)^{3/2}} \qquad \text{...(2.37)}$$

Putting $R^2 + z^2 = v^2$, and differentiating, we get (since z is to be treated as a constant for this integration), $2R\, dR = 2v\, dv$, and Eqn. (2.37) becomes

$$\overline{E} = \frac{q_s|z|\overline{u}_n}{2\varepsilon_0}\int_{R=0}^{R=a} \frac{v\, dv}{v^3}$$

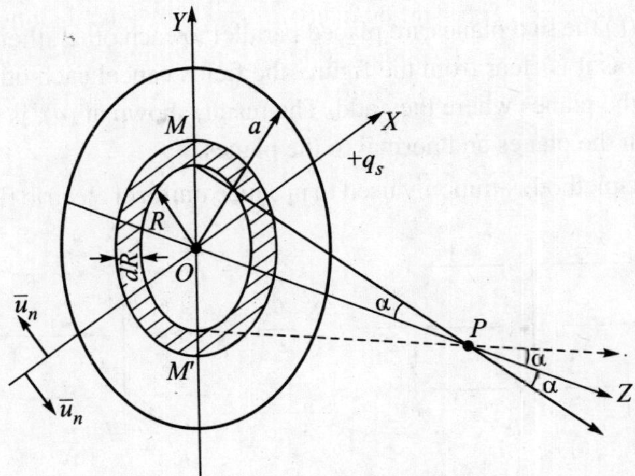

Fig. 2.13. Electric field intensity at a point on axis of uniformly charged circular disk

$$= \frac{q_s \,|\, z \,|\, \overline{\boldsymbol{u}}_n}{2\varepsilon_0}\left[\frac{v^{-1}}{-1}\right]_{R=0}^{R=a} = \frac{q_s \,|\, z \,|\, \overline{\boldsymbol{u}}_n}{2\varepsilon_0}\left[\frac{-1}{\sqrt{a^2+z^2}} - \frac{-1}{z}\right]$$

or
$$\overline{\boldsymbol{E}} = \frac{q_s}{2\varepsilon_0}\left[1 - \frac{|\, z \,|}{\sqrt{a^2+z^2}}\right]\overline{\boldsymbol{u}}_n \qquad ...(2.38)$$

This shows that the field is symmetrical on the two sides of the disk and is always directed along the outward normal to the plane.

2.15 ELECTRIC FIELD OF UNIFORMLY CHARGED INFINITE PLANE

In Eqn. (2.38) if the radius a of the disk is allowed to become infinitely large while the charge density remains the same, $(+q_s)$ C/m^2, then for the uniformly charged infinite plane

$$\overline{\boldsymbol{E}} = \frac{q_s}{2\varepsilon_0}\overline{\boldsymbol{u}}_n \text{ volt/metre} \qquad ...(2.39)$$

This is a remarkable result, because it does not depend on z at all. In other words, the electric field intensity is the same, both in magnitude and direction, everywhere. Such an electric field is called a uniform field. Notice, however, that the direction of the field is reversed on the opposite side of the plane.

2.16 ELECTRIC FIELD OF TWO UNIFORMLY BUT OPPOSITELY CHARGED PARALLEL INFINITE PLANES

In Fig. 2.14 the x-y plane is shown edgewise, so that the horizontal line represents the z-direction, normal to the plane. At (a) the uniform charge density being $(+\, q_s)$ C/m^2, the electric field everywhere is directed away form the plane, while at (b) with charge density $(-\, q_s)$ C/m^2, it is directed towards the plane.

When at (c) the two planes are placed parallel to each other, their fields overlap everywhere. But as it is clear from the figure, the fields cancel each other everywhere except between the planes where they add. The result, shown at (d), is a uniform field intensity between the planes and normal to the planes.

This is the method commonly used to produce uniform electric field in a region.

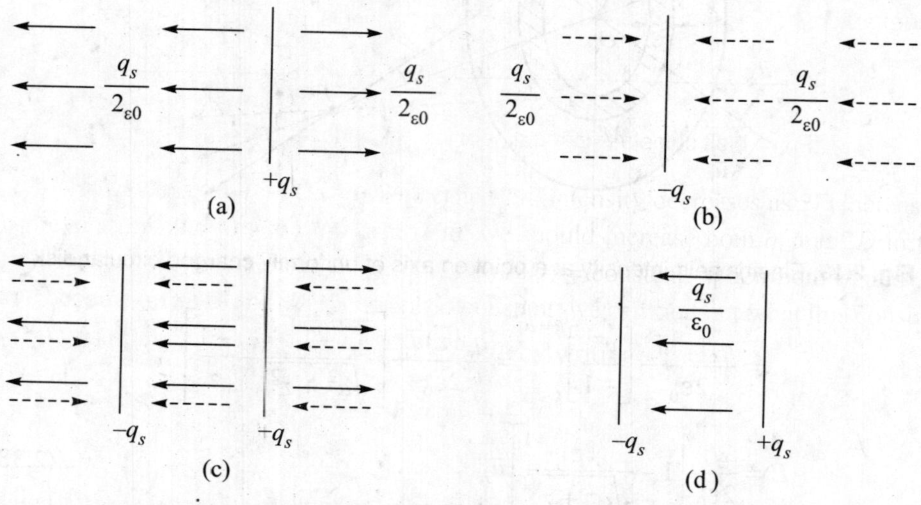

Fig. 2.14. Electric field intensity between two equally but oppositely charged parallel planes

2.17 ELECTRIC FIELD OF UNIFORMLY CHARGED SPHERICAL SURFACE

2.17.1 For Points Outside the Sphere

Uniform electric charge density of ($+\ q_s$) C/m^2 exists on the surface of a sphere of radius a metre. Let P be any point outside the sphere. Choose the centre of the sphere as the origin O and OP as the z-axis, so that $OP = z$. At a point M on the surface of the sphere, choose an infinitesimally small area ($r\ d\theta$) ($r \sin \theta\ d\phi$) where $\theta = m\angle MOP$ (see Fig. 2.15 where plane passing through the z-axis and point M is made the plane of the paper).

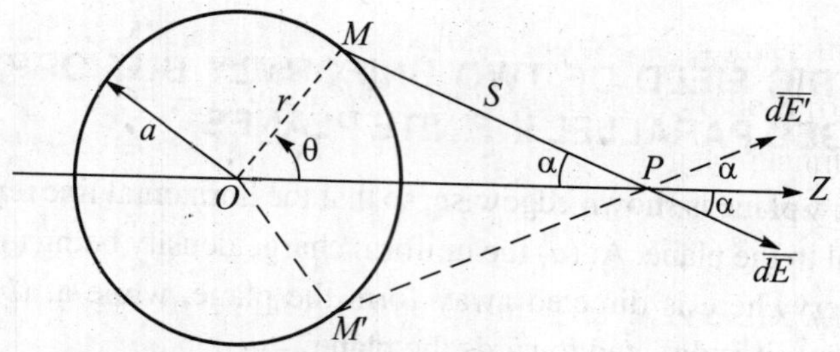

Fig. 2.15. Electric field intensity produced by uniformly charged sphere at a point outside the sphere

The charge ($q_s r^2 \sin\theta\, d\theta\, d\phi$) at M produces the electric field intensity $\overline{dE}$ at P making an angle α with the z-axis. A similar charge at a symmetrical point M' will produce an electric field intensity $\overline{dE'}$ of equal magnitude, but inclined at an angle α on the other side of the z-axis. Hence the resolved parts of these intensities at right angles to the z-axis will cancel each other, but the components parallel to the z-axis will add. The components of $\overline{dE}$ parallel to the z-axis is

$$\overline{E}_z = \left[\frac{q_s}{4\pi\varepsilon_0}\frac{r^2 \sin\theta\, d\theta\, d\phi}{s^2}\cos\alpha\right]\overline{u}_z \qquad ...(2.40)$$

where $MP = s$. Direct integration of this equation to cover the whole surface of the sphere is not possible because, besides the variation of θ and ϕ, α and s are also variable. For the triangle OMP, the Law of Cosines [Eqn. (1.31)] applied twice gives

$$s^2 = r^2 + z^2 - 2r\,z\cos\theta \qquad ...(2.41)$$

and

$$\cos\alpha = \frac{s^2 + z^2 - r^2}{2\,s\,z} \qquad ...(2.42)$$

Now r and z are constants when the integration is carried out over the surface of the sphere. The temptation to substitute the above values of s^2 and $\cos\alpha$ in Eqn. (2.40) to eliminate these two variables is to be resisted, because s^2 and s occur in Eqn. (2.42) and the expression in $\cos\theta$ becomes very complicated. It is easier to retain s and to eliminate θ and α from Eqn. (2.40)! Thus differentiating Eqn. (2.41), we have

$$2s\,ds = 2r\,z\sin\theta\, d\theta \qquad ...(2.43)$$

Now we first carry out the integration with respect to ϕ in Eqn. (2.40) and then eliminate $\cos\alpha$ and $\sin\theta\, d\theta$ using Eqns. (2.42) and (2.43) and write

$$\overline{E}_z = \frac{q_s r^2 \overline{u}_z}{4\pi\varepsilon_0}\int_{\theta=0}^{\theta=\pi}\left[\int_{\phi=0}^{\phi=2\pi} d\phi\right]\frac{\cos\alpha\sin\theta\, d\theta}{s^2}$$

$$= \frac{q_s r^2 \overline{u}_z}{4\pi\varepsilon_0}(2\pi)\int_{s=z-r}^{s=z+r}\frac{s^2+z^2-r^2}{s^2\,2sz}\,\frac{s\,ds}{rz}$$

$$= \frac{q_s r^2 \overline{u}_z}{4\pi\varepsilon_0}\,\frac{2\pi}{2rz^2}\int_{z-r}^{z+r}\left(1+\frac{z^2-r^2}{s^2}\right)ds$$

$$= \frac{q_s r^2 \overline{u}_z}{4\pi\varepsilon_0}\,\frac{2\pi}{2r\,z^2}4r = \frac{q_s(4\pi r^2)}{4\pi\varepsilon_0 z^2}\overline{u}_z$$

Since on the surface of the sphere $r = a$, $q_s(4\pi\, r^2) = q_s\,(4\pi\, a^2) = Q$, the total charge on the surface of the sphere. Therefore the above result can be written as

$$\overline{E}_z = \frac{Q}{4\pi\varepsilon_0 z^2}\overline{u}_z \qquad \text{V/m}, z > a \qquad ...(2.44)$$

But this is the same as Coulomb's law expression for a *point* charge Q located at the centre of the sphere. Hence, we get the important result that for points *outside* the sphere, the Coulomb's law can be applied simply by imagining that the total charge on the sphere is located at the centre of the sphere. It must be remembered that this short cut solution is valid if, and only if, the electric charge is *uniformly* distributed on the surface of the sphere.

2.17.2 For Points Inside the Sphere

Through any point P inside the sphere, we draw a straight line NPM intersecting the sphere in N and M. In Fig. 2.16 the plane passing through line MN and centre O of the sphere is made the plane of the paper.

With the point P as vertex and the line MN as axis, we draw a cone forming an infinitesimally small solid angle $d\Omega$ and extend it on both sides, so that the cone intercepts an area dA_1 around M and an area dA_2 around N out of the surface of the sphere. If $PM = s_1$ and $PN = s_2$, these lines give the directions of the unit vectors $\bar{u}_{s_1}$ and $\bar{u}_{s_2}$ (which are like the vector $\bar{u}_r$ in Fig. 1.23). The lines OM and ON give the directions of the normal unit vectors $\bar{u}_{n_1}$ and $\bar{u}_{n_2}$ for the areas dA_1 and dA_2 respectively. Hence by Eqn. (1.133) defining a solid angle, we have

$$\text{for } dA_1,\ d\Omega_1 = \frac{\bar{u}_{s_1} \cdot (\bar{u}_{n_1} dA_1)}{s_1^2}$$

$$\text{for } dA_2,\ d\Omega_2 = \frac{\bar{u}_{s_2} \cdot (\bar{u}_{n_2} dA_2)}{s_2^2} \qquad \text{...(2.45)}$$

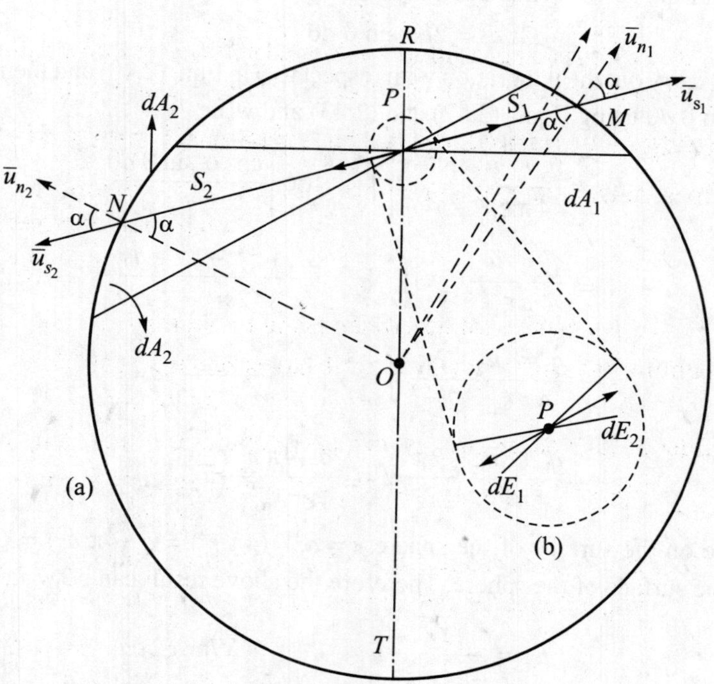

Fig. 2.16. Electric field intensity determination at a point inside a uniformly charged shpere

Since $OM = ON = a$, $m\angle OMN = \angle ONM = \alpha$ say. Then α is the angle between the vectors $\bar{u}_{s_1}$ and $\bar{u}_{n_1}$, and also the angle between $\bar{u}_{s_2} = \bar{u}_{n_2}$. Thus Eqn. (2.45) can be written as

$$\frac{dA_1}{s_1^2} = \frac{d\Omega}{\cos\alpha} = \frac{dA_2}{s_2^2} \qquad \text{...(2.46)}$$

The electric field intensities at P due to the charges ($q_s\, dA_1$) at M, ($q_s\, dA_2$) at N are given by (see enlarged view in Fig. 2.16 (*b*)

$$\overline{dE}_1 = \frac{q_s dA_1}{4\pi\varepsilon_0 s_1^2}(-\bar{u}_{s_1}),$$

$$\overline{dE}_2 = \frac{q_s dA_2}{4\pi\varepsilon_0 s_2^2}(-\bar{u}_{s_2}),$$

which by Eqn. (2.46) and the fact that $\bar{u}_{s_1} = -\bar{u}_{s_2}$, are equal and opposite, so that their resultant is zero. This is true for every direction in which the line *NPM* is drawn through *P*. Hence the electric field produced at *P* by the charge on the hemisphere *RMT* to the right of the plane *RPOT* is exactly cancelled by the charge on the hemisphere *RNT* to the left of the plane passing through *P* and *O*.

This gives the very important result that at *all points inside a uniform spherical* distribution of charge, the electric field intensity is zero. The electric field map for the uniformly charged sphere looks just like Fig. 2.2. for a point charge, except that the radial field lines are wiped out from the centre up to a distance *a* all round *O*.

2.18 ELECTRIC FIELD OF SPHERICAL SHELL WITH UNIFORM VOLUME CHARGE DENSITY

2.18.1 For Points Outside the Shell

Assume that a spherical shell of inner radius *a* metres and outer radius *b* metres has a uniform volume charge density of ($+q_v$) coulomb per cubic metre. To find the electric field intensity at a point *P* outside the spherical shell, we set up a spherical co-ordinate system as in Fig. 2.15 and at *M* at a distance *r* from *O*, we choose an infinitesimally small volume element $dr\,(r\,d\theta)\,(r \sin\theta\, d\theta)$. By the same reasoning as in Section 2.17, the electric field intensity at *P* due to the charge in this volume element is obtained as

$$\overline{dE}_z = \left[\frac{q_v}{4\pi\varepsilon_0}\,\frac{r^2 dr \sin\theta\, d\theta\, d\phi}{s^2}\cos\alpha\right]\bar{u}_z$$

The Eqns. (2.41), (2.42) and (2.43) are true in this case also. Hence using Eqns. (2.42) and (2.43) in the above equation, we get

$$\bar{E}_z = \frac{q_v \bar{u}_z}{4\pi\varepsilon_0}\int_{r=a}^{r=r}\left[\int_{\theta=0}^{\theta=\pi}\left\{\int_{\phi=0}^{\phi=2\pi} d\phi\right\} r^2 \frac{\cos\alpha}{s^2}\sin\theta\, d\theta\right] dr$$

$$= \frac{q_v \overline{u}_z}{4\pi\varepsilon_0} \frac{2\pi}{2z^2} \int_{r=a}^{r=r} \frac{r^2}{r}(4r)\, dr$$

using the results of integration with respect to s as in the previous section. Hence we get

$$\overline{E}_z = \frac{q_v\left(\frac{4}{3}\pi r^3 - \frac{4}{3}\pi a^3\right)}{4\pi\varepsilon_0 z^2} \overline{u}_z \qquad ...(2.47)$$

If the integration is performed with the limits for r as $r = a$ to $r = b$, we get

$$\overline{E}_z = \frac{q_v\left(\frac{4}{3}\pi b^3 - \frac{4}{3}\pi a^3\right)}{4\pi\varepsilon_0 z^2} \overline{u}_z = \frac{Q_{ab}}{4\pi\varepsilon_0 z^2}\overline{u}_z,\ z > b \qquad ...(2.48)$$

Since the quantity within the brackets is the volume of the shell, its product with q_v gives the total charge Q_{ab} within the shell. If $a = 0$, we get the shell becoming a charged sphere of radius b for which

$$\overline{E}_z = \frac{q_v\left(\frac{4}{3}\pi b^3\right)}{4\pi\varepsilon_0 z^2} \overline{u}_z = \frac{Q_b}{4\pi\varepsilon_0 z^2}\overline{u}_z,\ z > b \qquad ...(2.49)$$

The identical forms Eqns. (2.48) and (2.44) are not a result of a strange coincidence. They are the result of the principle of superposition. Equation (2.48) could have been derived from Eqn. (2.44) by considering the thick shell as made up of infinitesimally thin shells of radius r and thickness dr (like onion skins) and letting r vary from a to b.

2.18.2 For Points Inside the Shell

For these $z < a$. The process of reasoning is the same as in Fig. 2.16, except that instead of areas at M and N, we have the volumes $dA_1\, dr$ and $dA_2\, dr$, assuming the shell to be of radius r and thickness dr. Then the electric field intensities cancel out as before, and there is no electric field intensity for $z < a$ at any point inside the shell.

2.18.3 For Points Within the Shell

Now for the point P, $a < z < b$. For the part of the shell formed by the limits $r = z$ and $r = b$, the point P lies inside the shell and hence this part of the charge produces zero electric field intensity at the point P. For the shell with $r = a$ to $r = z$, the point P is outside the shell, and hence Eqn. (2.48) is applicable, and we get

$$\overline{E}_z = \frac{q_v\left(\frac{4}{3}\pi z^3 - \frac{4}{3}\pi a^3\right)}{4\pi\varepsilon_0 z^2} \overline{u}_z \text{ V/m},\ a < z < b \qquad ...(2.50)$$

If $a = 0$, this reduces to

$$\overline{E}_z = \frac{q_v z}{3\varepsilon_0}\overline{u}_z \quad \text{V/m},\ z < b \qquad ...(2.51)$$

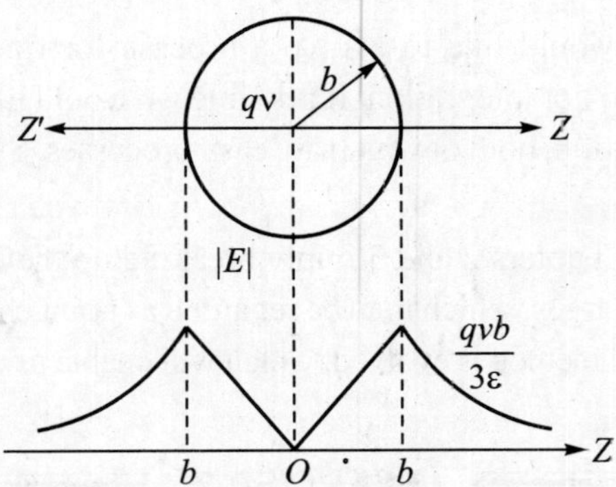

Fig. 2.17. Electric field intensity inside and outside a sphere of uniform volume charge density

showing that $|E_z|$ increases linearly with $|z|$ for $|z|$ varying from O to b inside a sphere having uniform volume charge density (Fig. 2.17). It reaches the maximum value of $\frac{q_v b}{3\varepsilon_0}$ at the surface of sphere and then begin to decrease according to the inverse square law as indicated by Eqn. (2.49). This variation is similar to the variation of the acceleration due to gravitation for points inside and outside the earth.

2.19 SUMMARY OF RESULTS AND COMMENTS

The three important and fundamental results established in this chapter are the formulas for the electric field intensity at P for special geometrical configuration of charges (ε is permittivity at point P) :

(*a*) For point charge $\overline{\boldsymbol{E}} = \frac{\overline{\boldsymbol{F}}}{Q_t} = \frac{Q}{4\pi\varepsilon r^2}\overline{\boldsymbol{u}}_r$ V/m Eqn. (2.11)

(*b*) For uniformly charged infinite line:

$$\overline{\boldsymbol{E}}_R = \frac{q_l}{2\pi\varepsilon R}\overline{\boldsymbol{u}}_R \text{V/m} \quad \text{Eqn. (2.27)}$$

(Cylindrical co-ordinates)

(*c*) For uniformly charged infinite plane:

$$\overline{\boldsymbol{E}} = \frac{q_s}{2\varepsilon}\overline{\boldsymbol{u}}_n \text{ V/m} \quad \text{Eqn. (2.39)}$$

The first equation shows $\overline{\boldsymbol{E}}$ as being inversely proportional to the square of the distance r; the second, inversely proportional to the distance R, and the third is not dependent on the distance at all.

The application of these formulas to find $\overline{\boldsymbol{E}}$ for a given charge distribution involves two big hurdles. The first is that, these give vector quantities which have to be resolved along mutually perpendicular unit vectors before we can proceed to replace vector addition by scalar addition (or integration). The second hurdle is that this summation involves a process of integration which is never too easy, but sometimes is not possible at all.

If possible we would like to evolve a process for finding $\overline{E}$, which neither involves vector addition nor integration. If possible we would like to have simple scalar addition and differentiation, both of which are easy processes. Such a method is evolved in Chapter 3.

However both Chapters 2 and 3 imply breaking up the charge distribution into infinitesimally small charges, which may be regarded as point charges. Hence in chapter 4 a much more powerful method is evolved, which will enable us to handle charge in bulk.

EXERCISE 2

1. Compute the force between two charges of one coulomb of each separated in vacuum by a distance equal to (*i*) the diameter of the earth (= 1.27×10^7 m), and (*ii*) one metre.

2. An electron has a negative electric charge e of 1.602×10^{-19} C and a mass m of 9.1066×10^{-31} kg, so that $\frac{e}{m} = 1.7592 \times 10^{11}$ C/kg. Calculate the ratio of electrostatic force of repulsion between two electrons in vacuum to the gravitational force of attraction between them. Assume that in Newton's law of gravitation, force $= G\frac{m_1 m_2}{r^2}$, m_1 and m_2 being the masses in kg, r the distance in metre and the gravitational constant $G = 6.67 \times 10^{-11}$ Nm^2/ kg^2.

3. Positive electric charges, each of magnitude $(4\pi\,\varepsilon_0)$ coulomb, are situated in the $x - y$ plane at the points $(-1, 0)$, $(0, 0)$ and $(1, 0)$ respectively, the metre being the unit of length. Find the magnitude and the direction of the electric field intensity at the point $P(1, 1)$.

4. In a square $ABCD$ of side one metre, electric charges of magnitude $(+\sqrt{4\pi\varepsilon_0})$ coulomb each are located at A and C, and of magnitude $(-\sqrt{4\pi\varepsilon_0})$ coulomb each at B and D. Find the magnitude and the direction of the resultant force on each charge due to other charges. Ascertain whether there are any null points where the electric field intensity is zero.

5. In cylindrical co-ordinates, for the planes $\phi = 0$, electric charges, each of magnitude $(+Q)$ coulomb are located on the z-axis at the point $z = 2a$ metres and $z = -2a$ metres, respectively, while a charge of magnitude $(-2Q)$ coulomb is located at the origin. Such an arrangement is called an electrical linear quadripole (= two dipoles "back to back"). Assuming that both $|z|$ and $|R|$ are $> 2a$, use binomial theorem to find the magnitude and direction of the electric field intensity at points on the z and R axes.

6. Two point charges, each of magnitude $(+Q)$ coulomb are located on the x-axis at $x = -3$ metres and $x = +3$ metres respectively, while a charge (kQ) is located on y-axis at $y = 9$ metres. Find the magnitude and sign of k in the three cases, in each of which a null point is produced on the y-axis by these three charges (that is, where electric field intensity is zero) at (*i*) $y_1 = -1$ m, (*ii*) $y_2 = 4$ m (*iii*) $y_3 = +10$ m.

7. In early experiments in electricity in previous centuries, pith balls suspended by strings were used. Two such pith balls, each having a mass m kg and a charge $(+\sqrt{4\pi\varepsilon_0})$ coulomb are suspended by two weightless strings, each of length L metres, from the same point. Show that the angle θ which each string makes with the vertical is given by $4mg\,L^2 \sin^3\theta = \cos\theta$.
8. Assume that three pith balls are suspended with three weightless strings from a common point. Each string is of length L metres and mass of each pith ball is m kg (weight = mg). It is found that when a charge of $(+\sqrt{4\pi\varepsilon_0})$ coulomb is placed on each pith ball, then, because of mutual repulsion, the three piths balls form an equilateral triangle of side L metres in the horizontal plane. Determine the relationship between L and mg.
9. For the uniformly charged short line in Fig. 2.7, find the electric field intensity for any point lying on the z-axis, such that $-\infty < z < -a$ or $+a < z < +\infty$. Can the answer be obtained by substituting appropriate value of R_1, α_1, and α_2 in Eqn. (2.20) and Eqn. (2.21)?
10. Find the equation to the direction lines for a uniformly charged semi-infinite line having a charge $(+q_l)$ coulomb per metre from $z = -\infty_1$ to $z = 0$.
11. Find the equation of the family of circles forming the direction lines shown in Fig. 2.11. [**Hint:** Use. Eqns. (2.32) and (2.33*b*)].
12. The direction lines for infinite parallel lines carrying equal and opposite uniform linear charges shown in Fig. 2.11 are segments of circles. On each circle, we can locate two points where tangent to the circle is parallel to the y-axis (or $E_x = 0$). Find the equation to the curves passing through such points.
13. Starting with Eqn. (2.27) for the infinite line, derive Eqn. (2.39) for the infinite plane. [**Hint:** Divide plane into strips parallel to the z-axis and of width dx].
14. The curved surface of a hemisphere carries a surface charge density of $(+q)$ C/m^2. Determine the magnitude and direction of the electric field intensity at the centre of the hemisphere.
15. The x-y plane represents an infinite plane sheet of charge density $(+q_s)$ C/m^2. If E is the magnitude of the electric field intensity on the z-axis at a point P where $OP = z$, show that one half $\left(=\dfrac{E}{2}\right)$ of it is due to the charge on the sheet enclosed within the circle $x^2 + y^2 = 3z^2$, and the other half is due to the charge lying outside this circle.
16. In a spherical charge distribution, the volume charge density varies with distance from the centre as

$$q_\upsilon = \begin{cases} +q_0\left(1 - \dfrac{r^2}{a^2}\right) \text{C/m}^3 & \text{for} \quad r \le a \\ 0 & \text{for} \quad r > a \end{cases}$$

Find the electric field intensity variation along the z-axis for the range $-2a < z < +2a$ and plot it as in Fig. 2.17. For what value of $\dfrac{r}{a}$ does this field intensity becomes maximum, and what is the maximum value?

CHAPTER

3 POTENTIAL AND ITS GRADIENT

3.1 LINE INTEGRAL OF A VECTOR

The electrical field intensity $\overline{E}$ is a vector quantity, and to find the resultant electric field intensity at a point due to a large number of point charges or a charge distribution involves a process of vector addition which is a laborious process. We therefore look around for something that will be proportional to E and will be a scalar and also be easily derived from $\overline{E}$, and also $\overline{E}$ should be easily derived from it. Since addition of vectors gives a vector only, we turn to multiplication which, in arithmetic, is an abridged process of addition. Of the three ways in which multiplication can be done using the vector $\overline{E}$, as given by Eqns. (1.70), (1.74) and (1.76), only the scalar or dot product of $\overline{E}$ and another vector will give a scalar result proportional to $\overline{E}$.

To decide what this other vector must be, we note that $\overline{E}$ is a force (newton per coulomb). In mechanics, the dot product of force $\overline{F}$ and a distance $\overline{d}$ gives the scalar result $Fd \cos \theta$, which is the work done. Hence in electrostatics also we select an infinitesimally small distance vector $\overline{ds}$ to make the dot product $\overline{E} \cdot \overline{ds} = E\, ds \cos \theta$. The vector $\overline{ds}$ is the small element of *any curved* or straight line as in differential or integral calculus. It must be remembered that this is *any line*, and should not be confused with the direction lines used in the previous chapter, because in the case of direction lines, $\overline{E}$ and $\overline{ds}$ are both in the same direction along the tangent to the direction line (see Step 5 of Section 2.6). However, in the present case, if any vectors $\overline{F}$ and $\overline{ds}$ are inclined at an angle θ, (Fig. 3.1), then we define

$$\text{Line integral} = \int_a^b \overline{F} \cdot \overline{ds} \qquad ...(3.1)$$

If, and only if, the vector $\overline{F}$ represents a force, (like $\overline{E}$, for instance), then only the line integral will give the work done when the point P is moved *along the line ab* from a to b. But if $\overline{F}$ is not a force but some other vector quantity, say a velocity, then the line integral will not have the units of work.

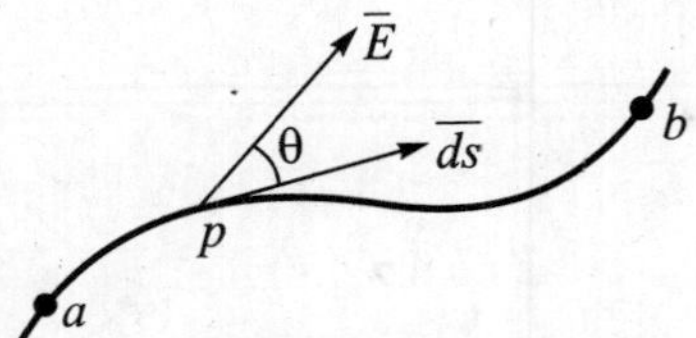

Fig. 3.1. Line integral

3.2 CONSERVATIVE AND NON-CONSERVATIVE VECTOR FIELDS

The usefulness of the concept of a line integral of a vector arises from the fact that it leads to the concept of energy (a scalar quantity) being uniquely associated with the position of a point in a vector field. This line integral evaluation always assumes that there are no irreversible processes associated with going from a to b, or in other words, the line integral from b to a is exactly equal and opposite of the line integral from a to b. An example of an irreversible process is the conversion of a part of the work done into heat due to friction. The direction of the force of friction while going from b to a will be the reverse of that obtained when going from a to b.

The unique value of the line integral implies that it is *only the positions of the point a and b* in the vector field which determine this unique value, and that *the value does not depend upon the path between a and b*. In such a field, the law of conservation of energy holds, and the field is therefore called a conservative field. Some of the vector fields we come across in nature, like the gravitational and the electrostatic fields, belong to this class of conservative fields.

Thus, in a gravitational field, Fig. 3.2 (a), when the mass m is shifted from the point a on the ground to the point b at a vertical height h metre above the ground level, work is done against the gravitational force mg (because angle $\theta > 90°$), and energy gets stored in the mass by virtue of its position at b being higher than at a. We say that the line integral or the energy stored in the mass m at point b is greater by an amount mgh than at the point a. Moreover this is independent of the path from a to b.

In the case of the oscillating simple pendulum [Fig. 3.2 (b)], where a and b are the lowest and the highest positions of the bob of mass m, the potential energy gain at b over that at a is mgh. On the down swing, the line integral from b to a gives the potential energy lost by the bob and this mgh gets converted into the kinetic energy $\frac{1}{2}mv^2$. If there is no frictional force (like air viscosity or friction at pivot) involved, this process of work done by the gravitational field (positive line integral, $\theta < \frac{\pi}{2}$) and work done against the gravitational field (negative line integral, $\theta > \frac{\pi}{2}$) can go on alternatively for ever, because the two line integrals are equal in magnitude but of opposite sign.

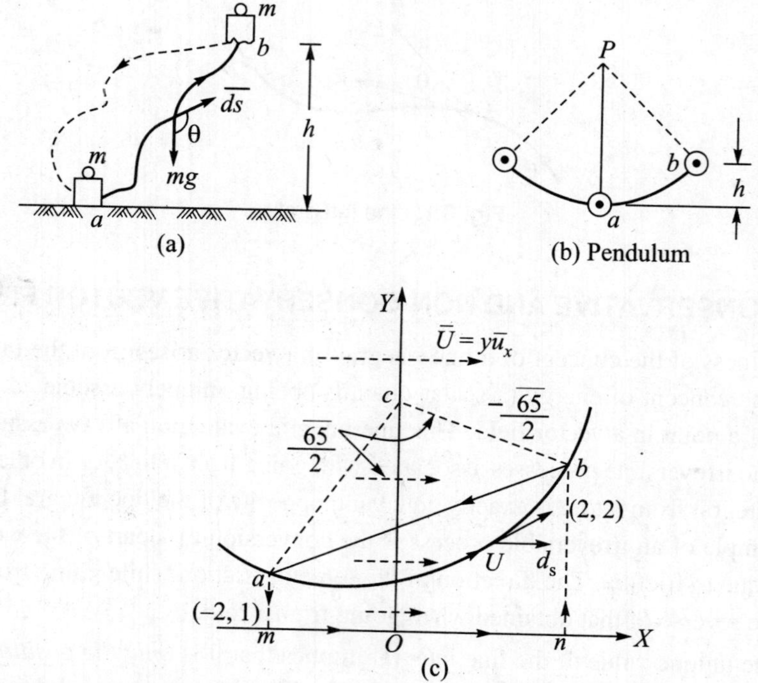

Fig. 3.2. *(a)*, *(b)* Conservative and *(c)* Non-Conservative vector fields

Because of the facts that

(i) the magnitudes of the line integrals between two points do not depend on the path taken, but are always equal and

(ii) that the sign of the line integral from b to a is opposite to that of the line integral from a to b, it follows that if the line integral is taken along a closed path – that is, from a to b, and then back from b to *a along the same or any other path* – then this line integral along a closed path written as

$$\oint \overline{E} \cdot \overline{ds} = \oint \overline{E} \cdot \overline{ds} = \int_a^b \overline{E} \cdot \overline{ds} \text{ along one path } +$$

$$\int_b^a \overline{E} \cdot \overline{ds} \text{ along the same or a different path} = 0 \qquad \text{...(3.2)}$$

The closed circle around the integral sign (with or without an arrow head) indicates integration around a closed path, the value of the integral being considered positive if the area enclosed is on the left hand side of a person as per Fig. 1.13. The line integral of a vector along a closed path [as in Eqn. (3.2)] is also called the circulation of the vector.

Next we consider another type of vector field. If water is flowing down a straight canal, because of viscous drag, the water layer at the bottom of the canal is stationary, but the velocity of water increases as we rise from the bottom, being a maximum at the top. Choosing the origin at the bottom and the x-axis in the direction of the water flow, and the y-axis vertical, we find [Fig. 3.2 (*c*)] that at a height y from the bottom, the velocity of water is $\overline{U}$ metre / second, where

$$\overline{U} = y\,\overline{u}_x \qquad \text{...(3.3)}$$

For this vector field, we find the line integrals from the point a (– 2, 1) to the point b (4, 4) along a circular path as given by

$$4x^2 + (2y - 9)^2 = 65 \qquad \text{...(3.4)}$$

and back from b to a along the straight line given by

$$2y = x + 4 \qquad \text{...(3.5)}$$

Using Eqns. (3.3) and (1.101), we have

$$\int \overline{U} \cdot \overline{ds} = \int (y\,\overline{u}_x)\,(\overline{u}_x\,dx + \overline{u}_y\,dy + \overline{u}_z\,dz) = \int y\,dx \qquad \text{...(3.6)}$$

Hence substituting the value of y from Eqn. (3.4)

$$\int_a^b \overline{U} \cdot \overline{ds} = \int y\,dx = \int_{-2}^{4} \left[\frac{\sqrt{65 - 4x^2} + 9}{2}\right] dx$$

$$= \int_{-2}^{4} \sqrt{\frac{65}{4} - x^2}\,dx + \frac{9}{2}\int_{-2}^{4} dx$$

Using Eqns. (1.52) and (1.56a)

$$\int_a^b \overline{U} \cdot \overline{ds} = \frac{1}{2}\left[x\sqrt{\frac{65}{4} - x^2} + \frac{65}{4}\sin^{-1}\frac{2x}{\sqrt{65}}\right]_{-2}^{4} + \frac{9}{2}[x]_{-2}^{4}$$

$$= \frac{1}{2}\left[4\sqrt{\frac{65}{4} - 16} - (-2)\sqrt{\frac{65}{4} - 4}\right] +$$

$$\frac{65}{8}\left[\sin^{-1}\frac{8}{\sqrt{65}} + \sin^{-1}\frac{4}{\sqrt{65}}\right] + \frac{54}{2}$$

or

$$\int_a^b \overline{U} \cdot \overline{ds} = \frac{63}{2} + \frac{65}{8}\left[\sin^{-1}\frac{8}{\sqrt{65}} + \sin^{-1}\frac{4}{\sqrt{65}}\right] \qquad \text{...(3.7)}$$

But $\int_a^b \overline{U} \cdot \overline{ds}$ along the straight line is obtained by putting the value of y from Eqn. (3.5) in Eqn. (3.6), so that

$$\int_a^b \overline{U} \cdot \overline{ds} = \int_4^{-2} \frac{x + 4}{2}\,dx$$

$$= \frac{1}{2}\left[\frac{x^2}{2} + 4x\right]_4^{-2} = -15 \qquad \text{...(3.8)}$$

Combining Eqns. (3.7) and (3.8), we find that

$$\oint \overline{U} \cdot \overline{ds} = \int_a^b \overline{U} \cdot \overline{ds} + \int_a^b \overline{U} \cdot \overline{ds} \neq 0 \qquad \text{...(3.9)}$$

In fact it is easy to see that by a proper choice of the path between the point *a* and *b*, the line integral can be made to have any value from 0 to ∞ !! Thus if *am* and *bn* are perpendicular to the *x*-axis, then line integral along the path *amnb* is found to be

$$\int_{amnb} \overline{U}\cdot\overline{ds} = \int_a^m \overline{U}\cdot\overline{ds} + \int_m^n \overline{U}\cdot\overline{ds} + \int_n^b \overline{U}\cdot\overline{ds}$$

= (zero, because angle $\theta = 90°$) +

(zero, because $\overline{U} = 0$ for $y = 0$) +

(zero, because angle $\theta = 90°$) = 0.

The $\overline{U}$ - vector field is therefore a non-conservative field, since its line integral around a closed path is not equal to zero. By proper choice of path, one can make the vector field to deliver energy. A fish in the canal knows this very well (without ever having evaluated a line integral !), and while it swims down the canal near the surface, it returns to the starting point by swimming back along the bottom of the canal !

3.3 ELECTRIC POTENTIAL, POTENTIAL ENERGY AND POTENTIAL DIFFERENCE

Consider the field map for the electrostatic field intensity $\overline{E}$ of a point charge + Q coulomb (Fig. 2.2). In spherical co-ordinates all direction lines are straight lines radiating from the point charge in all directions and extending up to infinity in all directions. For our starting point to evaluate the line integral in this case, we agree to adopt the standard starting point *a* at any point at infinity, and the path along the straight line joining this point to the point charge + Q.

If we place a test charge Q_t anywhere in this field, it will experience a force $\overline{F} = \dfrac{(+Q)\,Q_t}{4\pi \in_0 r^2}$ newton directed along a radial line *away* from + Q. If the test charge is initially placed at infinity, and moved along this radial line to the point P_1 at a distance r_1 from + Q (Fig. 3.3), then all the way from infinity to P_1 some external agency will have to do work against the force F, and thus the test charge will gain potential energy equal to the work done in this process, or

$$W_1 = -\int_\infty^{r_1} \overline{F}\cdot\overline{ds} \text{ joule}$$

Dividing both sides by the magnitude Q_t of the test charge, we get the work done per coulomb of test charge $= \dfrac{W_1}{Q_t}$

$$= -\int_\infty^{r_1} \left(\frac{\overline{F}}{Q_t}\right)\cdot\overline{ds} \text{ joule per coulomb}$$

This is called the potential at the point P_1 and is usually denoted by the symbol V_1, so that in general, for any point at a distance r from + Q, since $\overline{E} = \dfrac{\overline{F}}{Q_t}$

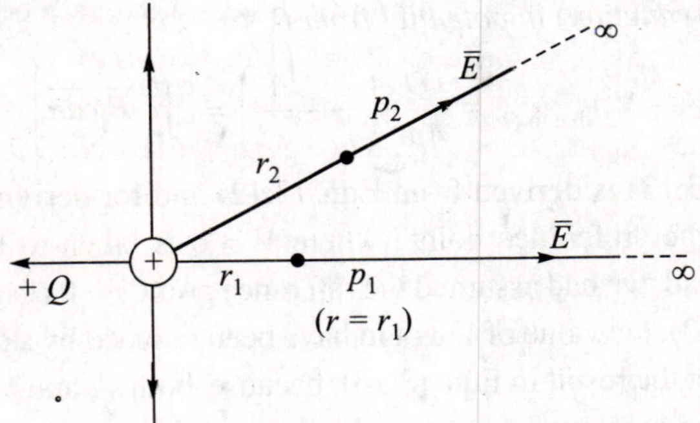

Fig. 3.3. Potential at a point and potential difference

$$\text{Electrostatic potential } V = -\int_{\infty}^{r} \overline{E} \cdot \overline{ds} \text{ joule/coulomb} \qquad ...(3.10)$$

which becomes the mathematical expression of the definition of the potential at a point in a field.

The negative sign in Eqn. (3.10) arises because work is done by an external agency *against* the coulomb force. But we could also say that it arises because of our mathematical convention. Thus *ds* being in the direction of the radial line, $\overline{ds} = \overline{dr}$. But *dr* is positive in the direction of increasing *r*, that is, away from + *Q*. For defining potential, since we move from infinity towards + *Q*, this displacement will be $(-\overline{u}_r\, dr)$ and the dot product which is to be integrated becomes

$$\int_{\infty}^{r} \left(\frac{F\,\overline{u}_r}{Q_t}\right) \cdot (-\overline{u}_r\, dr) = -\int_{\infty}^{r} (E\,\overline{u}_r) \cdot (-\overline{u}_r\, dr)$$

$$= -\int_{\infty}^{r} \overline{E} \cdot \overline{dr} = -\int_{\infty}^{r} \overline{E} \cdot \overline{ds} \qquad ...(3.11)$$

Hence, we get from Eqns. (3.10) and (3.11)

$$V = -\int_{\infty}^{r} \overline{E} \cdot \overline{ds} = -\int_{\infty}^{r} \overline{E} \cdot \overline{dr}$$

$$= \int_{\infty}^{r} \frac{(+Q)}{4\pi\varepsilon_0 r^2}\, dr = \frac{Q}{4\pi\varepsilon_0}\left[\frac{1}{r}\right]_{\infty}^{r}$$

$$= \frac{Q}{4\pi\varepsilon_0 r} \text{ J/C or volt} \qquad ...(3.12)$$

where the unit "joule per coulomb" is given the name "volt" in honour of the Italian physicist Alessandro Volta (1745–1827). The Eqn. (3.12) also shows why the units for the electric field intensity in Eqn. (2.12) are written as volt per metre.

Using Eqn. (3.12) we have from Fig. 3.3, the potentials at the points P_1 and P_2 at distance r_1 and r_2 from + *Q*.

$$V_1 = \frac{Q}{4\pi\varepsilon_0 r_1}$$

and

$$V_2 = \frac{Q}{4\pi\varepsilon_0 r_2}$$

so that *the fall (or reduction) in potential from P_1 to P_2 is*

$$V_{12} = (V_1 - V_2)_{\text{due to } Qn} = \frac{Q}{4\pi\varepsilon}\left(\frac{1}{r_1} - \frac{1}{r_2}\right) = \left[\int_{r_1}^{r_2} \overline{E} \cdot \overline{dr}\right] \quad ...(3.13)$$

Now Eqn. (3.13) is derived from Eqn. (3.12) and for deriving Eqn. (3.12), we had assumed that the "reference point" where $V = 0$ is taken to be at infinity. It is obvious that if instead, we had assumed the reference point $V = 0$ at some finite distance r_0, then in Eqn. (3.12), the value of V would have been reduced by a certain amount, but this would not affect the result in Eqn. (3.13), because both V_1 and V_2 would have been reduced by the same amount. In some problems we have to resort to this trick to get over some mathematical difficulty.

In Eqn. (3.12) the potential V gives the measure of the potential energy at the point P. Though we use the term "potential at a point", we really imply Eqn. (3.13) and mean that to be taken as the potential difference between that and a reference point for which V is *arbitrarily assigned to be zero.* It is only when the reference point is not specifically mentioned that it is understood to be at infinity, or at the ground (earth), just as altitudes in geography are with reference to mean sea level.

The Eqns. (3.12) and (3.13) are remarkable in two ways. First they are scalar relations derived from the field intensity vector. Secondly, they are dependent only on the position of the point (or points P_1 and P_2) in the electric field of the point charge $+ Q$. Moreover, V is inversely proportional to r whereas $\overline{E}$ is inversely proportional to the square of r. Hence as r increases, the rate of decrease of V is much less than the rate of decrease of $\overline{E}$.

At first sight, it may appear that the potential function defined by Eqn. (3.10) has not improved matters, because we have to know $\overline{E}$ first before V can be evaluated. It is just like shifting a load from one shoulder to another. But we see that by derving Eqns. (3.12) and (3.13) we are in a position to obtain V without knowing $\overline{E}$. Further the scalar nature of these equations easily permits us to take into account the effects of more than one point charge by the principle of superposition. Thus if $Q_1, Q_2, Q_3, ..., Q_n$ are the point charge, $r_{11}, r_{21}, r_{31}, ..., r_{n1}$ their distances to P_1 and $r_{12}, r_{22}, r_{32}, ...r_{n2}$ their distance to P_2, then

$$V_{12} = \sum_{n=1}^{n}(V_1 - V_2)_{\text{due to } Q_n} = \frac{Q_1}{4\pi\varepsilon_0}\left(\frac{1}{r_{11}} - \frac{1}{r_{12}}\right) + \frac{Q_2}{4\pi\varepsilon_0}\left(\frac{1}{r_{21}} - \frac{1}{r_{22}}\right) + ...$$

$$+ \frac{Q_n}{4\pi\varepsilon_0}\left(\frac{1}{r_{n1}} - \frac{1}{r_{n2}}\right) \quad ...(3.14)$$

If P_2 is at infinity, $r_{12} = r_{22} = r_{32} = ... = r_{n2} = \infty$, and Eqn. (3.14) reduces to

$$V_1 = \frac{1}{4\pi\varepsilon_0}\sum_{n=1}^{n}\frac{Q_n}{r_{n1}} \quad ...(3.15)$$

This equation shows very clearly the third remarkable aspect of the potential at a point. Even though Eqn. (3.12) was derived by a process of integration using the electric field intensity, Eqn. (3.15) shows that we do not have to know the value of the electric field intensity produced at P_1 by each point charge. All we need to know is the magnitudes and the signs of the point charges and their distances from P_1.

3.4 EQUIPOTENTIAL LINES ON POTENTIAL MAPS

In Eqn. (3.15), for a given charge distribution, the only quantities that are variable are V_1 and $r_{11}, r_{21}, r_{31}, ..., r_{nl}$. If one set of values for the distances gives a particular value for V_1, then it is possible for another set of values for the distances to give the same value for V_1. In fact there can be an infinitely large number of sets of values for the distances that give the same value for V_1. The three dimensional surface passing through the points defined by each of these sets of values of distances is such that at each point on it, the value of potential is the same value V_1. It is therefore called an equipotential surface.

By giving different values to V_1, we get different equipotential surfaces. If we take a section of these equipotential surfaces by a plane, we will get equipotential lines on the plane which thus become a two-dimensional potential map.

If two points a and b lie on the same equipotential surface, the same potential energy is associated with both of them, and hence, for any line joining them, the line integral $\int_a^b \overline{E} \cdot \overline{ds}$ is zero. Moreover, if we select this line in such a way that it wholly lies in the equipotential surface, then (by varying the length of the line from 0 to full length ab), we find that *at every point* of the line lying in the equipotential surface, $\overline{E} \cdot \overline{ds}$ is zero. Stated in other words, this implies that $\overline{E}$ *is always normal to the equipotential surface at every point.*

On the same map on a plane, if both the direction lines of the electric field intensity, and the equipotential lines are drawn, the everywhere these two sets of lines will intersect each other at right angles. In the mathematical theory of Functions of a Complex Variable this fact implies that the Cauchy-Riemann equations (5.26) are satisfied and that conformal transformations are possible. Without going into the mathematical intricacies at present, the physical aspect can be easily grasped by using the idea of rubber-sheet geometry. We imagine that on a plane, unstretched rubber sheet, electric field direction lines are drawn parallel to the x-axis, and the equipotential lines are drawn parallel to the y-axis. Thus these sets of lines are seen to intersect each other at right angles. Now if by applying suitable tensions at different points of the rubber sheet, these sets of lines are bent and made to coincide with the field direction lines and the equipotential lines, respectively, even then these sets will continue to intersect each other at right angles.

3.5 POTENTIAL GRADIENT

By writing Eqn. (3.10) in the differential form, we have at any point P

$$dV = -\overline{E} \cdot \overline{ds} \qquad \text{...(3.16)}$$

In this expression $\overline{ds}$ can have any direction we like, by $\overline{E}$ is always normal to the equipotential surface through P. Using Eqn. (1.92) we can also write

$$\text{gradient } V = \overline{\nabla} V = \frac{dV}{dn}\overline{u}_n \qquad \text{...(3.17)}$$

where $\overline{u}_n$ is the unit vector normal to the equipotential surface at P and dn is the projection of $\overline{ds}$ along the normal (PQ' in Fig. 1.14). Putting $\overline{E} = E\overline{u}_n$ in Eqn. (3.16) and combining with Eqn. (3.17), we get

$$dV = -E\overline{u}_n \cdot \overline{ds} = -E\,dn$$

or

$$E = -\frac{dV}{dn}$$

giving

$$\overline{E} = E\overline{u}_n = -\frac{dV}{dn}\overline{u}_n = -\overline{\nabla} V \qquad \text{...(3.18)}$$

This enables us to find the electric field intensity if the expression for the potential at the point P is known.

3.6 MAP WITH EQUIPOTENTIAL AND DIRECTION LINES

In the previous sections we have evolved a method of attack which answers the criticism made in Section 2.19, and is therefore easier than the Coulomb's Law approach. The steps in the procedure for preparing the map are as follows :

Step 1: Choose an infinitesimally small charge in the charge distribution, and also an arbitrary point P in space and knowing their co-ordinates in the system of co-ordinates selected, find the distance between them.

Step 2: Using Eqn. (3.12) write the infinitesimally small contribution dV made by the above charge to the potential at the point P. Then using Eqn. (3.15) in the integral form, find the total contribution of all charges and thus get the potential V at P. (Since the potential is a scalar function, only one integration is needed in this case instead of the three integrations for the resolved components of $\overline{dE}$ in Section 2.6).

The expression for V involves the data about the charge distribution, the magnitude of the total charge, and the co-ordinates of the point P. Hence by giving to V any constant value V_{const}, the equation defines the equipotential surface corresponding to that constant potential V_{const}.

Step 3: From the expression for V, using Eqn. (3.18) with $\overline{\nabla} V$ expressed in the appropriate co-ordinate system as given in Eqns. (1.90) (1.104) and (1.109), we get the components of $\overline{E}$ along three mutually perpendicular unit vectors at P. Putting these values in the system of Eqns. (2.14), (2.15) or (2.16), and solving the resultant differential equations, we get the equations to the direction lines of the electric field intensity in different planes. Since the gradient is a differentiation process, it is easier to get the value of $\overline{E}$ form V than to get it using Coulomb's law.

An important point to be noted in all field and potential maps is that, direction lines start from positive charges and end on negative charges. They are usually depicted as continuous lines with arrow head. The negative charges may be located at finite distances or assumed to be at infinity. On the other hand, equipotential surfaces are always closed surfaces enclosing a charge. Their section on a two-dimensional map forms equipotential lines, which are usually depicted as dotted lines without any arrow heads. Electric direction lines can never intersect each other because that would give rise to two tangents to the two curves at the point of intersection and field intensity at that point cannot be directed along two different directions. But in rare cases (see Fig. 5.5) a potential line may cross itself. The potential being a scalar quantity, the question of direction at point of intersection does not arise.

The step-by- step procedure is illustrated below for a few of the problems of Chapter 2, leaving the others for inclusion in exercises at the end of this Chapter.

3.7 EQUIPOTENTIAL FOR A POINT CHARGE + *Q*

Taking the origin of a system of spherical co-ordinates at the location of the charge + Q, we get from Eqn. (3.12)

$$\text{Potential } V = \frac{Q}{4\pi\varepsilon_0 r} \text{ volt} \qquad \text{by (Eqn. 3.12)}$$

For a constant value V_{const} for the potential, we get the equation to the equipotential surface as

$$r = \frac{Q}{4\pi\varepsilon_0 V_{\text{const}}} \qquad \text{...(3.19)}$$

which defines a system of concentric spheres with their centres at the origin. Next using Eqn. (1.109) with Eqn. (3.12) above, we get

$$\left.\begin{aligned} \overline{E}_r &= -(\overline{\nabla} V)_r = -\frac{\partial V}{\partial r} = \frac{Q}{4\pi\varepsilon_0 r^2}\overline{u}_r \\ \overline{E}_\theta &= 0 \\ \overline{E}_\phi &= 0 \end{aligned}\right\} \qquad \text{...(3.20)}$$

Curves for Eqns. (3.19) and (3.20) are shown in Fig. 3.4(a) with direction lines as continuous lines with arrow heads and equipotentials as dotted lines. Now Eqn. (3.12) when written as

$$rV = \frac{Q}{4\pi\varepsilon_0} = \text{Constant}$$

and compared with Eqn. (1.132) reveals that the curve is a rectangular hyperbola with r and V axes as asymptotes, as shown in Fig. 3.4(b). The combination of Figs. 3.4(a) and 3.4(*b*) looks like a contour map in geography, where at (*b*) we get a cross-section of the potential hill.

By drawing direction lines radiating with equal angular steps from + Q, and the equipotentials for equal voltage steps (and not equal increments or r), we are able to convey the additional information that the electric field intensity is greater in the region where the direction lines or the equipotentials are crowded together.

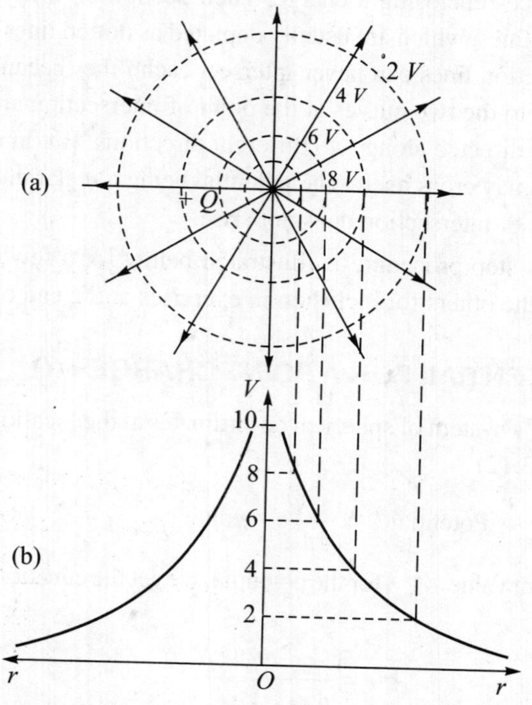

Fig. 3.4. (a) Direction lines and equipotentials for point charge + Q. and (b) Plot of v as function of r

3.8 EQUIPOTENTIALS FOR AN ELECTRIC DIPOLE

As in Fig. 2.4 assume that R and z are very large as compared to a. Then we can use the approximate values

$$R_1 \approx r + a \cos \theta$$

and

$$R_2 \approx r - a \cos \theta \qquad \text{...(3.21)}$$

and get the equation to the equipotential surface in spherical co-ordinates as follows :

$$V = \frac{Q}{4\pi\varepsilon_0(r - a\cos\theta)} + \frac{(-Q)}{4\pi\varepsilon_0(r + a\cos\theta)}$$

$$= \frac{(2aQ)\cos\theta}{4\pi\varepsilon_0(r^2 - a^2\cos^2\theta)} \approx \frac{(2aQ)\cos\theta}{4\pi\varepsilon_0 r^2}$$

Hence $$r_2 = \left[\frac{(2aQ)\cos\theta}{4\pi\varepsilon_0 V_{\text{const}}}\right]\cos\theta$$

$$= B\cos\theta \; (r >> a) \qquad ...(3.22)$$

where B is a constant. These equipotentials are shown in Fig. 2.5 as dotted lines.

The using Eqn. (1.109) for grad V in spherical co-ordinates in Eqn. (3.18), we get the values of the three components of $\overline{\boldsymbol{E}}$ at any point as

$$\left.\begin{aligned} \overline{\boldsymbol{E}}_r &= \left(\frac{2aQ}{4\pi\varepsilon_0}\right)\frac{2\cos\theta}{r^3}\,\overline{\boldsymbol{u}}_r \\ \overline{\boldsymbol{E}}_\theta &= \left(\frac{2aQ}{4\pi\varepsilon_0}\right)\frac{\sin\theta}{r^3}\,\overline{\boldsymbol{u}}_\theta \quad (r >> a) \\ \overline{\boldsymbol{E}}_\phi &= 0 \end{aligned}\right\} \qquad ...(3.23)$$

The reason for displaying the quantity $(2aQ)$ in these equations instead of cancelling out the common factors in numerator and denominator is that $(2aQ)$ is the moment of the electric dipole.

Finally putting these values of $\boldsymbol{E}_r$ and $\boldsymbol{E}_\theta$ in Eqn. (2.16), the equation of the direction line is obtained as

$$\frac{dr}{r\,d\theta} = \frac{E_r}{E_\theta} = \frac{2\cos\theta}{\sin\theta}$$

or $$\frac{dr}{r} = \frac{2\cos\theta\,d\theta}{\sin\theta} = \frac{2d\,(\sin\theta)}{\sin\theta}$$

Therefore $$r = A\sin^2\theta \; (r >> a) \qquad ...(3.24)$$

which is recognised as being the same as Eqn. (2.18), since in Fig. 2.4

$$r = \sqrt{R^2 + z^2}$$

and $$\sin^2\theta = \frac{R^2}{R^2 + z^2}.$$

3.9 EQUIPOTENTIALS FOR UNIFORMLY CHARGED SHORT LINE

Using the notation of Fig. 2.7, the potential dV at P contributed by the infinitesimally small charge $(+q_l\,dL)$ is given by

$$dV = \frac{q_l\,dL}{4\pi\varepsilon_0\sqrt{(z-L)^2 + R^2}} \qquad ...(3.25)$$

Since the co-ordinates z and R of the point P are to be regarded as constants for the purpose of integrating this equation over the length of the charge distribution, we can put Eqn. (3.25) in the form of Eqn. (1.56) by the substitution $z - L = u$. Integrating over the range $L = -a$ to $L = +a$, we get

$$V = \frac{+q_l}{4\pi\varepsilon_0} \log_e \left[\frac{(z+a)+\sqrt{(z+a)^2+R^2}}{(z-a)+\sqrt{(z-a)^2+R^2}} \right] \qquad ...(3.26)$$

The equation to the equipotential then becomes apparent, by putting V = constant potential V_{const}, as

$$\frac{(z+a)+\sqrt{(z+a)^2+R^2}}{(z-a)+\sqrt{(z-a)^2+R^2}} = e^{\frac{4\pi\varepsilon_0 V_{\text{const}}}{q_l}} = k \text{ (say)} \qquad ...(3.27)$$

With the notation of Fig. 2.7, this can be written as

$$\frac{z+a+R_2}{z-a+R_1} = k$$

Applying the algebraic operation of componendo and dividends to this

$$\frac{R_2+R_1+2z}{R_2-R_1+2a} = \frac{k+1}{k-1} = \frac{1}{e} \text{ (say)} \qquad ...(3.28)$$

From triangles F_2PT and F_1PT in Fig. 2.7, we write

$$R_2^2 - (z+a)^2 = R^2$$

and $$R_1^2 - (z-a)^2 = R^2$$

which give $$R_2^2 - R_1^2 = 4az$$

or $$R_2 - R_1 = \frac{4az}{R_2+R_1} \qquad ...(3.29)$$

Substituting (3.29) in (3.28) and cross-multiplying

$$e\,(R_2+R_1)^2 + 2\,(ze-a)\,(R_2+R_1) - 4az = 0$$

solving

$$R_2 + R_1 = \frac{(a-ze) \pm \sqrt{(ze-a)^2+4eaz}}{e} = \frac{2a}{e} \qquad ...(3.30)$$

This shows that the sum of the distances $PF_2 + PF_1$ = constant. Reference to Eqn. (1.125) shows that this is the property of an ellipse of which F_1 and F_2 are the foci. Hence the equipotential surface is an ellipse.

The equation to the ellipse can be derived from Eqn. (3.27) by cross-multiplication, collecting the radicals on the left hand side and the remaining terms on the right hand side. Squaring twice as in Section 2.10 to get rid of the radicals, it will be found that if the calculation is done carefully and accurately, a large number of common terms cancel out on the two sides of the equation, which can then be easily put in the form

$$\frac{z^2}{\left(a\frac{k+1}{k-1}\right)^2} + \frac{R^2}{\left(\frac{2\sqrt{ka}}{k-1}\right)^2} = 1 \qquad ...(3.31)$$

This is the desired equation to the equipotential line. Revolving this ellipse about the z-axis gives the ellipsoid of revolution in three-dimensions. These form egg-like shapes enclosed one within the other and forming layers as in an onion (Fig. 3.5). The constants for the ellipse are found from Eqn. (3.31) by using the Eqns. (1.124) to (1.126).

$$\text{Semi-major axis} = a\frac{k+1}{k-1} \qquad ...(3.32)$$

$$\text{Semi-major axis} = \frac{2\sqrt{ka}}{k-1} \qquad ...(3.33)$$

$$\text{Eccentricity } e = \frac{k-1}{k+1} \qquad ...(3.34)$$

as already suggested in Eqn. (3.28).

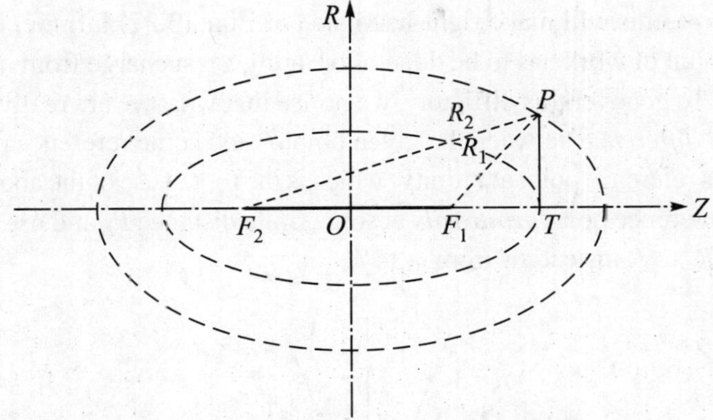

Fig. 3.5. Equipotential ellipsoids for uniformly charged short line

Focal distances OF_1 and OF_2

$$\pm \text{(semi-major axis) (eccentricity)}$$

$$\pm\, a\frac{k+1}{k-1}\,\frac{k-1}{k+1} = \pm a \qquad ...(3.35)$$

3.10 EQUIPOTENTIALS FOR UNIFORMLY CHARGED INFINITE LINE

It is not possible to apply the method of Section 2.11 and make *a* infinite in Eqn. (3.31) for the transition from the short line to the infinite line. A different approach starting from basic considerations is adopted below to evaluate *V* and also to highlight another difficulty.

We note that for any plane at right angles to the infinite line, the electric field map resembles that for a point charge, but *V* in one case varies as $\frac{1}{R}$ and in the other as $\frac{1}{r^2}$. Hence we can use the method of Sec. 3.3 and apply Eqn. (3.10) to Eqn. (2.27), thus giving.

For the infinite line

$$\overline{E}_R = \frac{q_l}{2\pi\varepsilon_0 R}\overline{u}_R \qquad \text{(by Eqn. 2.27)}$$

Therefore

$$V = -\int_{\infty}^{r} \overline{E}\cdot\overline{ds} = -\int_{\infty}^{R} \frac{q_l}{2\pi\varepsilon_0 R}\overline{u}_R\cdot\overline{ds} \qquad ...(3.36)$$

But $\overline{u}_R\cdot\overline{ds} = \overline{u}_R\cdot(\overline{u}_R\, dR + \overline{u}_\phi\, Rd\phi + \overline{u}_z\, dz) = dR$

Then

$$V = -\frac{q_l}{2\pi\varepsilon_0}\int_{\infty}^{R}\frac{dR}{R} = -\frac{q_l}{2\pi\varepsilon_0}\left[\log_e R\right]_{\infty}^{R} \qquad ...(3.37)$$

Since $\log_e \infty = \infty$, this will make right-hand side of Eqn. (3.37) infinite, implying that an infinite amount of work has to be done of bringing a test charge from infinity to any point in space. To get over this difficulty, we notice that what we are really interested in is the *potential difference* between the given point P and some reference point. Instead of choosing this reference point at infinity (which is the root cause of the above dilemma), we select the reference point *arbitrarily* at some finite distance R_0 and use the limits for integration as R_0 to R instead of from ∞ to R.

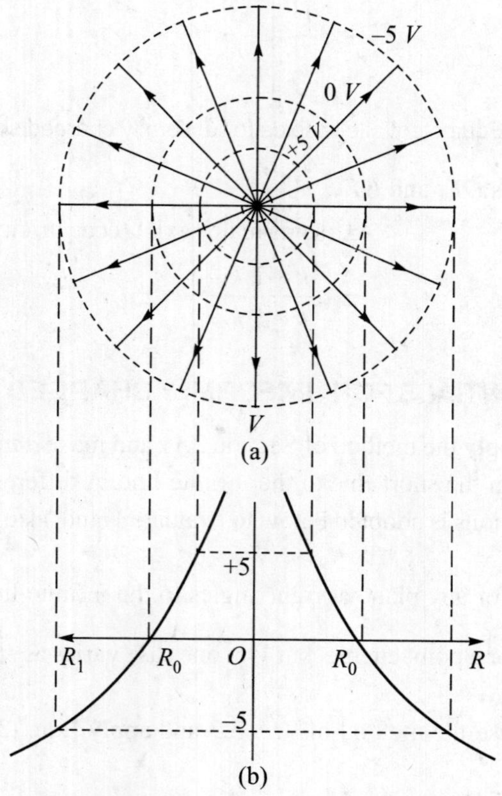

Fig. 3.6. (*a*) Direction lines and equipotentials and (*b*) Plot of V as a function of R, for uniformly charged infinite line

Then we get

$$V = -\frac{q_l}{2\pi \epsilon_0}[\log_e R]_{R_0}^{R} = \frac{q_l}{2\pi \epsilon_0}\log_e \frac{R_0}{R} \text{ volt} \quad ...(3.38)$$

This makes V positive if $R < R_0$ and negative if $R > R_0$. The equipotentials and the direction lines appear as shown in Fig. 3.6 (*a*); Eqn. (3.38) is plotted in Fig. 3.6 (*b*).

In three-dimensional space, the equipotential surface is obtained by moving Fig. 3.6 (*a*) and the plane of the paper parallel to itself, keeping the origin O always on the charged line. The equipotentials then form a family of coaxial cylinders of different radii, enclosing the charged line as their common axis. It should be carefully noted that though Fig. 3.4 (*a*) and Fig. 3.6 look alike, the curves in Fig. 3.4 (*b*) is different from that in Fig. 3.6 (*b*).

3.11 EQUIPOTENTIAL MAP FOR INFINITE PARALLEL LINES WITH EQUAL AND OPPOSITE UNIFORM LINEAR CHARGES

Using the notation of Fig. 2.9 and applying Eqn. (3.38), we get the potential at the point P as

$$V = V_+ + V_- = \frac{q_l}{2\pi \epsilon_0}\left(\log_e \frac{R_0}{R_1} - \log_e \frac{R_0}{R_2}\right) = \frac{q_l}{2\pi \epsilon_0}\log_e \frac{R_2}{R_1} \quad ...(3.39)$$

In this equation, since we have used the same value of R_0 in writing the values of both V_+ and V_- to get this term cancelled out, it implies that we have chosen the reference point of zero potential at equal distances R_0 from both the lines. Such points are located on the *y-z* plane at $x = 0$ (Fig. 3.7). Any point on this plane can be used as the reference point with $V = 0$, for whatever may be the value of R_0 for such a point, it cancels out and does not appear in the final value of V as seen from Eqn. (3.39). The *y-z* plane at $x = 0$ is therefore the equipotential surface for $V = 0$.

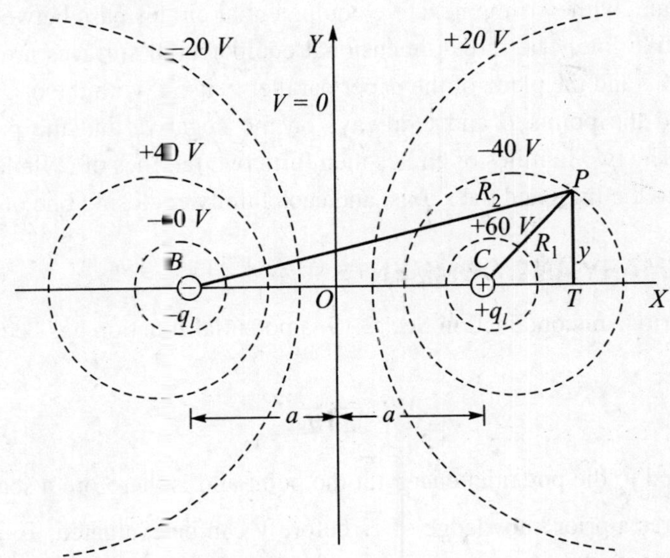

Fig. 3.7 Equipotential map for infinite parallel line

By giving to V the constant value V_{const}, we get the equation to the equipotential lines in Fig. 3.7 as

$$\frac{R_2}{R_1} = e^{\frac{2\pi \in_0 V_{const}}{q_l}} = k \text{ (say)} \qquad ...(3.40)$$

or

$$\frac{\sqrt{(x+a)^2 + y^2}}{\sqrt{(x-a)^2 + y^2}} = k,$$

since $OT = x$ and $TP = y$.

Squaring both sides and collecting terms, and then completing the square by adding a constant term to both sides, this can be put in the form

$$\left[x - a\left(\frac{k^2+1}{k^2-1}\right)\right]^2 + y^2 = \left(\frac{2ak}{k^2-1}\right)^2 \qquad ...(3.41)$$

This equation for the equipotential lines shows that these lines form a family of circles with centre at $\left(a\frac{k^2+1}{k^2-1}, 0\right)$ and radius $\frac{2ak}{k^2-1}$.

From Eqn. (3.40), we see that when $V_{const} = 0$, $k = e^0 = 1$, and this makes the radius of the circle infinite. That is how we get the y-axis as the equipotential line for $V_{const} = 0$. For positive values of V_{const}, $k > 1$, and the centre $(h, 0)$ of the circle is on the x-axis to the right of the positively charged line $(h > a)$. For negative value of V_{const}, $k < 1$, and the centre of the circle is on the x-axis to the left of the negatively charged line $(h < -a)$.

Unlike the family of circles for the direction lines, all of which pass through the two lines in Fig. 2.11, the family of equipotential lines in Fig. 3.7 do not intersect each other at all. Moreover *all* the positive equipotential circles pass between the origin and the positive line, while *all* the negative equipotential circles pass between the origin and the negative line. The three-dimensional equipotential surfaces are obtained by moving Fig. 3.7 and the plane of the paper parallel to itself, with the origin always on the z-axis and the points B and C always on the negative and the positive lines, respectively. The two families of circles then form two families of cylinders with their axes parallel to the lines and the z-axis, and each family enclosing one of the lines.

3.12 SUMMARY AND COMMENTS

To meet the criticism contained in Sec. 2.19, a potential function has been defined by the equation

$$V = -\int_{\infty}^{r} \overline{E} \cdot \overline{ds} \qquad \text{(By Eqn. 3.10)}$$

which is related to the potential energy at the point and is therefore a scalar quantity. But this requires a prior knowledge of $\overline{E}$ before V can be evaluated. To get over this difficulty, we derived the equation

$$V = \frac{Q}{4\pi\varepsilon_0 r} \text{ volt} \qquad \text{(By Eqn. 3.12)}$$

which does not require the knowledge of $\overline{E}$. This was then related to $\overline{E}$ by a space differentiation process as

$$\overline{E} = -\text{grad } V = -\overline{\nabla} V \quad V/m \qquad \text{(By Eqn. 3.18)}$$

Thus, we have succeeded in replacing the addition of vector quantities by the addition of scalar quantities only, as in Eqn. (3.15) and $\overline{E}$ is then derived by differentiation which is easier than integration.

However, in using Coulomb's Law as well as the potential function, a knowledge of r, the distance of the point P from the charge is necessary. Hence we now search for a method by which, at least in cases of problems involving simple geometries, we can get $\overline{E}$ without knowing r! Also Eqn. (3.15) still implies integration of distance functions. We would like to eliminate this. These topics will engage our attention in the next chapter.

EXERCISE 3

1. For the field of velocity vector $\overline{U} = y\,\overline{u}_x$ in Fig. 3.2 (*c*), evaluate the line integral $\int \overline{u} \cdot \overline{ds}$ along the paths

   (*i*) *acb* where *c* is the centre of the circle given by Eqn. (3.4) and

   (*ii*) *boa* where path is the *y*-parabola $x^2 = 4y$.

2. For the equal and like point charges of Fig. 2.3, find the potential V_0 at the points $(-3a, 0)$ and $(+3a, 0)$ on the *z*-axis. Next find the value of R for the points on the line $z = -a$ and $z = +a$ at which V has the same value V_0. Finally, find the value $|z| < a$ for the points on the *z*-axis for which the potential is V_0. Joining these points by a line which is always at right angles to the field direction lines of Fig. 2.3, obtain the shape of the equipotentials surrounding the two charges.

3. Electric charges of magnitude $(+Q)$ and $(-kQ)$ coulomb are located on the *x*-axis at $x = a$ and $x = -a$, respectively. Derive the equation to the equipotential line in the *x-y* plane for which the potential $V = 0$. Sketch the curves for the three cases for which

   (*i*) $k = \frac{1}{2}$ (*ii*) $k = 1$ and (*iii*) $k = 2$.

   For $k = \frac{1}{2}$, find the points where the equipotential intersects the *x*-axis and determine the value of $\overline{E}$ at the these points.

4. Derive the Eqns. (3.31) and (3.34) for the elliptic equipotential line for a uniformly charged short line of length $2a$.

5. (*a*) Starting with Eqn. (3.14), derive the Eqns. (2.28) and (2.29) for the components of the electric field intensity at any point in the case of the infinite parallel lines with equal and opposite uniform charges.

(*b*) For this infinite parallel line, prove that all positive equipotentials in Fig. 3.7 pass between 0 and the positive line, and that the positive line is enclosed by all of them.

6. For the linear electric quadripole of Ex. 2, Problem 5, show that for any point on the y-z plane (ϕ = constant) in spherical co-ordinates, the potential V at any distant point for which $r > 2a$, is given by

$$V = \frac{2a\,(2aQ)}{4\pi\varepsilon_0}\frac{(3\cos^2\theta - 1)}{r^3}$$

where (using notation of Fig 2.4) in the binomial expansion of $\frac{r}{R_1}$ and $\frac{r}{R_2}$ given by the Law of Cosines [Eqn. (1.31)], terms $\frac{a^3}{r^3}$ and higher are neglected.

7. For the uniformly charged circular loop of Fig. 2.12, find the potential at the point P on the z-axis at a distance z from the centre of the loop, and hence derive Eqn. (2.36).

8. For the uniformly charged circular disk of Fig. 2.13, find the potential on the axis of the disk at a point P whose distance from the centre of the disk is z, and hence derive Eqn. (2.38).

9. Show that for a spherical shell, having a uniform charge density given in spherical co-ordinates by

$$\text{Charge density} = \begin{cases} 0 & \text{for} \quad r < a \\ q_v & \text{for} \quad a < r < b \\ 0 & \text{for} \quad b < r \end{cases}$$

the potential at a distance $r > b$ from the centre is given by assuming that the total charge in the shell is located at the centre of the shell.

10. Assume that the electric charge density in the electron beam in a cathode-ray tube can be represented in cylindrical co-ordinates as

$$\text{Charge density} = \begin{cases} q_v\left(1 - \frac{R^2}{a^2}\right) & \text{for } 0 < R \leq a \\ 0 & \text{for } 0 < R < \infty \end{cases}$$

Find the following:

(*i*) Total charge Q in the beam per metre length.

(*ii*) Values of V and $\overline{E}$ at a point for which $R > a$.

(*iii*) Values of V and $\overline{E}$ at a point for which $R < a$.

(**Hint :** Assume beam to be infinitely long.)

11. Assume that, as a first approximation, the electric charge density in the nucleus of lithium can be represented in spherical co-ordinates as

$$\text{Charge density} = \begin{cases} q_v\left(1 - \frac{r^2}{a^2}\right) & \text{for } 0 < r \leq a \\ 0 & \text{for } 0 < r < \infty \end{cases}$$

Find the following :

(*i*) Total charge Q in the nucleus.

(*ii*) Values of V and $\overline{E}$ at a point for which $r > a$.

(*iii*) Values of V and $\overline{E}$ at a point for which $r < a$.

(*iv*) Hence find the value of $\frac{r}{a}$ for the point for which $\overline{E}$ has a maximum value.

CHAPTER

4 GAUSS'S LAW

4.1 FARADAY'S EXPERIMENT

The English physicist Michael Faraday (1791–1867) performed a very interesting experiment, the inference drawn from which forms the basis of the subject matter of this chapter. First he placed a charge $+Q$ on a small sphere S_{in} as in Fig. 4.1 (a). Next he surrounded the charged sphere by another hollow sphere S_{out} of larger size [Fig. 4.1 (b)] placed on an insulated stand, then he filled the space between the spheres by a dielectric. In the third step he connected only the outer sphere momentarily to the earth by closing the switch *SW*, and again opening it [Fig. 4.1 (c)]. Finally he removed the

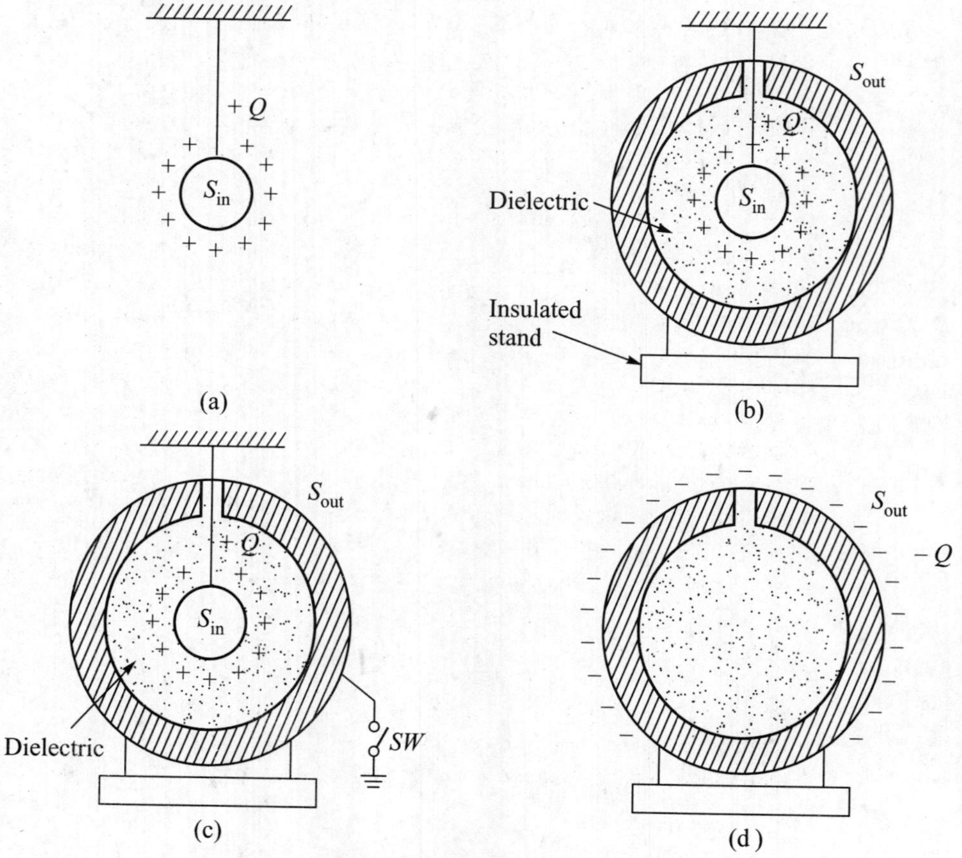

Fig. 4.1. Faradays experiment

inner charged sphere S_{in} and then proceeded to measure the charge on the outer sphere S_{out}. Whatever be the sizes of the inner and the outer spheres, and whatever be the dielectric (including the case of air as a dielectric), Faraday found that the charge on the outer sphere S_{out} was *always equal in magnitude* to the original charge Q placed on the inner sphere *but of opposite sign.*

Faraday therefore concluded that something was coming out of the inner sphere and reaching the outer sphere in all the cases, and that something depended only on the magnitude of the original charge placed on the inner sphere, and also that it did not depend on the sizes of the spheres or on the dielectric between the spheres. The phrase "coming out" (some books use "emanating") used in the previous sentence is not meant in the sense in which people come out of a cinema hall or water comes out of a fountain, because in these cases, after some time, the cinema hall or the water-tank feeding the fountain will become empty. The phrase is used in the sense in which roads come out of a large city in all directions and their number remains constant with time.

4.2 ELECTRICAL DISPLACEMENT OR FLUX

To this "something" that comes out from charge on the inner sphere to the surface of outer sphere, Faraday gave the name "electric displacement" or "electric flux". Thus electric displacement vector or "lines" of electric flux flow out of each positive charge. In SI units we say that the total electric flux ψ coming out of a charge $+ Q$ coulomb is

$$\text{Total electric flux } \psi = + Q \text{ coulomb} \qquad \text{...(4.1)}$$

showing that the unit for ψ in SI units is coulomb. If the charge is negative $(- Q)$, then the electric displacement vectors are reversed and they terminate on the negative charge, and $\psi = - Q$ coulomb.

Since the electric displacement or flux comes out of a charge $+ Q$, and passes outwards through any surface surrounding Q, we immediately begin to think of an electric displacement density or flux density $\bar{D}$ coulomb per (meter)^2 at different points of this surrounding or enclosing surface. This flux density is a vector quantity, because if we assume Q to be a point charge, then $\bar{D}$ will be along a straight line joining Q to the given point on the surface, and this direction will change from point to point on the surface.

Now to find the flux density, we have to measure a small area and also the amount of flux passing through it. This area itself is a vector quantity denoted by a vector along the normal to the area. However the directions of the vector $\bar{D}$ and the area vector may, in the general case, be different. In the very special case of a point charge $+ Q$, if we select the enclosing surface as a sphere of radius r with its centre at the location of the charge, we find that $\bar{D}$ and the area vector are in the same direction at *every point* on the sphere. Hence the magnitude of the electric flux density $|D|$

$$= \frac{\Psi}{4\pi r^2} \quad \text{or} \quad |D| = \frac{Q}{4\pi r^2} \qquad ...(4.2)$$

Comparing this equation with Eqn. (2.12) for the electric field intensity, we get the relation

$$\bar{D} = \varepsilon \bar{E} \text{ for isotropic media} \qquad ...(4.3)$$

The most important aspect to be noted about this equation is the fact that $\bar{D}$, ε and $\bar{E}$ all refer to *the same point on the surface* under consideration and ε is the value of permittivity at *this* point (and not at the location of the charge *Q*). Also Faraday's experiment has shown that $\bar{D}$ at the surface does not depend on the permittivity of the medium between the charge *Q* and the surface. This is the justification for the values of the permittivity used in Eqn. (2.11).

For anisoropic media, $\bar{D}$ and $\bar{E}$ will generally have different directions.

4.3 MIE PLATES

To find the magnitude and the direction of the vector $\bar{D}$ at any point, we preform a "thought experiment" of the type made popular by Albert Einstein (1879–1955). Its name implies that the experiment is not conducted in reality in a laboratory, but only in our imagination.

Two very minute disks, called Mie Plates, each of area *da* are placed together (that is, touching each other at all points) at the point *P* where $\bar{D}$ is to be measured. Then they are separated by a small distance to form a parallel-plate capacitor. Under the influence of the electric flux vector, the plates get charges $+\Delta q$ and $-\Delta q$ coulomb. The magnitude Δq of this charge depends on the magnitude of $\bar{D}$ at that point and also on the direction of $\bar{D}$ with respect to the normal to the plates. We remove (in thought only) the Mie plates from inside the dielectric and measure this induced charge Δq on them.

We repeat the experiment many times (again in thought only) until we are able to find that orientation of the Mie Plates which makes this induced charge Δq a maximum. Then

$$|D| = \frac{(\Delta q)_{\text{maximum}}}{da} \qquad ...(4.4)$$

and its direction is the normal from the positively charged plate towards the negatively charged plates. The vector nature of $\bar{D}$ and the fact that its magnitude does not depend on the dielectric between the charges like *Q* from which $\bar{D}$ emanates, and the point *P*, enables us to measure $\bar{D}$ as above using Mie plates.

4.4 LINES AND TUBES OF ELECTRIC DISPLACEMENT OR FLUX

Just as a direction line on an electric field map is a curve, such that the tangent to it at any point gives the direction of the electric field intensity at that point, similarly we can have a line of electric flux, the tangent to which will give the direction of the electric flux density $\overline{D}$ (or displacement density) at that point. Equation (4.3) shows that for an isotropic medium, these two (that is, direction lines for field intensity $\overline{E}$ and flux density $\overline{D}$) are identical.

The concept of *any vector* quantity (not electric displacement only) $\overline{D}$ flowing through an area a_{in} and normal to it, is the general form of "flux" (product of $\overline{D}$ and a_{in}). If we consider the lines of flux passing through each point on the periphery of area a_{in}, then these lines will form a surface in the shape of a tube which is called the "tube of flux". We take a normal section of this tube at another point and call the area of cross-section there as a_{out} (Fig. 4.2). Thus a_{out} is normal to the flux lines at that point. If we denote the surface area of the curved side of the tube as a_{side}, then the direction of the flux density $\overline{D}$ being *tangential* to this surface at every point, no flux "comes out" or crosses this surface of the tube from the inside to the outside.

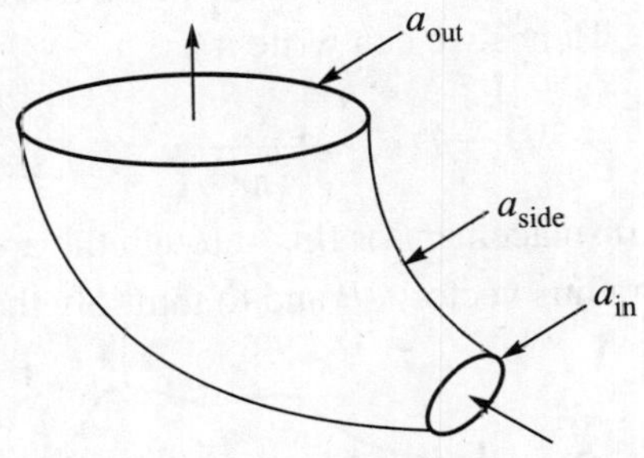

Fig. 4.2. Tube of flux

The total flux coming out of this part of the tube can be broken into three parts as

[Flux coming out of a_{in}] + [Flux coming out of a_{side}] + [Flux coming out of a_{out}].

It will be seen from the discussion in the next section that if there is no charge enclosed in the above part of the tube of flux, then the sum of the above three terms will always be zero. But the middle term pertaining to the curved side of the tube is already seen above to be equal to zero. Hence we get

[Flux coming out of plane a_{in}] + [Flux coming out of plane a_{out}] = 0

or [Flux coming out of a_{out}] = [Flux going in at a_{in}] ...(4.5)

This is a very interesting property of tubes of flux for any given vector $\overline{D}$, for it makes the vector behaves like water flowing through a tube of variable cross-section. Whatever amount goes in at one end, exactly the same amount comes out at the other end, and no water comes out through the sides of the tube. Though $\overline{\boldsymbol{D}}$ was considered to be *any* vector in this section, Eqn. (4.5) is true for electric flux density vector $\overline{\boldsymbol{D}}$ also, if the condition that "the total flux leaving the tube is zero" is satisfied.

On a field map, instead of tubes of flux, only the axial direction line of each tube is drawn. The cross-sectional areas of the tubes are selected in such a way that through each area, the same amount of flux passes. With this method of representation, besides the direction, additional information about the magnitude of the vector is visually presented. A closer spacing of the direction lines on the field map will indicate stronger field in that area.

4.5 GAUSS'S LAW

We begin with a point charge $+Q$ coulomb located at a point O in a linear, homogeneous, isotropic medium having permittivity ε. This point O is completely surrounding by an enclosing surface at *any shape* whatsoever, and P is a point on this surface at a distance $OP = r$ (Fig. 4.3). Consider an infinitesimally small area da surrounding P, which can be represented by a vector $da\,\overline{\boldsymbol{u}}_n$ where $\overline{\boldsymbol{u}}_n$ is the unit vector normal to the area da. The electric displacement density at P has a magnitude given by Eqn. (4.2) and its direction is along the radial line OP. Hence we can write it as

$$\overline{\boldsymbol{D}} = D\,\overline{\boldsymbol{u}}_r = \frac{Q}{4\pi r^2}\,\overline{\boldsymbol{u}}_r \qquad \text{...(4.6)}$$

To get the electric displacement or flux through the area da, we have to project the area da normal to the radius vector OP and to multiply the result by D. This gives

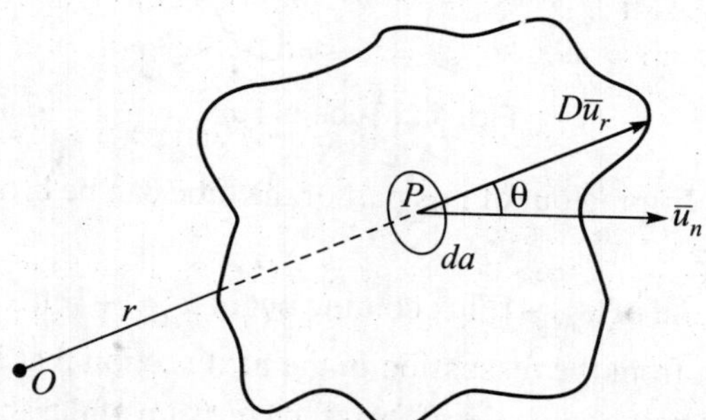

Fig. 4.3. Electric flux or displacement through a surface

$$d\psi = \overline{\boldsymbol{D}}\;.\;\overline{\boldsymbol{da}} = (D\,\overline{\boldsymbol{u}}_r)\,.\,(da\,\overline{\boldsymbol{u}}_n)$$
$$= D\,(da)\cos\theta \qquad \text{...(4.7)}$$

where θ is the angle between the vectors $\overline{\boldsymbol{u}}_r$ and $\overline{\boldsymbol{u}}_n$. But by Eqn. (1.133), the solid angle $d\Omega$ subtended by the area da at the point O is

$$d\Omega = \frac{\overline{\boldsymbol{u}}_r \cdot (da\overline{\boldsymbol{u}}_n)}{r^2} \text{ steradians} \qquad \text{...(1.133)}$$

Hence Eqn. (4.7) becomes on combining with Eqn. (4.6)

$$d\psi = Dr^2\, d\Omega = \frac{Q}{4\pi}\, d\Omega \qquad ...(4.8)$$

To get the total electric flux coming out of the surface, we integrate this equation over the whole surface and get

$$\psi = \oint_{\text{surface}} d\psi = \oint_{\text{surface}} \overline{D} \,.\, \overline{da}$$

$$= \frac{Q}{4\pi} \oint_{\text{surface}} d\Omega = \frac{Q}{4\pi}(4\pi) = Q \qquad ...(4.9)$$

where the integral sign with a circle and the word "surface" indicate an integration over *the entire surface enclosing* the point charge + Q, and permits us to use Eqn. (1.134) for the total solid angle as 4π.

Equation (4.9) is the same as Eqn. (4.1), and superficially it appears that we have achieved nothing in deriving Eqn. (4.9). But notice that though we started with Eqn. (4.2) which was based on the assumption of a *spherical* surface centered at O, we applied it to *any surface*, and obtained Eqn. (4.9) which shows that for finding the total displacement through a *surface completely enclosing O*, we do not need the distance from the point charge Q to the surface to be constant. The result is still consistent with Faraday's assumption of Eqn. (4.1), provided we are able to use the relation

$$\oint_{\text{surface}} d\Omega = 4\pi.$$

This encourages us to extend Eqn. 4.9 in steps as follows: Thus if there are many point charges $Q_1, Q_2, Q_3, ..., Q_n$, *all completely enclosed by the surface,* then

$$\text{Total electric displacement } \psi = \oint_{\text{surface}} \overline{D} \,.\, \overline{da}$$

$$= \sum_{n=1}^{n} Q_n \qquad ...(4.10)$$

As a corollary to this, it follows that if there are any charges situated outside this surface, then their contribution to the total electric displacement through this surface is nil.

The next and final step is to assume that instead of point charges, we have a charge distribution which is completely enclosed by the surface. Then in this volume enclosed by the surface, we choose an infinitesimally small volume dv over which we assume the charge density to be constant and equal to q_v coulomb per cubic metre, so that the charge in the volume dv becomes $q_v\, dv$ coulomb. The summation on the right hand side of Eqn. (4.10) can then be replaced by an integration of $q_v\, dv$ over the whole volume. This convert Eqn. (4.10) to the final form

$$\oint_{\text{surface}} \overline{D} \,.\, \overline{da} = \oint_{\text{surface}} \overline{D} \,.\, \overline{u}_n\, da$$

$$= \oint_{\text{volume}} q_v\, dv \qquad ...(4.11)$$

This equation is known as Gauss's law (in integral form), because it was obtained by the German mathematician Johann Friederich Karl Gauss (1777–1855). Stated in words, Gauss's law means that the *net outward* electric displacement through a *completely closed surface* is equal to the net charge contained in the volume enclosed by the surface.

The special aspects of Gauss's law to be noted are

(*i*) It relates a surface integral to a volume integral.

(*ii*) It is applicable *only if* the surface completely encloses the volume.

(*iii*) It does not specify any particular shape for the enclosing surface. Hence, we can choose any shape for the surface that will simplify our calculations.

(*iv*) It does not require a knowledge of the distances of the charges from the points on the surface. The only criterion is whether the charges are located *inside the enclosed volume,* because it does not apply to charges situated outside the enclosed volume.

(*v*) A *positive* value of the integral on the right hand side of Eqn. (4.11) indicates that the net flux is directed *outwards* (from the inside to the outside) through the enclosing surface. A negative value for the integral means that the net total flux is directed inwards.

Any *completely closed* surface to which Gauss's law is intended to be applied is therefore called a Gaussian surface.

4.6 PROCEDURE FOR APPLYING GAUSS'S LAW

Gauss's law is a very powerful tool for the study of electric fields. It is best suited for the analysis of those problems having such a geometry that because of symmetry considerations, the direction and magnitude (particularly zero magnitude) of $\overline{D}$ can be estimated purely from physical considerations (without performing any integration). When a conductor is placed in an electric field, the resulting situation is best analysed by the use of Gauss's law:

To apply Gauss's law, we have first to select a suitable Gaussian surface, which will simplify the solution by avoiding unnecessary, long calculations. This is done using the following considerations:

(*i*) If a plane can be located such that the charges are symmetrically distributed on the *two sides* of the plane along a straight line normal to the plane, then including such a plane as part of the enclosing surface will be useful, because there will be no electric displacement across this plane in either direction (inwards or outwards).

(*ii*) If a plane or a cylinder or a sphere can be chosen so that it is symmetrically situated with respect to the charge distribution, then the value of $\overline{D}$ on such a plane or cylinder or sphere will be the same at *every* point of this surface, and hence the left hand integral of Eqn. (4.11) becomes easy to evaluate.

(*iii*) Having located above types of surfaces, it is best to select as a first attempt, a surface such that the total charge enclosed by the entire surface is zero. (This may not be possible in every case, particularly if there are no negative charges anywhere in the system). Certain inferences can be drawn from this situation.

(*iv*) While retaining those surfaces across which the displacement is found to be zero, other parts of the enclosing surface are shifted to different locations and thus $\bar{D}$ is obtained at different points in space.

(*v*) Then using Eqn. (4.3) the value of the electric field intensity at different points in space is obtained. A few illustrative examples follow.

4.7 ELECTRIC FIELD OF INFINITE PLANE WITH UNIFORM CHARGE DENSITY

Assume that the uniform charge density of $+q_s$ coulomb per square metre is distributed on the plane $x = 0$. Then for any plane at right angles to this charged plane, because of symmetry of charge on both sides of the plane, there will be no electric displacement across this perpendicular plane in either direction. In other words, $\bar{D}$ is everywhere normal to the plane $x = 0$. Further, because of symmetry, $\bar{D}$ must have the same *magnitude* at equal distances from the plane $x = 0$.

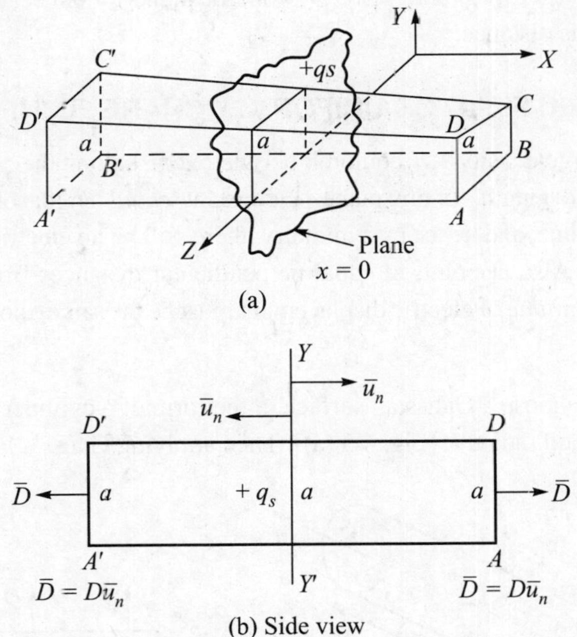

Fig. 4.4. Electric field of uniformly charged infinite plane

Hence we choose a Gaussian surface in the form of a rectangular parallelopiped [Fig. 4.4 (a)] with planes $ABCD$ and $A'B'C'D'$ parallel to the plane $x = 0$ and at equal distances on the two sides of the plane $x = 0$. Then applying Gauss's law to this surface, we have, if area $ABCD$ is equal to a,

(Net outward flux) = (charge enclosed)

or (Flux through face $A'B'C'D'$) + (Flux through four plane sides perpendicular to plane $x = 0$) + (Flux through face $ABCD$)

$$= aq_s$$

or $$(a\,D) + (\text{zero}) + (a\,D) = aq_s$$

so that $$D = \frac{q_s}{2}.$$

Notice that the direction of the vector $\overline{D}$ is in opposite directions on the two sides of the plane $x = 0$ [Fig. 4.4(b)]. This fact can be expressed mathematically in two different ways as follows:

$$\overline{D} = \frac{q_s}{2}\,\overline{u}_n = \frac{q_s x}{2\,|x|}\,\overline{u}_x$$

Then applying Eqn. (4.3) we get the electric field intensity as

$$\overline{E} = \frac{\overline{D}}{\varepsilon_0} = \frac{q_s}{2\varepsilon_0}\,\overline{u}_n$$

$$= \frac{q_s x}{2\varepsilon_0\,|x|}\,\overline{u}_x \quad \text{V/m} \qquad ...(4.12)$$

which is the same as Eqn. (2.39). This result remains the same whatever may be the distance of the planes $ABCD$ and $A'B'C'D'$ from the plane $x = 0$, because Eqn. (4.12) is independent of this distance.

4.8 ELECTRIC FIELD OR UNIFORMLY CHARGED INFINITE LINE

Let the linear charge density $+\,q_l$ coulomb per metre exist along the z-axis. Then since the line is of *infinite* length, for any plane z = constant, *equal* lengths of line exist on the two sides of the plane, and hence by symmetry, there will be no electrical displacement across this plane. Also at points at equal perpendicular distances from the line, (R = constant), the *magnitude* of electric displacement must be the same, though the direction will be along $\overline{u}_R$.

Hence we choose a Gaussian surface in the form of a cylinder of length L with its axis on z-axis and radius R [Fig. 4.5 (a)]. Then applying Gauss's law,

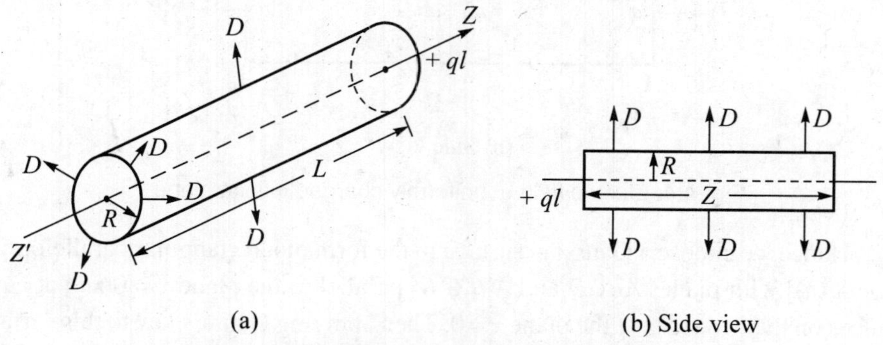

Fig. 4.5. Electric field of uniformly charged infinite line

(Outward flux through the two plane end faces) + (Outward flux through the curved surface of the cylinder) = (Charge enclosed)

or $$0 + D(2\pi RL) = q_l L$$

or $$D = \frac{q_l}{2\pi R}$$

And using Eqn. (4.3) $$\overline{\boldsymbol{E}} = \frac{q_l}{2\pi\varepsilon_0 R}\,\overline{\boldsymbol{u}}_R \text{ V/m} \qquad ...(4.13)$$

which is the same as Eqn. (2.27).

4.9 ELECTRIC FIELD OF UNIFORMLY CHARGED SPHERICAL SHELL

In spherical co-ordinates, let the uniform charge density be $+q_v$ C/m^3 for $a < r < b$ and zero everywhere else [Fig. 4.6 (a)]. By symmetry, it is seen that at all points on *any* sphere whose centre is at the origin O, the outward electrical displacement will always be normal to the surface of the sphere, and also that it will have the same magnitude at all points on the surface of the sphere. The solution is divided into three distinct cases.

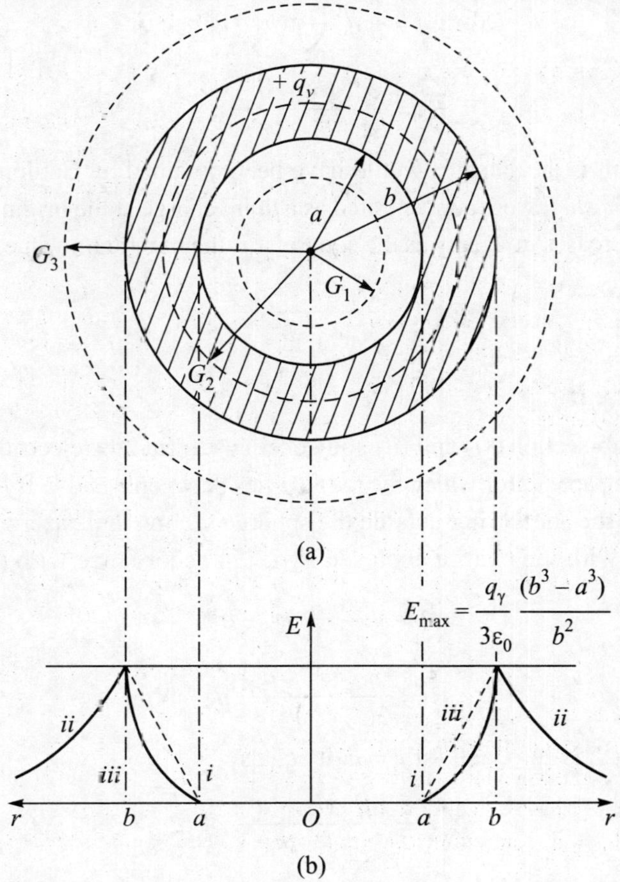

Fig. 4.6. Electric field of uniformly charged spherical shell

4.9.1 (i) r < a

A sphere G_1 of radius r less than a, with center on O, will not enclose any charge. This makes the right hand side of Eqn. (4.11) equal to zero, and shows that there is no electric field inside the shell, or $\overline{E} = 0$.

4.9.2 (ii) r > b

A Gaussian surface in the form of a sphere G_3 of radius r greater than b, and with centre at O, will enclose the total charge in the shell. Since the surface area of a sphere is $4\pi r^2$ this total charge is found by integrating

$$\int_a^b + q_v (4\pi r^2)\, dr$$

as $$q_v \frac{4}{3}\pi\,(b^3 - a^3) \text{ coulomb.}$$

This gives the right hand side of Eqn. (4.11). Equating this to the left hand side of Eqn. (4.11), noting that D is constant over the surface of the sphere G_3, we get

$$D(4\pi r^2) = +\, q_v \frac{4}{3}\pi(b^3 - a^3)$$

so that $$\overline{E} = \frac{q_v}{3\varepsilon_0}\frac{(b^3 - a^3)}{r^2}\,\overline{u}_r \quad \text{V/m} \qquad ...(4.14)$$

This result is the same as would have been obtained by Coulomb's Law if *all the charges in the shell* were concentrated as a *point* charge at the origin O. Eqn. (4.14) shows that E decreases inversely as the square of r, the maximum value, at $r = b$, being

$$E_{max} = \frac{q_v}{3\varepsilon_0}\frac{(b^3 - a^3)}{b^2}$$

4.9.3 (iii) *a* < *r* < *b*

The results of sub-sections (*i*) and (*ii*) above help us to find the electric field intensity on a Gaussian surface G_2 for which the radius r lies between a and b. By (*i*) the portion of the charge of the shell lying outside the sphere G_2 contributes zero displacement. Hence we deal with the charge enclosed by G_2 in accordance with (*ii*). Then

$$D(4\pi r^2) = +\, q_v \frac{4}{3}\pi(r^3 - a^3)$$

so that $$\overline{E} = \frac{q_v}{3\varepsilon_0}\frac{(r^3 - a^3)}{r^2}\,\overline{u}_r \quad \text{V/m} \qquad ...(4.15)$$

For $r = a$, E is zero, and for $r = b$, it equals

$$E_{max} = \frac{q_v}{3\varepsilon_0}\frac{(b^3 - a^3)}{b^2}$$

This variation is shown in Fig 4.6 (*b*).

4.10 ELECTRIC FIELD OF A CYLINDRICAL ELECTRON BEAM

In a cathode-ray tube or a television kinescope (picture tube), electrons emerge from an electron gun in the form of a cylindrical beam in which all electrons may be assumed to be travelling along parallel straight lines at uniform, constant speed. In other words the mutual repulsion of the electrons in the transverse direction that tends to cause the beam diameter to increase progressively, is neglected. At first sight one may hesitate to accept this as a problem in electrostatics, because the electrons are in motion, but when we consider that a place vacated by an electron is taken up by another, we can regard this as a case of uniform, static distribution of charge throughout the volume of a cylinder. We further simplify the problem by assuming that the electron beam is infinitely long, which is approximately correct in practice, since the beam length is very much greater than the beam diameter.

Let the electronic charge density be $(-q_v)$ C/m^3 in an infinitely long cylinder of radius a metre, having its axis on z-axis and assume that the electrons are travelling in the negative direction of the z-axis with a constant velocity $(-v_0)$ metre/second. [Fig. 4.7 (*a*)]. This is equivalent to a direct current of I amperes flowing in the positive direction of the z-axis, where

$$I = (-q_v)(-v_0)(\pi a^2) \qquad \text{...(4.16)}$$

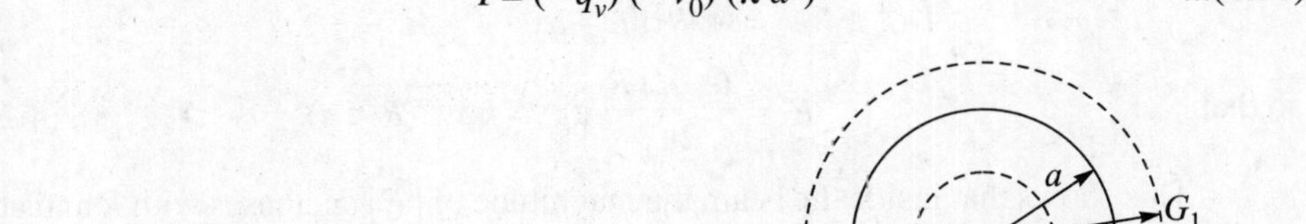
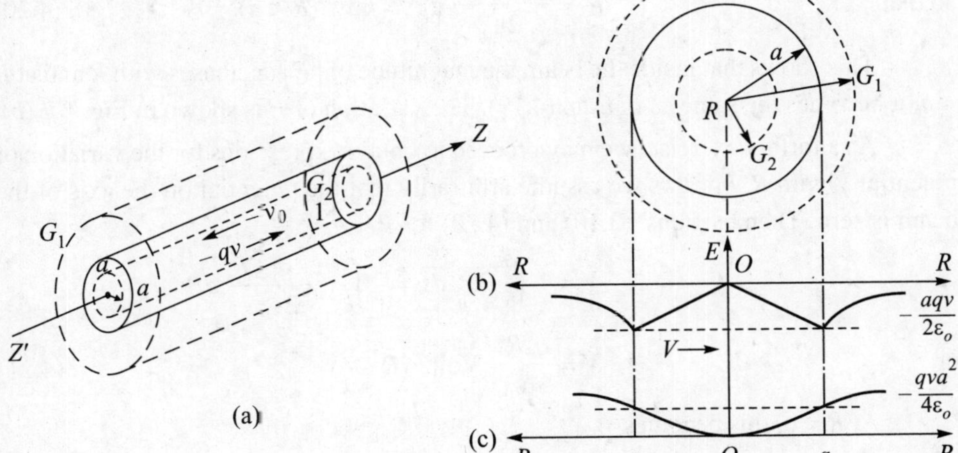

Fig. 4.7. Electric field of a cylindrical electron beam

This follows from the fact that all charges contained in a cylinder of length v_0 and cross-sectional area (πa^2) will flow in one second past a plane at right angles to the beam. Further notice that the same Eqn. (4.16) would have been obtained if the charge density were $(+q_v)$ C/m^3 and the charges were moving with velocity $(+v_0)$ m/s in the positive direction of the z-axis.

The charge per unit length of the cylindrical beam is therefore

$$(\pi a^2)(1)(-q_v) = \frac{I}{(-v_0)} \qquad \text{...(4.17)}$$

As in Section 4.8, because of symmetry reasons, the electric displacement will be only at right angles to the z-axis. Hence if a Gaussian surface in the form of a cylinder with axis on z-axis and radius $R > a$ is selected, D will be constant on its curved surface but zero through its end faces. If the length of cylinder G_1 is L, Gauss's law gives

$$D(2\pi RL) = (-q_v)(\pi a^2)L$$

so that

$$\overline{\boldsymbol{E}} = \frac{(-q_v)\pi a^2}{2\pi\varepsilon_0 R}\overline{\boldsymbol{u}}_R, \qquad \text{V/m } (R > a) \qquad ...(4.18)$$

Thus for $R > a$, the electric field is the same as given by Eqn. (4.13), if we assume that the linear charge density given by Eqn. (4.17) is situated on the axis only, and not over the volume of the cylindrical beam. The negative sign in Eqn. (4.18) shows that the field is directed towards the axis (because the electronic charge is negative).

The maximum negative magnitude of $|E|$ occurs when $R = a$ as

$$|E_{max}| = \frac{aq_v}{2\varepsilon_0} \text{ V/m} \qquad ...(4.19)$$

For $R < a$, choosing the Gaussian surface G_2 as in Fig. 4.7 (*a*), and remembering that the charge *outside* G_2 does not contribute to the displacement D, we get

$$D(2\pi RL) = (\pi R^2)L(-q_v)$$

so that

$$\overline{\boldsymbol{E}} = \frac{(-q_v)R}{2\varepsilon_0}\overline{\boldsymbol{u}}_R, \text{ V/m} \quad (R < a) \qquad ...(4.20)$$

This shows that inside the beam the magnitude of $|E|$ decreases with R and at $R = a$, it becomes the same as in Eqn. (4.19). The variation of E is shown in Fig. 4.7 (*b*).

As a further exercise, we may proceed to obtain expressions for the variation of potential V with R. For this we assume arbitrarily that the potential on the axis of the beam is zero. Then by Eqns. (3.10) and (4.20) for $R < a$,

$$V = -\int_0^R \overline{\boldsymbol{E}} \cdot \overline{ds} = -\int_a^R \frac{(-q_v)R}{2\varepsilon_0}dR$$

or

$$V = \frac{q_v R^2}{4\varepsilon_0} \text{ Volt}, \ (R \le a) \qquad ...(4.21)$$

At $R = a$, this becomes

$$V_a = \frac{q_v a^2}{4\varepsilon_0} \text{ Volt} \qquad ...(4.22)$$

Again for $R > a$, using Eqns. (3.10), (4.22) and (4.18), we have

$$V = -\int_0^R \overline{\boldsymbol{E}} \cdot \overline{ds}$$

$$= -\int_0^a \overline{\boldsymbol{E}} \cdot \overline{ds} - \int_a^R \frac{(-q_v)\pi a^2}{2\pi\varepsilon_0 R}dR$$

or

$$V = \frac{q_v a^2}{4\varepsilon_0} + \frac{q_v a^2}{2\varepsilon_0}\log_e\frac{R}{a} \text{ Volt}, \ (R \ge a) \qquad ...(4.23)$$

The variation of V with R is shown in Fig. 4.7 (c). Eqn. (4.21) is similar to Eqn. (1.22c) and shows that it is a parabola open upwards, for points inside the beam, but the curvature reverses at the surface of the beam, and the curve ceases to be a parabola.

4.11 ELECTRIC FIELD OF A UNIFORM VOLUME CHARGE DISTRIBUTION IN CARTESIAN CO-ORDINATES

Assume that the uniform volume charge distribution extends to infinity in the x- and y-directions, but is restricted in the z-direction, so that

$$\text{Volume charge density} = \begin{cases} 0 & \text{for} \quad -\infty < z < -a \\ -q_v & \text{for} \quad -a < z < 0 \\ +q_v & \text{for} \quad 0 < z < +a \\ 0 & \text{for} \quad +a < z < +\infty \end{cases}$$

as shown in Fig. 4.8 (a), in which the x-axis is perpendicular to the plane of the paper and coming out of the paper.

It is obvious that because of symmetry in planes parallel to the x-y plane, there cannot be any component of D anywhere parallel to the x-y plane. The solution is obtained in two parts.

4.11.1 (i) For | z | > a

(i) This appears to be very similar to the case of two infinite parallel planes with equal and opposite uniform surface charge densities on them as discussed in Sec. 2.16, except that the charges are in the form of thick slabs instead of planes.

We take a Gaussian surface G_1 in the form of a rectangular parallelopiped with four sides parallel to the z-axis, and the two ends having area of a_0 square metres each. The end faces are located in the region which is free of charge as shown in Fig. 4.8 (a). Then since the charges extend to infinity in the x- and y-directions, because of symmetry, there is no electric displacement across any of the four side faces. Let the outward displacement across the end faces be as shown. Then the net outward displacement is $D_1(2a_0)$. The total charge enclosed is

$$aa_0 (+q_v) + aa_0 (-q_v) = 0.$$

Equating these by Gauss's law, we get $D_1 = 0$ as in Section 2.16 for region outside the oppositely charged plates.

4.11.2 For | z | < a

For this case, we keep the lower end face of the parallelopiped fixed, so that we know this end face and the four side faces (total five faces out of six) have zero displacement coming out of them. The length of the parallelopiped in the z-direction is then reduced, so that the upper end face lies in the region $0 < z < +a$ [Gaussian surface G_2 in Fig. 4.8 (a)]. Then applying Gauss's law to G_2.

(Zero displacement through five faces) + $D_2\, a_0$

$$= |z|\, a_0\, (+q_v) + a\, a_0\, (-q_v)$$

so that

$$\overline{E} = \frac{\overline{D}_2}{\varepsilon_0} = \frac{q_v(|z| - a)}{\varepsilon_0}\, \overline{u}_z \text{ V/m} \qquad ...(4.24)$$

Since $|z| < a$, the right hand side is negative indicating that $\overline{E}$ is directed in the negative direction of the z-axis. The reason for this is that the amount of negative charge enclosed by G_2 is more than the positive charge, and hence D_2 is actually directed into G_2 [and not out of G_2 as indicated in Fig. 4.8 (a)].

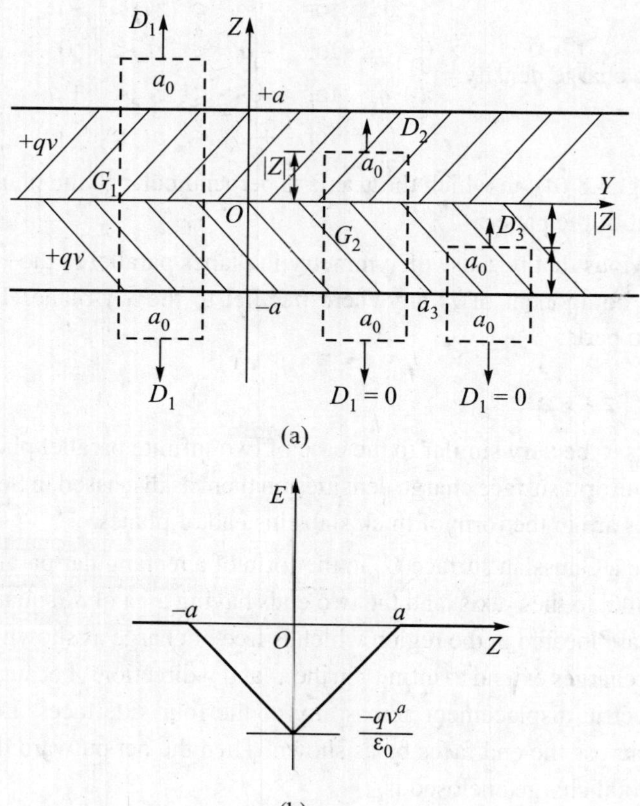

Fig. 4.8. Electric field of a uniform volume charge in cartesian co-ordinates

As $|z|$ decreases to zero, the amount of positive charge enclosed by G_2 goes on diminishing while the enclosed negative charge remains a constant. This maintains $\overline{E}$ negative all through, but the magnitude of E increases linearly, being a maximum when $|z| = 0$.

and

$$E_{\max} = -\frac{q_v a}{\varepsilon_0}\, \overline{u}_z \text{ V/m.}$$

When z becomes negative, but $|z| < a$, we get the Gaussian surface G_3 for which

$$D_3\, a_0 = -\, q_v[a - |z|]$$

and
$$\overline{E} = \frac{\overline{D}_3}{\varepsilon_0} = \frac{-\, q_v[a - |z|]}{\varepsilon_0}\,\overline{u}_z$$

$$= \frac{q_v[|z| - a]}{\epsilon_0}\,\overline{u}_z \text{ V/m.}$$

which is the same as Eqn. (4.24). As z varies from 0 to $-a$, E varies linearly from its maximum negative value E_{max} found above, to the value zero at $z = -a$. This is because as the length of the parallelopiped continues to decrease from $z = 0$ to $z = -a$, $|z|$ increases from 0 to $|z| - a$ varies from $-a$ to zero. The variation of E with z is shown in Fig. 4.8 (*b*), the negative value of E being in the direction of ($-\overline{u}_z$).

4.12 CONDUCTORS, CONDUCTIVITY AND CURRENT DENSITY

In all the discussion so far, we have only talked about static electric charges and dielectrics, but there has been no conductor present anywhere. Metals are good conductors of electricity. To get a crude picture of conductivity of metals, we note that in metals, the atoms are so closely packed together that the outermost orbits of valence electrons of neighbouring atoms almost intertwine. The valence electrons of one atom often leave the atom and begin to revolve around the outermost valence electron orbit of another atom. Since this happens to every atom, every atom continues to have its normal quota of valence electrons at all times, though no valence electron remains with any one atom for a long time.

Expressed in another way, inside the metal, such valence electrons experience attractive forces towards the positively charged nuclei of many neighbouring atoms, so that the resultant force on them is negligibly small. Such electrons are then not tied down to one atom for a long time, but are free to move about in the metal. Such free electrons can therefore be thought of as providing an electron gas in metals, in which the electrons have random velocities; they collide with each other and their velocities undergo a change of magnitude and direction. In short the electron gas and its electrons provide a picture very similar to that of a gas and its molecules in accordance with the kinetic theory of gases.

At the surface of the metal, however, the nuclei of the atoms near the surface being all on one side of the free electron, tend to pull these surface electrons back into the metal, so that these electrons cannot leave the metal. If they could leave the metal like water evaporating from a surface, then very soon the metal would have begun to show a positive charge, because of loss of electrons. But such spontaneous charging of a metal never occurs.

Let a long conducting wire of metal be connected to a battery through a switch. When the switch is turned on, an electric current is produced in the wire. The current starts flowing throughout the length of the wire at the same instant. The magnitude of the current at every cross-section along the length of the wire is the same, that is, there is no accumulation of charge at any point. This leads a student to make the common mistake of imagining that electrons leave the negative terminal of the battery, race through the wire at lightning speed and reach the positive terminal of the battery, and since electrons carry negative charge, we say that a positive current flows from the positive terminal of the battery through the wire to the negative terminal.

This picture is in error in a very vital respect, because in reality, the electrons do not move with lightning speed in the wire, but their velocity along the wire is only a few centimetres per second. An analogy will make the point clear. Assume that on a vast plain, thousands of children are standing, forming a single, long closed chain within arm's length of each neighbour. The time is on a dark night. Let each child hold a lighted candle in his right hand. At a signal (equivalent to the closing of the switch in the electrical circuit), each child transfers the lighted candle from his right to his left hand, and thence to his left hand neighbour's right hand. This process is continued. Then an observer in a helicopter looking at the plain will notice points of light moving continuously in a closed circuit. There is no accumulation of light points at any part of the circuit. But if the observer concentrates his attention on a single light point, he discovers that it is really moving very slowly. It is the *simultaneous movement of every light point* towards the left that gives the illusion of lightning speed.

Ohm's law, given by German physicist, Professor Georg Simon Ohm (1787–1854) in 1827 as

$$I = \frac{1}{R} V \qquad \text{...(4.25)}$$

where I is in ampere, R in ohm and V in volt, is the basic law of current electricity. From this we get the idea of resistivity or specific resistance ρ of a conductor as the resistance between the opposite faces of a unit cube (of side one metre). Then the resistance R of a conductor of length L and cross-sectional area A is given by

$$\left.\begin{aligned} R &= \frac{\rho L}{A} \text{ ohm} \\ \text{or} \quad \rho &= \frac{RA}{L} \text{ ohm-metre} \end{aligned}\right\} \qquad \text{...(4.26)}$$

It is obvious that if both A and L are made equal to unity (that is, if we have a cube of side one metre), then the resistivity becomes the resistance between the opposite faces of this unit cube.

Silver is the best conductor of electricity with $\rho = 1.629 \times 10^{-8}$ ohm-metre, annealed copper with $1.7241 \times 10^{-8}\ \Omega - \text{m}$ a close second best. For aluminium which is commonly used in India nowadays for power distributions, the value is 2.828×10^{-8} $\Omega - \text{m}$. For most conductors (metals, alloys and graphite), the resistivity lies in the range 10^{-8} to $10^{-6}\ \Omega - \text{m}$.

Insulators or dielectrics, on the other hand, have their atoms far apart and they do not have any electron gas. Their resistivity is therefore high in the range 10^{11} to 10^{15} $\Omega - \text{m}$. Fused quartz is the best insulator with $\rho = 2.5 \times 10^{17}\ \Omega - \text{m}$.

In between these extremes lie the group called semi-conductors like Germanium with $\rho = 0.60\ \Omega - \text{m}$ and silicon with $\rho = 1500\ \Omega - \text{m}$.

Energy level diagrams give an explanation for these differences. In a solid, the energy levels of individual atoms spread out into bands because of the proximity of neighbouring atoms. Between the lower energy band called the valence electron band, and the higher energy band called the conduction band, the gap is very large in the case of insulators, is zero for metals, and is small for semi-conductors.

Writing Ohm's law in the form of Eqn. (4.25) is not suitable for our purpose. In the first place, all the quantities in it — current, resistance and voltage – are scalar quantities, whereas we would like to have vector quantities on both sides of the equation. Moreover, this form of Ohm's law is suitable for analysing an electrical circuit that is spread out in space. In the vector calculus approach used for field theory, we need equations that are applicable at a *point* in space. We therefore proceed first to write Ohm's law in a form suited to our needs.

First we decide to use conductance G in our equations instead of the resistance R, and conductivity σ in place of resistivity ρ so that Eqn. (4.26) becomes

$$\left.\begin{aligned} G &= \frac{1}{R} = \frac{\sigma A}{L} \text{ siemens} \\ \text{and} \quad \sigma &= \frac{1}{\rho} = \frac{GL}{A} \text{ siemens per metre} \end{aligned}\right\} \quad ...(4.27)$$

The name given to the unit of conductance in the past was mho (ohm spelt backwards), but in SI units, the name siemens (symbol : *S*) has been given in honour of the German physicist Dr Werner Siemens (1816–1892). Along with mho, the symbols Ω^{-1} or ℧ have been discarded and should not be used. For copper the conductivity has the value 5.800×10^7 S/m.

The advantage of this change will become apparent when we define the capacitance parameter later and proceed to determine it for various shapes of conductors. Thus the Eqn. (4.27) for G will be found to be of the same form as the capacitance of a parallel-plate capacitor Eqn. (4.34).

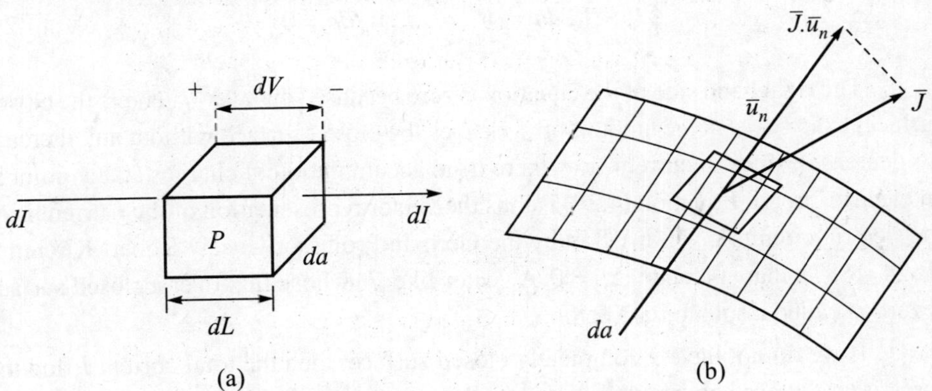

Fig. 4.9. Ohm's law in vector notation

Next we proceed to apply Eqns. (4.25) and (4.27) to a point P inside a conductor [Fig. 4.9 (*a*)], where we take an infinitesimally small volume dv having length dL in the direction of the current dI, and cross-sectional area da at right angles to the current dI. If dV is the potential difference between the end faces of volume dv, we get

$$dI = \frac{I}{R} dV = \sigma \frac{da}{dL} dV$$

or
$$\frac{dI}{da} = \sigma \frac{dV}{dL} \quad \text{or} \quad \bar{J} = \sigma \bar{E} \qquad \text{...(4.28)}$$

where $\bar{J}$ is the *volume current density* in amperes per *square metre of area at right angles to the direction of the current*, and $\bar{E}$ is the electric field intensity at P in volt per metre. (Note that in the figure, dV is the voltage decrease (drop) in the direction of the current, and hence the negative sign is not written before it). If we are dealing with a two-dimensional current sheet, then the *surface current density* $\bar{J}_s$ would be in ampere per *metre width at right angles to the direction of the current.* Eqn. (4.28) in Ohm's law in vector form applicable to a point in space. Thus by introducing via Eqn. (4.27) area da and length dL, both of which are vector quantities, we have been able to convert the current and voltage (scalar quantities) of Eqn. (4.25) into the *vector quantities—current density* and *voltage gradient*—of Eqn. (4.28).

We immediately see the similarity between Eqns. (4.3) and (4.28), but there is a vital difference. The electric displacement was compared to roads leaving a city, but an electric current is actual charge entering the volume dv at the left-hand face and leaving it at the right-hand face in Fig. 4.9 (*a*). This actual movement of charge involves expenditure of energy, but in case of D coming out of Q, there is no expenditure of energy.

In case I and da are not at right angles to each other, we have to take the resolved part of the current density vector $\bar{J}$ in the direction of the unit vector $\bar{u}_n$ normal to the area da [Fig. 4.9 (*b*)] just as in the case of D. Hence we get for a completely enclosed Gaussian surface, the Gauss's Law for current density

$$\oint_{\text{surface}} \bar{J} . \overline{da} = \oint_{\text{surface}} \bar{J} . \bar{u}_n \, da = 0 \qquad \text{...(4.29)}$$

The right hand side of the equation is zero because whatever $\bar{J}$ enters the closed surface, exactly the same amount of $\bar{J}$ leaves the closed surface without any increase (or decrease), since in current flow there is no accumulation of charges at any point in an electric circuit. Equation (4.29) is thus the vector representation of the Current Law for electric circuits given in 1849 by the German Professor Gustav Robert Kirchhoff (1824–1887), that at a node, $\Sigma i = 0$. A vector like $\bar{J}$ whose flux over a closed surface is zero is called a solenoidal vector.

If we do not have a completely closed surface, then the total current I flowing across that surface can be evaluated as

$$I = \oint_{\text{surface}} \overline{J} \,.\, \overline{u}_n \, da \qquad ...(4.30)$$

4.13 COROLLARIES ABOUT CONDUCTORS DEDUCED FROM GAUSS'S LAW

Even though we do not have conductors with infinite conductivity, it is found that the conductivity of actual conductors is sufficiently high, and therefore the re-arrangement of charges in a conducting region as described in this section is almost instantaneous. Certain corollaries can be deduced from Gauss's law for regions where conductors are present. These are discussed below with proof.

Corollary I: In electrostatics, a conductor must be an equipotential region

This is a direct consequence of Eqn. (4.28), which shows that a current must flow if there is a potential gradient, or conversely, since electrostatic implies that charges are not moving (no current), therefore the potential gradient must be zero. To get a physical picture, when a certain amount of electric charge is placed at some points in a conductor, the mutual repulsion between these charges will give rise to an electric field intensity $\overline{E}$ in the conductor, which will produce a current given by Ohm's law. This current must flow so long as there is an electric field intensity $\overline{E}$. The flow of current causes the charges to get distributed and the current flow ceases only when the electric field intensity at every point of the conductor becomes zero. Since the potential difference between any two points in the conductor can be obtained by the line integral of $\overline{E} \,.\, \overline{ds}$ between them, and $\overline{E}$ is everywhere zero, it follows that all points of the conductor are at the same potential, or the region occupied by the conductor is an equipotential region.

Corollary 2: In electrostatics, charges can only reside on the surface of the conductor, and not inside a conductor

By Corollary 1, when the motion of the charges in the above example has stopped, $\overline{E}$ and therefore $\overline{D}$, is zero everywhere in the conductor region. Hence if we select a Gaussian surface G_1 just inside the surface of the conductor [Fig. 4.10 (*a*)], D is zero at all points on the surface G_1. This makes the left hand surface integral of Gauss's law equal to zero, and hence by Gauss's law, the volume integral on the right hand side must also be zero. Stated in words this means that there are no (stationary) electric charges inside the conductor. Even if a charge is somehow released inside a conductor, it must flow out almost instantaneously to the surface of the conductor. This shows that in electrostatics, charges can only reside on the surface of a conductor.

From these two corollaries, it is seen that when electric charges are placed on a conductor, the charges get arranged on the surface of the conductor in such a way that the *net or resultant electric field in the region of the conductor is everywhere zero.* It must be noted however, that this does not mean that the charge density (C/m^2) on the surface is uniform everywhere. Only if the conductor has an infinite plane as its surface, or if the conductor is in the form of a sphere, will this charge density on the surface be uniform. In particular for a conductor tapering towards a point, the charge density will be very high in the neighbourhood of the sharp pointed end. This is the basis for the use of such sharply pointed metal rods, as lightning conductors on tops of tall buildings, for protecting the building from lightning, a practice started by Benjamin Franklin in 1749.

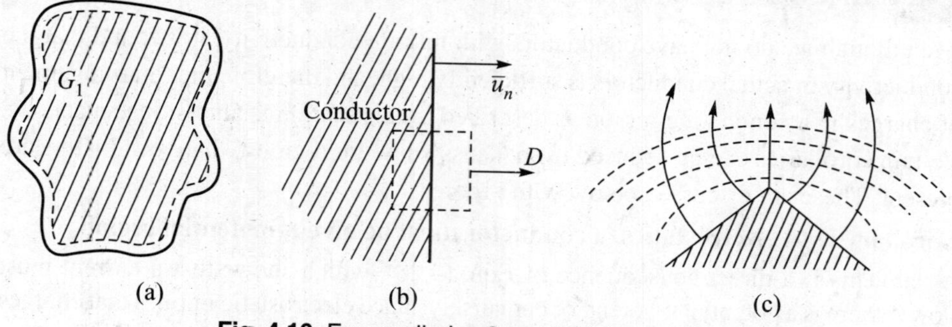

Fig. 4.10. For corollaries 2 and 3 of Gauss's law

This corollary also implies that when an uncharged conductor is introduced in an electrostatic field, the field must become zero in the region occupied by the conductor. This requires that induced charges must appear on the surface of the conductor, and they must arrange themselves in such a way that the electric field produced by them exactly neutralizes the original field in the region of the conductor.

Corollary 3 : In electrostatics, the electric field intensity at all points on the surface of a conductor must be normal to the surface

This follows from the fact that we are dealing with electrostatic problems, in which the movement of charges has ended and all charges are at rest. If the electric field at any point on the surface of the conductor is not normal to the surface but is inclined, then it can be resolved along the surface to give a tangential component $E_{\text{tangential}}$, and the material being a conductor, this $E_{\text{tangential}}$ will produce a current along the surface. But this contradicts our original assumption that all charges are at rest. Hence there cannot be any tangential component of $\overline{E}$ at any point of the surface.

The expression for the field intensity at any point on the surface of the conductor can be found from Fig. 4.10 (*b*). If $(+\,q_s)$ C/m^2 is the surface charge density at any point on the surface of the conductor, and we select a Gaussian surface in the form of a rectangular parallelopiped whose four sides are normal to the surface of the conductor, and the end faces of cross-sectional area *da* square metres are located just inside and just outside the conductor surface, then by Corollary 2, there is no *D* across the left hand face located inside the conductor. By this Corollary 3, there is also no *D* coming out of the four side faces, as this would result in a tangential component. Thus *D* comes only out of the end face just outside the conductor.

Hence by Gauss's Law

$$D\,da = (+\,q_s)\,da$$

so that
$$\overline{E} = \frac{q_s}{\varepsilon_0}\,\overline{u}_n \text{ V/m} \qquad ...(4.31)$$

At corners or over sharply curved surfaces, the charge density is higher than over a plane surface, and hence the electric field intensity is greater there as shown by the closer spacing of the equipotentials (dotted lines) in Fig. 4.10 (*c*). If this electric field intensity exceeds a critical value (approximately 30 kV/cm), then the air can become ionized, producing a corona discharge with a pale-voilet light near the surface of the conductor. For this reason, the radius of curvature of any part of the surface of high voltage bus-bars, switches or transmission lines is not permitted to be small. On the other hand, the ends of lightning conductors are made sharply pointed to produce ionization of air to make the air conducting, so that the lightning stroke is diverted through the conductor to the earth instead of striking any part of the building.

Corollary 4 : A completely closed cavity inside a conductor cannot enclose any net charge

Inside a closed cavity in the conductor [Fig. 4.11(*a*)], suppose we deliberately put some charges at different points (not on the conductor surface) in an attempt to disprove this corollary. Let the sum of these charges be + Q.

Now we choose a Gaussian surface G_1 in such a way that every point on it lies totally inside the conductor, and which completely encloses the cavity. Since there is no field inside the conductor (Cor. 2), it follows that the left hand side of Gauss's law must be zero. This implies that the right hand side is also zero, or that the *net* charge enclosed by G_1 must be zero. But since we had put a total charge + Q inside the cavity, the net charge within G_1 can become zero only if there is an exactly equal and opposite charge – Q induced on the inner surface of the conductor forming the cavity. In other words, the sum of the charges placed and the charges induced must always be zero, thus proving the truth of this Corollary.

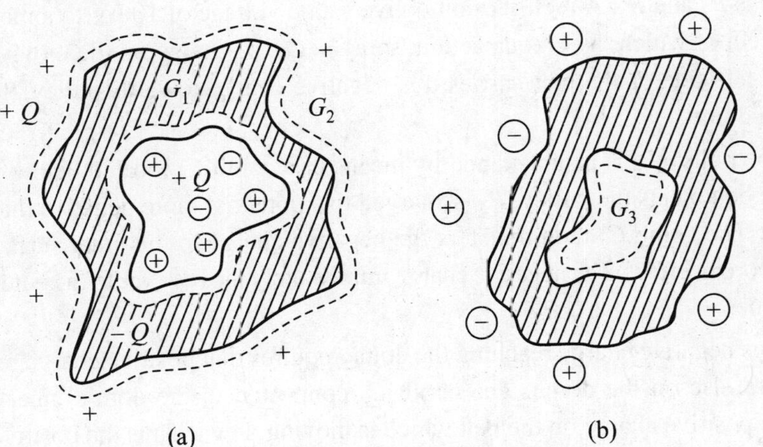

Fig. 4.11. Corollaries 4 and 5 of Gauss's law

Cor. 5 : If a charge is placed inside a cavity completely enclosed within a conductor, then an exactly equal and like charge must be induced on the outer surface of the conductor.

This is the same situation as that of a charge-free conductor placed in an electric field. In Cor. 4 we have seen the movement of charges in the conductors to make it an equipotential surface, by having an equal and opposed induced charge to appear on the inner surface of the conductor in the cavity. Since the conductor as a whole must have a net zero charge, there must be an equal and like charge induced in the conductor at some other place. By Cor. 2, the inner and the outer surfaces of the conductor are the only surfaces where the induced charges reside. Therefore, this equal and like charge must be on the outer surface.

To give a more direct proof using Gauss's law, we select the Gaussian surface G_2 [Fig. 4.11 (*a*)] totally lying just outside the conductor and completely enclosing it. Then the electric displacement through G_2 must be equal to the total charge $+Q$ deliberately placed in the cavity. But Cor. 4 makes the net charge within the conductor cavity as zero. Hence this charge $+ Q$ must be on the outer surface of the conductor, which is the only possible place for it to be.

Van de Graaff Generator. In Section 2.17 it was seen that there is no electric field inside a spherical charge distribution. Even if the surface is not completely closed, only a very weak field can exist inside such a cavity, because of mutual cancellation of the effects of major part of the charge distribution. A second concept that has been established in Cor 5 is that, a charge placed inside a cavity gets transfered to the outer surface of a hollow conductor. These two concepts were combined together by Prof. R. J. Van de Graaff of Massachusetts Institute of Technology, to build in 1933 an electrostatic generator, giving a voltage of several million volts and a direct current of a few milliamperes. This was used for research in nuclear physics to accelerate charged particles.

The generator consists of a very large sized conducting dome supported on an insulated hollow column (see Fig. 4.12). Through his column runs a rubber belt around two pulleys P_1 and P_2. A rectifier unit delivers a dc voltage of 10 kV to a metal comb C_1, the teeth of which, by corona action, spray negative charge on the moving rubber belt. The left side of the belt carries this negative charge towards pulley P_2 located inside the dome.

The pulley P_2 is insulated and by means of a comb C_2 located below the point where the belt meets the pulley, P_2 is charged to a potential more negative than that of the dome. The comb C_3 then transfers the negative charge to the inner surface of the dome by corona effect. The negative charge immediately moves over to the outer surface of the dome.

This negative charge reaching the dome requires that positive charge must shift somewhere else on the dome. The comb C_4 connected to the dom's inner surface, sprays the positive charge on the belt which is moving down. Thus the positive charge is carried down to the pulley P_1 which is earthed. Hence the positive charge leaks away to the earth.

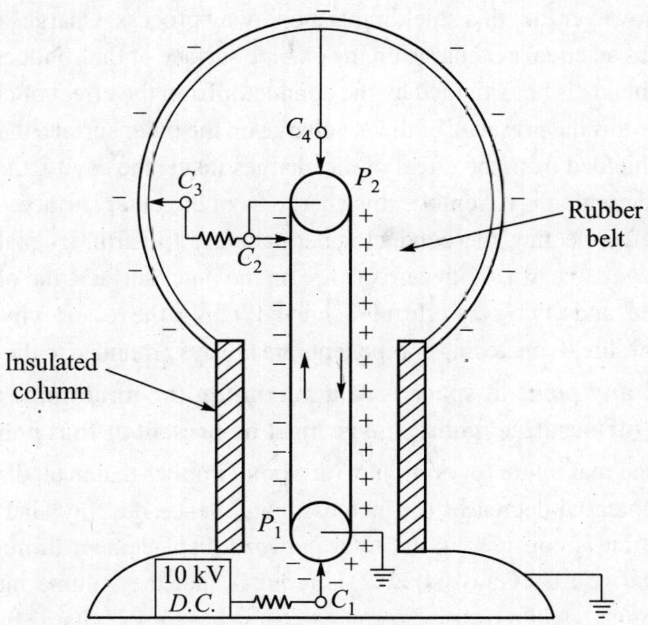

Fig. 4.12. Van de Graaff generator

Corollary 6 : Electric charges located outside a conductor cannot produce an electric field inside a completely enclosed cavity within the conductor.

We proceed to prove this by the method of reductio ad absurdum. We assume that the corollary is not true and that an electric field does get produced inside the cavity. Then the electric field lines must meet the inner surface of the conductor in the cavity at right angles to the surface at all points by Cor. 3. Hence we can choose an equipotential surface as shown in Fig. 4.11 (*b*), which will naturally be normal to the electric field lines and located just inside the cavity. This equipotential surface will thus be parallel to the inner surface of the cavity and will therefore form a closed surface. We can therefore choose it to be a Gaussian surface. Applying Gauss's law to it tells us that the electric displacement through it is equal to the charge enclosed by this equipotential surface. But we had not placed any charge inside the cavity (charges were placed outside the conductor). Hence the enclosed charge is zero. We have thus an electric field and an electric displacement through the surface as assumed, but the enclosed charge is zero. This contradicts Gauss's law and therefore our original assumption of the existence of an electric field inside the cavity is incorrect. Thus the corollary is true after all, and both D and E are zero in the cavity.

This is a very interesting result. It forms the basis of what is called the electrostatic shielding of electronic devices and apparatus placed inside such closed cavities, from the effect of any external charges or electric fields. Metals shields enclosing IF transformers in radio sets are examples of this. Even a precision measurements laboratory itself can be located inside such a Faraday Cage to eliminate effects of external or stray fields on the measurements.

Notice however that this shielding is a one-way process. Charges placed *inside* a cavity appear as an equal *net* charge on the *outside* surface of the conductor by Cor. 5, and hence the outside is not shielded by the conductor from the effect of charges inside the cavity. Since it is the presence of the net charge on the outer surface that does not let the outside be shielded from the effect of the charges inside the cavity, the only way to shield the outside region is to remove this charge from the outer surface. This is easily done simply by connecting the conductor permanently to earth, to enable the outer charge to leak away to earth. With zero charge on the outer surface, the outer region is thus also shielded, and a two-way effect is achieved. This is the reason why in electronic equipment, metal shields enclosing components are always grounded to the metal chasis.

Corollary 7: If any point in space is at a maximum (or minimum) of potential, then a positive (or negative) point charge must be present at that point

A potential maximum (or minimum) at a point implies that in all directions from that point, the potential decreases (or increases) and hence if a tiny Gaussian surface enclosing the point is considered, the total *outward* displacement through it will be positive (or negative). By Gauss's law, this requires that the volume integral on the right hand side must yield a positive (or negative) value for the charge enclosed, thus proving the corollary.

4.14 CAPACITANCE

Two conductors of any shape, insulated from each other, form a system which has the ability or capacity to store electric charge. Hence the system is called a capacitor. Notice that in the definition, there is no mention of the distance separating the two conductors. In particular, if we remove the second conductor to an infinite distance away from the first conductor, we are left with a capacitor apparently having only one conductor. However, the presence of the second conductor at infinity is still necessary for the single conductor to be able to store charge. The second conductor at infinity is also needed as a reference point at zero voltage to measure the voltage or potential (really the potential difference from the reference point) of the first conductor.

It is found that if an infinitesimally small charge dQ is added to the first conductor of the capacitor, its voltage with respect to the second conductor rises by an amount dV and the two are proportional to each other, no matter what their amounts are, or what the value of the initial charge on the conductor is, before dQ is added to it. This means that $dQ = C\,dV$ where C is the constant of proportionality, called the capacitance of the capacitor. Integrating this we get

$$Q = CV \qquad (\text{if } V = 0 \text{ when } Q = 0) \qquad \text{...[4.31 (b)]}$$

This shows that if we have two conductors of any shape, we can determine the capacitance of the system by finding the voltage difference V between them (by the methods already studied) when a known charge Q is placed on one conductor of the capacitor, and finding C as the ratio Q/V.

It will always be found that for the charge to be stored in a capacitor, if a charge (+ Q) is located on the first conductor, then there must be a charge (– Q) on the second conductor, so that the forces of attraction hold the charges on the conductors, and prevent the like charges on one conductor from moving away from each other by mutual repulsion and going to other parts of the circuit.

4.15 PARALLEL PLATE CAPACITOR

Two parallel conducting plates, with area A square metres for *one side of each plate*, have a dielectric of permittivity ε between them [Fig. 4.13 (*a*)]. If charges (+ Q) and (– Q) coulomb are placed on the two plates respectively, then the surface charge density on the positive plate is

$$q_s = +(Q/A)\ \text{C/m}^2.$$

Let the plates be separated by a distance d metres.

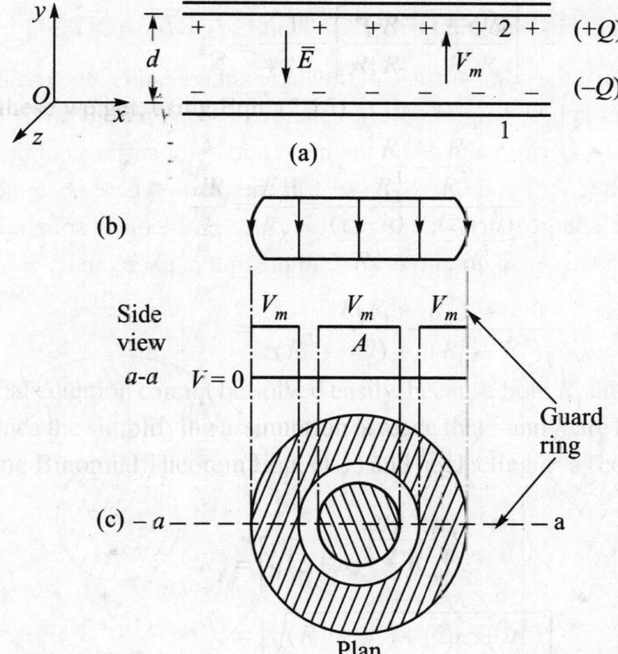

Fig. 4.13. (a) Parallel plate capacitor, (b) Fringing of field at edges and (c) Guard ring

Assuming that these plates are *infinite planes* with charge density q_s as found above, we can use the method of Sec. 4.7 to apply Gauss's law and the principle of Sec. 2.16 to obtain the electric field intensity between the plates as

$$\overline{\boldsymbol{E}} = \frac{-q_s}{\varepsilon}\overline{\boldsymbol{u}}_y = \frac{-Q}{\varepsilon A}\overline{\boldsymbol{u}}_y \ \text{V/m} \quad ...(4.32)$$

Hence the potential difference between the capacitor plates 1 and 2 is obtained as

$$V_{r1} = -\int_1^2 \overline{\boldsymbol{E}} \cdot \overline{\boldsymbol{ds}} = \frac{Qd}{\varepsilon A} \ \text{Volt} \quad ...(4.33)$$

Therefore using Eqn. (4.31*b*), the capacitance is

$$C = \frac{Q}{V_m} = \frac{\varepsilon A}{d} \text{ farad} \qquad ...(4.34)$$

The assumption of infinite extent of the parallel plates on which Eqn. (4.32) is based, is required to ensure that all field lines are parallel to ($-\bar{u}_y$). The assumption is reasonably true if the length and breadth of the plates is large as compared to the separation *d* between the plates. The cause of the error is the bending of the electric field direction lines [Fig. 4.13 (b)] at the edge of the plates, called the fringing of the field. To eliminate this bending and to have the direction lines normal to the planes, the upper plate is surrounded by a guard ring which is maintained at the same potential as the upper plate, though it is separated from the upper plate by a narrow insulated strip [Fig. 4.13 (c)]. This shifts the fringing to the outer edge of the guard ring, from where it does not disturb the uniform and parallel electric field in the area *A* of the capacitor.

4.16 CONCENTRIC SPHERES FORMING A CAPACITOR

Let r_1 be the radius of the outer surface of the inner sphere, and r_2 the radius of the inner surface of the outer sphere (Fig. 4.14). If a charge (+ *Q*) is placed on the inner sphere, charges (– *Q*) and (+ *Q*) are induced on the inner and outer surfaces of the outer sphere in accordance with Cor. 5 and 6 of Sec. 4.13. If the outer sphere is connected to earth [Fig. 4.14 (b)], the charge (+ *Q*) on the outer surface of the outer sphere leaks away to earth. Hence we can assign a potential $V = 0$ to the outer sphere and $V = V_m$ to the inner sphere.

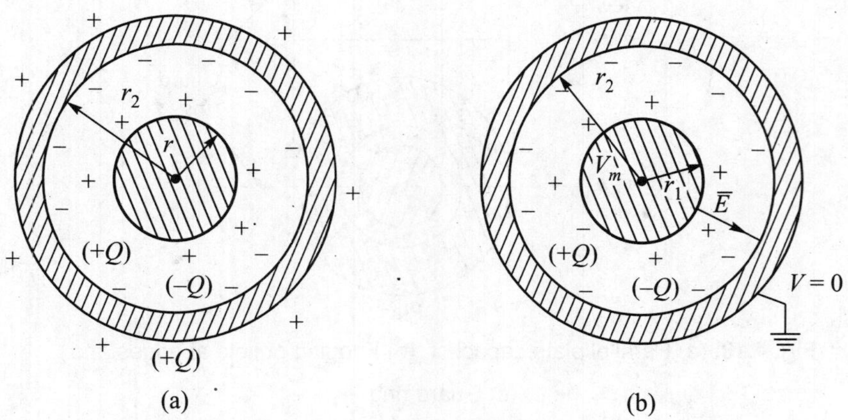

Fig. 4.14. Concentric spherical capacitor

If $r < r_1$, the point, being inside the conductor forming the inner sphere, is on an equipotential surface at potential V_m, while the electric field is zero [Fig. 4.15 (a) and (b)].

If $r_1 < r < r_2$, then we can choose a sphere of radius *r* as the Gaussian surface and by Gauss's law, get

$$D(4\pi r^2) = \text{Charge enclosed} = +Q$$

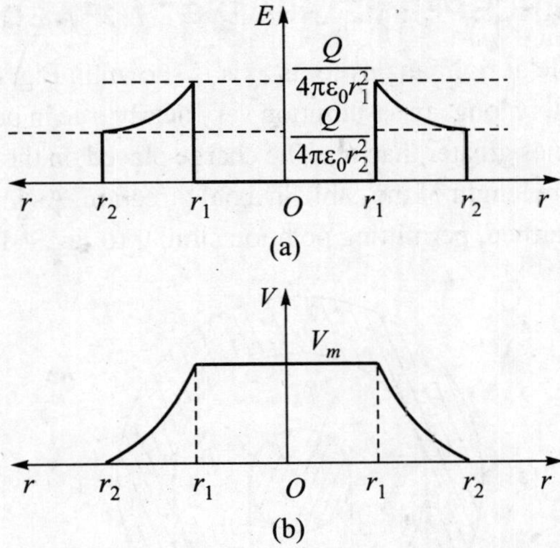

Fig. 4.15. (a) Electric field intensity and (b) Potential in concentric spherical capacitor

or
$$\overline{E} = \frac{Q}{4\pi\varepsilon r^2}\,\overline{u}_r \quad \text{V/m} \qquad ...(4.35)$$

where ε is the permittivity of the dielectric between the spheres. The electric field intensity decreases from the value

$$\frac{Q}{4\pi\varepsilon\, r_1^2} \text{ at } r = r_1$$

to
$$\frac{Q}{4\pi\varepsilon\, r_2^2} \text{ at } r = r_2$$

as shown in Fig. 4.15 (b). Also the potential at a point at a distance r from the centre, where $r_1 < r < r_2$, is given by

$$V = -\int_{r_2}^{r} \overline{E}\cdot\overline{dS} = -\frac{Q}{4\pi\varepsilon}\int_{r_2}^{r}\frac{dr}{r^2}$$

$$= \frac{Q}{4\pi\varepsilon}\left(\frac{1}{r} - \frac{1}{r_2}\right) \text{ volt} \qquad ...(4.36)$$

It has the value $V_m = \frac{Q}{4\pi\varepsilon}\left(\frac{1}{r_1} - \frac{1}{r_2}\right)$ at $r = r_1$,

decreasing to zero at $r = r_2$ as shown in Fig. 4.15 (*c*).

Hence the capacitance is obtained as

$$C = \frac{Q}{V_m} = \frac{4\pi\varepsilon}{\dfrac{1}{r_1} - \dfrac{1}{r_2}} \text{ farad} \qquad ...(4.37)$$

If r_2 is made equal to infinity, the capacitance of the sphere of radius r_1 is seen to be ($4\pi\,\varepsilon\, r_1$) farad.

4.17 CAPACITANCE PER METRE LENGTH OF A CO-AXIAL CABLE

A section of the cable at right angles to its axis is shown in Fig. 4.16 (*a*). The cable is assumed to be infinitely long, an assumption very nearly true in practice if the length of the cable is many times greater than R_2. The charge placed on the inner conductor is (+ q_l) coulomb per metre length of the cable in axial direction. As in previous section, the outer conductor is earthed, permitting notation similar to Fig. 4.14.

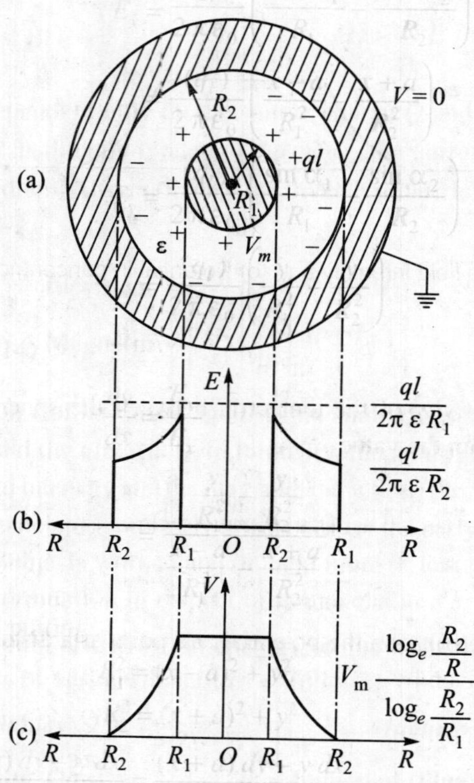

Fig. 4.16. (a) Co-axial cable (b) Electric field intensity and (c) Potential variation in it.

If $R_1 < R < R_2$, then we can choose a cylinder of radius R and length L, and with its axis on axis of cable, as the Gaussian surface to which Gauss's law is applied to give

$$D(2\pi R)L = q_l L$$

or

$$\overline{E} = \frac{q_l}{2\pi\varepsilon R}\overline{u}_R \text{ V/m} \qquad \text{...(4.38)}$$

and

$$V = -\int_{r_2}^{r} \frac{q_l dR}{2\pi\varepsilon R} = \frac{q_l}{2\pi\varepsilon} \log_e \frac{R_2}{R} \text{ Volt} \qquad \text{...(4.39)}$$

At $R = R_1$, $V = V_m$. Therefore

$$V_m = \frac{q_l}{2\pi\varepsilon} \log_e \frac{R_2}{R_1} \qquad \text{...(4.40)}$$

so that
$$V = \frac{V_m}{\log_e \frac{R_2}{R_1}} \log_e \frac{R_2}{R} \qquad ...(4.41)$$

The capacitance per metre length is given as

$$C = \frac{q_l}{V_m} = \frac{2\pi\varepsilon}{\log_e \frac{R_2}{R_1}} \text{ F/m} \qquad ...(4.42)$$

The variation of E and V with R is depicted in Fig. 4.16 (b) and (c). (Compare Eqn. (4.42) with the solution to Ex. 4, Prob. 12).

4.18 CAPACITANCE PER METRE LENGTH OF TWO-WIRE LINE

A two-wire transmission line used in telephony, radio engineering or electric power distribution usually consists of two metal conductors in the form of cylinders of radius R metre each, parallel to each other with a separation of D metres between their centres (Fig. 4.17). Usually D is very much greater than R. If a uniform charge density of $(+q_l)$ C/m exists on one conductor and $(-q_l)$ C/m on the other, the charges will distribute themselves to make the surface of each conductor an equipotential surface.

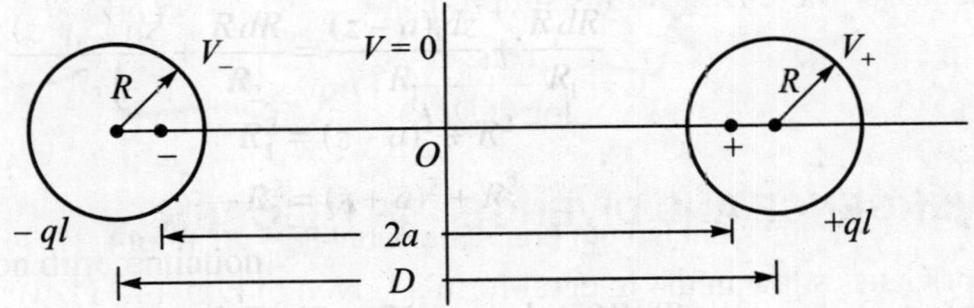

Fig. 4.17. Two-wire transmission line

This situation reminds us of the equipotential map of parallel line of infinite length (Fig. 3.7), in which the equation to the equipotentials was given as Eqn. (3.41). With the notation of Eqn. (3.41), we equate the radii of similar equipotential cylinders and the distance between their centres with R and D respectively, to get

$$\frac{2ak}{k^2-1} = R \quad \text{and} \quad 2a\left(\frac{k^2+1}{k^2-1}\right) = D \qquad ...(4.43)$$

Dividing to eliminate $2a$,

$$\left(\frac{R}{D}\right) = \frac{k}{(k^2+1)} \quad \text{or} \quad Rk^2 - Dk + R = 0$$

This gives
$$k = \frac{D + \sqrt{D^2 - 4R^2}}{2R} \qquad ...(4.44)$$

We are justified in writing Eqn. (4.43) though the actual line is never infinitely long, because we assume that the length of the line in the axial direction is very much greater than D.

For the sake of symmetry, if we assume that the plane midway between the lines is at zero potential ($V = 0$), then the potentials V_+ and V_- of the two cylinders of Fig. 4.17 can be written using Eqns. (3.14) and (3.15) as

$$V_+ = \frac{+q_l}{2\pi\varepsilon_0}\log_e k$$

and

$$V_- = \frac{-q_l}{2\pi\varepsilon_0}\log_e k.$$

The potential difference between the two cylinders is

$$V_m = V_+ - V_- = \frac{+q_l}{\pi\varepsilon_0}\log_e k.$$

Hence the capacitance of the transmission line per metre length is

$$C = \frac{q_l}{V_m} = \frac{\pi\varepsilon_0}{\log_e k}$$

$$= \frac{\pi\varepsilon_0}{\log_e\left[\dfrac{D+\sqrt{D^2-4R^2}}{2R}\right]} \text{ F/m} \qquad \text{...(4.45)}$$

For the usual case where $D \gg R$, this simplifies to

$$C \approx \frac{\pi\varepsilon_0}{\log_e (D/R)} \text{ F/m} \qquad \text{...(4.46)}$$

4.19 DIVERGENCE AND DIVERGENCE THEOREM

By the use of Gauss's law in the integral form as given in Eqn. (4.11), we have solved half the problem posed in Sec. 3.12, because with its help we have been able to obtain the value of $\overline{E}$ at a point without knowing its distance from the charge distribution. However, the other half of the problem still remains, because Eqn. (4.11) has to be applied to finite and comparatively large areas and volumes, while we would prefer to have a relation similar to that for potential, $\overline{E} = -\overline{\nabla} V$ which is applicable at a *point.*

In this search we take a look again at Eqn. (1.94), and apply that concept to Gauss's law. Consider an infinitesimally small volume Δv in which the average charge density is $(q_v)_{av}$ C/m^3. Applying Gauss's law to it gives

$$\oint_{\text{surface}} \overline{D}\cdot\overline{u}_n\, da = (+q_v)_{av}\,\Delta v$$

Dividing both sides by Δv and proceeding to take the limit, we get [see Eqn. (1.94)]

$$\lim_{\Delta v\to 0}\left[\frac{\oint_{\text{surface}} \overline{D}\cdot\overline{u}_n\, da}{\Delta v}\right] = \text{div}\,\overline{D} = \overline{\nabla}\cdot\overline{D} = q_v \qquad \text{...(4.47)}$$

where q_v C/m^3 is the volume charge density at that point. This familiar trick of differential calculus gives a finite value for the ratio of the surface integral and the enclosed volume even when, in the limit, both have a zero value : Eqn. (4.47) is Gauss's law in differential vector form. In words it states that the net outward electric flux per unit volume is equal to the volume charge density at that point. The internationally used symbol for volume charge density is the Greek letter ρ(rho). Hence we write ρ C/m^3 instead of q_v C/m^3 and write Gauss's law in differential vector form as

$$\overline{\nabla} \,.\, \overline{D} = \rho \quad \text{C/m}^3 \qquad \text{...(4.48)}$$

It is necessary to remind ourselves that in Eqns. (4.47) and (4.48) we have a *dot operation* by del on $\overline{D}$ as indicated or defined in detail by the integral and the limit at the left of Eqn. (4.47), and *not a dot product of* $\overline{\nabla}$ *and* $\overline{D}$. There is no angle θ between $\overline{\nabla}$ and $\overline{D}$ and no term like cos θ can appear in the resulting scalar or dot product. The units for $\overline{D}$ are coulomb/m^2 and of q_v or ρ are C/m^3. Therefore this shows that the del dot operation involves a differentiation with respect to distance, so as to obtain the same dimensions on both sides of these equations.

In the rest of the book, we shall be using Gauss's law in the differential vector form most of the time, but it must be remembered that this form is not of much use in solving problems discussed in the earlier sections of this chapter. As the saying goes "Where a needle is needed, a sword is of no use", though both are made of steel !

By substituting the value of q_v from Eqn. (4.47) into Eqn. (4.11), we get

$$\oint_{\text{surface}} \overline{D} \cdot \overline{u}_n \, da = \oint_{\text{volume}} (\overline{\nabla} \cdot \overline{D}) \, dv \qquad \text{...(4.49)}$$

In this equation a surface integral operation on a vector $\overline{D}$ is related to a volume integral operation on the *same* vector $\overline{D}$. Hence this equation is true for any vector. It is called the Divergence theorem and enables us to shift from a surface integration to a volume integration, or vise versa.

4.20 APPLYING GAUSS'S LAW IN DIFFERENTIAL VECTOR FORM

Two examples are given below to illustrate how Gauss's law in differential vector form can be used to solve problems of a certain type:

(*i*) Find the charge distribution which will give the following electric field intensity in cartesian co-ordinates:

$$\overline{E} = \begin{cases} \dfrac{z^3}{2\varepsilon_0 \, |z|} \overline{u}_z & \text{for } |z| < a \\[2ex] \dfrac{a^2 |z|}{2\varepsilon_0 \, z} \overline{u}_z & \text{for } |z| > a \end{cases}$$

Using Eqn. (1.93) we write divergence $\overline{D}$ in the expanded form in cartesian co-ordinates as

$$\text{div } \overline{D} = \overline{\nabla} \cdot \overline{D} = \frac{\partial D_x}{\partial x} + \frac{\partial D_y}{\partial y} + \frac{\partial D_z}{\partial z}$$

In this problem $\overline{D}$ $(= \varepsilon_0 \overline{E})$ does not depend on x or y. Hence the first two terms on the right hand side of the expanded form of divergence expression will always be zero. In other words, the volume charge density will be the same everywhere on any x-y plane for which z is a constant.

For $0 < z < a$, $z = +|z|$. Hence we get

$$\overline{D} = \varepsilon_0 \overline{E} = \frac{z^3}{2|z|}\overline{u}_z = \frac{z^3}{2z}\overline{u}_z = \frac{z^3}{2}\overline{u}_z$$

By Gauss's law,

$$\overline{\nabla} \cdot \overline{D} = \frac{\partial D_z}{\partial z} = z \quad \text{or} \quad \rho = +|z|$$

For $-a < z < 0$, $z = -|z|$. Now we have

$$\overline{D} = \frac{z^3}{2|z|}\overline{u}_z = \frac{z^3}{2(-z)}\overline{u}_z = \frac{-z^2}{2}\overline{u}_z$$

which gives

$$\frac{\partial D_z}{\partial z} = -z = +|z| \quad \text{or} \quad \rho = +|z|$$

For $|z| > a$, $z = \pm|z|$ makes

$$\overline{D} = \frac{a^2|z|}{2z}\overline{u}_z = \pm\frac{a^2}{2}\overline{u}_z, \text{ (a constant)}$$

Hence $\dfrac{\partial D_z}{\partial z} = 0$, or $\rho = 0$.

Putting these results together, we have the answer

$$\rho = \begin{cases} |z|\, C/m^3 & \text{for } |z| < a \\ 0 & \text{for } |z| > a \end{cases}$$

(*ii*) Find the charge distribution which will give the following electric field intensity in cartesian co-ordinates:

$$\overline{E} = \begin{cases} -\dfrac{3q_0}{\varepsilon_0}\overline{u}_z & \text{for } -\infty < z < -a \\ -\dfrac{q_0}{\varepsilon_0}\overline{u}_z & \text{for } -a < z < +a \\ +\dfrac{3q_0}{\varepsilon_0}\overline{u}_z & \text{for } +a < z < +\infty \end{cases}$$

Here it is obvious that D_x and D_y are zero as in the above example. But now we find that everywhere D_z is constant (independent of z) which will make $\frac{\partial D_z}{\partial z} = 0$ and $\rho = 0$ everywhere !! Obviously there is something wrong in our thinking here, because an electric field intensity cannot exist without any charges anywhere.

The cause of the error in our thinking is soon identified as our failure to note that though D is constant and independent of z everywhere, the value of this constant is different in the three regions. If z is varied from $-\infty$ to $+\infty$, this constant undergoes sudden changes at $z = -a$ and at $z = +a$. These are the infinite parallel planes where the charges are located !!

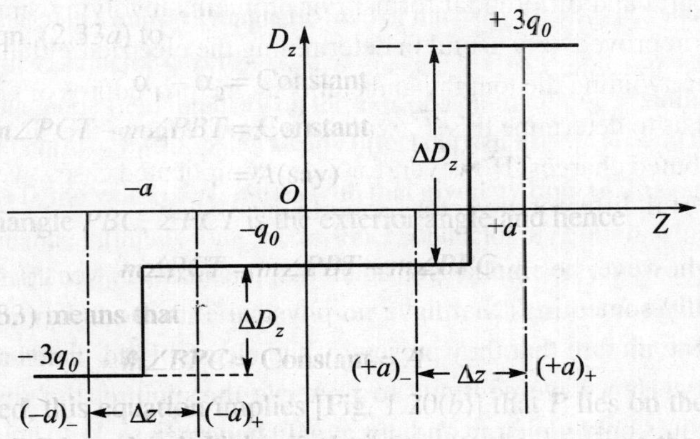

Fig. 4.18. Showing how to deal with a step in a function

To solve the problem, we employ a very useful mathematical trick illustrated in Fig. 4.18 in an exaggerated form for clarity. We split the point $z = -a$ into two points $z = (-a)_-$ and $z = (-a)_+$ as shown. These are the values of z at $z = -a$ when we approach the point $z = -a$ from the negative and positive sides, respectively. The change $(-a)_-$ to $(-a)_+$ is then denoted as Δz, while the change in the value of D_z at the point is denoted as ΔD_z.

Then at $z = -a$,

$$\frac{\partial D_z}{\partial z} = \frac{\Delta D_z}{\Delta z} = \frac{(-q_0) - (-3q_0)}{(-a)_+ - (-a)_-} = \frac{+2q_0}{\Delta z} = \rho$$

Therefore $+2\,q_0 = \rho\Delta\, z$

Since ρ is the *volume* charge density, $\rho\,\Delta\, z$ is the charge enclosed in a volume of area one square metre and thickness $\Delta\, z$. As we make $\Delta\, z \to 0$, this volume becomes zero, but the surface area remains one square metre all the while. Hence the value of $\rho\,\Delta\, z$ when $\Delta\, z = 0$, becomes the surface charge density q_s C/m². Thus at $z = -a$ we have

$$q_s = (+2q_0) \text{ C/m}^2.$$

Using a similar trick at $z = +a$, we get

$$\frac{\partial D_z}{\partial z} = \frac{(+3q_0) - (-q_0)}{(+a)_+ - (+a)_-} = \frac{+4q_0}{\Delta z} = \rho$$

giving, at $z = +a$, $q_s = (+4q_0)$ C/m^2. The results are combined to give the surface charge density

$$q_s = \begin{cases} +2q_0 & \text{at } z = -a \\ +4q_0 & \text{at } z = +a \\ 0 & \text{everywhere else.} \end{cases}$$

4.21 SUMMARY AND COMMENTS

We introduced an auxillary vector $\overline{D}$ to separate the influence of electric charge and the medium on the electric field intensity at a point. This enabled us to obtain Gauss's law in the integral and differential forms. For problems involving simple geometries, the integral form proved very useful in determining the electric field intensity in a very elegant manner avoiding the lengthy and cumbersome procedures of Chapters 2 and 3. It also enabled us to determine this $\overline{E}$ without a knowledge of the distance of the point from the distributed charges. However, for more complicated cases, we had to develop the more powerful differential vector form of Gauss's law.

It must however be emphasised again that the auxillary vector $\overline{D}$ and the tubes of flux are really somethings that have no physical existence; they exist only in our imagination! The picture that they present of the electric field yields results that tally with the reality. Hence we continue to preserve the picture, but remind ourselves frequently that it is only a picture and not a real phenomenon.

Before taking the final step towards developing the most powerful mathematical tool for the study and analysis of electrostatic fields, we step aside in the next chapter to draw inspiration from the science of geometrical optics, and to borrow some ideas from geographers and mathematicians working in the field of Theory of Functions of Complex Variables.

EXERCISE 4

1. In spherical co-ordinates, the permittivity varies as follows:

$$\text{Permittivity} = \begin{cases} \varepsilon_1 & \text{for } 0 < r < a \\ \varepsilon_2 & \text{for } a < r < \infty \end{cases}$$

A charge $+Q_1$ coulomb is located at the origin. Another charge $+Q_2$ coulomb is located at a distance b metres from the origin where $b > a$. Find the coulomb forces on Q_1 and Q_2.

2. In cartesian co-ordinates, the permittivity varies as follows:

$$\text{Permittivity} = \begin{cases} \varepsilon_1 & \text{for } -\infty < x < -a \\ \varepsilon_2 & \text{for } -a < x < +b \\ \varepsilon_3 & \text{for } +b < x < +\infty \end{cases}$$

On the x-axis, two charges are located : $+Q_1$ coulomb at $x = -c$ where $|c| > |a|$, and $+Q_2$ coulomb at $x = +d$ where $d > b$. Find the Coulomb forces on Q_1 and Q_2.

3. A uniform charge density $+ q_v$ C/m^3 exists throughout a sphere of radius b metres. Using Gauss's law, find the value of $\overline{E}$ inside and outside the sphere and plot the variation of E with distance from the centre of the sphere.

4. Uniform charge densities of $(+ q_s)$ and $(- q_s)$ C/m^2 exist on two concentric spherical surfaces of radii a and c metres $(c > a)$, respectively. The permittivity of the dielectric varies, being ε_1 for $0 < r < b$ and ε_2 for $b < r < \infty$ where $a < b < c$. Determine the magnitude of the electric field intensity in different regions as r varies from zero to infinity.

5. *(a)* Two infinite, parallel planes have uniform surface charge densities of $(+ q_s)$ and $(- q_s)$ C/m^2, respectively on them. Find the electric field intensity in the region between the planes and the region outside the planes.

(b) Does the electric field intensity depend on the distance d between the planes?

(c) Assume the potential of the negatively charged plane to be zero. Will the potential at a point at a fixed distance d_1 from the negatively charged plane vary if the distance d between the planes is varied? Explain.

6. In cartesian co-ordinates, a variable volume charge density exists between the planes $z = - a$ and $z = + a$ given by

$$q_v = \begin{cases} z \text{ C/m}^3 & \text{for } |z| < a \\ 0 & \text{for } |z| > a \end{cases}$$

Determine the electric field intensity at different points on the z-axis as z varies from $- \infty$ to $+ \infty$.

7. *(i)* Three infinitely long lines, parallel to the z-axis, cross the x-axis at the points $x = x_1, x_2$ and x_3 respectively and have uniform, linear charge densities of $+ ql_1$, $+ ql_2$ and $+ ql_3$ respectively. Show that the equation to the direction line of the electric field can be written as

$\alpha_1\, ql_1 + \alpha_2\, ql_2 + \alpha_3\, ql_3 = \text{Constant}$

for any plane z = constant, where α_1, α_2 and α_3 are the angles shown in Fig. Ex. 4 Prob. 7 for a point P on a direction line.

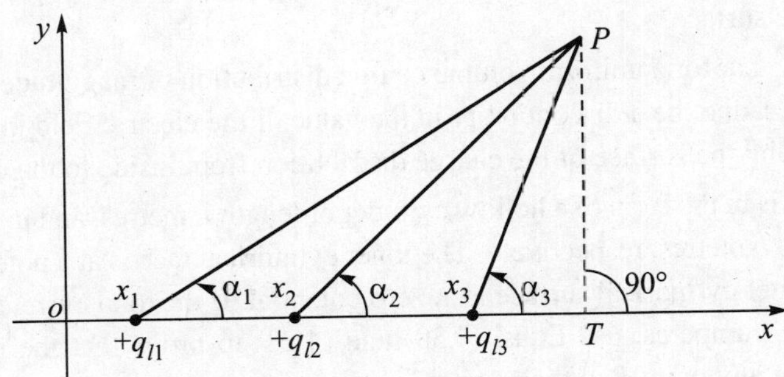

Fig. Ex. 4. Prob. 7

(ii) Hence putting appropriate values for the different x, α and q_l, derive Eqn. (2.33).

[**Hint:** By symmetry, there is no displacement parallel to the z-axis. Draw PT perpendicular to the x-axis, and consider a rectangle on the plane x = constant, passing through P, with sides PT and Δz. Let D_1', D_2' and D_3' be the displacements through this rectangle (PT) Δz, due to lines q_{l_1}, q_{l_2} and q_{l_3}. Therefore, total D' through the rectangle $= D_1' + D_2' + D_3'$, by principle of superposition. If D' is held a constant as P moves, then by the concept of tubes of flux (Fig. 4.2), P will trace a direction line.]

8. Apply the method of the above problem to the case of a dipole with point charges $-Q$ and $+Q$ (Fig. 2.4) and show that the equation to the direction line is

$$\cos\alpha_1 - \cos\alpha_2 = \text{constant.}$$

[**Hint :** Use result of Ex. 1, Prob. 27.]

9. Two infinitely long cylindrical surfaces have their axes parallel to the z-axis, the axes being separated by a distance h metres. A uniform volume charge density of $(+q_v)$ C/m^3 exists only in the region between the two surfaces. The intersection of these surfaces with the plane $z = 0$ gives the curves

$$x^2 + y^2 = b^2 \quad \text{and} \quad (x - h)^2 + y^2 = a^2$$

where $a > \frac{b}{2}$ and $h < a$. If the centre of the smaller circle is denoted as the point A, show that the electric field intensity at any point on the charge-free portion of the x-axis is given by

$$\bar{E} = \frac{q_v(\overline{OA})}{2\varepsilon_0},$$

O being the origin of the co-ordinates. [**Hint:** Use results of Sec. 4.10]

10. Using Gauss's law, obtain the value of the electric field intensity for $r < a$ and for $r > a$ in Ex. 3, Prob. 11 and show that the maximum value of E occurs when $\frac{r}{a} = \frac{\sqrt{5}}{3}$.

11. (*i*) Show that for a completely closed surface charge distribution of uniform density $(+q_s)$ C/m^2, there must be a discontinuity (that is, a sudden jump in magnitude) in the value of the electric field intensity, when passing from inside to the outside of the surface.

(*ii*) Show that for a uniform volume charge distribution of magnitude $(+q_v)$ C/m^2, there cannot be a discontinuity in the value of the electric field intensity when crossing the surface of the charge distribution from inside to the outside.

12. A resistor is in the form of a hollow cylinder of length L metres, and inner and outer radii a and b metres, respectively. The inner cylindrical face is at a potential V volts and the outer cylindrical surface is at zero potential. If the total current through the resistor is I amperes, use Eqns. (4.30) and (4.28) to prove that the electric field intensity at any point P in the resistor is

$$\bar{E} = \frac{(I/L)}{2\pi\sigma R}\bar{u}_R$$

where R is the perpendicular distance of P from the axis of the cylinder. Hence show that conductance of the resistor between its inner and outer cylindrical surfaces is

$$\frac{G}{L} = \frac{2\pi\sigma}{\log_e \dfrac{b}{a}} \text{ siemens per metre.}$$

[Compare these results with Eqns. (2.27) and (4.42).]

13. At a point P on a completely closed conducting shell, the surface charge density is $(+ q_s)$ C/m^2, and hence by Cor. 3, Sec. 4.13, the magnitude of the electric field intensity at P is $\dfrac{+q_s}{\varepsilon_0}$. If an infinitesimally small pinhole is made in the conductor at P, show that the electric field intensity in the hole will become $\dfrac{+q_s}{2\varepsilon_0}$.

14. A conductor in the form of a sphere of radius a has a cavity of any shape inside it. If a charge $+ Q$ coulomb is placed anywhere inside the cavity, find the electric field intensity at any point at a distance r from the centre of the sphere, if $r > a$.

15. In the parallel plate capacitor of Fig. 4.13 (*a*), if there are two dielectric slabs of permittivities ε_1 and ε_2 between the plates, the thickness of the slabs being d_1 and d_2 (where $d_1 + d_2 = d$), find the capacitance of the capacitor, and hence derive the formula for finding the capacitance of two capacitors in series.

16. A charge of one coulomb is placed on a conducting sphere of the size of the earth, located in space. Find the potential of the sphere, assuming that the diameter of the earth is 1.27×10^7 m.

(This shows the enormous effect that one coulomb of charge has in electrostatics, though for power engineers, a current of one ampere (equal to one coulomb per second) is insignificant.)

17. Find the charge distribution which will give the following electric field intensity in cylindrical co-ordinates.

$$\overline{E}_R = \left(\frac{1-\epsilon^{-R}}{\varepsilon_0 R}\right)\overline{u}_R \text{ for } 0 < R < \infty\,; \quad \overline{E}_\phi = 0\,;\ \overline{E}_z = 0$$

18. Find the charge distribution which will give the following electric field intensity in spherical co-ordinates:

$$\overline{E}_r = \begin{cases} 0 & \text{for } 0 < r < a \\ \dfrac{+q_0\, a^2}{\varepsilon_0\, r^2} & \text{for } a < r < b\,; \ \overline{E}_\theta = 0;\ \overline{E}_\phi = 0 \\ 0 & \text{for } b < r < \infty \end{cases}$$

19. In cylindrical co-ordinates, the following potential distribution exists:

$$V = \begin{cases} \dfrac{\varsigma_0\, a}{\varepsilon_0} \log_e \dfrac{b}{a} & \text{for } 0 < R < a \\ \dfrac{\varsigma_0\, a}{\varepsilon_0} \log_e \dfrac{b}{R} & \text{for } a < R < b \\ 0 & \text{for } b < R < \infty \end{cases}$$

Find the charge distribution.

CHAPTER

5 Method of Images and Conformal Transformations

5.1 GEOMETRICAL OPTICS

While studying the behaviour of rays of light in the presence of plane and convex mirrors, it is found that light intensity at any point P is dependent first, on the direct ray from the light source to the point P and, secondly also on the ray reflected from the mirror. To determine this reflected ray the method used is to find the location of the *image* of the light source. Then we assume that *the mirror is removed,* and a ray travels in a straight line from the image to the point P. This ray is drawn *only in front* of the mirror, because no ray exists behind the mirror.

These ideas from geometrical optics are borrowed in electrostatics to study problems involving point or line charges in front of a conducting plane, cylinder or sphere. This is possible because, like light travelling in straight lines, the Coulomb's Law force also acts along the straight line joining the two point charges. This is the principle of the Method of Images for the study of electrostatic problems involving conductors.

In electrostatic fields, conductors have two effects. As the Corollaries of Gauss's law show, the surface of a conductor must be an equipotential surface. Also a charge is induced on the surface of the conductor, and at every point on the surface of the conductor, the electric field intensity is $\overline{E} = \dfrac{q_s}{\varepsilon_0}\overline{u}_n$, Eqn. (4.31) normal to the surface. Though the potential on the conductor is everywhere the same, the charge density q_s can vary from point to point on the surface. This variation of q_s makes it inconvenient to deal with the conductor placed in an electric field. We would like to have a method by which we can *get rid of the conductor altogether,* and yet produce the same electric field intensity at the surface previously occupied by the conductor, but using *one single charge or line charge only.*

5.2 IMAGE OF POINT CHARGE IN INFINITE CONDUCTING PLANE

Consider a point charge $+Q$ located at a distance a in front of an infinitely large conducting plane (Fig. 5.1). By Cors. 1 and 3 of Gauss's law, this plane, being a conductor, must be an equipotential surface, and the electric field lines due to the charge $+Q$ must meet the surface of the plane at right angles at all points on the plane.

But this is the very special feature noted in the last paragraph of Sec. 2.9 about the electric field map of a dipole shown in Fig. 2.5, where the plane shown by the dotted line is at zero potential and all direction lines are perpendicular to it. Hence the field map of Fig. 2.5 will not be altered if at the position of the dotted line, we place a conducting plane of infinite extent and regard the potential of this plane as zero.

But now this conductor acts like a screen. It has a total charge (– Q) induced on its surface as indicated by Cor. 4 of Gauss's law. This is depicted in Fig. 5.1, where the direction lines to the left of the plane are shown dotted because they do not exist.

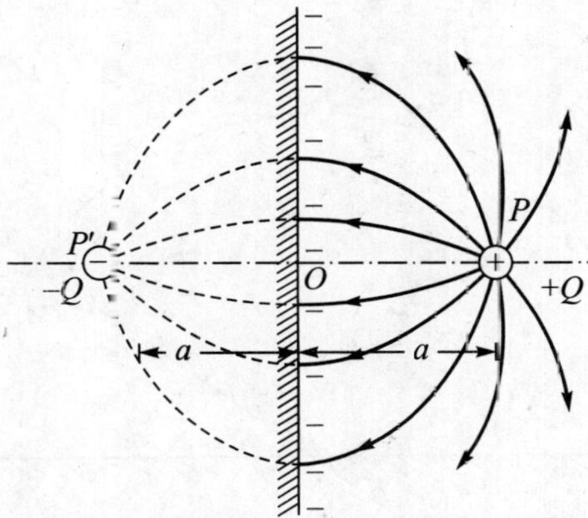

Fig. 5.1. Image of point charge in infinite conducting plane

If there is any difficulty in imagining how a plane can form a closed cavity as required for applying corollary of Gauss's law, it should be remembered that this plane extends to infinity in *all* directions, and infinity can be thought of as a single point! Recall the variation of tan θ in trigonometry as θ varies from 0 to 2π, and how we regard $-\infty$ and $+\infty$ as being the same point to maintain the continuity of the curve for tan θ. Alternatively, in this case of conducting plane, we can regard the right-hand side of Fig. 5.1 enclosed in a *conducting* hemisphere of infinite radius, with the mouth of the hemisphere closed by the infinite conducting plane. Since we had assumed in Sec. 3.3 the potential due to a point charge to be zero at infinite distance, this conducting hemisphere at zero potential is acceptable.

We are thus able to outline the procedure for obtaining the field map for a point charge + Q coulomb located at P at a distance a metres from the plane. We draw a perpendicular PO on the plane and produce it to P' making $P'O = OP = a$ metre. At the point P' we place an image charge (– Q) coulomb (*equal* in magnitude, but *opposite* in sign to the point charge at P), and *remove the conducting plane*. Then the two charges form a dipole whose field direction lines can be plotted by the method already described, regarding the potential of the plane as zero. However, these lines are drawn only on the right hand side of the plane, because the conducting plane acts as a screen, and there is no electric field on the left-hand side of the plane. The image charge is only used as a device for calculating the field; it has no physical existence. We also note that the image charge is of the same magnitude as the given point charge, just as in optics, the image in a plane mirror is of the same size as the object. The two images are also located as far behind the plane as the charge or object is in front of it.

5.3 MULTIPLE IMAGES

The purpose of introducing an image charge as in the previous section, is to make the region occupied by the conductor *an equipotential surface even in the absence of the conductor.* With only one plane, one image charge for *each* point charge is needed as seen in Sec. 5.2. But if we have one point charge + Q at point O and two conducting infinite planes A and B as in Fig. 5.2, then it is not enough to have an image charge – Q

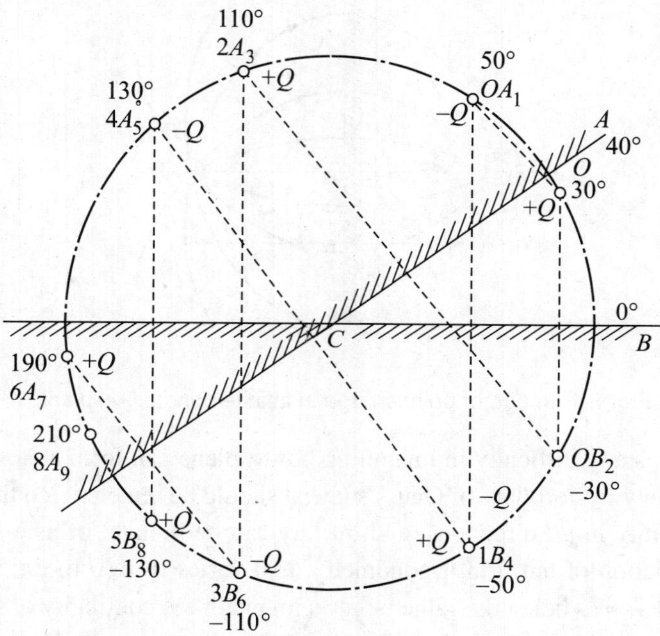

Fig. 5.2. Multiple images

at point O_{A_1} and a second image charge – Q at point O_{B_2} is needed. This notation O_{A_1} means "the image of charge at O in conductor A is at point 1". The charges + Q at O and – Q at OA_1 make the plane A an equipotential surface, but the presence of the second image charge – Q at OB_2 disturbs this and plane A is no longer equipotential with these three charges. To maintain plane A as equipotential, the effect of image charge – Q at OB_2 has to be compensated for by another "image of image charge" + Q at $2A_3$.

A similar argument holds for maintaining plane B as an equipotential surface. The process has to be repeated over and over again, so long as an image charge lies in front of (that is, on the same side as the original charge + Q at O) each plane. In Fig. 5.2 the back side of the planes has been marked by short lines.

It will be found that all these images lie on a circle with centre at C, the point where the planes A and B intersect, and with radius = CO. The two planes are maintained as equipotentials by the pairs of charges shown in Table 5.1.

Table 5.1. Pairs of charges

Plane A as an equipotential			*Plane B as an equipotential*		
$+Q$ at O	and	$-Q$ at OA_1	$+Q$ at O	and	$-Q$ at OB_2
$-Q$ at OB_2	and	$+Q$ at $2A_3$	$-Q$ at OA_1	and	$+Q$ at $1B_4$
$+Q$ at $1B_4$	and	$-Q$ at $4A_5$	$+Q$ at $2A_3$	and	$-Q$ at $3B_6$
$-Q$ at $3B_6$	and	$+Q$ at $6A_7$	$-Q$ at $4A_5$	and	$+Q$ at $5B_8$
$+Q$ at $5B_8$	and	$-Q$ at $8A_9$		—	—

The process ends because both points $6A_7$ and $8A_9$ are behind both the planes *A* and *B*. The angular positions of the images are shown in Fig. 5.2 taking *CB* as the reference line at 0°.

In general these last image points lying behind the planes do not coincide. However, if the angle *ACB* between the planes is an integral submultiple of 180° (that is, if the ratio 180°/ ($m\angle ACB$) is an integer), then the positions of the last two images coincide. When this happens, it must be noted that the magnitude of the charge at the common last image point will be only *Q* (and not 2*Q*!), because the *same* image charge *Q* (with appropriate sign) will make *both* planes as equipotential (see Ex. 5, Probs. 1 and 2). Finally, if the two conducting planes are parallel to each other, all the images will lie on the normal to the planes passing through the point *O* and will form two infinite series of charges located behind both the planes.

5.4 IMAGE OF UNIFORMLY CHARGED INFINITE LINE IN INFINITE PLANE

In Fig. 5.1, if the point charge + *Q* at *P* moves parallel to the infinite plane along a straight line, the image charge – *Q* will likewise have to move parallel to the plane, because for maintaining the equipotential nature of the conducting plane, equal and opposite charges are needed at equal distances on the two sides of the plane.

Hence the electric field map for a uniformly charged infinite straight line in front of an infinite conducting plane is obtained by locating a parallel infinite line with uniform charge of same magnitude but opposite sign, at equal distance behind the conducting plane, then removing the conducting plane, and obtaining the electric field as in Fig. 2.9 and Sec. 2.12, but plotting only that half which is on the same side of the plane as the given charged line. In this analysis, the potential of the plane is taken as zero.

5.5 IMAGE OF POINT CHARGE IN GROUNDED CONDUCTING CIRCULAR RING

For this problem, we make use of an interesting geometrical proposition called the Theorem of Inverse Points. Let a charge + *Q* be located at the point *P* at a distance *D* from the centre *C* of the earthed conducting ring of radius *r* (Fig. 5.3). We have to find out the magnitude and location of an image charge (– *Q*′) such that, with the conducting ring removed the charges (+ Q) and (– Q') will make the circle an equipotential line. The discussion so far has shown that the image charge is always located on the other side of the conductor. Since in this case the point charge + Q is outside the conducting ring, the image charge must be within the circle. By symmetry it must also be located on the line CP at B where $CB = d$.

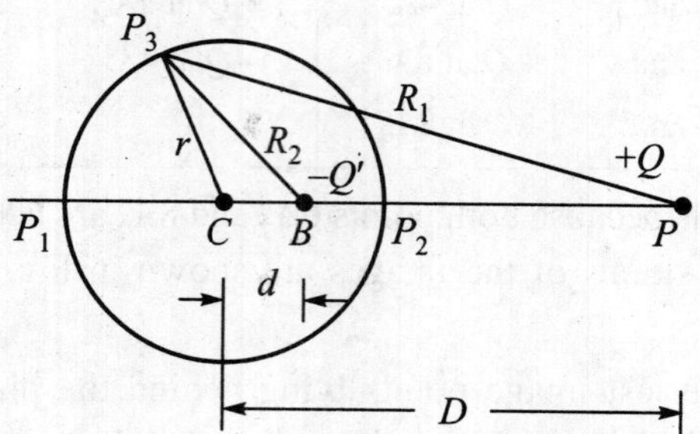

Fig. 5.3. Charge and image charge at inverse points

If PC (produced) intersects the circle at P_1 and P_2, then the potentials at these points due to the two charges + Q and (– Q') must be equal to zero, because the ring is earthed, or

$$V_{P_1} = V_{P_2} = 0.$$

Therefore

$$\frac{Q}{4\pi\varepsilon_0(D+r)} + \frac{-Q'}{4\pi\varepsilon_0(r+d)} = \frac{Q}{4\pi\varepsilon_0(D-r)} = \frac{-Q'}{4\pi\varepsilon_0(r-d)} = 0$$

or $$r + d = \frac{Q'}{Q}(D+r) \quad \text{and} \quad r - d = \frac{Q'}{Q}(D-r) \qquad ...(5.1)$$

Solving $$Q' = Q\left(\frac{r}{D}\right) \text{ and } d = \frac{r^2}{D} \qquad ...(5.2)$$

The points P and B which satisfy the relation $Dd = r^2$ are called inverse points in geometry, and this relation is the Theorem of Inverse Points.

With this magnitude of the image charge and this location, P_1 and P_2 are found to be at zero potential. Now we have to verify whether Eqn. (5.2) make *every* point on the circle to be at zero potential. Let P_3 be *any* point on the circle and let $PP_3 = R_1$ and $BP_3 = R_2$. Then

Potential at $$P_3 = V_{P_3} = \frac{1}{4\pi\varepsilon_0}\left[\frac{Q}{R_1} - \frac{Q'}{R_2}\right] \qquad ...(5.3)$$

By Law of Cosines, Eqn. (1.31), applied to triangles PCP_3 and BCP_3,

$$R_1^2 = D^2 + r^2 - 2Dr\cos(\angle PCP_3)$$
$$= D^2 + r^2 + 2Dr\cos(\angle P_1CP_3)$$

and $$R_2^2 = d^2 + r^2 - dr\cos(\angle BCP_3)$$

$$= \frac{r^4}{D^2} + r^2 + 2\frac{r^2}{D} r \cos (\angle P_1CP_3)$$

$$= \frac{r^2}{D^2}\left[D^2 + r^2 + 2Dr \cos (\angle P_1CP_3)\right]$$

Putting $Q' = Q\left(\frac{r}{D}\right)$ and these values of R_1 and R_2 in Eqn. (5.3), it is found that $V_{P_3} = 0$, thus confirming that Eqns. (5.2) do make the entire ring at zero potential.

Another very useful relation can be established by nothing that in Fig. 5.3, the triangle PCP_3 and P_3CB are similar, because two of their sides are proportional to each other by Eqn. (5.2) and the included angle $\angle PCP_3 = \angle P_3CB$, or

$$\frac{P_3C}{PC} = \frac{r}{D}, \quad \frac{CB}{CP_3} = \frac{d}{r}$$

Therefore,

$$\frac{r}{D} = \frac{d}{r} = \frac{P_3C}{PC} = \frac{CB}{CP_3} = \frac{P_3B}{PP_3} = \frac{R_2}{R_1} = k \text{ (say)} \quad \text{...(5.4)}$$

Since D and r are given, it is obvious that k is a constant, and as $r < D$, $k < 1$.

The reciprocal relationship of inverse points contained in Eqns. (5.2) and (5.4) indicate that if $(-Q')$ is the magnitude and B the location of the *given* charge, then its *image* is of magnitude $Q = \frac{+Q}{k}$ and it is located at P. Since $k < 1$, we find that the magnitude of the image charge Q now is greater than the given charge Q'. This means that for a given charge Q, the magnitude of its image Q' in a *convex* conducting surface is smaller, but if charge Q' is given inside the ring, then the magnitude of its image in the *concave* conducting surface (as seen from inside the ring) is larger. This tallies with geometrical optics, in which a convex mirror produces a smaller image, and a concave mirror produces a larger image than the object.

The ring was grounded to make its potential $V = 0$. This was done to simplify the derivation of Eqns. (5.2) and (5.4). But even if V is not equal to zero, and the ring is not grounded, these equations still make the ring an equipotential surface, but at a potential other than zero.

5.6 IMAGE OF A UNIFORM LINE CHARGE IN A PARALLEL CONDUCTING CYLINDER

The problem considered in the previous section is only a theoretical one, because it is two-dimensional, whereas in practice, we have to deal with three-dimensional problems in space. There are two methods that we can use to extend the results of Eqn. (5.4) from two to three dimensions.

One way is to imagine that the plane of the paper with Fig. 5.3 on it moves parallel to itself, so that the point P becomes the location of an infinitely long uniform linear charge density of ($+ q_l$) C/m. Then the ring traces out a cylinder of radius r with its axis C parallel to the line charge. Since Eqn. (5.4) is true for every position of the plane, it means that the electric field can be determined by placing an infinitely long image charge at B, *removing the conducting cylinder*, and proceeding to calculate the electric field intensity and potentials at different points by the methods of Secs. 2.12 and 3.11, if we know the magnitude of the charge density at B.

A look at Fig. 3.7 shows that it has a line charge ($+ q_l$) C/m, and the equipotential surfaces in the form of cylinders surrounding a parallel line charge of density ($- q_l$) C/m. The cylinder on the left side whose radius is r metres can thus be selected as the conducting cylinder. For this cylinder to be an equipotential, Eqn. (3.40) shows that R_2/R_1 should be equal to a constant k. But Eqn. (5.4) shows that this is also the condition giving the inverse point B, where the image charge is to be located as stated above. However Fig. 3.7 shows that the charge density at this location is ($- q_l$) C/m [and not smaller as given by Eqn. (5.2)]. Thus in this case, the location of the image charge is at B as given by Eqn. (5.4), but its magnitude is ($- q_l$) C/m. This tallies with geometrical optics, where the length of the image in the axial direction of the cylindrical mirror is the same as that of the object. The reason why the image charge density is the same as the given charge density (like the image of a point charge in a plane mirror) is because, there is no curvature in the direction of the axis and ($- q_l$) C/m is measured in the direction of the axis. The curved side of the cylinder only determines the location of the image charge in accordance with Eqn. (5.4). The cylindrical conductor acquires an induces charge of ($- q_l$) C/m.

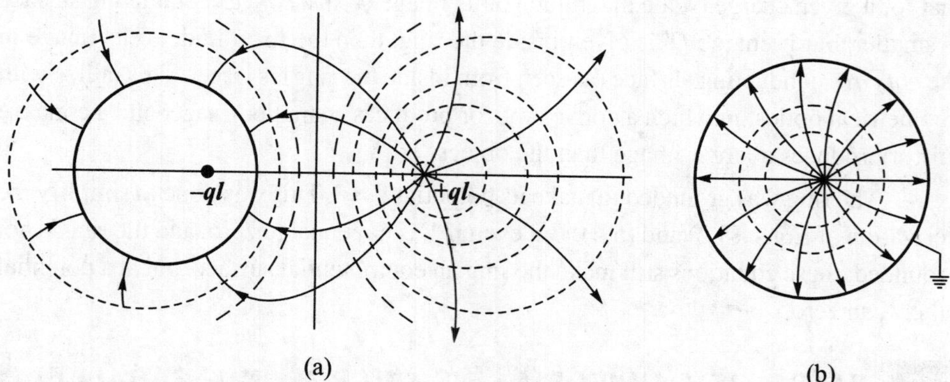

Fig. 5.4. Electric field and potential lines with uniform infinite line charge located (a) outside and (b) inside infinite conducting cylinder

If the cylinder is earthed, its potential will become zero. The direction lines and equipotentials will then be as shown in Fig. 5.4 (a), which has the same general appearance as Figs. 3.7 and 2.11.

Notice that all direction lines meet the conducting surface of the cylinder at right angles as required by the Cor. 3 of Gauss's law. There are no potential lines inside the hollow cylinder in accordance with Cor. 6. The same diagram will be true by Cor. 1 and 2 even if the cylinder is not hollow but a solid conducting cylinder.

In Fig. 5.4 (b), the positive charged line is assumed to be at B inside the hollow cylinder. Its image charge $(-q_l)$ will be at P outside the cylinder [Fig. 5.4 (*a*)], but now the field is plotted only *inside* the cylinder, because there is no field outside as the conductor is earthed, and hence all induced charges on the outer surface of the conducting cylinder (Cor. 5 of Gauss's law) have leaked away to earth.

5.7 IMAGE OF POINT CHARGE IN CONDUCTING SPHERE

In this second extension of Eqn. (5.4) from two to three dimensions, we revolve Fig. 5.3 about axis PC. Then at P and B, we continue to have point charges, but the ring becomes a sphere with centre at C and radius r.

For each position of the revolving plane, Eqns. (5.2) and (5.4) are true. Hence in this case, the image of a point charge $+Q$ located at P will be a charge $-Q' = -\frac{r}{D}Q$ located at B. Then the electric field and the equipotentials outside the sphere can be obtained by removing the sphere, and using the methods of Secs. 2.9 and 3.9. These are shown in Fig. 5.5 for a hollow conducting sphere of radius r.

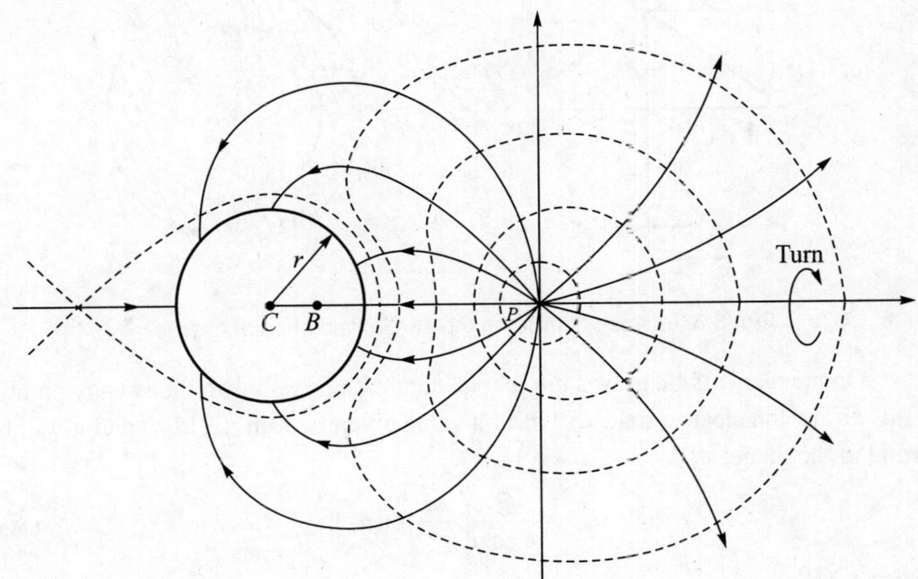

Fig. 5.5. Direction lines and equipotentials for a point charge $+Q$ at P in front of a conducting hollow sphere of radius r. A charge $(-Q)$ is induced on the outer surface of the sphere

The diagram is to be revolved about the axis CP to get the three-dimensional surfaces. It is seen that the conducting sphere is the only equipotential surface which has a truly spherical shape. Other equipotentials tend to become more and more spherical as they get nearer to the point P, but they are nowhere truly spherical, because of the negative charge at B.

5.8 INFINITE CONDUCTING PLANE PLACED IN FIELD OF A POINT CHARGE

When an infinite conducting plane is introduced in the electric field of a point charge $+Q$ located at a point O, it acts as a shield causing the electric field behind (that is, on the other side from point O) the plane to vanish. There is also a charge induced at all points of the plane, whether the plane is grounded or not. If the plane is grounded, its potential $V = 0$. But if the plane is not earthed, V can have any value depending on which point is taken as the reference point with zero potential.

The charge density induced at different points on the plane is not uniform. To find it, we first draw a perpendicular OT from O to the plane [Fig. 5.6 (*a*)] and produce it to O' making $OT = TO' = a$. Then we place an image charge $(-Q)$ at O' and remove the conducting plane. The electric field at any point is now found by the method of Section 2.9.

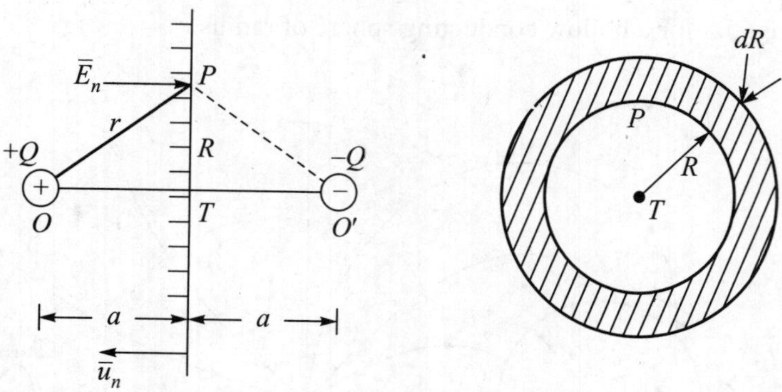

Fig. 5.6. Induced charge on a plane in field of point charge

On the plane, if we draw a circle with centre T and radius R, then at any point P on this circle, the electric field will have a value given by Eqn. (2.17 *a*) and it will be normal to the plane:

$$\overline{E}_n = \frac{Q}{4\pi\varepsilon_0}\left(\frac{2a}{r^3}\right)(-\overline{u}_n) \qquad ...(5.5)$$

where $r = OP$.

The negative sign is needed with the unit vector $\overline{u}_n$, because this unit vector is always the *outward* drawn normal, but the field is directed towards the plane. By Eqn. (4.31), this field is related to the surface charge density (q_s) coulomb/m^2 at P as

$$\overline{E}_n = \frac{q_s}{\varepsilon_0}\overline{u}_n = -\frac{Qa}{2\pi\varepsilon_0 r^3}\overline{u}_n \qquad ...(5.6)$$

Therefore $$q_s = -\frac{Qa}{2\pi r^3}\ \text{C/m}^2 \qquad ...(5.7)$$

This shows that

(*i*) a negative charge density is induced on the plane and

(*ii*) the charge density is maximum at point T, where $r = a$ and it decreases as we move away on the plane from T. The charge enclosed in a circular strip of radius R and width dR (Fig. 5.6 *b*) is

$$\text{Charge in circular strip} = q_s\,(2\pi R)\,dR$$

$$= -\frac{Qa}{2\pi r^3}(2\pi\, RdR) \qquad ...(5.8)$$

To find the total charge induced on the plane, we have to integrate this expression for values of R from $R = 0$ to $R = \infty$.

$$\text{Total charge induced on the plane} = \int_{R=0}^{R=\infty}\frac{(-\,Qa\,RdR)}{r^3} \qquad ...(5.9)$$

To integrate, we note that $r^2 = R^2 + a^2$ so that $2r\,dr = 2R\,dR$

Hence total charge induced on the plane

$$= \int_{R=0}^{R=\infty}\frac{(-\,Qa)r\,dr}{r^3} = \int_{R=0}^{R=\infty}(-\,Qa)r^{-2}\,dr$$

$$= -\,(Qa)\left[\frac{-1}{r}\right]_{R=0}^{R=\infty} = -\,Q \qquad ...(5.10)$$

This shows that the *total* charge induced on the plane is equal in magnitude to the given charge but of *opposite* sign. Equation (5.7) also shows that the charge density is maximum at the point T and it decreases rapidly as we move away from T.

5.9 CAPACITANCE OF A CAPACITOR FORMED BY A CONDUCTING SPHERE NEAR A CONDUCTING INFINITE PLANE

By definition these two conductors, insulated from each other, form a capacitor. Since both are conductors, the sphere forms one equipotential surface and the plane forms another equipotential surface. In Secs. 5.7 and 5.2, we have handled these surfaces separately and seen where image charges of proper magnitude and sign can be located, to make each surface an equipotential surface when the conductor is removed. But in this problem, we have both surfaces together. We cannot make *both* surfaces as equipotential by *one* image charge.

Hence we adopt the method of successive approximations, by trying to find series of image charges which will make, first the sphere, and then the plane, as equipotential, repeating this cycle over and over again. This step-by-step process is described below.

To fix our ideas, we assume that the plane is grounded ($V = 0$) and the sphere is at a potential V with respect to the plane. Then if we can find the charge Q on the sphere which has given it this potential, we can derive the capacitance $C = \frac{Q}{V}$ from Eqn. (4.31 *a*). Of course since both the sphere and the plane will not be equipotential surfaces *simultaneously*, after each step, we shall get only an approximate answer, the accuracy of which can be made better by increasing the number of steps.

Let a be the radius of the sphere (Fig. 5.7) and $\frac{D}{2}$ the distance of its centre from the plane where $\frac{D}{2} > a$. *Remove both the conductors.*

Step 1: Put a charge Q_1 at the *centre* of the sphere. Note that even if the charge is not placed at the centre of a conducting sphere, for points *outside* the conducting sphere, the charged sphere acts as if the charge is located at its centre and the sphere is removed. But in the present case, we are placing the charge *after* removing the conducting sphere. Hence, if the charge is not placed at the centre, the spherical surface will not become an equipotential surface.

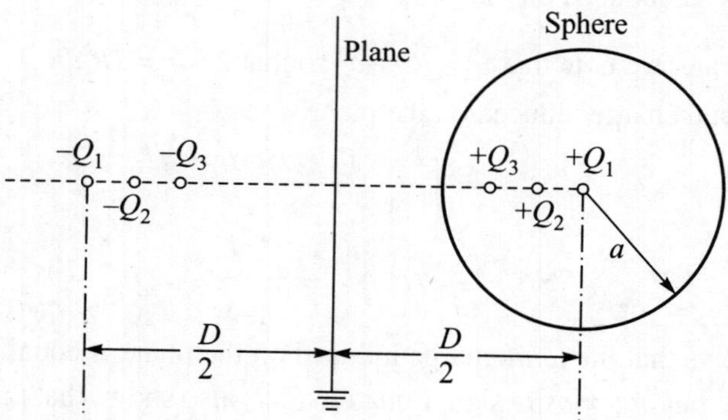

Fig. 5.7. Capacitor formed by sphere and infinite plane

With Q_1 at the centre of the sphere, all points equidistant from it will form an equipotential surface (Fig. 3.4). Thus the sphere has been made an equipotential surface. But the points on the plane are not at equal distances from Q_1. Hence the plane is not an equipotential surface.

Step 2: We place an image charge $(-Q_1)$ behind the plane on the normal to the plane through Q_1, the distance of the image charge being $\frac{D}{2}$ from the plane. Now the charges $(+Q_1)$ and $(-Q_1)$ together make the plane an equipotential surface. Notice that the charge $(+Q_1)$ alone at the centre had made the sphere an equipotential surface, but now, because of the presence of the charge $(-Q_1)$, the sphere is no longer an equipotential surface (see Fig. 2.5).

Step 3: Using Eqns. (5.2), we find that if we locate a charge $(+Q_2) = -\frac{a}{D}(-Q_1)$ $= +rQ_1$, (where $r = \frac{a}{D}$) at a distance $d = \frac{a^2}{D} = ar$ from the centre of the sphere, then the charge $(-Q_1)$ and its image charge $(+Q_2)$ *together* will make the surface of the sphere an equipotential surface, which was the position at the end of Step 1. Also because of the charge $(+Q_2)$, the plane is not an equipotential at the end of Step 3.

Step 4: Since the charge $(+Q_2)$ is at a distance $\left(\frac{D}{2}-ar\right)$ from the plane, to make the plane an equipotential surface, we place a charge $(-Q_2)$ at a distance $\left(\frac{D}{2}-ar\right)$ behind the plane. These charges $(+Q_2)$ and $(-Q_2)$ *together* make the plane an equipotential surface, but the sphere is no longer an equipotential. This was the position at the end of Step 2.

The process can be continued and the results tabulated as in Table 5.2, we see that at the end of an even number of steps, the plane is an equipotential, because of the pairs of charges $(+Q_1, -Q_1)$, $(+Q_2, -Q_2)$, ..., etc., and the sphere is not an equipotential.

On the other hand, at the end of an odd number of steps, the sphere is an equipotential, due to charge $(+Q_1)$, and pairs of charges $(-Q_1, +Q_2)$, $(-Q_2, +Q_3)$, $(-Q_3, +Q_4)$, ..., etc, because as indicated by Eqn. (5.1), the pairs of charges shown in the brackets, contribute zero potential to the sphere, as they satisfy Eqn. (5.2). Hence, at the end of any odd number of steps, the potential of the sphere is only due to the charge $(+Q_1)$ placed at the centre of the sphere, and by Eqn. (3.12) it is

$$V = \frac{Q_1}{4\pi\varepsilon_0 a} \text{ volt} \qquad \text{...(5.11)}$$

Although at the end of an odd (or even) number of steps, the sphere (or the plane) becomes an equipotential surface *exactly*, the other surface is not an exact equipotential, but tends to become more and more nearly equipotential, as the number of steps is increased. This implies that by taking a sufficiently large number of steps, this deviation from equipotentialness of the other surface can be made as small as we please. The total charge on each conductor to produce this condition is $(+Q)$ coulomb on the sphere $(-Q)$ coulomb on the plane, where

$$Q = Q_1 + Q_2 + Q_3 + Q_4 + \text{... to infinity}$$

Using the values of Q_2, Q_3, Q_4, ..., etc., from Table 5.2, and expressing them in terms of Q_1, we get

$$Q = Q_1\left[1 + r + \frac{r^2}{(1-r^2)} + \frac{r^3}{(1-r^2)\left(1-\dfrac{r^2}{1-r^2}\right)} + \frac{r^4}{(1-r^2)\left(1-\dfrac{r^2}{1-r^2}\right)\left[1-\dfrac{r^2}{1-\dfrac{r^2}{1-r^2}}\right]} + \ldots\right] \qquad \text{...(5.12)}$$

Table 5.2. Charges and their locations

Behind the Plane (on side opposite to sphere)			*In front of the Plane (on same side as the sphere)*		
Step No.	*Charge*	*Distance from centre of sphere*	*Step No.*	*Charge*	*Distance from centre of sphere*
			1	$+Q_1$	0
2	$-Q_1$	D	3	$+Q_2 = +rQ_1$	ar
4	$-Q_2$	$D - ar$	5	$+Q_3 = \frac{+r}{1-r^2}(rQ_1)$	$\frac{ar}{1-r^2}$
6	$-Q_3$	$D - \frac{ar}{1-r^2}$	7	$+Q_4 = \frac{r}{(1-r^2)} \cdot \frac{r}{\left(1 - \frac{r^2}{1-r^2}\right)} \cdot (rQ_1)$	$\frac{ar}{1 - \frac{r^2}{1-r^2}}$
Result at end of even step: Plane equipotential, but not sphere.			Result at end of odd step: Sphere equipotential, but not plane.		

Equation (5.12) enables us to write the expression for the capacitance in the form of an infinite series as

$$C = \frac{Q}{V} = 4\pi\varepsilon_0\, a\left[1 + r + \frac{r^2}{1-r^2} + \frac{r^3}{(1-r^2)\left(1 - \frac{r^2}{1-r^2}\right)} + \frac{r^4}{(1-r^2)\left(1 - \frac{r^2}{1-r^2}\right)\left(1 - \frac{r^2}{1 - \frac{r^2}{1-r^2}}\right)} + \ldots\right] \text{ farad} \qquad \ldots(5.13)$$

As the infinite conducting plane moves away from the sphere, D increases and $r = (a/D)$ decreases. In the limit, when $D \to \infty$, $C \to 4\pi\, \varepsilon_0\, a$ as found from Eqn. (4.37). The effect of the plane is thus seen to cause an increase in the capacitance of the sphere.

The electric field direction lines and equipotentials are as shown in Fig. 5.8.

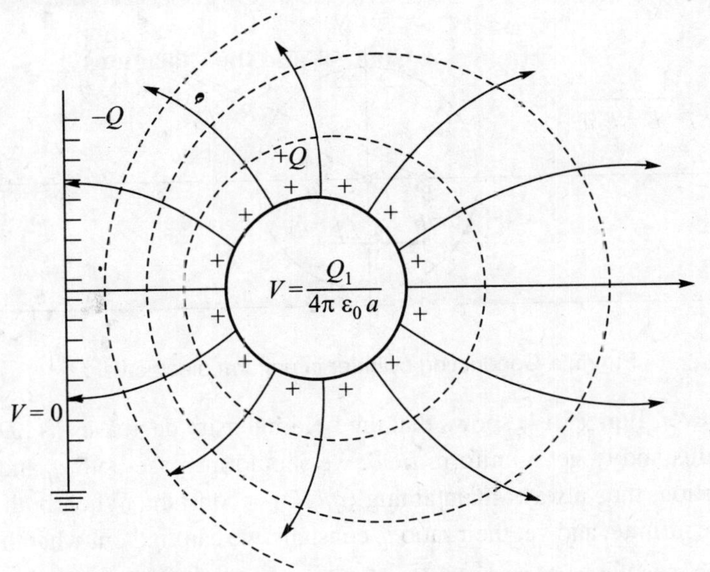

Fig. 5.8. Direction lines and equipotentials for capacitor formed by sphere and infinite plane

5.10 EFFECT OF PLACING CONDUCTING CYLINDER ACROSS UNIFORM ELECTRIC FIELD

Let $\overline{\boldsymbol{E}} = E_u \overline{\boldsymbol{u}}_x$ be the uniform electrostatic field, and let the infinitely long conducting cylinder of radius a be placed with its axis coinciding with the z-axis. The surface of a conductor must be an equipotential surface by Cor. 1 of Gauss's law. Since this equipotential surface is cylindrical, we base our solution on the problem discussed in Sec. 3.11 and Fig. 3.7.

In Sec. 2.16 we had produced the uniform electric field by using two oppositely charged infinite parallel planes. Here we shall first obtain the uniform electric field by using two oppositely charged parallel lines. We denote by $2D$ the distance between these two oppositely charged infinite parallel lines with charge densities $(+\,q_l)$ and $(-\,q_l)$ coulomb/metre on them (Fig. 5.9). The lines are parallel to the z-axis and intersect the x-axis at P' and P, respectively. At the origin O, the field produced by these infinite line charges will add to give the resultant field

$$\frac{(+\,q_l)}{2\pi\varepsilon_0\,D}\overline{\boldsymbol{u}}_x + \frac{(-\,q_l)}{2\pi\varepsilon_0\,D}(-\,\overline{\boldsymbol{u}}_x) = \frac{(2q_l)}{\pi\varepsilon_0(2D)}\overline{\boldsymbol{u}}_x \qquad \text{...(5.14)}$$

A study of Fig. 2.11 reveals that if the separation between the lines is increased, the radii of circles representing the field lines will become larger and larger, and the arcs of circles crossing the y-axis will become less and less curved. In the limit, if the distance $2D$ becomes infinitely large, with P' at $-\infty$ and P at $+\infty$, the circular arcs will become parallel straight lines, and thus the electric field will consist of parallel straight lines *everywhere*.

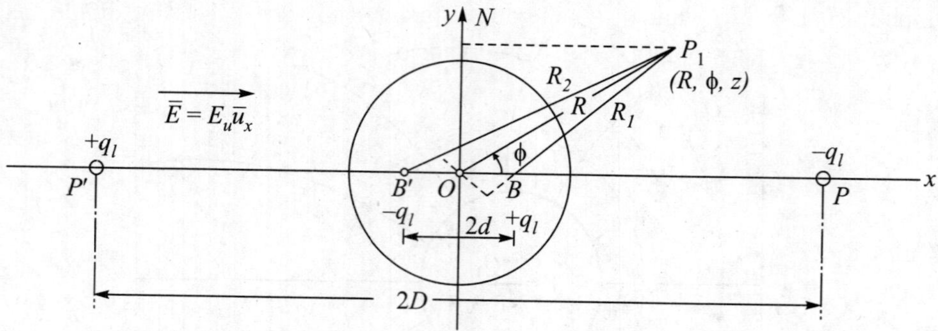

Fig. 5.9. Conducting cylinder in uniform electric field

However, Eqn. (5.14) shows that the field intensity decreases as $2D$ increases. To prevent this and to get a uniform field, we should increase both q_l and $2D$ in the same proportion, thus always maintaining $(q_l/2D)$ = constant. When both q_l and $2D$ have become infinite, and yet their ratio is constant, we can find out what this constant must be to give us the uniform field $\overline{\boldsymbol{E}} = E_u\,\overline{\boldsymbol{u}}_x$ by writing

$$\overline{\boldsymbol{E}} = E_u\overline{\boldsymbol{u}}_x = \frac{2q_l}{\pi\varepsilon_0(2D)}\overline{\boldsymbol{u}}_x$$

so that

$$\left(\frac{q_l}{2D}\right) = \frac{\pi\varepsilon_0}{2}E_u \qquad ...(5.15)$$

In this way we have, in the limit, produced the desired uniform electric field using two infinitely long line charges.

The significance of this method of producing the uniform electric field will become obvious, when we next tackle the problem of using the method of images to make the cylindrical surface as an equipotential, so that we may ultimately put a conducting cylinder there. For this, by Sec. 5.6, to compensate for the line charge $(-q_l)$ C/m at P, we put a line charge $(+q_1)$ at B, where $OB = d = \dfrac{a^2}{D}$. Similarly for $(+q_l)$ at P', we have $(-q_l)$ at B', so that $BB' = 2d$. These four line charges at P, P', B and B' make the cylinder an equipotential surface. Using the relation $d = (a^2/D)$ in Eqn. (5.15)

$$(2d\,q_l) = 4a^2\left(\frac{q_l}{2D}\right) = 2a^2\,\pi\,\varepsilon_0\,E_u \qquad ...(5.16)$$

Notice that as q_l and $2D$ tend to become infinite to produce the uniform field $E_u\overline{\boldsymbol{u}}_x$, the distance $2d$ between the image charges tends to become zero. In the limit the left hand side of Eqn. (5.16) becomes $(0 \times \infty)$, an indeterminate quantity, but the right-hand side remains a constant, which is the condition that makes the cylinder an equipotential surface.

Before going to the limiting value infinity for q_l and $2D$, we will write an expression for the potential V_i produced by the two image line charges giving a line dipole (or two-dimensional electric dipole of the same type as three-dimensional dipole with point charges (+ Q) and (– Q) and dipole moment $P = 2aQ$). The line dipole has moment (q_l $2d$). At any point P_1 with cylindrical co-ordinates (R, ϕ, z) as in Fig. 5.9, if $BP_1 = R_1$ and $B'P_1 = R_2$ then, as in Eqn. (3.39), we have the potential at P_1 as

$$V_i = \frac{q_l}{2\pi\varepsilon_0} \log_e \frac{R_2}{R_1} \qquad ...(5.17)$$

Putting $OP_1 = R$ and noting that ultimately $R \gg 2d$, we write

$$R_2 = R + d\cos\phi \text{ and } R_1 = R - d\cos\phi$$

so that

$$\frac{R_2}{R_1} = \frac{R + d\cos\phi}{R - d\cos\phi} = \left(1 + \frac{d\cos\phi}{R}\right)\left(1 - \frac{d\cos\phi}{R}\right)^{-1}$$

$$\approx \left(1 + \frac{d\cos\phi}{R}\right)\left(1 + \frac{d\cos\phi}{R}\right)$$

(by Binomial Theorem)

$$\approx 1 + \frac{2d}{R}\cos\phi$$

Putting this value in Eqn. (5.17) and simplifying by using Eqn. (1.14) and neglecting higher powers of $\left(\frac{2d}{R}\right)$, and using Eqn. (5.16), we get the potential at the point P_1 due to image line charges at B and B' as

$$V_i = \frac{q_l}{2\pi\varepsilon_0} \log_e\left(1 + \frac{2d}{R}\cos\phi\right)$$

$$= \frac{(q_l\, 2d)\cos\phi}{2\pi\varepsilon_0 R} = \frac{(2a^2\pi\varepsilon_0 E_u)\cos\phi}{2\pi\varepsilon_0 R}$$

$$= \frac{a^2\cos\phi}{R} E_u \qquad ...(5.18)$$

Note that as $q_l \to \infty$, $2D \to \infty$ and $2d \to 0$, the value of V_i remains unchanged.

The original line charges at P and P' produce the uniform field $\overline{E} = E_u \overline{u}_x$. To obtain the potential V_0 at P_1 due to this field, we observe that for deriving Eqn. (3.39) on which Eqn. (5.17) is based, the reference potential on the plane $x = 0$ is assumed to be zero. The perpendicular distance of P_1 from the plane $x = 0$ is $NP_1 = R\cos\phi$ and this is parallel to $\overline{E}$. Hence the work done in going from the plane $x = 0$ to P_1 along this line NP_1 is $= E_u R\cos\phi$. This is equal to the potential difference $V_{x=0} - V_0 = 0 - V_0$ Therefore

$$V_0 = -E_u R\cos\phi \qquad ...(5.19)$$

Adding these voltages to get the resultant voltage V,

$$V = V_i + V_0 = \frac{a^2\cos\phi}{R} E_u - R\cos\phi\, E_u$$

or

$$V = \left(\frac{a^2}{R} - R\right)\cos\phi\, E_u \qquad ...(5.20)$$

Using the expression for the gradient V in cylindrical co-ordinates given in Eqn. (1.104), we get the components of the electric field intensity at P_1 as

$$\overline{E}_R = -\frac{\partial V}{\partial R}\overline{u}_R = \left[\left(\frac{a^2}{R^2}+1\right)E_u \cos\phi\right]\overline{u}_R \qquad ...(5.21)$$

$$\overline{E}_\phi = -\frac{1}{R}\frac{\partial V}{\partial \phi}\overline{u}_\phi = \left[\left(\frac{a^2}{R^2}-1\right)E_u \sin\phi\right]\overline{u}_\phi \qquad ...(5.22)$$

$$\overline{E}_z = 0$$

5.11 CONFORMAL TRANSFORMATIONS

Let z and w be two complex numbers which, when separated into their real and imaginary parts, can be written as

$$z = x + jy$$

and

$$w = u + jv, \quad j = \sqrt{-1} \qquad ...(5.23)$$

If w is some function of z, or

$$w = F(z) = u(x, y) + jv\,(x, y) \qquad ...(5.24)$$

then u and v must be real functions of x and y. In general, this implies that when z is given (that is, when x and y are given), we can have one or more values of w. Also the differentiation of w with respect to z, defined as

$$\frac{dw}{dz} = \lim_{\Delta z \to 0} \frac{F(z+\Delta z) - F(z)}{\Delta z} \qquad ...(5.25)$$

can have, in general, an infinite number of different values depending upon the manner in which x and y are tending towards zero along different paths in the z-plane. But if we desire that for a given value of z, we should have one single or unique value of $\frac{dw}{dz}$ then it is found by mathematicians Cauchy and Riemann that the following conditions must be satisfied:

$$\frac{\partial u}{\partial x} = \frac{\partial v}{\partial y} \quad \text{and} \quad \frac{\partial u}{\partial y} = -\frac{\partial v}{\partial z} \qquad ...(5.26)$$

where $\frac{\partial u}{\partial x}, \frac{\partial u}{\partial y}, \frac{\partial v}{\partial x}, \frac{\partial v}{\partial y}$ are continuous functions of x and y.

Equation (5.26) is referred to as the Cauchy-Riemann conditions.

It is not possible to show the relationship between x, y, u and v (four variables) in a three-dimensional space. To get around this difficulty we plot z and w on two separate planes (Fig. 5.10) called the z and the w planes. For each point z defined by x and y on the z-plane, there is a corresponding point on the w-plane. If changes of x and y cause the point z to describe a curve on the z-plane, there will be a corresponding curve described by the point w on the w-plane, but the shapes of the two curves may be different, depending upon the nature of $F(z)$. This process is called the mapping of the curve in the z-plane on to the w-plane. Usually $\frac{dw}{dz}$ is complex and may be written as

$$\frac{dw}{dz} = a\, e^{j\phi} = a \angle\phi \qquad \text{...(5.27)}$$

which shows that (Fig. 5.10) a segment Δw is obtained from a segment Δz by multiplying it by a and rotating it through an angle $\angle\phi$. This means that for any area in the z-plane, the corresponding area in the w-plane has linear dimensions that are a times as large, and the orientation of the area is turned through an angle ϕ. This similarity of form has given the name conformal transformation, to Eqn. (5.24) when the Cauchy-Riemann are satisfied and the function is said to be analytic.

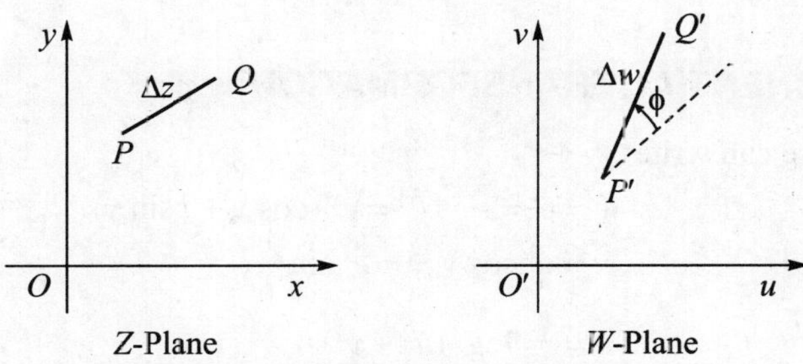

Fig. 5.10. Conformal transformation

5.12 ORTHOGONAL PROPERTIES OF CONJUGATE FUNCTIONS

For the analytic function of Eqn. (5.24) satisfying Eqn. (5.26), let us define two families of curves by the relations

$$u(x, y) = C_1 \quad \text{and} \quad v(x, y) = C_2 \qquad \text{...(5.28)}$$

where C_1 and C_2 are arbitrary constants. By giving different values to C_1, we get different curves of the first family, and similarly different values of C_2 give members of the second family of curves. Let the curves of Eqn. (5.28) intersect at a point (x_1, y_1). Then the slopes of the tangents of these curves at (x_1, y_1) may be written as

$$\left[\frac{dy}{dx}\right]_{u=\text{const}} = -\frac{\frac{\partial u}{\partial x}}{\frac{\partial u}{\partial y}} \quad \text{and} \quad \left[\frac{dy}{dx}\right]_{v=\text{const}} = -\frac{\frac{\partial v}{\partial x}}{\frac{\partial v}{\partial y}} \qquad \text{...(5.29)}$$

Since Eqn. (5.26) is satisfied, it follows that

$$\left[\frac{dy}{dx}\right]_{u=\text{const}} \times \left[\frac{dy}{dx}\right]_{v=\text{const}} = -1 \qquad \text{...(5.30)}$$

Now two straight lines $y = mx + c$ and $y = m'x + c'$ intersect at right angles if $mm' = -1$. Hence Eqn. (5.30) shows that the curves $u(x, y) = C_1$ and $v(x, y) = C_2$ intersect at right angles *everywhere* on the w-plane, just as the families of straight line $x = C_3$ and $y = C_4$ on the z-plane intersect at right angles *everywhere* on the z-plane.

Thus the orthogonal property of the curves on the z-plane is preserved when they are transferred into curves on the w-plane. The functions of Eqn. (5.28) are called conjugate functions.

It is this special feature of the conformal transformation, which is of interest in the study of electric fields, because as pointed out in Sec. 3.4, the direction lines and equipotentials are always orthogonal to each other at every point in space. The conformal transformation enables us to alter an unfamiliar field configuration into a familiar type, like the ones discussed in the previous chapters for which the solutions are easily found.

Another field using orthogonal curves is geography, in which lines of longitude and parallels of latitudes are everywhere orthogonal. This orthogonality is preserved when the curved surface of the globe is depicted on a two-dimensional map by any of the map projection methods used by cartographers.

5.13 EXPONENTIAL TRANSFORMATION $w = e^z$

For this case we can write

$$w = e^z = e^{(z+jy)} = e^x (\cos y + j \sin y) \qquad \text{...(5.31)}$$

and

$$u = e^x \cos y,\ v = e^x \sin y,$$

$$x = \frac{1}{2} \log_e (u^2 + v^2)$$

$$y = \tan^{-1} \frac{u}{v} \qquad \text{...(5.32)}$$

If we have a rectangle on the z-plane (Fig. 5.11) formed by the lines $x = 0, x = 1$, $y = 0$ and $y = \frac{\pi}{4}$, then we get the rectangle $OABC$ on the z-plane transformed into a sector $O''A'B'C'$ on the w-plane. If the side BC were shifted to B_1C_1 at $y = 2\pi$, we would get the complete dotted circles $u^2 + v^2 = 1$ and $u^2 + v^2 = e^2$. Since $\frac{v}{u} = \tan y = \tan (y + 2n\pi)$,

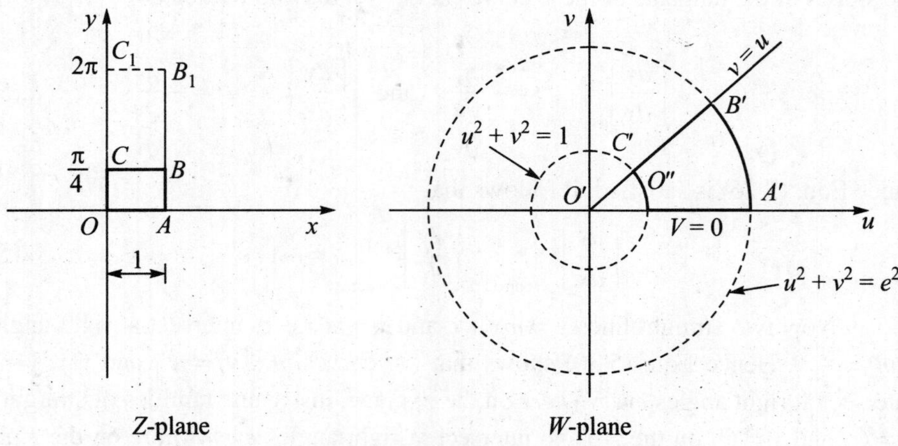

Fig. 5.11. Exponential transformation $w = e^z$

it follows that as side C_1B_1 is shifted from $y = 0$ to $y = \infty$, the section $O''A'$ of the radius vector will go round and round the point O' (origin) on the w-plane, but will always lie between the circles of radii 1 and e.

On the z-plane as the width of the strip is increased by shifting side AB from $x = 0$ to $x = \infty$, the radius of the circle on the w-plane increases from 1 to ∞. Thus the entire half of the z-plane corresponding to positive values of x is mapped on to the w-plane *outside* the circle of unit radius. The remaining half of the z-plane for negative values of x is plotted inside the circle $u^2 + v^2 = 1$ on the w-plane. However, the point O' (for which $w = 0$) has no finite value of z corresponding to it, because for it, $z = -\infty$.

5.14 LOGARITHMIC TRANSFORMATION $w = \log_e z$

This is the inverse of the exponential transformation, because now

$$w = \log_e z$$

or
$$z = e^w = e^u (\cos v + j \sin v) \quad \text{...(5.33)}$$

so that
$$x = e^u \cos v,\ y = e^u \sin v,$$

$$u = \log_e \sqrt{x^2 + y^2}\,,$$

$$v = \tan^{-1} \frac{y}{x} \quad \text{...(5.34)}$$

These relations show that Fig. 5.11 is applicable to this case, if w, u and v are replaced by z, x and y respectively, and vice versa.

If necessary, to fit boundary conditions in a problem, we can introduce constants into Eqns. (5.31) and (5.33) and write them as

$$w = C_1 e^{(z + C_2)}$$

and
$$w = C_1 \log_e z + C_2\,, \quad \text{...(5.35)}$$

without altering the general nature of the transformations. As an illustration we apply the logarithmic transformation to the case of a coaxial cable.

If the planes in Fig. 5.11 are moved parallel to themselves, so that origins O and O' describes the z-axis for each plane, then with one plane we shall get an infinitely long box with rectangular section (like OAB_1C_1), and with the other, a coaxial cylinder (see Fig. 5.12). Let R_1 be the outer radius of the inner cylinder and R_2 the inner radius of the outer cylindrical conductor. Then we assume the transformation $w = C_1 \log_e z + C_2$, which transforms these cylinders on the z-plane to a rectangle on the w-plane. With the notation of Fig. 5.12, we can express z in the polar form as

$$z = x + jy = Re^{j\phi}$$

where
$$R = \sqrt{x^2 + y^2}\,,\ \phi = \tan^{-1} \frac{y}{x}$$

Then
$$w = C_1 (\log_e R + j\phi) + C_2,$$

or
$$u = C_1 \log_e R + C_2,\ v = C_1\phi \quad \text{...(5.36)}$$

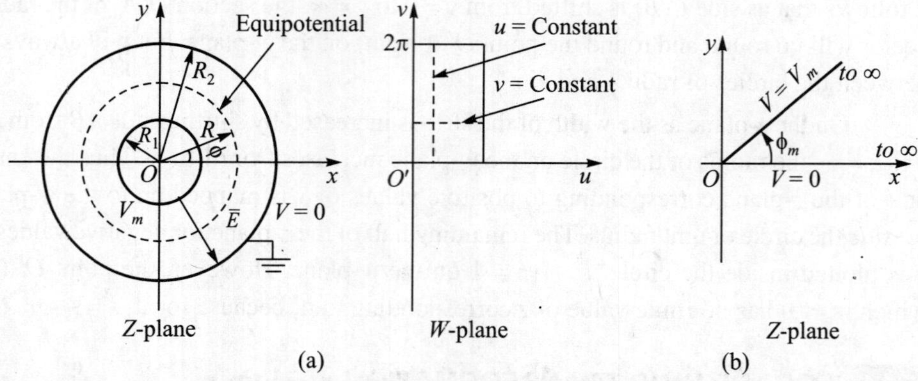

Fig. 5.12. Logarithmic transformation

5.14.1 (i) *u* = Constant as Equipotential

If we choose u = constant to represent an equipotential surface then, as seen from Eqn. (5.36), it makes the cylinder R = constant on the z-plane as an equipotential surface. Assume that the inner cylinder is at potential V_m volts and the outer cylinder, being assumed to be grounded, is at zero potential. Substituting these values in Eqn. (5.36) for u

$$V_m = C_1 \log_e R_1 + C_2$$
$$0 = C_1 \log_e R_2 + C_2$$

Solving we get $$C_2 = -\frac{V_m}{\log_e \dfrac{R_2}{R_1}},$$

$$C_2 = V_m \frac{\log_e R_2}{\log_e \dfrac{R_2}{R_1}}, \qquad \text{...(5.37)}$$

Then Eqn. (5.36) becomes

$$V = -\frac{V_m \log_e R}{\log_e \dfrac{R_2}{R_1}} + \frac{V_m \log_e R_2}{\log_e \dfrac{R_2}{R_1}}$$

or $$V = V_m \frac{\log_e (R_2/R)}{\log_e (R_2/R_1)} \qquad \text{...(5.38)}$$

This is the same as Eqn. (4.41). Then using Eqn. (1.104), we get

$$\overline{\boldsymbol{E}} = -\frac{\partial V}{\partial R}\overline{\boldsymbol{u}}_R = \frac{V_m}{R \log_e \dfrac{R_2}{R_1}}\overline{\boldsymbol{u}}_R,$$

$$R_1 < R < R_2 \qquad \text{...(5.39)}$$

Since u = constant is an equipotential, it follows from the orthogonal properties of u and v, that v = constant will give the direction of electric flux lines or the electric intensity. At any point for which $R_1 < R < R_2$, we can write in Cartesian co-ordinates

$$E_x = -\frac{\partial u}{\partial x}, \quad E_y = -\frac{\partial u}{\partial y} \qquad \text{...(5.40)}$$

Then using the Cauchy-Riemann Eqns. (5.26) and (5.40) and (4.3), we can change the total differential dv as follows:

$$dv = \frac{\partial v}{\partial x}dx + \frac{\partial v}{\partial y}dy = -\frac{\partial u}{\partial y}dx + \frac{\partial u}{\partial x}dy$$

$$= E_y\,dx - E_x\,dy$$

or

$$\varepsilon\,dv = D_y\,dx - D_x\,dy \qquad \text{...(5.41)}$$

If we consider two direction lines v and $v + dv$ as in Fig. 5.13, then they form a tube of flux as shown in Fig. 4.2. The Eqn. (5.41) indicates that the flux *entering* the volume shown by dotted lines is given by the right-hand side of the equation. Now a positive value of the flux $d\psi$ indicates the flux *leaving* the volume as shown in Fig. 5.13. This means that the right-hand side of Eqn. (5.41) is $(-\,d\psi)$. (There is no flux across the direction lines, and total flux leaving volume must be zero as no charge is enclosed).

Hence integrating

$$\int \varepsilon\,dv = -\int d\psi \text{ and } \varepsilon v = -\,\psi \qquad \text{...(5.42)}$$

where the constant of integration is made zero by arbitrarily choosing $\psi = 0$ for $v = 0$.

Combining Eqn. (5.42) with Eqns. (5.36) and (5.37)

$$\psi = -\,\varepsilon v = -\,\varepsilon\,C_1\phi = \frac{V_m\,\varepsilon\phi}{\log_e \dfrac{R_2}{R_1}} \qquad \text{...(5.43)}$$

If ψ_m is the total electric flux through any cylinder of radius R lying between R_1 and R_2, the length of the cylinder in the z-direction being one metre, then it corresponds to an angle $\phi = 2\pi$. But by Gauss's law, this ψ_m must then be equal to the charge q_l coulomb/metre placed on the inner conductor, which has produced the potential V_m on it. Hence from Eqn. (5.43) we can write, for the coaxial cable

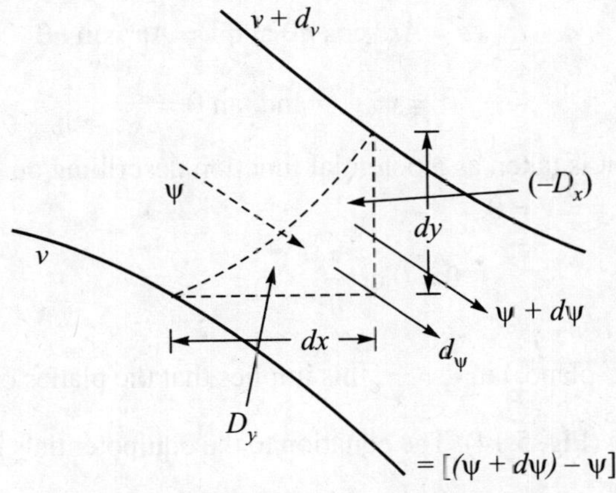

Fig. 5.13. Interpretation of Eqn. (5.40)

$$\text{Capacitance per metre} = \frac{q_l}{V_m} = \frac{\Psi_m}{V_m} = \frac{2\pi\varepsilon}{\log_e \dfrac{R_2}{R_1}} \text{ (F/m)} \qquad ...(5.44)$$

This is the same as Eqn. (4.42).

5.14.2 (ii) *v* = Constant as Equipotential

If instead of u, we make v = constant as an equipotential, then as shown in Fig. 5.12 (*b*), on the z-plane this makes the two radial lines as equipotentials. This is the case of two conducting semi-infinite planes extending from the origin to infinity, but *insulated from each other along the z-axis.*

Now we have $V = 0$ for $\phi = 0$ and $V = V_m$ for $\phi = \phi_m$ so that equation $v = C_1\phi$ becomes

$$V_m = C_1\phi_m \text{ or } C_1 = \frac{V_m}{\phi_m}$$

and
$$V = \frac{V_m}{\phi_m}\phi \qquad ...(5.45)$$

and
$$\overline{E}_R = 0,$$

$$\overline{E}_\phi = \frac{1}{R}\frac{\partial V}{\partial \phi}\overline{u}_\phi = -\frac{V_m}{R\phi_m}\overline{u}_\phi \qquad ...(5.46)$$

$$\overline{E}_z = 0$$

5.15 POWER FUNCTION TRANSFORMATION *w* = *Az*ⁿ

This transformation is best handled by expressing z in the polar form as $z = r(\cos\theta + j \sin\theta)$ and using De Moivre's theorem, Eqn. (1.37). Then assuming that A and n are real numbers, we write

$$w = Ar^n (\cos\theta + j\sin\theta)^n$$
$$= Ar^n (\cos n\theta + j \sin n\theta)$$

so that
$$u = Ar^n \cos n\theta \text{ and } v = Ar^n \sin n\theta \qquad ...(5.47)$$

and
$$r^2 = x^2 + y^2 \text{ and } \tan\theta = \frac{y}{x} \qquad ...(5.48)$$

If v = constant is taken as a potential function describing an equipotential, then it is obvious that for $v = V = 0$,

$$\theta = 0 \text{ or } \frac{\pi}{n} \qquad ...(5.49)$$

Since on the z-plane, $\tan\theta = \dfrac{y}{x}$, this implies that the planes $\phi = 0$ and $\phi = \dfrac{\pi}{n}$ are *both* at zero potential (Fig. 5.14). The equation to the equipotentials is then obtained by putting $n = \dfrac{\pi}{\phi}$ in Eqn. (5.47) as

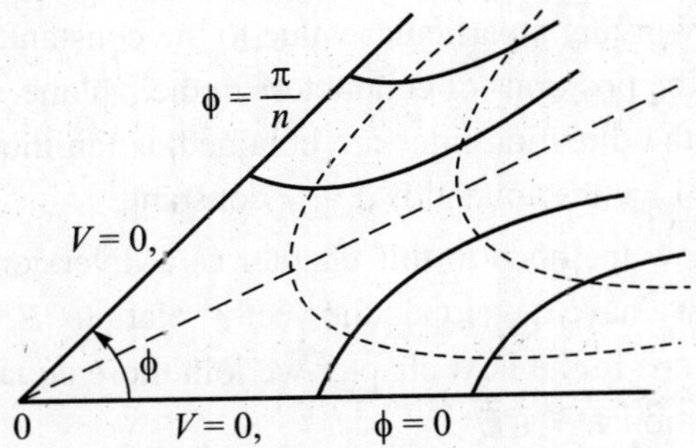

Fig. 5.14. Power function transformation

$$v = Ar^{\frac{\pi}{\phi}} \sin \frac{\pi\theta}{\phi} = \text{Constant} \quad ...(5.50)$$

The direction lines are obtained by putting u = constant and they are

$$u = Ar^{\frac{\pi}{\phi}} \cos \frac{\pi\theta}{\phi} = \text{Constant} \quad ...(5.51)$$

The equipotentials and the direction lines are symmetrical with respect to the bisector of the angle ϕ. Since the two planes forming the angular space are at zero potential, there must be some point on this bisector, which is at a potential other than zero, and therefore it produces this electric field in the wedge formed by the two planes.

5.16 SUMMARY AND COMMENTS

The two methods of approach to the problems of electric field outlined in this chapter are based on the requirement that a conductor must be an equipotential surface. In the method of images, we tried to find an *additional* charge distribution which, along with the original (given) charge distribution, will produce an equipotential surface at the correct potential at the places occupied by the conductors. When this situation is achieved, we remove the conductors, and proceed to find the field produced by the original and the additional (image) charges. However, this field is plotted *only* on that side (front) of the conductor on which the original charge is situated. The reason for this is that the conductor acts as an electrostatic screen, preventing the effect of the original charges being felt behind the conductor.

Considering the vast size of the earth, as compared to any electrical devices like antennas above the ground, the earth can be regarded as an infinite conducting plane, and the method of images can be used to plot the electric field above the ground.

The need for having an equipotential at the location of the conductor is achieved in the method of conformal transformations by mathematical manipulation between z- and w-planes instead of using image charge!! On the w-plane, the lines u = constant and v = constant form a grid consisting of families of mutually perpendicular straight lines parallel to the axis. The fact that if the Cauchy-Riemann conditions are satisfied by the transforming function, the functions $u(x, y)$ = constant and $v(x, y)$ = constant give curves on the z-plane, which are orthogonal to each other, enables us to choose either u or v as potential function. If by giving a particular value to the constant, this function can be made to coincide with the positions of conductors in the z-plane, then the equations to the equipotentials and the direction lines are obtained as families of curves on the z-plane in the form $u(x, y)$ = constant and $v(x, y)$ = constant.

Having learnt new methods in this chapter as a diversion, we go back to the main stream, in which we have developed the vector relations $\overline{E} = -\overline{\nabla}V$ and Gauss's law in the form $\overline{\nabla} \cdot \overline{D} = \rho$. In the next chapter we join these equations into one single equation using the equation $\overline{D} = \varepsilon\overline{E}$.

EXERCISE 5

1. In Cartesian co-ordinates, two conducting infinite planes are located at $x = 0$ and $y = 0$. If a charge + $(4\pi\,\varepsilon_0)$ coulomb is located at the point $\left(\frac{1}{2}, \frac{1}{2}, 0\right)$, find the electric field intensity at the point (1, 1, 0), metre being the unit of length.
2. In Fig. 5.2, if the angle $ACB = 45°$ and a charge $(+\,Q)$ coulomb is located at O where angle $OCB = 40°$, find the locations of the image charges and their magnitudes and signs.
3. For the oppositely charged infinite, parallel lines, use the data contained in Eqns. (3.40) and (3.41) to verify Eqns. (5.4).
4. In Fig. 5.4 (*b*), show that if the cylinder were not earthed, the field would continue to be the same inside the cylinder, but there would also be a field outside the cylinder. Sketch this field giving reasons for the shape and directions of the electric field lines.
5. With the notation of Fig. 5.3, assume that the sphere is not grounded, but is charged to a potential V volt. If a point charge $(+\,Q)$ coulomb is located at P, find the magnitude and locations of the point charges which will be needed to analyse the field at points lying outside the sphere, assuming that the conducting sphere is removed for this analysis.
6. A conducting sphere of radius a metre is located with its centre at the origin of spherical co-ordinates. If a positive charge $(+\,Q)$ is placed at a distance D from the centre of the sphere and $D > a$, then using the method of Section 5.8, find an expression for the induced charge density at any point on the sphere. Integrating this over the whole surface of the sphere, find the total charge induced on the sphere.
7. A conducting cylinder of infinite length and of radius a metre is located with its axis along the z-axis. If a uniform, line charge of density $(+\,q_l)$ C/m, and of infinite length, is located parallel to the z-axis at a distance D from the axis of the cylinder, then using the method of Section 5.8, find the expression for the induced charge density per metre length at any point on the surface of the cylinder. Integrating this over the circumference of the cylinder, find the total charge density induced per metre length of the cylinder.

8. Using the method of images, find the capacitance per metre length of two infinitely long, parallel cylindrical conductors, each of radius a metre, and with their axes separated by a distance D metre.

9. (*a*) Show that the capacitance of two conducting spheres, each of radius a metre, with their centres D metre apart ($D \gg 2a$), is given, to a first approximation, as

$$C \approx \frac{2\pi \varepsilon_0 a}{1 - \dfrac{a}{D}} \approx 2\pi \varepsilon_0 a \quad \text{farad}$$

(*b*) If in the above example, $D = 2a$, but there is an infinitesimally small insulating gap between the two spheres, show that the capacitance now is exactly given by

$$C = 8\pi \varepsilon_0 a \log_e 2 \quad \text{farad}$$

[**Hint:** Assume their potentials to be $-\dfrac{V_m}{2}$ and $+\dfrac{V_m}{2}$ with respect to the central plane, due to line charges $-q_l$ and $+q_l$ C/m. Use Sec. 5.6 to locate the image charges needed to make cylinder an equipotential, remembering that for points outside a cylinder, an enclosed charge acts as if it is located on the axis. Relate $\dfrac{V_m}{2}$ to respective image line charges by Eqn. (3.39).]

10. Using the hyperbolic cosine transformation $z = a \cosh w$, show that the curves u = constant represent confocal ellipses, and v = constant represent confocal hyperbolas. Show that in the limit when $u = 0$, the ellipse degenerates into a straight line, and the ellipses and hyperbolas are the equipotentials and the direction lines shown in Figs. 3.5 and 2.8.

11. Using the results of Ex. 5, Prob. 10 above, find the capacitance per metre (along z-axis) of a cylindrical capacitor which is formed by a flat, plane strip of width $2a$ lying symmetrically on the plane $y = 0$, and the section of the elliptical cylinder being an ellipse $u = \text{constant} = \cosh^{-1} \dfrac{a}{d}$.

12. For the power function transformation $w = \sqrt{2}z^{\frac{1}{2}}$, use the potential function as u = constant. Hence show that for $u = 0$, the potential function degenerations into the semi-infinite line of Ex. 2, Prob. 9. Verify that $v = \text{constant} = \sqrt{c}$ gives the same equation for the direction line as was obtained in Ex. 2, Prob. 9.

[**Hint:** Write the transformation as $z = \dfrac{w^2}{2}$ and equate real and imaginary parts.]

13. For the power function $w = \dfrac{1}{z}$, find the equation to the family of straight lines in the w-plane, which transforms into a family of circles in the z-plane, and vice versa.

CHAPTER

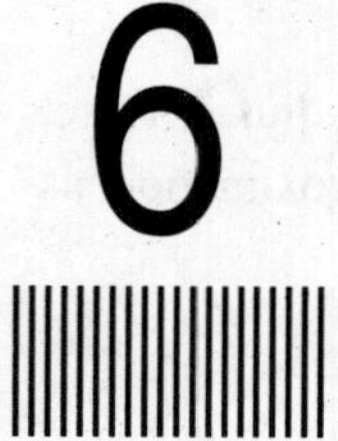

6 POISSON'S EQUATION AND LAPLACE'S EQUATION

6.1 POISSON'S EQUATION

In a linear, homogeneous and isotropic medium, the permittivity ε is a constant and a real, scalar quantity. With this assumption, we combine Eqns. (4.3), (3.18) and (4.48) and get

$$\overline{D} = \varepsilon\,\overline{E}, \quad \overline{E} = -\overline{\nabla} V$$

and

$$\nabla \cdot \overline{D} = \rho$$

or

$$\overline{E} = \frac{\overline{D}}{\varepsilon} \text{ and gradient } V = \overline{\nabla} V = -\overline{E} = -\frac{\overline{D}}{\varepsilon}$$

Taking divergence of both sides

$$\text{div (grad } V) = \text{div}\left(-\frac{\overline{D}}{\varepsilon}\right)$$

or

$$\overline{\nabla} V \cdot (\overline{\nabla} V) = -\frac{1}{\varepsilon}(\overline{\nabla} \cdot \overline{D}) = -\frac{\rho}{\varepsilon}$$

or

$$\overline{\nabla}^2 V = -\frac{\rho}{\varepsilon} \qquad \text{...(6.1)}$$

This equation was derived by the French mathematician Simeon Denis Poisson (1781–1840) and is known as Poisson's equation. It is a result that is true at every point in space. The left hand side is a shorthand notation for the longer expressions given by Eqns. (1.98), (1.107) or (1.112), depending upon the system of co-ordinates chosen. Hence Eqn. (6.1) is really a partial differential equation of the second order. It relates some aspect of the auxilliary scalar quantity V (or the potential at that point) which we had introduced for the study of the electric field, with another scalar quantity, the flux D of a vector or the electric displacement [equal to charge as per Eqn. (4.1)] throughout the space. Any relation that satisfies the boundary conditions, that is, the given charge distribution in space and the potentials of the equipotential surfaces formed by the presence of conductors, and also Eqn. (6.1) is a solution of Poisson's equation. It is found that for a given charge distribution and conductor location is space, there is one, *and only one*, possible solution of Poisson's equation, indicating that the resultant potential distribution is unique. This Poisson's equation is the most powerful mathematical tool available for the study of problems in electrostatics. All the solutions that we have so far obtained in this book for different charge distributions and conductor locations are really solutions of Poisson's equation, though we have obtained the solutions using simpler mathematical tools.

6.2 CHILD'S LAW FOR SPACE-CHARGE LIMITED CURRENT IN A HIGH VACUUM DIODE

Solving partial differential equations with three variables is a very tricky business. To bring the problem within the scope of this elementary textbook, we shall take up a one variable Poisson's equation for solution by assuming that a high vacuum doide is formed with cathode and anode in the form of infinite, parallel planes at $y = 0$ and $y = d$, respectively, and having potentials $V = 0$ and $V = V_p$ (Fig. 6.1), where V_p is a constant.

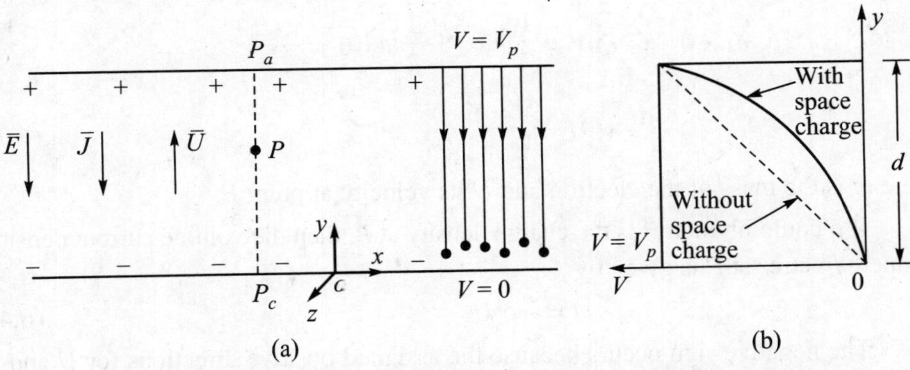

Fig. 6.1. (a) For space-charge limited current analysis for high vacuum diode (b) Potential variation between cathode and anode

Since there is no variation of V in the x- and z- directions, $\frac{\partial V}{\partial x}$ and $\frac{\partial V}{\partial z}$ are equal to zero, and Poisson's equation becomes

$$\frac{\partial^2 V}{\partial y^2} = -\frac{\rho}{\varepsilon} \qquad ...(6.2)$$

To solve this equation, three further assumptions are made:

(*i*) Only electrons are present in space between the cathode and the anode.

(*ii*) The number of electrons available near the cathode is very much greater than number demanded by the anode to form the plate current.

(*iii*) The initial velocity of all electrons forming the plate current is zero, when they are at the cathode.

Apparently since there is a current flowing, this cannot be a problem in electrostatics. Hence, we examine this aspect first.

The field intensity $\overline{E}$ between the infinite parallel planes will be uniform everywhere when there are no electrons present (that is, with cold cathode and not thermionic emission of electrons) and the potential will vary linearly from cathode to anode as shown by the dotted line in Fig. 6.1 (b). However, when a very large cloud of electrons is formed near the cathode, all direction lines starting from the anode end up on electrons (negative charges) and none reach the surface of the cathode. Hence $E = -\frac{\partial V}{\partial y} = 0$ at the surface of the cathode ($y = 0$) when the current is space-charge limited. This is shown at right hand end of Fig. 6.1 (a) and in Fig. 6.1 (b).

Under the influence of the field intensity $\overline{E}$, an electron with negative charge ($-e$) coulomb travels from point P_c on cathode towards point P_a on anode in a straight line perpendicular to the planes. Let the potential at any point P be denoted as V. Then the law of conservation of energy requires that the loss of potential energy of the electron while going from P_c to P must equal the gain in its kinetic energy, or

(Potential energy at P_c) – (P.E. at P) = (Kinetic energy at P) – (K.E. at P_c)

or
$$(-e) \times 0 - (-e)\, V = \frac{1}{2} m U^2 - \frac{1}{2} m\,(0)^2$$

or
$$U = \sqrt{\frac{2eV}{m}} \qquad \text{...(6.3)}$$

where m is the mass of the electron and U its velocity at point P.

If ρ coulomb/metre^3 is the charge density at P, then the volume current density J ampere /metre^2 at P is given by

$$J = -\rho\, U \qquad \text{...(6.4)}$$

The negative sign occurs because the assigned positive directions for U and J are opposite to each other as indicated by arrows in Fig. 6.1 (a).

Now J must be a constant everywhere, because of Kirchhoff's current law [Eqn. (4.29)], as there cannot be any accumulation of charge anywhere. This means that when one electron moves from a region, its place is taken by another electron, and thus *the charge density at a point does not change with time*. This constancy of ρ with time enables us to treat this as a problem in electrostatics.

Though the charge density at any point does not vary with time, it must be remembered that charge density decreases as the point P is shifted from cathode towards the anode. This is because the electrons are accelerated by the electric field and therefore their velocity U increases from P_c to P_a. Since the product of ρ and U is constant by Eqn. (6.4), it follows that ρ must decrease from P_c to P.

Then combining Eqns. (6.4) and (6.3), we get

$$-\rho = \frac{J}{U} = J\sqrt{\frac{m}{2\text{eV}}}$$

and this changes Poisson's equation (6.2) into

$$\frac{d^2V}{dy^2} = J\left(\frac{1}{\varepsilon_0}\sqrt{\frac{m}{2e}}\right) V^{-\frac{1}{2}} = J\, K_1\, V^{-\frac{1}{2}} \qquad \text{...(6.5)}$$

where K_1 is the constant term within brackets. Though J is also a constant, it is not included in K_1, because we are interested in finding the volume of J in terms of the anode potential V_P. To solve the differential equation [Eqn. (6.5)], multiply both sides by $2\left(\frac{dV}{dy}\right) dy$ and integrate. Then

$$\int 2\left(\frac{dV}{dy}\right)\left(\frac{d^2V}{dy^2}\right) dy = 2K_1 J \int V^{-\frac{1}{2}}\left(\frac{dV}{dy}\right) dy$$

$$= 2K_1 J \int V^{-\frac{1}{2}}\, dV$$

Therefore, $$\left(\frac{dV}{dy}\right)^2 = 2K_1 J \frac{V^{\frac{1}{2}}}{1/2} + C_1$$

But at $$y = 0,\ V = 0 \text{ and } \frac{dV}{dy} = 0.$$

Hence $$C_1 = 0$$

Now $$\frac{dV}{dy} = \sqrt{4K_1 J}\, V^{\frac{1}{4}}$$

so that $$\int V^{-\frac{1}{4}}\, dV = \sqrt{4K_1 J} \int dy$$

or $$\frac{4}{3} V^{\frac{3}{4}} = \sqrt{4K_1 J}\, y + C_2$$

Since at $y = 0$, $V = 0$, therefore $C_2 = 0$

Also at $y = d$, $V = V_p$.

This gives $$J^{\frac{1}{2}} = \left(\frac{4}{3\sqrt{4K_1}\, d}\right) V_p^{\frac{3}{4}} = \sqrt{K}\, V_p^{\frac{3}{4}} \text{ say,}$$

where $\sqrt{K}$ is the constant term within brackets

Squaring $$J = K\, V_p^{\frac{3}{2}} \qquad \text{...(6.6)}$$

This shows that unlike Ohm's law, in a high vacuum diode, the current density is proportional to the 3/2 power of the anode-to-cathode voltage. This Eqn. (6.6) was derived by C.D. Child in 1911 and is known as Child's law.

6.3 LAPLACE'S EQUATION

In a region where the charge density is zero, the Poisson's equation becomes

$$\nabla^2 V = 0 \qquad \text{...(6.7)}$$

This is known as Laplace's equation. It is the most powerful mathematical tool available for the study of electrostatic problems for the region, where there is no charge present. In practical problems, since the charges are confined to small regions while major part of the space is charge-free, it is obvious that Laplace's equation has far greater utility than Poisson's equation.

There can be an infinite variety of solutions for the second order partial differential Eqn. (6.7). Out of these, we have to select the one which satisfies the boundary conditions of equipotential surfaces at certain potentials in the regions occupied by the conductors. Very often the solutions involve complicated integrals for which exact solutions are not easy to find. In such cases, approximate solutions can be found using a computer or a graphical method.

Solutions of Laplace's equation, which are continuous through the second derivative, are called harmonic functions. Thus depending upon the co-ordinate system used, we get rectangular harmonics, cylindrical harmonics and spherical harmonics as solutions of the Laplace's equation.

6.4 PRODUCT SOLUTION OF LAPLACE EQUATION BY SEPARATION OF VARIABLES IN CARTESIAN CO-ORDINATES

In Cartesian co-ordinates, Laplace's equation is

$$\frac{\partial^2 V}{\partial x^2}+\frac{\partial^2 V}{\partial y^2}+\frac{\partial^2 V}{\partial z^2}=0 \qquad \text{by Eqn. ...(1.98)}$$

Assume a product solution with variables separated in the form

$$V = X(x)\, Y(y)\, Z(z) \qquad \text{...(6.8)}$$

in which $X(x)$ is a function involving x only, function $Y(y)$ involves y only and function $Z(z)$ involves z only. If we use the abbreviation X' for $\frac{\partial X(x)}{\partial x}$ and X'' for $\frac{\partial^2 X(x)}{\partial x^2}$, and similarly for others, we get by partial differentiation of Eqn. (6.8)

$$\left.\begin{aligned} \frac{\partial V}{\partial x} &= Y\,Z\,X' & \frac{\partial^2 V}{\partial x^2} &= Y\,Z\,X'' \\ \frac{\partial V}{\partial y} &= Z\,X\,Y', & \frac{\partial^2 V}{\partial y^2} &= Z\,X\,Y'' \\ \frac{\partial V}{\partial z} &= X\,Y\,Z', & \frac{\partial^2 V}{\partial z^2} &= X\,Y\,Z'' \end{aligned}\right\} \qquad \text{...(6.9)}$$

where $X(x)$, $Y(y)$, $Z(z)$ have been abbreviated to X, Y, Z. Substituting Eqn. (6.9) in Eqn. (1.98) above, and dividing throughout by $X\,Y\,Z$, we get

$$\frac{X''}{X}+\frac{Y''}{Y}+\frac{Z''}{Z}=0 \qquad \text{...(6.10)}$$

Thus we have succeeded in separating the variables, because the first term of Eqn. (6.10) involves x only, the second term involves y only and the third term involves z only. The complete separation into three equations, each involving one variable only, is achieved with a beautifully precise knife-like argument.

Equation (6.10) must be true for *all* values that x, y and z may have. Also if, say y is a constant, then the second term *must be constant*, but the first and third terms *may vary* as x and z are varied, because they do not involve y.

We extend this argument and maintain two variables, say y and z, as constant and allow only x to vary. Then the second and third terms are constant, and only the first term *may vary* as x is varied. But Eqn. (6.10) shows that the sum of all the three terms is always a constant (equal to zero). Hence, even when x is varied, the first term must always be equal to one constant only, say k_x^2.

By a similar argument, we arrive at the conclusion that even when y and z are varied, the second and third terms also must be constants, say k_y^2 and k_x^2. Then the variables get completely separated because Eqn. (6.10) splits up into three independent equations

$$\left.\begin{aligned}\frac{1}{X}\frac{d^2X}{dx^2} &= k_x^2,\\ \frac{1}{Y}\frac{d^2Y}{dy^2} &= k_y^2,\\ \frac{1}{Z}\frac{d^2Z}{dz^2} &= k_z^2\end{aligned}\right\} \qquad ...(6.11)$$

where
$$k_x^2 + k_y^2 + k_z^2 = 0 \qquad ...(6.12)$$

But the sum of three constants cannot be equal to zero unless at least one of them (and at most two of them) is negative. Since $\sqrt{-1} = j$, it follows that at least one (and at most two) of the constants k_x, k_y and k_z must be imaginary.

An obvious solution to a differential equation like

$$\frac{\partial^2 X}{\partial x^2} = k_x^2 X$$

is $X = e^{k_x x}$ or $e^{-k_x x}$ or sums and differences of the two. In general the solution can be written as

$$X = A_x \cos h\, k_x x + B_x \sin h\, k_x x \qquad ...(6.13)$$

and the complete solution for Eqn. (6.8) becomes

$$V = [A_x \cos h\, k_x x + B_x \sin h\, k_x x]\, [A_y \cos h\, k_y y + B_y \sin h\, k_y y]\, [A_z \cos h\, k_z z + B_z \sin h\, k_z z] \qquad ...(6.14)$$

Of the constants k_x, k_y and k_z, at least one (or at most two) is imaginary. For that imaginary constant, by Eqn. (1.39), the hyperbolic sine and cosine terms change into simple trigonometric sin and cos terms which, unlike sin h and cos h, are *periodic* functions. This term then should be used for that direction in which the voltage has repeated zeros. The sin h and cos h terms are useful for those directions, in which the boundary extends to infinity. Finally, it may be noted that in Eqn. (6.11), if any separation constant, say k_x, becomes zero, we do not get any sinusoidal or hyperbolic sinusoidal terms at all, but a linear relation $X = ax + b$.

6.5 RECTANGULAR HARMONIC SOLUTION

A simple two-dimensional problem is solved to illustrate the method developed in the previous section. Or rather we have chosen such boundary conditions for this problem, that the restrictions imposed on A, B, k in the previous section could be applied here. Let $y = 0$ and $y = b$ be two conducting planes extending from $x = 0$ to $x = \infty$ in the x-direction, and from $z = -\infty$ to $z = +\infty$ in the z-direction. A conducting strip of width from $y = 0$ to $y = b$ lies in the plane $x = 0$ and extends from $z = -\infty$ to $z = +\infty$. The system thus forms a U-shaped trough as shown in Fig. 6.2. If an electric field exists in this trough, we wish to find the equation to the equipotential surface, for which the voltage is V_0 at the point for which $x = a$, $y = b/2$ and z = any value, assuming the conducting trough to be at $V = 0$.

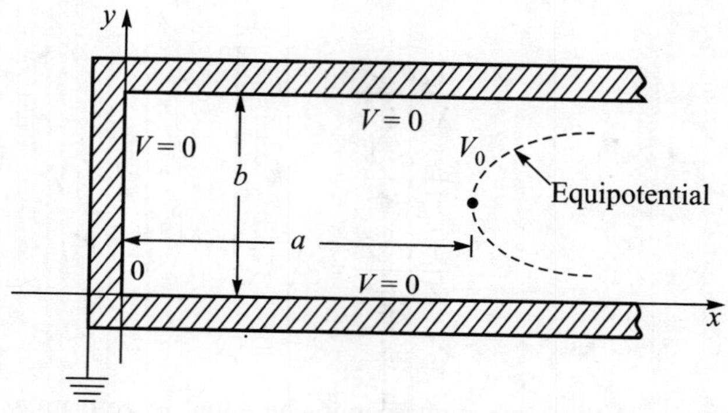

Fig. 6.2. Rectangular harmonics solutions

We notice that the problem involves only the two variables x and y, as there is no variation of potential in the z-direction. From Eqns. (6.14) and (6.12), we see that the solution will be of the form

$$V = [A_x \cos h\, k_x x + B_x \sin h\, k_x x]\,[A_y \cos h\, k_y y + B_y \sin h\, k_y y] \quad \text{...(6.15)}$$

and

$$k_x^2 + k_y^2 = 0 \quad \text{...(6.16)}$$

The constants A_x, A_y, B_x, B_y, k_x and k_y are to be determined from the boundary conditions.

As before, either k_x^2 or k_y^2 must be negative. Since $V = 0$ at $y = 0$ and again $V = 0$ at $y = b$, this presence of repeated zero values of V in the y-direction indicate that k_y^2 should be negative. Hence putting $k_x = k$ and $k_y = jk$ [so as to satisfy Eqn. (6.16)], Eqn. (6.15) becomes

$$V = [A_x \cos h\, kx + B_x \sin h\, k x]\,[A_y \cos k y + B_y j \sin k y]$$

Using the boundary condition that at $y = 0$, $V = 0$, we get

$$0 = [A_x \cos h\, kx + B_x \sin h\, kx]\, A_y \quad \text{...(6.17)}$$

Since Eqn. (6.17) must be true for all values of x, therefore $A_y = 0$.

Next we use in Eqn. (6.15) the condition that for $x = 0$, $V = 0$, we get

$$0 = A_x\,[B_y\, j \sin k\, y] \qquad \text{...(6.18)}$$

Since Eqn. (6.18) must be true for all values of y, where $0 < y < b$, therefore $A_x = 0$. The required solution then reduces to

$$V = (B_x\, B_y\, j) \sin h\; k\, x \sin k\, y$$

or

$$V = C_1 \sin h\; kx \sin k\, y \qquad \text{...(6.19)}$$

where C_1 is a constant.

The third boundary condition used is that at $y = b$, $V = 0$. Then Eqn. (6.19) will be true for all values of x if $\sin kb = 0$ or

$$k = \frac{\pi}{b} \qquad \text{...(6.20)}$$

Thus Eqn. (6.19) now becomes

$$V = C_1 \sin h\left(\frac{\pi\, x}{b}\right) \sin\left(\frac{\pi\, y}{b}\right) \qquad \text{...(6.21)}$$

Lastly, we determine the constant C_1 using the condition that at $x = a$, $y = \frac{b}{2}$, $V = V_0$. Hence

$$V_0 = C_1 \sin h\left(\frac{\pi\, a}{b}\right) \sin\left(\frac{\pi}{2}\right)$$

or

$$C_1 = \frac{V_0}{\sin h\left(\frac{\pi\, a}{b}\right)}$$

The final solution of Laplace's equation for the boundary conditions of this problem becomes

$$V = \frac{V_0}{\sin h\left(\frac{\pi\, a}{b}\right)} \sin h\left(\frac{\pi\, x}{b}\right) \sin\left(\frac{\pi\, y}{b}\right) \qquad \text{...(6.22)}$$

Putting $V = V_0$ we get the equation to the equipotential as

$$\sin h\left(\frac{\pi\, x}{b}\right) \sin\left(\frac{\pi\, y}{b}\right) = \sin h\left(\frac{\pi\, a}{b}\right) \qquad \text{...(6.23)}$$

The shape of the equipotential is shown in Fig. 6.2. We could place a conductor of the shape of the equipotential surface and put a charge on it to make its potential V_0. Then Eqn. (6.22) will give the value of the voltage at any point in the trough between the conductors at zero potential and this conductor at potential V_0.

6.6 SERIES SOLUTION USING RECTANGULAR HARMONICS

What will happen in the above example if the conductor at potential V_0, instead of having the shape of the equipotential [Eqn. (6.23)], is in the form of a plane $x = a$ as shown in Fig. 6.3? It is assumed that this plane extends from $z = -\infty$ to $z = +\infty$, and from $y = (0+)$ to $y = (b-)$, that is, it does not touch the grounded planes at $y = 0$ and at $y = b$, but is insulated from them by an infinitesimally thin layer of dielectric.

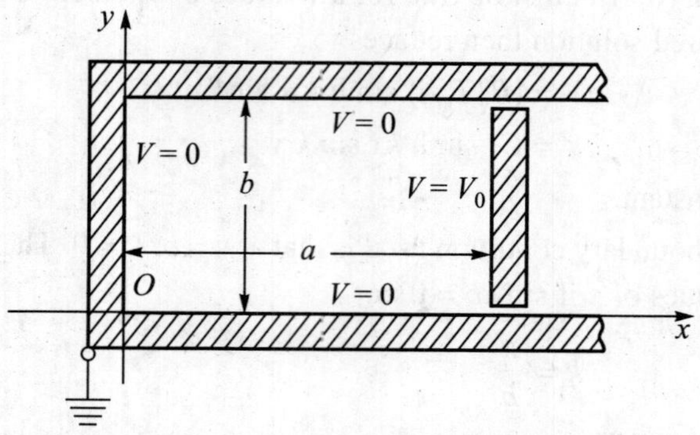

Fig. 6.3. Series solution with rectangular harmonics

Now one single term like Eqn. (6.22) will not suffice, because for $x = a$, y is not only equal to $\frac{b}{2}$ but y can have all values from 0 to b. Hence the solution begins in the same way as in the previous section, and all equations from (6.15) to (6.19) are the same as before.

However, now V has certain values for variation of y from $y = 0$ to $y = b$ (in this case, it is a constant $= V_0$). What values we give to V *outside* this range $y = 0$ to $y = b$ for other values of y in order to get a solution is immaterial, so long as $V = V_0$ *within* this range? This suggests two things. First, we modify Eqn. (6.20) so that sin kb may be equal to zero for values of y other than $y = b$, and write

$$k = \frac{n\pi}{b} \qquad \text{...(6.24)}$$

The second modification is the use of half-range Fourier series (see Sec. 1.10) with odd function symmetry as shown in Fig. 6.4. For this we modify Eqn. (6.21) and write

$$V = \sum_{n=1}^{\infty} C_n \sin h\left(\frac{n\pi\, x}{b}\right) \sin\left(\frac{n\pi\, y}{b}\right) \qquad \text{...(6.25)}$$

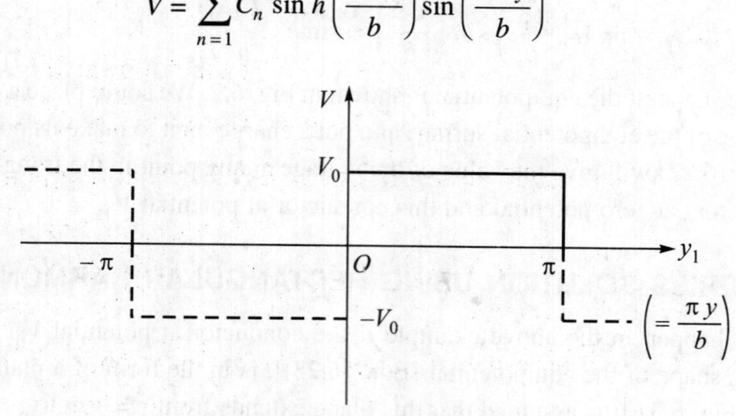

Fig. 6.4. Odd function symmetry

Now we apply the fourth boundary condition that $V = V_0$ for $x = a$ and $0 < y < b$ and get

$$V_0 = \sum_{n=1}^{\infty} C_n \sin h\left(\frac{n\pi\, a}{b}\right) \sin\left(\frac{n\pi\, y}{b}\right) \text{ for } 0 < y < b \qquad ...(6.26)$$

Also following the method of Sec. 1.10 and using Fig. 6.4, we write

$$\begin{aligned} F_0(y_1) &= V_0, & 0 < y_1 < \pi \\ F_0(y_1) &= -V_0, & -\pi < y_1 < 0 \\ F_0(y_1) &= F_0(y_1 + 2\pi k_1), & k_1 \text{ is any integer} \end{aligned} \qquad ...(6.27)$$

Then using Eqn. (1.64)

$$a_n = \frac{1}{\pi}\int_{-\pi}^{\pi} F_0(y_1) \cos ny_1 \, dy_1 = 0 \qquad ...(6.28)$$

and

$$\begin{aligned} b_n &= \frac{1}{\pi}\int_{-\pi}^{\pi} F_0(y_1) \sin ny_1 \, dy_1 \\ &= \frac{1}{\pi}\left[\int_{-\pi}^{0} (-V_0) \sin ny_1 \, dy_1 + \int_{0}^{\pi} (V_0) \sin ny_1 \, dy_1\right] \\ &= \frac{V_0}{\pi}\left\{\left[\frac{\cos ny_1}{n}\right]_{-\pi}^{0} - \left[\frac{\cos ny_1}{n}\right]_{0}^{\pi}\right\} \end{aligned}$$

$$\left.\begin{aligned} &\text{If } n \text{ is even, } b_n = \frac{V_0}{n\pi}[(1-1)-(1-1)] = 0 \\ &\text{If } n \text{ is odd, } b_n = \frac{V_0}{\pi}\left[\frac{1-(-1)}{n} - \frac{(-1)-(1)}{n}\right] = \frac{4\,V_0}{n\,\pi} \end{aligned}\right\} \qquad ...(6.29)$$

With the values of a_n and b_n as given by Eqns. (6.28) and (6.29), we can write

$$V_0 = \sum_{n=1}^{\infty} b_n \sin \frac{n\pi\, y}{b}$$

where

$$b_n = \begin{cases} 0 & \text{when } n \text{ is even} \\ \dfrac{4V_0}{n\pi} & \text{when } n \text{ is odd} \end{cases} \qquad ...(6.30)$$

Comparing Eqns. (6.26) and (6.30)

$$C_n \sin h\left(\frac{n\pi\, a}{b}\right) = \frac{4V_0}{n\pi} \text{ for } n \text{ odd,} \qquad ...(6.31)$$

$$C_n = 0 \text{ for } n \text{ even}$$

Putting this value of C_n in Eqn. (6.25) gives the potential at any point in the rectangular box of Fig. 6.3 as

$$V = \sum_{n \text{ odd}}^{\infty} \frac{4V_0}{n\pi} \frac{1}{\sin h\left(\dfrac{n\pi\, a}{b}\right)} \sin h\left(\frac{n\pi\, x}{b}\right) \sin\left(\frac{n\pi\, y}{b}\right) \qquad ...(6.32)$$

6.7 CYLINDRICAL HARMONICS

Assuming that there is no variation of V in the z-direction, we reduce the general Laplace's equation in cylindrical co-ordinates as given by Eqn. (1.107) to a two-dimensional form as

$$\frac{1}{R}\frac{\partial}{\partial R}\left(R\frac{\partial V}{\partial R}\right)+\frac{1}{R^2}\frac{\partial^2 V}{\partial \phi^2}=0 \qquad \text{...(6.33)}$$

We assume a product solution

$$V = f(R)\ F(\phi) \qquad \text{...(6.34)}$$

in which $f(R)$ is a function of R only $F(\phi)$ is some other function of ϕ only.

Then
$$\frac{\partial V}{\partial R} = F(\phi)\, f'(R)$$

and
$$\frac{\partial^2 V}{\partial \phi^2} = f(R)\ F''(\phi)$$

Substituting these values in Eqn. (6.33) and multiplying by $\dfrac{R^2}{f(R)\,F(\phi)}$

$$\frac{R^2}{f(R)}\frac{1}{R}\frac{d}{dR}[R\,f'(R)]+\frac{1}{F(\phi)}\frac{d^2F(\phi)}{d\phi^2}=0$$

Thus the variables are separated. Next by using the same argument as given for deriving

Eqns. (6.11) and (6.12), we get

$$\frac{R}{f(R)}\frac{d}{dR}[R\,f''(R)]=k^2$$

and
$$\frac{1}{F(\phi)}\frac{d^2F(\phi)}{d\phi^2}=-\,k^2 \qquad \text{...(6.35)}$$

Then carrying out the differentiation, the first equation in Eqn. (6.35) can be written as

$$R^2\frac{d\,[f'(R)]}{dR}+Rf'(R)=k^2 f(R) \qquad \text{...(6.36)}$$

6.7.1 (i) If $k = 0$

Equation (6.36) becomes

$$R\frac{d\,[f'(R)]}{dR}+f'(R)=0$$

or
$$\frac{d\,[f'(R)]}{f'(R)}+\frac{dR}{R}=0$$

Integrating,

$$\log_e [f'(R)] + \log_e R = \log_e A,$$

where A is a constant. This gives

$$\log_e [f'(R)] = \log_e \left[\frac{A}{R}\right]$$

or

$$f'(R) = \frac{d\, f(R)}{dR} = \frac{A}{R}$$

Therefore,

$$f(R) = A \log_e R + B \qquad ...(6.37)$$

When $k = 0$, $\dfrac{d^2 F(\phi)}{d\phi^2} = -k^2 = 0$ gives the solution

$$F(\phi) = C\phi + D \qquad ...(6.38)$$

Thus for $k = 0$,

$$V = (A \log_e R + B)(C\phi + D) \qquad ...(6.39)$$

6.7.2 (ii) If k ≠ 0

Equation (6.36) obviously has a solution

$$F(R) = (A_k R^k + B_k R^{-k}) \qquad ...(6.40)$$

Also the second equation of Eqn. (6.35) is the standard equation for simple harmonic oscillations of the form

$$\frac{d^2 F(\phi)}{d(\phi)^2} = -k^2 F(\phi)$$

for which the solution is

$$F(\phi) = (C_k \cos k\phi + D_k \sin k\phi) \qquad ...(6.41)$$

Putting in Eqns. (6.40) and (6.41) in Eqns. (6.34), we have another possible solution as

$$V = (A_k R^k + B_k R^{-k})(C_k \cos k\phi + D_k \sin k\phi) \qquad ...(6.42)$$

Equations (6.39) and (6.42) can be used to fit the boundary conditions of the problem by assigning suitable values to the constants. These are called the cylindrical (or circular) harmonic solutions of Laplace's equation.

6.8 CONDUCTING CYLINDER IN UNIFORM FIELD

To illustrate the application of cylindrical harmonics, we shall seek the effect of placing a conducting cylinder of radius a with its axis coinciding with the z-axis, the cylinder being infinitely long. It is assumed that in the absence of the cylinder, there is a uniform electric field $\overline{\boldsymbol{E}} = E_u \overline{\boldsymbol{u}}_x$ present *everywhere* (see Sec. 5.10).

Because of the cylindrical symmetry about the z-axis, we seek a solution of the Laplace's equation using cylindrical harmonics to satisfy the following boundary conditions:

(*i*) At very great distances from the axis of the cylinder (that is, for $R \to \infty$), the electric field intensity will be unaffected by the presence of the cylinder, and its value will be $E_u \overline{\boldsymbol{u}}_x$. Consider the uniform electric field shown in Fig. 6.5 (*a*). The equipotential surface will be a plane at right angles to the x-axis at points far away from the cylinder. Assume that the cylinder (of radius a) is earthed. Then if the potential at point P at a distance R from the origin O is V, it is equal to the potential difference between P' (on the same equipotential, and x-axis) and the point O' (where the cylinder crosses the x-axis). That is

$$V = -\int_{O'}^{P'} \overline{\boldsymbol{E}} \cdot \overline{\boldsymbol{ds}} = -E_u \int_{O'}^{P'} \overline{\boldsymbol{u}_x} \cdot \overline{\boldsymbol{ds}}$$

$$= -E_u\,(O'P') \approx E_u\,(OP'), \text{ as } R \gg a$$

or

$$V \approx -E_u x = -E_u R \cos\phi, \text{ as } R \to \infty \qquad \text{...(6.43)}$$

(*ii*) The potential, when $R = a$ must be equal to zero.

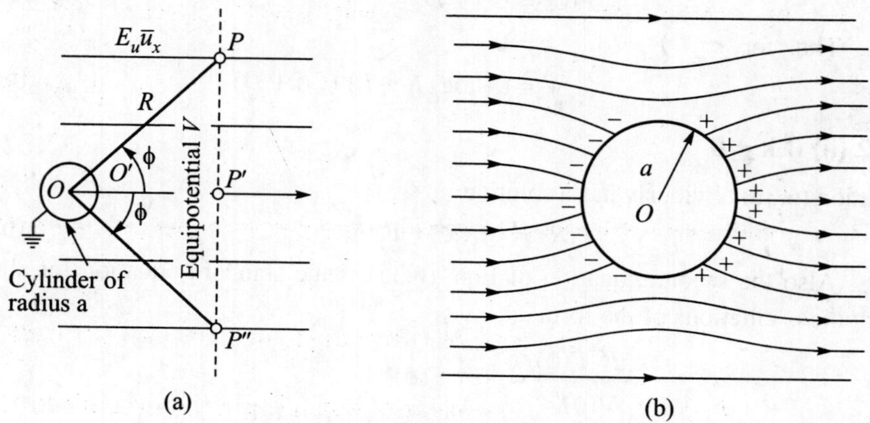

Fig. 6.5 (*a*) Equipotential in uniform field
(*b*) Field direction lines near conducting cylinder

The boundary condition (*i*) or Eqn. (6.43) indicates that a solution in the form of Eqn. (6.39) is not applicable in this case. Then choosing a solution in the form of Eqn. (6.42), it is obvious that k must be equal to 1, giving the solution in the form

$$V = \left(A_1 R + \frac{B_1}{R}\right)(C_1 \cos\phi + D_1 \sin\phi) \qquad \text{...(6.44)}$$

Comparing Eqns. (6.43) and (6.44), $A_1\,C_1 = -E_u$. Also since the potential at point P (R, ϕ) is the same as at the point P'' $(R, -\phi)$ as shown in Fig. 6.5(a), it follows that $D_1 = 0$. Then applying the second condition that the potential at $R = a$ must be equal to zero, we get from

$$V = A_1 C_1 R \cos\phi + \frac{B_1 C_1}{R} \cos\phi$$

or

$$0 = -E_u a \cos\phi + \frac{B_1 C_1}{a} \cos\phi$$

Therefore $B_1 C_1 = a^2 E_u$

Hence, the final form of this solution of Laplace's equation in cylindrical co-ordinates becomes

$$V = \left(\frac{a^2}{R} - R\right) E_u \cos\phi \qquad \text{...(6.45)}$$

From this, using Eqn. (1.104), the components of the electric field in the neighbourhood of the cylinder are given by

$$\left.\begin{aligned}\overline{E}_R &= -\frac{\partial V}{\partial R}\overline{u}_R = \left[\left(\frac{a^2}{R^2}+1\right)E_u\cos\phi\right]\overline{u}_R\\ \overline{E}_\phi &= -\frac{1}{R}\frac{\partial V}{\partial \phi}\overline{u}_\phi = \left[\left(\frac{a^2}{R^2}-1\right)E_u\sin\phi\right]\overline{u}_\phi\\ \overline{E}_z &= 0\end{aligned}\right\} \quad ...(6.46)$$

which are the same as Eqns. (5.21) and (5.22).

6.9 SPHERICAL HARMONIC SOLUTION FOR CONDUCTING SPHERE IN UNIFORM FIELD

Because of the spherical shape of the equipotential corresponding to the surface of the conductor, we use Laplace's equation in the spherical co-ordinates with the origin at the centre of the sphere, as given in Eqn. (1.112)

$$\frac{1}{r^2}\frac{\partial}{\partial r}\left(r^2\frac{\partial V}{\partial r}\right)+\frac{1}{r^2\sin\theta}\frac{\partial}{\partial\theta}\left(\sin\theta\frac{\partial V}{\partial\theta}\right)+\frac{1}{r^2\sin^2\theta}\frac{\partial^2 V}{\partial\phi^2}=0$$

We make the simplifying assumption that the electric field has polar axial symmetry, with the result that there is no variation in the ϕ direction. Thus the last term in the above equation can be put equal to zero. Before attempting to separate the variables r and θ in the first two terms, we modify the equation by the substitution $u = \cos\theta$, so that $\sin^2\theta = (1 - u^2)$ and $du = -\sin\theta\, d\theta$. Then we write Laplace's equation as

$$\frac{1}{r^2}\frac{\partial}{\partial r}\left(r^2\frac{\partial V}{\partial r}\right)+\frac{1}{r^2}\frac{\partial}{\sin\theta\,\partial\theta}\left(\sin^2\theta\frac{\partial V}{\sin\theta\,\partial\theta}\right)=0$$

or

$$\frac{\partial}{\partial r}\left(r^2\frac{\partial V}{\partial r}\right)+\frac{\partial}{\partial u}\left[(1-u^2)\frac{\partial V}{\partial u}\right]=0 \quad ...(6.47)$$

As before we assume a product solution

$$V = f(r)\,F(u) \quad ...(6.48)$$

Substituting this in Eqn. (6.47) and dividing throughout by $F(r)\,F(u)$, we get

$$\frac{1}{f(r)}\frac{d}{dr}\left[r^2\frac{d\,f(r)}{dr}\right]+\frac{1}{F(u)}\frac{d}{du}\left[(1-u^2)\frac{dF(u)}{du}\right]=0 \quad ...(6.49)$$

The variables have now been separated and we can therefore write the two separate equations

$$\frac{1}{f(r)}\frac{d}{dr}\left[r^2\frac{df(r)}{dr}\right]=k \quad ...(6.50)$$

$$\frac{1}{F(u)}\frac{d}{du}\left[(1-u^2)\frac{dF(u)}{du}\right]=-k \quad ...(6.51)$$

Equation (6.50) can be expanded as

$$r^2 f''(r) + 2r f'(r) = k f(r) \qquad ...(6.52)$$

First we try a solution like $f(r) = r^n$. Then

$$r^2 n (n-1) r^{n-2} + 2rnr^{n-1} = kr^n \qquad ...(6.53)$$

This shows that r^n can be a solution if $k = n(n+1)$. If we replace n by $-(n+1)$, we get $k = -(n+1)(-n-1+1) = -(n+1)(-n)$ which is the same value as before. Thus if $k = n(n+1)$, both r^n and $r^{-(n+1)}$ are solutions, and hence combining them, we get the solution

$$f(r) = Ar^n + Br^{-(n+1)} \qquad ...(6.54)$$

Substituting $k = n(n+1)$ in Eqn. (6.51) and writing P_n in place of $F(u)$ we have

$$\frac{d}{du}\left[(1-u^2)\frac{dP_n}{du}\right] + n(n+1) P_n = 0 \qquad ...(6.55)$$

As before, replacing n by $-(n+1)$ does not change the form of Eqn. (6.55), hence $P_{-(n+1)}$ will also be a solution of Eqn. (6.55), because

$$P_{-(n+1)} = P_n \qquad ...(6.56)$$

Equation (6.55) was first obtained by the French mathematician Adrien-Marie Legendre and it is known as Legendre's equation. The functions P_n are in the form of polynomials of u (or $\cos\theta$) of degree n, and are referred to as Legendre polynomials of degree n. The final solution of Laplace's equation then becomes

$$V = \sum_{n=0}^{\infty} (A_n r^n + B_n r^{-(n+1)}) P_n \qquad ...(6.57)$$

where the values of P_n for some values of n are given below:

Legendre Polynomials

$$\left.\begin{aligned}
n=0, &\quad P_0 = 1\\
n=1, &\quad P_1 = u = \cos\theta\\
n=2, &\quad P_2 = \tfrac{1}{2}(3u^2-1) = \frac{1}{4}(3\cos 2\theta + 1)\\
n=3, &\quad P_3 = \tfrac{1}{2}(5u^3 - 3u) = \frac{1}{8}(5\cos 3\theta + 3\cos\theta)\\
n=4, &\quad P_4 = \tfrac{1}{8}(35u^4 - 30u^2 + 3)\\
n=5, &\quad P_5 = \tfrac{1}{8}(63u^5 - 70u^3 + 15u)
\end{aligned}\right\} \qquad ...(6.58)$$

These are that spherical harmonics.

The uniform electric field parallel to the polar axis (see Fig. 1.9) is $\overline{\boldsymbol{E}} = E_u \overline{\boldsymbol{u}}_z$ and we have the boundary conditions that have to be satisfied as

(*i*) when $r = a$, $V = 0$, the surface of the grounded conductor being chosen as the zero of the potential.

(*ii*) as before, when $r \to \infty$, $V = -E_u r\cos\theta$.

In the neighbourhood of the sphere, the electric field will depend on both r and θ. Hence $|n|$ cannot be equal to zero, because P_0 does not involve θ. Therefore, we try the next solution with $n = 1$ and write

$$V = (A_1 r + B_1 r^{-2}) \cos\theta$$

Applying the first boundary condition that $V = 0$ for $r = a$,

$$0 = A_1 a + \frac{B_1}{a^2}$$

or

$$B_1 = -A_1 a^3$$

Next applying the second boundary condition, for $r \to \infty$,

$$-E_u r \cos\theta = A_1 \left(r - \frac{a^3}{r^2} \right) \cos\theta$$

$$-E_u \cos\theta = A_1 \left(1 - \frac{a^3}{r^3} \right) \cos\theta$$

$$= A_1 \cos\theta, \text{ when } r \to \infty$$

Therefore $\quad A_1 = -E_u$

Hence the final solution becomes

$$V = \left(\frac{a^3}{r^3} - 1 \right) E_u \, r \cos\theta \qquad \text{...(6.59)}$$

Then using Eqn. (1.109), the electric field intensity components are

$$\left.\begin{aligned} \bar{E}_r &= -\frac{\partial V}{\partial r}\bar{u}_r = \left[\left(\frac{2a^3}{r^3} + 1 \right) E_u \cos\theta \right] \bar{u}_r \\ \bar{E}_\theta &= -\frac{1}{r}\frac{\partial V}{\partial \theta}\bar{u}_\theta = \left[\left(\frac{a^3}{r^3} - 1 \right) E_u \sin\theta \right] \bar{u}_\theta \\ \bar{E}_\phi &= 0 \end{aligned}\right\} \qquad \text{...(6.60)}$$

The difference between Eqns. (6.60) and (6.46) should be carefully noted, though the field map of Eqn. (6.60) shown in Fig. 6.8 looks very much like Fig. 6.5 (*b*) turned counter-clockwise through 90°.

6.10 GRAPHICAL SOLUTION OF LAPLACE'S EQUATION

Graphical solutions of Laplace's equation in two dimensions can be obtained in two different ways as follows, when a sketch of the conductors is drawn *to scale* and the potentials of different conductors are denoted on it.

6.10.1 Finite Differences Method

In this method, the region where the potentials at different points are required has a graph-paper-like square grid pattern drawn with each side of the square having a length h, which may be chosen to be as small as we please, depending upon the accuracy of the answer desired. Then the following difference equations can easily be written from Fig. 6.6, where V_0 is the potential sought at the point O, and V_1, V_2, V_3 and V_n are the potentials at distances h from O in the x and y directions, and a, b, c, d are points at distances $\frac{h}{2}$ from O. Then

$$\left(\frac{\partial V}{\partial x}\right)_a = \frac{V_1 - V_0}{h}, \quad \left(\frac{\partial V}{\partial x}\right)_c = \frac{V_0 - V_3}{h}$$

$$\left(\frac{\partial V}{\partial y}\right)_b = \frac{V_2 - V_0}{h}, \quad \left(\frac{\partial V}{\partial y}\right)_d = \frac{V_0 - V_4}{h}$$

Therefore,

$$\left(\frac{\partial^2 V}{\partial x^2}\right)_0 = \frac{1}{h}\left[\left(\frac{\partial V}{\partial x}\right)_a - \left(\frac{\partial V}{\partial x}\right)_c\right] = \frac{1}{h^2}[(V_1 - V_0) - (V_0 - V_3)]$$

and

$$\left(\frac{\partial^2 V}{\partial y^2}\right)_0 = \frac{1}{h}\left[\left(\frac{\partial V}{\partial y}\right)_b - \left(\frac{\partial V}{\partial y}\right)_d\right] = \frac{1}{h^2}[(V_2 - V_0) - (V_0 - V_4)]$$

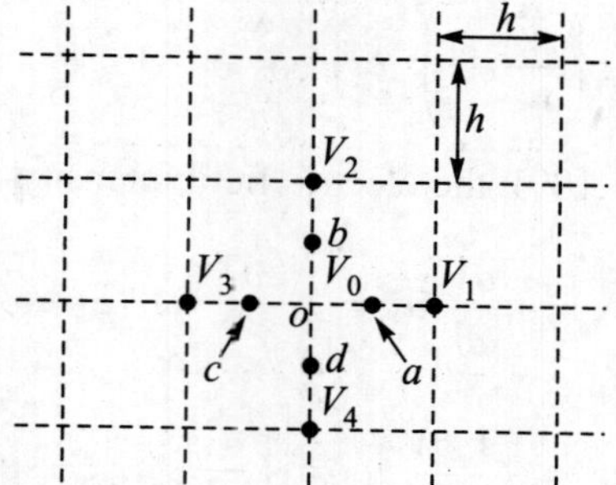

Fig. 6.6. Finite differences solution of Laplace's equation

Putting these values in Laplace's two-dimensional equation in cartesian co-ordinates, we get

$$\left[\frac{\partial^2 V}{\partial x^2} + \frac{\partial^2 V}{\partial y^2}\right] = \frac{1}{h^2}[V_1 + V_2 + V_3 + V_4 - 4V_0] = 0$$

Therefore $$V_0 = \frac{1}{4}[V_1 + V_2 + V_3 + V_4] \qquad ...(6.61)$$

To illustrate the method, the potentials for Fig. 6.2 are obtained by setting up a grid as shown in Fig. 6.7 and assuming $a = b$, $h = \frac{a}{4}$, and $V_0 = 100$ V.

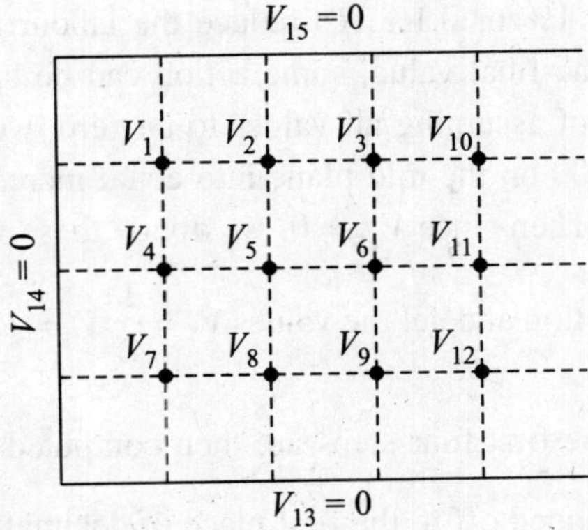

Fig. 6.7. Potentials in Fig. 6.2

As seen from the diagram and the notation of Fig. 6.2, V_{11} = 100 V, and $V_{13} = V_{14} = V_{15} = 0$. Also by symmetry $V_7 = V_1$, $V_8 = V_2$, $V_9 = V_3$ and $V_{12} = V_{10}$. To evaluate the potentials V_1 to V_9, we must know the potentials at the surrounding points V_{10} to V_{15}. Of these, all are known except V_{10} and V_{12}. To determine their values by the finite differences method will require knowledge of potential at distance h to the right of points V_{10} and V_{12}. To avoid this extra calculation, we note that the co-ordinates of the point marked V_{12} are $\left(a, \frac{b}{4}\right)$. Hence using this value in Eqn. (6.22), the potential

$$V_{10} = V_{12} = \frac{100}{\sin h\left(\frac{\pi a}{b}\right)} \sin h\left(\frac{\pi a}{b}\right) \sin\left(\frac{\pi b}{b4}\right)$$

$$= \frac{100}{\sqrt{2}} = 70.7 \text{ V}$$

The computation is best done in the form of a table such as Table 6.1. Only the values from V_1 to V_6 are tabulated as V_7, V_8, V_9 have the same values as V_1, V_2, V_3. At step 1 we can assume all these values to be zero, or some other values. Then we use Eqn. (6.61) for each point and get the potentials from the equations:

$$\left.\begin{aligned}
V_1 &= \tfrac{1}{4}(V_2 + V_{15} + V_{14} + V_4) &&= \tfrac{1}{4}(V_2 + V_4), \\
V_2 &= \tfrac{1}{4}(V_3 + V_{15} + V_1 + V_5) &&= \tfrac{1}{4}(V_1 + V_3 + V_5) \\
V_3 &= \tfrac{1}{4}(V_{10} + V_{15} + V_2 + V_6) &&= \tfrac{1}{4}(V_2 + V_6 + V_{10}) \\
V_4 &= \tfrac{1}{4}(V_5 + V_1 + V_{14} + V_7) &&= \tfrac{1}{4}(2V_1 + V_5) \\
V_5 &= \tfrac{1}{4}(V_6 + V_2 + V_4 + V_8) &&= \tfrac{1}{4}(2V_2 + V_4 + V_6) \\
V_6 &= \tfrac{1}{4}(V_{11} + V_3 + V_5 + V_9) &&= \tfrac{1}{4}(2V_3 + V_5 + V_{11})
\end{aligned}\right\} \quad \text{...(6.62)}$$

The important point to be noted is that, for each of these computations, the *most recently computed values* of the voltages must be used. By these successive approximations, we approach closer to the final value at each step until we come to a final set of values while will satisfy Eqn. (6.62) to the *desired degree of accuracy.*

Usually this is taken as 1% or 0.1%. To reduce the labour and to make the values converge fast towards the final value, some action can be taken at different stages. Thus at Step 1, instead of assuming all values to be zero, we can divide the change from $V_{14} = 0$ to $V_{11} = 100$ on the mid-plane into equal increments, and put $V_4 = 25$, $V_5 = 50$ and $V_6 = 75$ V. Then since $V_{15} = 0$, we divide these values by 2 to get equal increments in the y-direction and get the values $V_1 = \frac{1}{2} V_4 = 12.5$, $V_2 = \frac{1}{2} V_5 = 25$ and $V_3 = \frac{1}{2} V_6 = 37.5$ V. The first four steps are then computed using Eqn. (6.62). The computed values are rounded off to the first place of decimal.

Table 6.1. Finite differences solution of Laplace's equation

	Steps							*Using*
Voltages	*1*	*2*	*3*	*4*	*5*	*6*	*Final*	*Eqn. (6.22)*
V_1	12.5	12.5	11.0	9.2	7.4	6.1	5.9	5.3
V_2	25.0	25.0	22.4	18.8	15.2	15.3	15.2	14.1
V_3	37.5	42.7	37.1	35.1	33.1	33.3	33.2	32.0
V_4	25.0	18.8	14.5	11.8	9.1	8.5	8.3	7.5
V_5	50.0	36.0	28.7	25.0	21.3	21.7	21.4	19.9
V_6	75.0	55.4	50.7	48.8	46.9	47.1	47.0	45.2

After four steps it is observed that all voltages are decreasing at each step. Also the rate of decrease is small from steps 3 to 4 in comparison to the change from steps 2 to 3. This is always the case and it implies that more steps are required in the later phase of computation to obtain the same amount of change than in the earlier phase. To reduce the work therefore, step 5 is not written using Eqn. (6.22), but the same constant multiple of (change from steps 3 to 4) is made from steps 4 to 5. In this case, the common multiplying factor is chosen as unity. Thus for V_1, since $11 - 9.2 = 1.8$, hence in step 5 we write $9.2 - 1.8 = 7.4$, and so on for other voltages. Now step 6 is computed from step 5 using Eqn. (6.62) and the step shows that most of the voltages changes are less than one volt, providing an answer with 1% accuracy.

These computations are obviously best performed using a computer. The final answer is shown in next column, and is seen to be very close to the values arrived at in step 6.

The exact values, using Eqn. (6.22) are shown in the last column. Why are they different from the final values given in the previous column? This is because the so-called final answer is only a solution of the simultaneous equations (6.62), and these equations like Eqn. (6.61), are based on the assumption that $\frac{\partial V}{\partial x}$ and $\frac{\partial V}{\partial y}$ are varying linearly over the interval h. This will be nearer the truth, only if the size of the interval is made very small. Using very small steps and a computer, answers close to the exact values can be obtained with the desired degree of accuracy.

6.10.2 Graphical Solution Using Drawing Board

This method is a graphical way of implementing the underlying idea of the method of conformal transformation, namely, that if the equipotentials and the direction lines form a square grid pattern on one plane, then they will continue to form curvilinear square grid pattern on the other plane. In the graphical pattern, on the same map drawn to scale, the undisturbed uniform field gives the square grid pattern, while in the portion modified by boundary conditions, the curvilinear squares are to be found.

As an illustration we proceed to find graphically the effect of placing a conducting sphere in a uniform electric field $\overline{E} = E_s \overline{u}_z$. The Eqn. (6.60) shows that for $r > 3a$, the field is almost undisturbed, the error caused by the term $\frac{a^3}{r^3}$ being less than 4%. In Fig. 6.8 the sphere is located at the centre of a square of side $6a$. To save space, different stages of the graphical process are shown in the four quadrants around the centre.

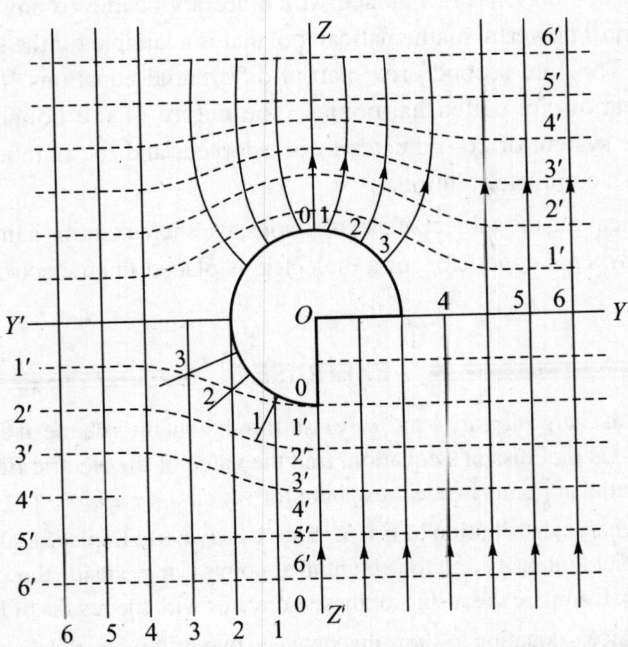

Fig. 6.8. Field lines and equipotentials around sphere placed in uniform electric field

In the fourth quadrant, the uniform field is depicted with equally spaced direction lines with arrow-heads, and dotted equipotential lines intersecting them at right angles and forming a grid of squares. In the first three quadrants, the sphere is shown in position at the centre. In the third quadrant, a tentative beginning is made by starting the direction lines and equipotentials from the outer boundary of the square. The central direction line $Z'Z$ and the central equipotential $Y'Y$ will remain unchanged throughout, on account of symmetry in the field.

The direction lines 0,1 and 2 are intercepted by the sphere and 3 is very close to the sphere and will therefore bend to terminate on the sphere. Hence from the surface of the sphere in the third quadrant, equally spaced short lines 1, 2, 3 are drawn normal to the sphere as required by Cor. 3 of Gauss's law. The equipotentials $1'$, $2'$ intercepted by the sphere have to bend round the sphere. Hence all the equipotential $1'$ to $6'$ have to be accommodated within a-b on the z'-axis, as shown by short lines $1'$ to 6 near the z'-axis.

The equipotentials and direction lines near the sphere appear to form curvilinear rectangles rather than curvilinear squares. To correct for this, the spacing between them is adjusted by shifting (in second quadrant) the direction lines towards the z-axis as judged by the eye, to make curvilinear squares of the type in Fig. 5.11 w-plane. The final result is depicted in the first quadrant, and the complete map is obtained by having mirror images of these curves repeated in the second, third and fourth quadrants.

6.11 SUMMARY AND COMMENTS

The equations of Poisson and Laplace which are applicable to any point in space, represent the most powerful mathematical tool that is available for the study of fields in electrostatics. They are second order partial differential equations. The solutions of Laplace's equation are called harmonics. The nature of the boundary conditions determines the system of co-ordinates to be adopted, and the number of harmonics needed to give a complete solution.

Before leaving the subject of electrostatics, we shall examine a little more closely what happens when a conductor or a dielectric is placed in an electrostatic field.

EXERCISE 6

1. A uniform charge density $+\rho$ C/m^3 exists throughout the volume of a sphere of radius b metre. Using Poisson's equation, find the value of the electric field intensity and the potential at points inside the sphere for which $0 < r \leq b$.
2. For the charge distribution of Ex. 6, Prob. 1 above, use Laplace equation to find the electric field intensity and the potential at points lying outside the sphere for which $b < r < \infty$. Compare the results of these exercises with the results of Ex. 4, Prob. 3.
3. Use Laplace's equation to show that between two infinite, parallel conducting planes at voltages V_m and 0 respectively, the potential at any point P is proportional to its distance from the second plate at voltage 0. Comparing this value with the result of Sec. 2.16, find the formula for the capacitance of a parallel plate capacitor.
4. Using Laplace's equation in cylindrical co-ordinates, obtain an expression for $\overline{E}$ at any point P in a coaxial cable, lying between the inner and outer conductor. Using Gauss's law, obtain another expression for $\overline{E}$ at the same point. Comparing these two values of $\overline{E}$, derive an expression for the capacitance per metre length of a coaxial cable.

5. Using the method of Ex. 6, Prob. 4 above, find the capacitance of a capacitor formed by two concentric conducting spheres of radii r_1 and r_2, where $r_2 > r_1$.

6. Two infinite conducting planes, both at potential $V = 0$, are located at $y = 0$ and $y = \pi$. Between these is the third plane at $x = 0$, which extends from $y = (0+)$ to $y = (\pi -)$, but is insulated from the planes at potential $y = 0$. If the third plane at $x = 0$ has a potential $V = \sin y$ on it, determine the potential at any point in the trough formed by the three planes. Assume that for $x = \pm\infty$, $V = 0$.

7. Using the method of Section 5.10, show that Eqns. (6.60) for the field when a conducting sphere is placed in a uniform field can be obtained for points outside the sphere by superposing on the uniform field $\overline{E} = E_u\,\overline{u}_z$, the field due to a dipole of moment $(4\pi\,\varepsilon_0\,E_u\,a^3)$ located at the centre of the sphere. [The dipole produces a field which is superimposed on the uniform field $\overline{E} = E_u\,\overline{u}_z$.]

8. Use Eqns. (6.39) and (6.42) to show that by appropriate choice of constants in them, they can give the electric field due to
   (*i*) uniformly charged infinite line, and
   (*ii*) coaxial cable of infinite length.

9. Show that in spherical co-ordinates, $V = \log_e\left(\cot\dfrac{\theta}{2}\right)$ satisfies the Laplace's equation, and is therefore a solution of Laplace's equation.

10. The surface described by a constant value of $\theta = \theta_1$ in spherical co-ordinates consists of two cones with the polar z-axis as their common axis, and having their vertices at the origin. If the infinite cones towards the positive and negative directions of the z-axis are made of conductors, and the cones are insulated from each other by an infinitesimally small gap between their vertices, show that the capacitance of the resulting capacitor is

$$C = \frac{\pi\,\varepsilon_0}{\log_e\left(\cot\dfrac{\theta_1}{2}\right)}$$

CHAPTER

7 CONDUCTORS AND DIELECTRICS

7.1 CONDUCTORS

The ability of the metals to conduct electric current easily has been explained in Sec. 4.12.

Table 7.1. Conductivity of materials

Material	*Conductivity (S/m)*	*Uses of group*
Silver	6.1×10^7	As conductor material
Copper	5.8×10^7	
Gold	4.1×10^7	Surface plating on conductors
Aluminium	3.5×10^7	and waveguides
Tungsten	1.8×10^7	Lamp filament
Brass	1.5×10^7	
Nickel	1.3×10^7	
Constantin	2.0×10^6	For resistance boxes
Mercury	1×10^6	Switches
Nichrome	8.9×10^5	Heater element
Germanium	2.2	In semiconductor device
Silicon	1.6×10^{-3}	With doping, increases 10 to 100 times
Sea water	4	Ground wave propagation of
Good soil	5 to 10×10^{-3}	radio waves
Fresh water	10^{-3}	
Distilled water	2×10^{-4}	
Dry earth	10^{-5}	
Glass	10^{-10} to 10^{-14}	
Porcelain	2×10^{-11}	Insulators
Mica	10^{-11} to 10^{-15}	
Fused Quartz	4×10^{-18}	

It was also noted that the conductivity of materials, expressed in siemens/metre (S/m) varies over a wide range, the values for silver, the best conductor of electricity and for fused quartz, the worst conductor of electricity differing by a factor of 10^{25}, as seen from Table 7.1.

While in the majority of materials, electrons with their negative charge are the carriers of electric charge, which produces the electric current, in some cases other carriers of electric charge are available. Atoms of elements of Group IV of periodic classification of elements like carbon, silicon and germanium form covalent or double bonds to make a crystal lattice structure. Under thermal agitation of the lattice, an electron may leave the covalent bond leaving a vacancy or hole behind. Similarly, if a few (about one in 10^8 atoms) of the Group IV atoms in the lattice are replaced by Group III elements like boron, aluminium, gallium or indium, then since these elements have only three valence electrons instead of four, one of the covalent bonds has a hole in it. When a new electron-hole pair is created, and this electron goes and occupies an earlier created hole to complete the covalent bond there, then this process results in the shifting of a hole from the old position to the new one. Thus holes also contribute to the total current, because their movement involves the movement of negative electronic charge in the positive direction.

In liquids because of hydrolysis, positively charged ions are available as charge carriers. One or more of these three – electrons, holes and positive ions – help to carry current in materials. Since even insulators conduct some electric current, it is obvious that the terms conductors and insulators are rather loosely applied only to indicate which aspect is given importance in a particular case. Thus we normally use the word conductor to mean material having conductivity in the megasiemens/metre range, but for radio antennas, the earth is also treated as a good conductor ! When Poly-vinyl-chloride (PVC) coating is provided between the conductors of a twin-flexible wire, we call PVC an insulator. But when Titanium dioxide (TiO_2) is used between aluminium foils to provide greater capacitance in the same volume, we call TiO_2 a dielectric.

7.2 EFFECT OF PLACING A CONDUCTOR IN ELECTRIC FIELD

Normally an electric conductor is electrically neutral. The distribution of positive charges located in the nuclei of the atoms and the negative charges of the surrounding electrons is uniform throughout the entire volume. Everywhere these positive and negative charges cancel each other exactly, making the entire conductor electrically neutral. When the conductor is placed in an electric field, these positive and negative charges shift in the direction of the electric field and in the opposite direction respectively, and almost instantaneously (in time less than one attosecond or 10^{-18} second), this movement of charges ceases, and an equilibrium condition is reached. The electric field theory so far developed indicates that in electrostatics, the charges reside only on the surface of the conductors, and there is no electric field intensity inside the conductor. This is the equilibrium condition attained. The surface charges are so distributed, that they produce a secondary electric field intensity $\overline{E}_s$ which, when added to the original field intensity $\overline{E}_0$, makes $\overline{E}_0 + \overline{E}_s = 0$ in the entire region occupied by the conductor.

This can be illustrated by the simplest example of a conducting slab, infinite in extent in the *y*- and *z*-directions, and having a thickness *d* metre in the *x*-direction, placed normal to the uniform electric field $\overline{E}_0 = E_u\ \overline{u}_x$ as shown in Fig. 7.1 (a). The movement of the charges in the conductor results in a charge distribution as shown in Fig. 7.1 (b). This gives a secondary field

$$\overline{E}_s = \frac{q_s}{\varepsilon_0}\ (-\ \overline{u}_x) \qquad \text{...(7.1)}$$

This added to the original field makes the resultant

$$\overline{E} = \overline{E}_0 + \overline{E}_s = 0 \qquad \text{...(7.2)}$$

or
$$E_u\ \overline{u}_x - \frac{q_s}{\varepsilon_0}\ \overline{u}_x = 0$$

Therefore
$$q_s = \varepsilon_0\ E_u \qquad \text{...(7.3)}$$

(a) $\overline{E}_0 = E_u\overline{u}_x$ (b) $\overline{E}_s = \frac{q_s}{\varepsilon_0}(-\overline{u}_x)$

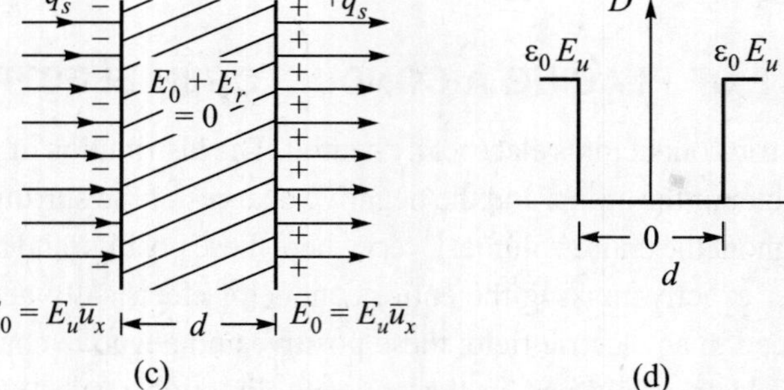

(c) (d)

Fig. 7.1. Conducting slab in uniform electric field

This gives the magnitude of the surface charge density. The field lines on the left of the slab terminate on the negative charges on the left surface, and the direction lines on the right start from the positive charges on the right surface, as shown in Fig. 7.1 (*c*).

Equation (7.3) can also be derived if we plot D against x as in Fig. 7.1 (*d*), and then apply the method of Fig. 4.18 to account for the abrupt change in D at $x = -\frac{d}{2}$ and at $x = +\frac{d}{2}$.

A similar surface charge distribution, but one which was not uniform, was found in Sec. 5.8 for the case of an infinite plane conducting sheet placed in the electric field of a point charge.

For the case of the conducting sphere placed in a uniform field, we had obtained the Eqns. (6.60) for the field outside the sphere. At the surface of the sphere, the field must be normal to the surface at all points, showing that only E_r has a value when $r = a$, and

$$[E_r]_{r=a} = \left[\left(1+\frac{2a^3}{a^3}\right) E_u \cos\theta\right] \bar{u}_r$$

$$= (3E_u \cos\theta)\, \bar{u}_r$$

Near the surface of a conductor, field intensity is given by Eqn. (4.31) as

$$[E_r]_{r=a} = \frac{q_s}{\varepsilon_0}\, \bar{u}_n .$$

Hence $$q_s = 3\varepsilon_0\, E_u \cos\theta \qquad \text{...(7.4)}$$

7.3 DIELECTRICS

So far as the poor conductors of electricity are concerned, we are not interested in their ability to prevent movement of electric charges through them. We are here interested in their ability to change the electric field intensity at a point. For this reason, we refer to them as dielectrics, and not as insulators.

The absence of an abundant supply of free electrons is therefore the basic difference between a dielectric and a conductor. We express this in other words, and say that in dielectrics, the electrons are tightly bound to the positively charged nuclei and are unable to leave their parent atoms. Hence, when the dielectric is placed in an electric field, a flow of electrons to produce induced charges on the surface is not possible. Thus the complete cancellation of the electric field that takes place inside a conductor does not occur in a dielectric, and inside a dielectric, we do have an electric field, though it is weaker than the field outside the dielectric for reasons discussed below.

7.4 ELECTRIC POLARIZATION

In an atom there is a positively charged nucleus around which a number of electrons revolve within a spherical surface. The centre of these negative charges in the sphere coincides with the positively charged nucleus, and since these positive and negative charges are equal in magnitude, the atom as a whole is electrically neutral. When two or more such similar atoms get together to form a molecule, the centres of the positive nuclei and the negative electrons still coincide, and the molecule exhibits no electrical effects outside. Such a molecule is referred to as a non-polar molecule.

However, if two dissimilar atoms form an ionic bond and become a molecule, things are different. For instance when a potassium atom which has one electron in its outermost shell combines with an atom of chlorine which has a deficiency of one electron in its outermost shell to make a KCl molecule, the outermost electron of potassium fits into the vacancy in the outermost shell of chlorine, making that shell completely filled. But by this process, the chlorine configuration has excess negative charge and the potassium configuration has less amount of negative charge. Thus a dipole is created with a positive charge $+q$ near centre of potassium configuration and a negative charge $-q$ near centre of chlorine configuration. Such a molecule is said to have a dipole moment $\bar{p} = qd$ where d is the distance between the charges $+q$ and $-q$. Naturally this dipole moment is a permanent feature of the KCl molecule. Such molecules are called polar molecules.

Under certain circumstances, a non-polar molecule may temporarily acquire a dipole moment. This happens when it is placed in an electric field. The field causes the centres of the positive and the negative charges to shift slightly, but in opposite directions, thus providing two charges $+q'$ and $-q'$ with distance d' between them. This quantity $q'd'$ is referred to as the induced dipole moment. We can explain this shift in another way by noting that the electrons, under the influence of the external electric field, spend more time on one side of the nucleus than on the other side, and this loss of symmetry is interpreted as a shift in the centre of the negative charge.

In the case of polar molecules, in the absence of an external electric field, the dipole moments of individual molecules are oriented in a random manner, so that the overall result of millions of molecules is to have zero dipole moment, [Fig. 7.2 (a)]. When the material is placed in an electric field, the individual molecules experience a torque as shown in Fig. 7.2 (b). This torque causes the molecules to align themselves in the direction of the field as shown in Fig. 7.2 (c). The degree of alignment achieved depends upon the freedom with which a molecule can turn around its axis.

In the case of a material like carnauba wax, this freedom is greater when the wax is in a molten state, but is almost zero when the wax is in a solid state. Hence, it is possible to melt the wax, keep it in an electric field so that the molecules align themselves, as shown in Fig. 7.2 (c), and while still under the influence of the electric field, allow the wax to cool and to solidify. Then the molecules get set in the aligned position, and the wax piece becomes an electret – the electrical equivalent of a bar magnet. In an electric field, a freely suspended electret will align itself with the field just as a magnetic needle aligns itself with a magnetic field.

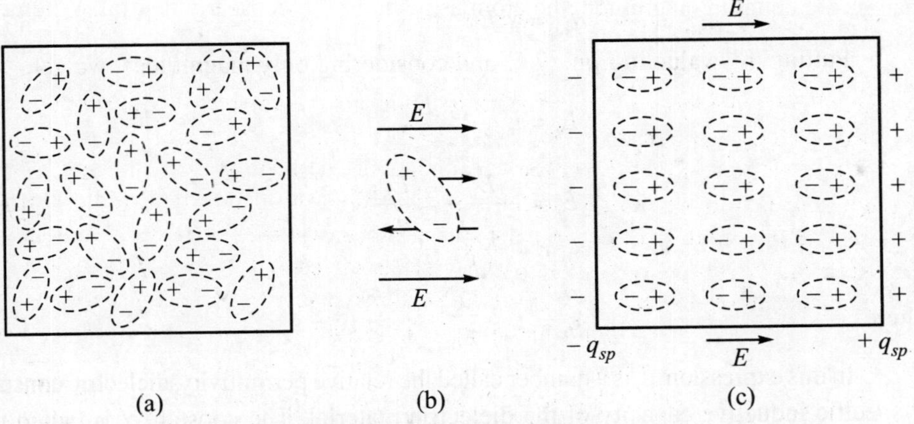

Fig. 7.2. Polar molecules in electric field

7.5 EFFECT OF PLACING A DIELECTRIC IN ELECTRIC FIELD

The basic concepts outlined in the previous section bring out the essential difference in the effect produced by an electric field inside a conductor and inside a dielectric. In a conductor, however weak the electric field might be, the electrons, being free, quickly move and surface charges appear to give a secondary field inside, which exactly cancels the external field in the region inside the conductor. On the other hand, the first effect of the electric field on a dielectric having non-polar molecules is to change the shape of the molecule from a sphere to an ellipsoid, the elongation being in the direction of the field due to the displacements of the centres of the positive and the negative charges. The elongation therefore depends upon the strength of the field *inside* the dielectric. The second effect is that the dipoles tend to align themselves in the direction of the field, the torque again being proportional to the field *inside* the dielectric.

Hence if $+q_{sp}$ and $-q_{sp}$ C/m^2 are the polarisation surface charges which appear at the surface of the dielectric because of the free ends of the dipoles near the surface [Fig. 7.2 (*b*)], we have the secondary electric field intensity $\overline{\boldsymbol{E}}_{sp}$ when the dielectric is placed in the uniform electric field $\overline{\boldsymbol{E}}_0 = E_u\,\overline{\boldsymbol{u}}_x$ as

$$\overline{\boldsymbol{E}}_{sp} = \frac{q_s p}{\varepsilon_0}(-\overline{\boldsymbol{u}}_x) \qquad \text{...(7.5)}$$

However, the total electric field $\overline{\boldsymbol{E}}_t$ inside the dielectric is not zero but is given by

$$\overline{\boldsymbol{E}}_t = \overline{\boldsymbol{E}}_0 + \overline{\boldsymbol{E}}_{sp} = \left(E_u - \frac{q_{sp}}{\varepsilon_0}\right)\overline{\boldsymbol{u}}_x \qquad \text{...(7.6)}$$

The difference between a conductor and a dielectric lies in the fact that in a dielectric, the surface charge density q_{sp} does not depend on E_u but on E_t, or if χ is a constant

$$q_{sp} = \chi\, E_t \qquad \text{...(7.7)}$$

Putting this value in Eqn. (7.6) and considering only magnitudes, we get

$$E_t = E_u - \frac{\chi\, E_t}{\varepsilon_0}$$

or

$$E_t = \frac{E_u}{1 + \dfrac{\chi}{\varepsilon_0}} = \frac{E_u}{\varepsilon_r} \qquad \text{...(7.8)}$$

where

$$\varepsilon_r = 1 + \frac{\chi}{\varepsilon_0}, \qquad (\varepsilon = \varepsilon_r\, \varepsilon_0) \qquad \text{...(7.9)}$$

In this expression, ε_r is a number called the relative permittivity, dielectric constant or specific inductive capacity of the dielectric material. The constant χ is called the dielectric susceptibility with units as Farad per metre. It is a constant for a homogeneous, linear, isotropic medium. Equation (7.8) shows that a feedback type of phenomenon takes place in the system before an equilibrium condition is reached. This is shown as a block diagram in Fig. 7.3. The applied uniform electric field $\overline{\boldsymbol{E}}_0 = E_u\, \overline{\boldsymbol{u}}_x$ acts on the dielectric producing a polarisation of its molecules with dipole moment $\overline{\boldsymbol{p}} = q''\, d''$. This is the picture at the microscopic level. If we consider the total volume of the dielectric containing millions of such dipole moments, we get the macroscopic picture. Now we define a polarisation vector $\overline{\boldsymbol{P}}$, which gives the dipole moment per unit volume (m^3) of the dielectric. Obviously P depends upon the ultimate field intensity E_t in the dielectric (and not on the applied external field E_0), and is given by

$$\overline{\boldsymbol{P}} = P_0\, \overline{\boldsymbol{u}}_x = \chi E_t\, \overline{\boldsymbol{u}}_x = q_{sp}\, \overline{\boldsymbol{u}}_x \qquad \text{...(7.10)}$$

where χ is the dielectric susceptibility. The units for $\overline{\boldsymbol{P}}$ are coulomb per square metre.

This polarisation gives rise to the surface charge, and the latter gives the secondary field which is directed opposite to the external field. The two fields add algebraically to give the total field which is less than E_0. This reduces the original polarisation produced by E_0, which reduces the surface charge which reduces the opposing secondary field, which increases the total field. The process repeats over and over round the feedback loop until an equilibrium is reached ultimately with the various quantities having the values given by Eqns. (7.5) to (7.10).

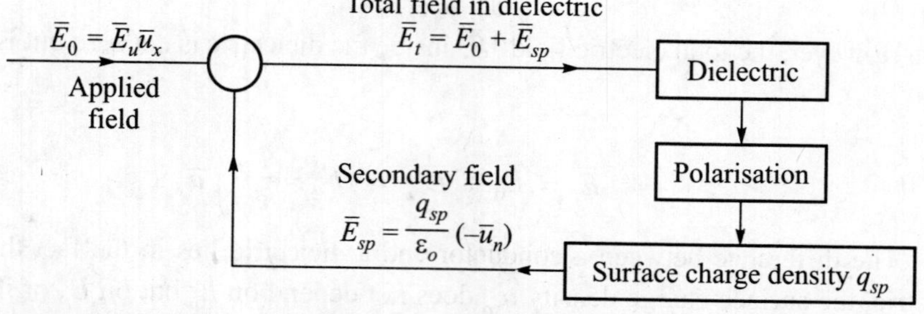

Fig. 7.3. Feedback loop showing effect of placing dielectric in electric field

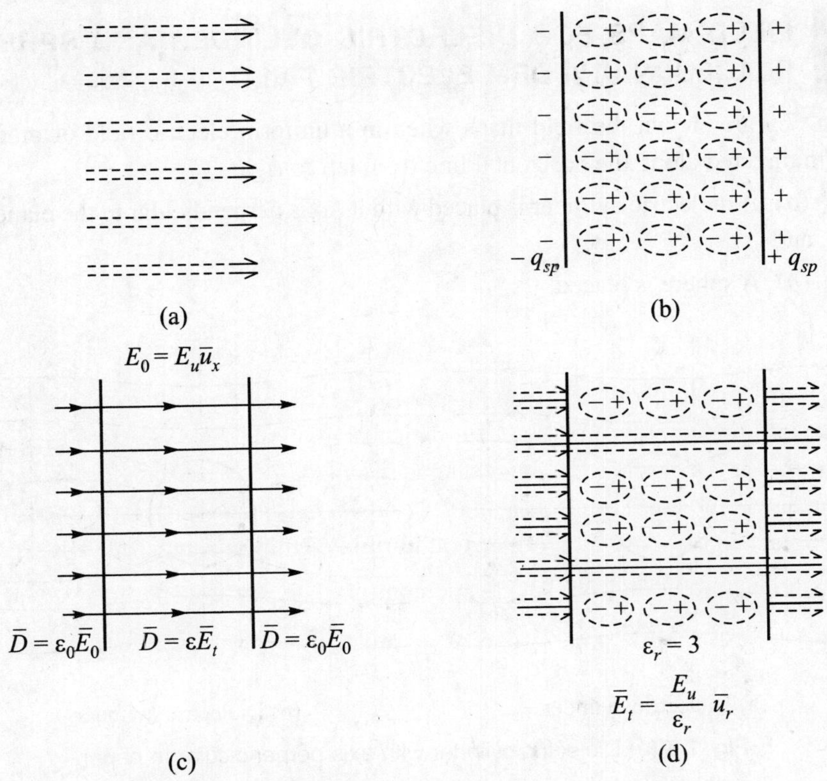

Fig. 7.4. Showing effect of placing dielectric in electric field

The phenomenon is shown in a different manner in Fig. 7.4. The applied field $\overline{\boldsymbol{E}}_0 = E_u \, \overline{\boldsymbol{u}}_x$ [Fig. 7.4 (a)] produces polarisation and surface charges q_{sp} [Fig. 7.4 (b)]. The presence of the dielectric makes no difference to the electric displacement vector $\overline{\boldsymbol{D}}$, which has the same magnitude in the dielectric as it has outside [Fig. 7.4 (c)]. In Fig. 7.4 (d) is shown the fact that some of the field direction lines terminate on the surface charges and do not continue inside the dielectric, while a lesses number are able to continue through the dielectric showing a reduction in field intensity inside the dielectric. In Fig. 7.4 (d), ε_r is taken as 3 (= 1 + 2) as per Eqn. (7.9). Hence for every two lines terminating on the surface charges, one direction line passes through. This shows the basic difference between D lines and E lines with a dielectric. Since the surface charges are in the nature of a "reaction" of the dielectric to the action of the electric field, the D lines ignore the surface charges and do not terminate on them. But the electric field lines terminate on negative surface charges on the left and originate again from the positive surface charges on the right hand side.

7.6 FIELD MAPS FOR DIELECTRIC CYLINDER AND SPHERE PLACED IN UNIFORM ELECTRIC FIELD

In Fig. 7.5 are shown the field maps when in a uniform electric field of magnitude E_u V/m directed along the horizontal line from left to right,

(*a*) A dielectric cylinder is placed with its axis perpendicular to the plane of the paper, and

(*b*) A sphere is placed.

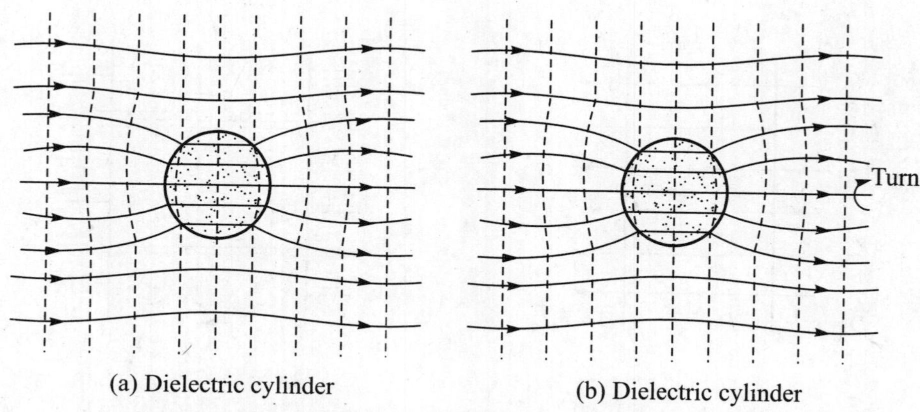

Fig. 7.5 (a) Dielectric cylinder with axis perpendicular to paper (b) Dielectric sphere, both placed in uniform electric field

Notice that in both cases, there is a uniform field inside the dielectric as shown by the parallel direction lines there. The equipotentials are more widely spaced inside the dielectric, which shows a weaker electric field intensity there. The direction lines do not meet the surface of the dielectric at right angles to the surface, as they did in the case of conductors. On either side of the dielectric, there is a weakening of the electric field near the vertical axis, and a corresponding strengthening of the field on the horizontal axis. Finally to obtain the three-dimensional picture, Fig. 7.5 (*a*) is to be moved parallel to itself so that the centre of the circular section describes the axis of the cylinder coming out of the paper, but Fig. 7.5 (*b*) is to be rotated around the horizontal line through the centre of the sphere as shown by arrow marked TURN. These figures should be compared with Fig. 6.5 (*b*) and the differences noted.

7.7 TABLE OF DIELECTRIC CONSTANTS

In general it is found that the dielectric constant is not constant, but it varies with frequency. Hence in physical tables, the dielectric constant is often given at frequencies like 50 Hz, 1 kHz, 1 MHz and 1 GHz. In Table 7.2 however nominal values are given for selected materials to show the range of variation of ε_r.

Table 7.2. Dielectric constant of materials

Material	*Dielectric constant*	*Material*	*Dielectric constant*
Air at N.T.P.	1.000 590	Ruby Mica (Muscovite)	5.4
Paraffin wax	2.25	Porcelain	5.87
Rubber	2.4	Aluminium oxide	8.8
Bees wax	2.53	Ethyl alcohol	24.5
Shellac	3.47	Distilled water	81
Poly-vinyl-chloride (PVC)	3.52	Titanium dioxide (TiO_2)	100
Fused silica	3.78		1100
Bakelite	4.67	Barium Titanate	to
Corning 7060 Pyrex glass	4.84	($BaTiO_3$)	10,000

7.8 BOUNDARY CONDITIONS IN ELECTRIC FIELDS

Certain conditions are associated with the boundary between two media, whether they are two dielectrics, or one dielectric and one conductor, or two conductors meeting at a common surface. These arise because certain laws like Gauss's law (or Kirchhoff's current law for electric currents) and Law of Conservation of Energy must hold good at the boundary.

Consider [Fig. 7.6 (a)] two media, 1 and 2, where, at their common boundary, the flux vectors D_1 and D_2 in the two media are shown at point P on the boundary. Let D_1 and D_2 make angles θ_1 and θ_2 respectively, with the normal to the boundary at the point P.

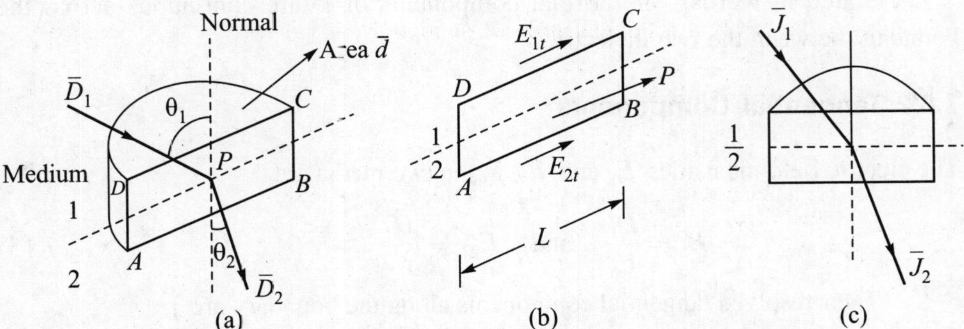

Fig. 7.6. Boundary conditions between medium 1 and medium 2

7.8.1 Normal Components

Surrounding the point P situated on the boundary, a Gaussian surface is selected in the form of a short cylinder with flat ends parallel to the boundary at P. This is often described as a pillbox, because it resembles the small box in which people keep medicinal pills or tablets. Half the box is shown in Fig. 7.6 (*a*), the point *P* being at the centre of the rectangle *ABCD*. Now the box is allowed to shrink in size so that $AD \to 0$ and $BC \to 0$, but all the time, we assume that the boundary lies between the upper and lower flat surfaces. Hence *CD* always lies in medium 1 and *AB* in medium 2, while *P* is always between *AB* and *CD*.

In the limit, when $AD = 0$ and $BC = 0$, there will be no flux through the curved surface of the cylinder, as this area will be zero. Hence when applying Gauss's Law, we are concerned only with the components of D_1 and D_2 normal to the boundary. Denoting these components as D_{1n} and D_{2n}, we can write

$$D_{1n} = D_1 \cos \theta_1 \quad \text{and} \quad D_{2n} = D_2 \cos \theta_2 \qquad ...(7.11)$$

D_{1n} enters the Gaussian surface and is therefore to be counted as negative for applying Gauss's law, while D_{2n} is outwards through the Gaussian surface, and therefore the total outward flux is $= (D_{2n} - D_{1n})\, a$, where a is the area of a flat end of the Gaussian surface. This must be equal to the charge enclosed. If there is charge density of $+q_s$ C/m^2 at *P*, then the charge on area a is aq_s coulomb. Equating these,

$$D_{2n} - D_{1n} = q_s \qquad ...(7.12)$$

Only two different cases can arise:

(*i*) Let medium 1 be the conductor and medium 2 a dielectric. Then since there cannot be an electric field inside a conductor, $D_1 = 0$. Therefore $D_{1n} = 0$. Then Eqn. (7.12) gives

$$D_{2n} = q_s, \text{ the surface charge density on conductor at } P \qquad ...(7.13)$$

This checks with Cor. 3 of Gauss's law and Eqn. (4.31)

(*ii*) If both the media are dielectrics, and also there is no surface charge density on the boundary separating them, then by Eqn. (7.12)

$$D_{2n} - D_{1n} = 0 \quad \text{or} \quad D_{2n} = D_{1n} \qquad ...(7.14)$$

Stated in words, "the normal components of *D* are continuous across the boundary between the two dielectrics."

7.8.2 Tangential Components

The electric field intensities $\overline{E}_1$ and $\overline{E}_2$ in the two media are

$$\overline{E}_1 = \frac{\overline{D}_1}{\varepsilon_1} \quad \text{and} \quad \overline{E}_2 = \frac{\overline{D}_2}{\varepsilon_2} \qquad ...(7.15)$$

Their resolved tangential components along the boundary are

$$E_{1t} = E_1 \sin \theta_1 \quad \text{and} \quad E_{2t} = E_2 \sin \theta_2 \qquad ...(7.16)$$

Since an electrostatic field is a conservative field, it means that [Fig. 7.6 (*b*)] the line integral $\int \overline{E} . \overline{ds}$ along the closed path *ABCDA* must always be zero. Hence if $AB = CD = L$, then in the limit with $AD = BC = O$.

$$\oint \overline{E} \cdot \overline{ds} = \oint_A^B \overline{E} \cdot \overline{ds} + \int_C^D \overline{E} \cdot \overline{ds} = 0.$$

or $$E_{2t} L + (-E_{1t}) L = 0.$$

or $$E_{2t} = E_{1t} \qquad ...(7.17)$$

Stated in words, "the tangential components of E are continuous across the boundary between two media."

Then combining Eqns. (7.11) and (7.14),

$$D_2 \cos \theta_2 = D_1 \cos \theta_1 \qquad ...(7.18)$$

Similarly combining Eqns. (7.16) and (7.17),

$$E_2 \sin \theta_2 = E_1 \sin \theta_1 \qquad ...(7.19)$$

Dividing Eqn. (7.19) by Eqn. (7.18) and using Eqn. (7.15), we get

$$\frac{E_2 \sin \theta_2}{D_2 \cos \theta_2} = \frac{E_1 \sin \theta_1}{D_1 \cos \theta_1},$$

or $$\frac{\tan \theta_1}{\tan \theta_2} = \frac{\varepsilon_1}{\varepsilon_2} \qquad ...(7.20)$$

7.8.3 Boundary between Two Conductors

The problem of the boundary between two conductors cannot arise in the case of electrostatics, because in both media, $\overline{\boldsymbol{D}}$ and $\overline{\boldsymbol{E}}$ will be zero. Hence the boundary between two conductors becomes important only in case there is steady direct current flow across the boundary. If the normal components of the current densities $\overline{\boldsymbol{J}}_1$ and $\overline{\boldsymbol{J}}_2$ ampere/metre^2 in the two media are J_{1n} and J_{2n}, then since there is no accumulation of charge at the point P on the boundary by Kirchhoff's current law, Eqn. (4.29) applied to the Gaussian surface [Fig. 7.6 (*c*)], it follows that

$$J_{2n} = J_{1n} \qquad ...(7.21)$$

and by Ohm's law, this means that

$$\sigma_2 E_{2n} = \sigma_1 E_{1n} \qquad ...(7.22)$$

7.9 SUMMARY AND COMMENTS

In this chapter a very simple and elementary view of the effects of electric fields on conductors and dielectrics is presented. In some books, the point of view of Lorentz is followed. He defined $\overline{\boldsymbol{D}}$ in a dielectric as

$$\overline{\boldsymbol{D}} = \varepsilon_0 \overline{\boldsymbol{E}}_t + \overline{\boldsymbol{P}}$$

and the polarisation $\overline{\boldsymbol{P}}$ as

$$\overline{\boldsymbol{P}} = \varepsilon_0 \chi \overline{\boldsymbol{E}}_t,$$

calling χ the electric susceptibility. Obviously this makes χ a dimensionless constant. This is mentioned so that the reader may not get confused when consulting other books.

The boundary conditions that have been established for the electric field are useful in determining constants of integration when solving differential equations, and in particular, when we shall deal later with time varying electric and magnetic fields.

Having completed our study of electrostatic fields, we shall now proceed to study magnetostatic fields.

EXERCISE 7

1. Show that when an infinitely long, conducting cylinder is placed at right angles to a uniform electric field $\bar{E} = E_u\, \bar{u}_x$, the charge density induced on its surface is given by

$$q_s = 2\varepsilon_0 E_u \cos\theta \quad \text{C/m}^2$$

2. In a parallel plate capacitor formed by planes $z = 0$ and $z = d$, the permittivity of the dielectric varies according to the relation

$$\varepsilon = \left(\frac{\varepsilon_2 - \varepsilon_1}{d}\right) z + \varepsilon_1.$$

Find the capacitance of the capacitor per square metre area of the plates.

3. In an infinitely long co-axial cable, assume that the permittivity of the dielectric varies inversely as the radius. Find the capacitance per unit length of the cable.

4. A thin electret in the form of a disk of radius a and thickness t is made of a dielectric of permittivity ε. Assume that its permanent polarisation is uniform and has a value $\bar{P} = P_0 \bar{u}_z$, where the z-axis is the axis of the disk. Find the values of $\bar{E}$ and $\bar{D}$ at

   (*i*) any point on the z-axis lying outside the disk ($|z| > \frac{t}{2}$) and

   (*ii*) for any point inside the disk.

5. Dipoles, each of moment $\bar{p}$, are located at the corners of a tiny cube, with all the $\bar{p}$ vectors in the same direction. Such tiny cubes with the dipoles at their corners are arranged in the form of a three dimensional lattice to fill completely the volume of a sphere of radius R_0, whose centre is at the centre of one of the cubes. Show that the resultant field intensity at the centre of the sphere is zero.

   [**Hint :** Use Eqn. (3.23).]

CHAPTER 8

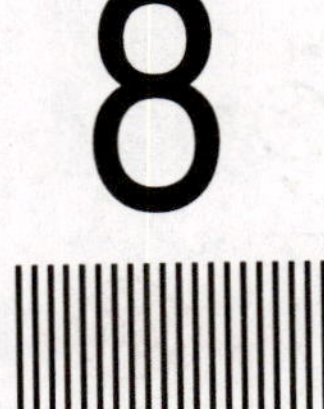

MAGNETOSTATICS

8.1 MAGNETIC FIELD PRODUCED BY ELECTRIC CURRENT

Similar to his law for electric charges, Coulomb had also obtained experimentally his law for magnetic poles, which was also an inverse square law. However, we shall abandon this high school concept of magnetism, and use a modern approach consistent with the SI system of units that we have decided to use. It is found that electric charges at rest produce the electrostatic fields which we have studied in the previous chapters. Electric charges in motion constitute an electric current, and such charges in motion produce a magnetic field, or simply that an electric current produces a magnetic field. Here we do not consider the individual electrons that are mostly responsible for the electric current because, as seen in Sec. 4.12, these individual charges have random motion and a superimposed drift motion. Instead we think of them in bulk, because the statistical average is substantially constant.

A current of I amperes flowing in a straight conductor as shown in Fig. 8.1 produces a magnetic field around it directed along a circle in a plane normal to I in the direction of the curved arrow. (One is taught in schools that this is the direction in which the north pole (more correctly, the north-seeking pole) of a magnetic compass needle will be urged to move). This curved arrow is the direction in which a right-handed screw will have to be turned to make it progress in the direction of the current I. Another easy way to get the direction of the magnetic field is to hold the conductor in the right hand with the thumb pointing in the direction of the current, then the fingers of the right hand give the direction of the magnetic field.

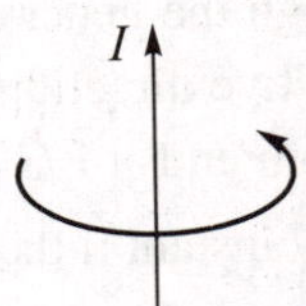

Fig. 8.1. Magnetic field of electric current

8.2 MAGNETIC FLUX

In Fig. 8.2 (a) near a straight conductor carrying a steady direct current I amperes, a coil is placed and ends of the coil are connected to a ballistic galvanometer BG. If a pulse of current flows through the galvanometer, and the time for which the pulse of current flows is small compared to the period of oscillation of the galvanometer, then the deflection or throw of the galvanometer is proportional to the total charge Q, which flows through the galvanometer circuit. In other words, the ballistic property of the galvanometer has been used to perform an integration with respect to time and to give

$$\int_0^t i\,dt = \int_0^t \frac{dq}{dt}\,dt = Q$$

As seen from Fig. 8.1, the coil being in the plane of the paper, the direction of the magnetic field near the coil is going into the paper. This is indicated by the crosses shown inside the coil in Fig. 8.2 (a) according to the convention that if an arrow (as at b) goes into the paper, the feathers near its tail will look like a cross. (If the field were coming out of the paper, it would have been shown by an array of dots instead of crosses, to represent the pointed end of the arrow head).

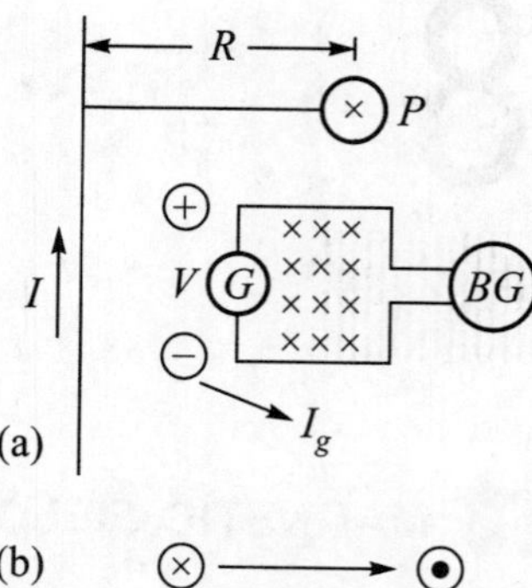

Fig. 8.2. Relation between magnetic flux and induced voltage

Let us imagine that at time $t = 0$, the current in the conductor is zero, and that it increases to a value I at time t, where the duration 0 to t is small compared to the period of one oscillation of the ballistic galvanometer. Then, as discovered by Faraday in 1831, the changing magnetic field linking with the coil would induce an electromotive force in the coil, which is depicted in the diagram as generator G in the coil circuit. Before stating whether this induced emf or voltage V should be called positive or negative, we must first decide which polarity of V we would call positive. We *assign* this positive direction by noting that to produce a magnetic field going into the paper (in other words, in the *same direction* as the field produced by the positive current I), the current through the generator G must flow from bottom to top of the coil. Hence the *assigned* positive direction for voltage V is as shown by the signs + and – *in circles* at top and bottom ends of G or V.

At any instant if the voltage in the coil-galvanometer circuit is V, a current $I_g = \frac{V}{R_g}$ will flow in this circuit, and as the current I is being established in the straight conductor during the time 0 to t, the total charge that will flow through the ballistic galvanometer will be

$$Q = \int_0^t I_g\,dt = \int_0^t \frac{V}{R_g}\,dt = \frac{1}{R_g}\int_0^t V\,dt \qquad ...(8.1)$$

when the current $I = 0$, I_g is also zero. At time t, the current in the straight conductor becomes I and remains I thereafter. Since only a changing magnetic field can induce a voltage in a stationary coil, the current I_g stops flowing after time t. During the time interval 0 to t, the magnetic flux linking the coil changes from 0 the value Φ say. Faraday showed that this magnetic flux is always equal to the integral of the induced voltage V with respect to time or

$$\Phi = -\int_0^t V\,dt \qquad ...(8.2)$$

The negative sign arises from the fact that the actual direction of flow of current I_g when current I is being established (changing from 0 to I) is as shown in Fig. 8.2 (a), that is, it is opposed to the *assigned* positive direction for V.

If the current I were flowing at time $t = 0$, and it was reduced to zero by time t, the flux Φ would have got reduced to value 0 and the change of magnetic flux from time 0 to t would be

$$0 - \Phi = \int_0^t V\, dt$$

because the current I_g would then be flowing in the direction opposite to that shown in Fig. 8.2 (a), that is, it would be flowing in the positive direction of V.

In both the above cases, the actual direction of flow of the induced current I_g is such that it tries to maintain that value of the flux which existed before the current I began to change at time $t = 0$. This is Lenz's law enunciated in 1834 by the Russian physicist Henri Frederic Emile Lenz (1804 - 1885). In both cases, Eqn. (8.2) is obtained. Hence differentiating this equation, we write

$$V = -\frac{\partial \Phi}{\partial t} \quad \text{...(8.3)}$$

This equation is called the Faraday's law of induction.

The Eqn. (8.2) shows that the units for magnetic flux Φ are volt-second. The old terms for the unit of magnetic flux were "lines" and "Maxwells" but the name in SI units is "Weber" (abbreviated Wb) in honour of the German physicist, Wilhelm Eduard Weber (1804 – 1891), one weber being equal to 10^8 lines or 10^8 maxewells.

8.3 MAGNETIC FLUX DENSITY OR MAGNETIC INDUCTION

The magnetic flux density, denoted by $\overline{\boldsymbol{B}}$, could be defined as the magnetic flux per unit area perpendicular to the direction of the magnetic flux. The old unit for magnetic flux density was called Gauss (Gs), one Gauss being equal to one line or one maxwell per square centimetre. In the MKS system of units, the unit for magnetic flux density was weber per square metre or Wb/m^2. It is obvious that one Wb/m^2 is equal to 10^4 gauss.

In the SI system of units the magnetic flux density is called the magnetic induction and its unit is called the tesla (T), so named in honour of the Serbian born Nikola Tesla (1856 – 1943) who migrated to America and there invented the rotating magnetic field, and three phase alternating current and the induction motor. In SI system, the tesla is defined first as the induction of such a field in which, on each metre length of a conductor carrying a steady direct current of one ampere and arranged perpendicular to the direction of the vector of magnetic induction, a mechanical force of one newton will act. The unit weber of magnetic flux is then derived from the tesla as the magnetic flux produced by an induction of one tesla through an area of one square metre perpendicular to the vector of magnetic induction. It is seen that one tesla = one Wb/m^2 = 10^4 gauss.

Obviously, the magnetic induction $\overline{B}$ is a vector quantity and we can obtain the magnetic flux Φ from it by the relation

$$\Phi = \int_{\text{surface}} \overline{B} \cdot \overline{da} = \int_{\text{surface}} \overline{B} \cdot \overline{u}_n \, da \qquad \text{...(8.4)}$$

where $\overline{da}$ is the area vector.

8.4 MAGNETIC FIELD INTENSITY

Experiments conducted to measure the magnetic flux density around a straight conductor carrying a steady direct current indicate that the value of the magnetic induction $\overline{B}$ is dependent on the magnitude of the current I in the conductor and the medium surrounding the conductor, but is inversely proportional to the perpendicular distance R of the point P from the conductor. The Coulomb's law shows a spherical symmetry about the point charge and hence a factor 4π is used in it in the denominator. In the case of this magnetic field, there is a cylindrical symmetry about the current carrying conductor. Hence, we write [see Fig. 8.2 (a)]

$$\overline{B} = \frac{\mu I}{2\pi R} \overline{u}_\phi \quad \text{Wb/m}^2 \qquad \text{...(8.5)}$$

where μ is a constant called the permeability of the medium, and $\overline{u}_\phi$ is the unit vector going into the paper at the point P.

For vacuum the permeability is denoted by the symbol μ_0. The ratio $\dfrac{\mu}{\mu_0}$ is then a number denoted as μ_r, the relative permeability of the medium. Hence

$$\mu = \mu_r \, \mu_0$$

or

$$\mu_r = \frac{\mu}{\mu_0} \qquad \text{...(8.6)}$$

To find the magnitude and the units for m0, we use a result proved by Maxwell [see eqn. (11.7b)] that the velocity of light in vacuum » 3×10^8 m/s $= \dfrac{1}{\sqrt{\mu_0 \varepsilon_0}}$. Putting the approximate value of ε_0 as $\dfrac{10^{-9}}{36\,\pi}$, we get

$$\mu_0 = \frac{1 \times 36\,\pi}{9 \times 10^{16} \times 10^{-9}}$$

$$= 4\mu \times 10^{-7} \quad \text{H/m} \qquad \text{...(8.7)}$$

It has been decided to make the value of μ_0 *arbitrarily as exactly equal* to $4\,\pi \times 10^{-7}$. Even if more accurate measurements later give us a value for the velocity of light (or electomagnetic waves) in vacuum different from 3×10^8 m/s, it is the value of ε_0 which we shall adjust [see derivation of Eqn. (2.9)]. and *not* the value of μ_0.

Equation (8.2) shows that the magnetic flux Φ has dimensions of volt-second. This makes the dimensions of the magnetic induction as (volt-second)/(metre)^2. Using these values in Eqn. (8.5), the dimensions of the permeability are

$$[\mu] = \left[\frac{BR}{I}\right] = \left[\frac{\text{(Volt-Second)}}{\text{(Metre)}^2}\,\frac{\text{(Metre)}}{\text{(Ampere)}}\right]$$

We simplify this by ncting that in circuit theory, the volt-ampere relation for an inductance is $e = L\,\dfrac{di}{dt}$, which gives the dimensions

$$[\text{Volt}] = \left[\text{Henry}\,\frac{\text{Ampere}}{\text{Second}}\right]$$

Using this relation in above, the dimensions of μ are found to be henry per metre as indicated in Eqn. (8.7) in anticipation.

Dividing Eqn. (8.5) by the permeability μ, we write

$$\overline{H} = \frac{\overline{B}}{\mu} = \frac{I}{2\pi R}\,\overline{u}_\phi \quad \text{ampere per metre} \qquad ...(8.8)$$

where

$$\overline{B} = \mu\,\overline{H} \quad \text{weber per (metre)}^2 \qquad ...(8.9)$$

This quantity $\overline{H}$ is called the magnetic field intensity. Like the magnetic induction $\overline{B}$, it is also a vector quantity, but the units of $\overline{H}$ are ampere per metre. At every point in space, the directions of $\overline{B}$ and $\overline{H}$ are the same, since μ is only a scalar quantity.

8.5 BIOT-SAVART LAW OR AMPERE'S CURRENT LAW

In 1820, the French physicists, Jean Baptiste Biot (1774–1862) and Felix Savart (1791–1841) announced, based on their experiments, a formula for determining the infinitesimally small magnetic field intensity $\overline{dH}$ produced at a point P by an infinitesimally small segment of current $I\,\overline{ds}$, and this is known as the Biot-Savart law. In the same year, another French physicist, Andre-Marie Ampere (1775–1836), experimenting independently, also gave the same law which is therefore also referred to as Ampere's current law, but in general, it is commonly known as Biot-Savart law. It states (Fig. 8.3) that

$$\overline{dH} = \frac{I\,(\overline{ds} \times \overline{u}_r)}{4\pi\, r^2} \text{ A/m} \qquad ...(8.10)$$

or

$$\overline{dH} = \frac{(I\,ds)\sin\theta}{4\pi\, r^2}\,\overline{\mu}_\phi \text{ A/m} \qquad ...(8.11)$$

where r is the distance of point P from $\overline{ds}$. The vector $\overline{dH}$ is directed into the paper in the direction of the unit vector $\overline{u}_\phi$.

Integrating over the entire length of the current element, we get

$$\overline{H} = \oint \frac{I\,(\overline{ds} \times \overline{u}_r)}{4\pi\, r^2} \text{ A/m} \qquad ...(8.12)$$

where the circle on the integral sign indicates that the integration is to be performed over the entire length of the closed circuit or loop in which the current I flows.

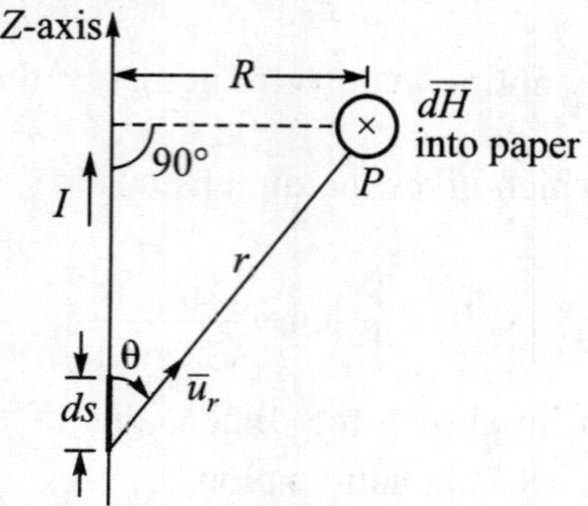

Fig. 8.3. Biot-Savart law

8.6 FORCE ON CURRENT ELEMENT IN MAGNETIC FIELD

It was also found experimentally that if a current element $I\overline{dL}$ is located at a point in a magnetic field where the magnetic induction is $\overline{B}$ weber/(metre)^2, then the current element experiences a force $\overline{dF}$ which is given by

$$\overline{dF} = I\ \overline{dL} \times \overline{B} \text{ newton} \qquad ...(8.13)$$

where I is in amperes, $\overline{dL}$ in metres and $\overline{B}$ in tesla or Wb/m^2. The cross-product of the vectors on the right hand side indicates that the force is at right angles to the plane containing the vectors $\overline{dL}$ and $\overline{B}$.

8.7 FORCE BETWEEN CURRENT ELEMENTS

In Eqn. (8.13) if we assume that the magnetic induction $\overline{B}$ (= μ $\overline{H}$) is produced by another current as per Eqn. (8.12), then we get an expression for the force between two current carrying conductors, which would be the magnetic equivalent of Coulomb's law for force between two point charges.

In Fig. 8.4 I_1 and I_2 are the steady direct currents flowing in two infinitely long, parallel conductors. At a point P_1 on conductor 1, the magnetic field intensity due to current I_2 is given by Eqn. (8.8) and the magnetic induction $\overline{B}_2$ is given by Eqn. (8.9). Hence

$$\overline{B}_2 = \frac{\mu_0\, I_2}{2\pi\, R}\, \overline{u}_\phi$$

as shown. Therefore the force $\overline{dF}_1$ on the current element $I_1\ \overline{dL}_1$ is

$$\overline{dL}_1 = I_1\ \overline{dL}_1 \times \overline{B}_2 \qquad ...(8.14)$$

or force per unit length of conductor 1

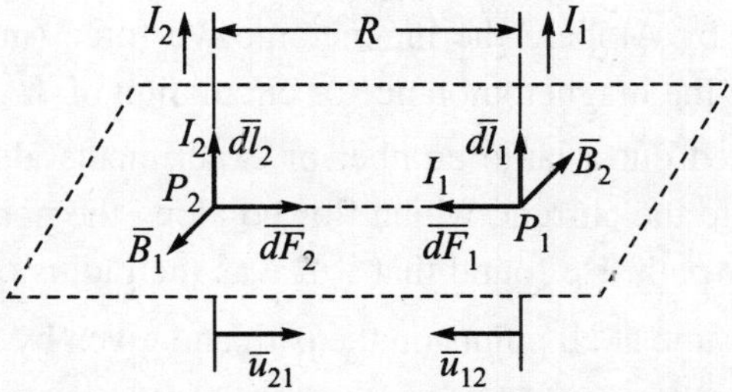

Fig. 8.4. Force between current in parallel conductors

$$= \frac{\overline{dF_1}}{\overline{dL_1}} = \frac{\mu_0 \, I_1 \, I_2}{2\pi \, R} \overline{u}_{12} \quad \text{N/m} \qquad ...(8.15)$$

At P_1 the direction of $\overline{dF}_1$ is towards the current I_2. The product $I_1 I_2$ in Eqn. (8.15) shows that the magnitude to the force acting per metre length of conductor 2 at P_2' is equal in magnitude to $\overline{dF}_1$ but its direction is towards conductor 1, so that we can write at P_2.

Force per unit length of conductor 2

$$= \frac{\overline{dF_2}}{\overline{dL_2}} = \frac{\mu_0 \, I_1 \, I_2}{2\pi \, R} \overline{u}_{21} \quad \text{N/m} \qquad ...(8.15a)$$

Notice that in both these cases the trick used is the same as was used in defining the electric field intensity in Section 2.4. Even though $\overline{dF} \to 0$ when $\overline{dL} \to 0$, their ratio always remains a constant. This shows that unlike the case of electric charges, "like (in same direction) currents attract, but opposite currents repel".

Equation (8.15) forms the basis of the definition of the Ampere in SI units. Thus if $I_1 = I_2 = 1$ ampere and $R = 1$ metre, then

$$\frac{\overline{dF_1}}{\overline{dL_1}} = \frac{\mu_0}{2\pi} = \frac{4\pi \times 10^{-7}}{2\pi} = 2 \times 10^{-7} \text{ N/m.}$$

If equal, steady and direct currents I_1 and I_2 flowing in infinitely long, parallel conductors separated from each other in vacuum by a distance of one metre, experience a force of 2×10^{-7} N/m length of each, then each current has a strength of one ampere.

It is obvious that the total force on the conductor is obtained by integrating Eqn. (8.14) to get

$$\overline{F}_1 = \oint I_1 \, (\overline{dL}_1 \times \overline{B}_2) \qquad ...(8.16)$$

the integration being done around the entire closed path of the current.

8.8 LINE INTEGRAL OF MAGNETIC FIELD INTENSITY

Using the notation of Eqn. (3.2) we can write an expression for the line integral of the magnetic field intensity $\overline{H}$ along a given line from point a to point b, as the line integral

$$\mathcal{F} = \int_a^b \overline{H} \cdot \overline{ds} \qquad ...(8.17)$$

where $\mathcal{F}$ was called by Ampere the magnetomotive force (mmf) between a and b. Others have called it the magnetomotance or circulation of $\overline{H}$.

Ampere carried out a large number of experiments along circular paths in a plane perpendicular to the current, which flowed along the normal to the plane at the centre of the circular path. He found that if R was the radius of the circular path, the value of $\overline{H}$ was the same at all points on the path and given by Eqn. (8.8), from which he concluded that the line integral around a closed path was

$$\mathcal{F}_{\text{circle}} = \oint \overline{H} \cdot \overline{ds} = \oint \frac{I}{2\pi R} ds$$

$$= \frac{I}{2\pi R}(2\pi R) = I \qquad \text{...(8.18)}$$

Equation (8.18) is referred to as Ampere's circuital law or ampere's work law. In words it may be stated as

$$\text{Magnetomotive force} = \text{Ampere-turns (At)} \qquad \text{...(8.19)}$$

the ampere-turns being the product of the current in amperes with the number of complete turns in the coil carrying the current. In Eqn. (8.18) we have made only one complete turn around the current along the circular path, whose perimeter is $2\pi R$. If we had made N complete turns, we would have got the length of the path as $N(2\pi R)$. Or if there were N turns in the coil, the current in each being I, then the magnetic field intensity would have been $N\overline{H}$. In the both these cases.

$$\mathcal{F}_{\text{circle}} = N I \text{ ampere-turns.}$$

It must be remembered that the line integral of Eqn. (8.18) has come out to be positive, because we have gone round the circular path in the direction of $\overline{H}$. If we had travelled in the opposite direction, the integral would have given a negative value – I. The rule for determining the sign is : If a right handed screw, when turned in the direction from a to b, progresses in the direction of the current, then the positive sign is to be taken.

The fact that Eqn. (8.18) differs from Eqn. (3.2) should be carefully noted. Since the line integral along a closed circular path in the magnetic field around a current is equal to the current enclosed (and not equal to zero), it follows that the magnetic field is not a conservative field, and the line integral depends on the path from a to b. However, Ampere found experimentally that for closed paths, the line integral equals the current enclosed even if the current is not at the exact centre of the path.

8.9 MAGNETIC FLUX THROUGH COMPLETELY ENCLOSED SURFACE

A very important difference between the electrostatic and the magnetostatic field must have become obvious by now. In electrostatic fields, the flux lines start from positive charges and end on negative charges. But in magnetostatic fields, the magnetic flux lines form continuous loops, they have no beginning and no end. Therefore, for any completely enclosed Gaussian surface that we may choose (Fig. 8.5), whether it encloses the current carrying conductor or not, whatever flux enters the surface, exactly the same flux also leaves it. The flux lines do not stop or end at any point within any Gaussian surface. It follows therefore that the surface integral of the magnetic induction $\overline{B}$ taken over the entire Gaussian surface will always be zero, because

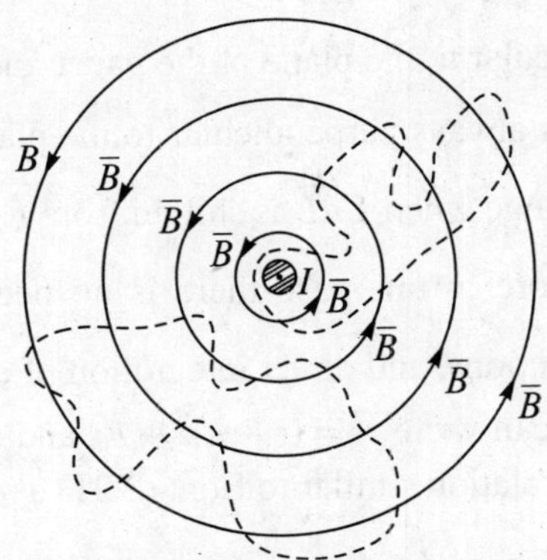

Fig. 8.5. Magnetic flux through Gaussian surfaces

Magnetic flux entering surface = Magnetic flux leaving surface or

$$\oint \overline{B} \cdot \overline{da} = \oint \overline{B} \cdot \overline{u}_n \, da = 0 \qquad \text{...(8.20)}$$

This looks like Gauss's law, Eqn. (4.11) applied to the restriced charge-free regions, where charge density $\rho = 0$.

8.10 MAGNETIC FIELD INTENSITY DUE TO A SHORT, STRAIGHT CURRENT ELEMENT

Obviously the solution to this problem is best expressed in cylindrical co-ordinates. Notice the similarity with the electrostatics case of Sec. 2.10. The notation of Fig. 8.6 is therefore made the same as that in Fig. 2.7. Observe the similarity in expressions for the electric and the magnetic field intensities.

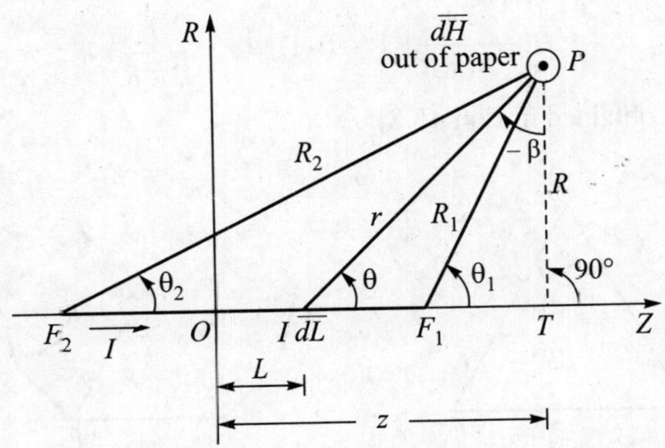

Fig. 8.6. Magnetic field intensity due to short line

$$\overline{dE}_R = \frac{ql\, dL \sin\alpha}{4\pi\varepsilon_0 r^2}\,\overline{u}_R \qquad (2.19)$$

$$\overline{H} = \frac{I\, dL \sin\theta}{4\pi\, r^2}\,\overline{u}_\phi \qquad (8.11)$$

Here $\overline{u}_\phi$ is perpendicular to the plane of the paper and coming out of the paper at point P. The vector $\overline{u}_\phi$ is always perpendicular to the plane containing the vectors $\overline{dL}$ and $\overline{r}$. As $\overline{dL}$ and $\overline{r}$ change when $I\,\overline{dL}$ is shifted from F_2 to F_1, $\overline{dH}$ remains in the same direction $\overline{u}_\phi$. Therefore in this case there is no need to resolve $\overline{dH}$ in two mutually perpendicular directions, and direct integration of $\overline{dH}$ is possible.

As in Sec. 2.10, we can write $r^2 = (z - L)^2 + R^2$ and substitute $z - L = R \cot\theta$ and by integration obtain a relation similar to Eqn. (2.21) as

$$\overline{H} = \frac{I}{4\pi R}[\cos\theta_2 - \cos\theta_1]\,\overline{u}_\theta \qquad ...(8.21)$$

Unlike the electrostatics case, it is not necessary in this magnetostatics problem to carry out elaborate calculations for obtaining the equation to the direction lines. Equation (8.21) shows that the magnetic field intensity is always directed at P coming out of the paper as indicated by the unit vector $\overline{u}_\phi$. If we revolve Fig. 8.6 about the z-axis, the point P will describe a circle in the plane at right angles to the z-axis. The vector $\overline{u}_\phi$ will always lie in this plane and be tangential to the circle at every point. Hence this circle is the direction line passing through the point P. The direction lines in three dimensions therefore form a family of circles in planes perpendicular to the z-axis, with centre on the z-axis and radius $R = (T\,P)$ a constant.

It is obvious that the result of Eqn. (8.21) can be extended to an infinite line if we make $\theta_2 \to 0$ and $\theta_1 \to \pi$. Thus for infinite straight conductor carrying steady, direct current

$$\overline{H} = \frac{I}{4\pi R}[(1) - (-1)]\,\overline{u}_\phi = \frac{I}{2\pi R}\,\overline{u}_\phi$$

as experimentally obtained in Eqn. (8.8).

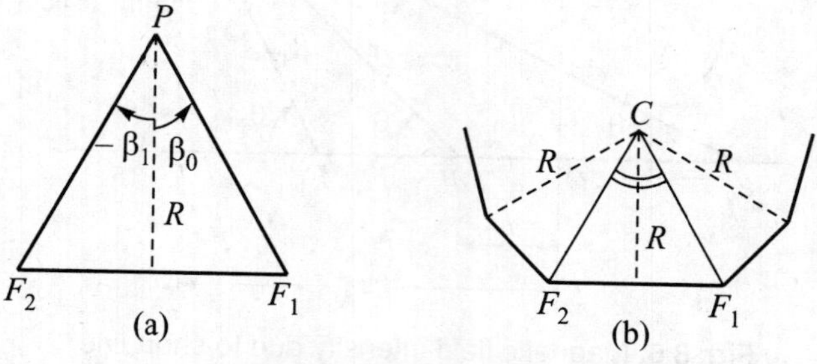

Fig. 8.7. Special case of Fig. 8.6

For certain types of problems, particularly when the point P is situated on the R-axis and there is a symmetry about the R-axis [see Fig. 8.7(a)], a solution with angles at P is better than one with angle with the current-carrying conductor as was done in Ex. 12, Prob. 13. Then (Fig. 8.6) calling the angle APT at P as $(-\beta)$ we can substitute

$$(z - L) = R \tan(-\beta),$$

and write $$-dL = -R \sec^2 \beta \, d\beta,$$

and $$r^2 = R^2 + (z - L)^2$$

$$= R^2 \left[1 + \left(\frac{z - L}{R}\right)^2\right] = R^2 [1 + \tan^2 \beta] = R^2 \sec^2 \beta$$

and $\sin\theta = \cos\beta$. This changes Eqn. (8.21) as

$$\overline{dH} = \frac{I (R \sec^2 \beta \, d\beta) \cos\beta}{4\pi R^2 \sec^2 \beta} \overline{\mu}_\phi = \left[\frac{I}{4\pi R} \cos\beta \, d\beta\right] \overline{\mu}_\phi$$

Hence for the point P in Fig. 8.7 (a)

$$\overline{H} = \frac{I \overline{\mu}_\phi}{4\pi R} \int_{-\beta_0}^{\beta_0} \cos\beta \, d\beta = \left[\frac{I}{2\pi R} \sin\beta_0\right] \overline{u}_\phi \qquad \text{...(8.22)}$$

If a conductor is bent into the form of a regular polygon of N sides, then [see Fig. 8.7(b)] angle $F_2CF_1 = 2\beta_0 = \frac{2\pi}{N}$ so that $\beta_0 = \frac{\pi}{N}$. If a steady, direct current flows round this polygon whose sides are at a perpendicular distance R from the centre C, then each side will contribute a magnetic field intensity $\overline{H}$ as given by Eqn. (8.22) and the resultant or

Total $\overline{H}$ at C due to N sides

$$= N\left[\frac{I}{2\pi R} \sin\left(\frac{\pi}{N}\right)\right] \overline{\mu}_\phi \qquad \text{...(8.23)}$$

If N is increased indefinately to infinity, the polygon will become a circle. Then writing Eqn. (8.23) in the form

$$\overline{H} = \frac{I}{2R} \frac{\sin\left(\frac{\pi}{N}\right)}{\left(\frac{\pi}{N}\right)} \overline{\mu}_\phi$$

and noting that as $N \to \infty$,

$$\lim_{N \to \infty} \frac{\sin\left(\frac{\pi}{N}\right)}{\left(\frac{\pi}{N}\right)} = 1 \qquad \text{...(8.24)}$$

Therefore for circular loop

$$\overline{H} = \frac{I}{2R} \overline{\mu}_\phi \quad \text{A/m at centre} \qquad \text{...(8.25)}$$

8.11 MAGNETIC FIELD INTENSITY DUE TO CURRENT IN A CONDUCTOR OF FINITE DIAMETER

Assume that a steady, direct current I flows in a cylindrical conductor, infinite in length, and of radius a metre. Further assume that the current density J amperes per square metre [Fig. 8.8 (a)] is uniform over the cross-section of the conductor, so that

$$J = \frac{I}{(\pi a^2)}$$

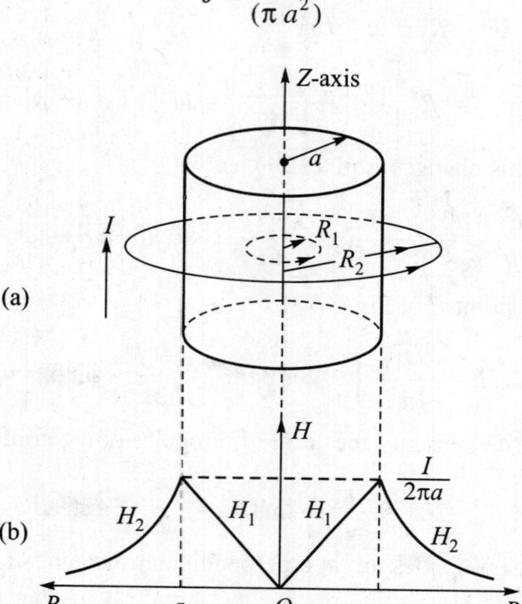

Fig. 8.8. Magnetic field intensity due to infinitely long conductor of finite diameter

8.11.1 (i) For $R < a$

In this case we have the line integral evaluated along a circle of radius R_1 so that

$$\oint \overline{\boldsymbol{H}_1}\, \overline{\boldsymbol{ds}} = H_1\,(2\pi R_1)$$

The current enclosed is

$$J\left(\pi R_1^2\right) = \frac{I R_1^2}{a^2}$$

Equating these, as in Eqn. (8.18),

$$H_1 = \frac{I}{2\pi a^2} R_1 \qquad \text{A/m} \qquad \ldots(8.26)$$

At $R_1 = 0$, $H_1 = 0$ and at $R_1 = a$, $H_1 = \dfrac{I}{2\pi a}$. Between these limits, H_1 increases linearly with R_1 as shown in Fig. 8.8 (b).

8.11.2 (ii) For $R > a$

In this case since a circle of radius R_2 (> a) encloses the entire current I, we get

$$H_2\,(2\pi\, R_2) = I$$

or
$$H_2 = \frac{I}{2\pi\, R_2} \quad \text{A/m} \qquad \text{...(8.27)}$$

Compare this result with comments on Eqn. (2.44). It is seen that for points lying outside the conductor, the total current may be assumed to be flowing along the axis of the cylinder and the result written by using Eqn. (8.8). The variation of H_2 is shown in Fig. 8.8 (*b*).

8.12 MAGNETIC FIELD INTENSITY DUE TO INFINITE PLANE SHEET OF CURRENT

In cartesian co-ordinates (Fig. 8.9), the plane sheet is at $z = +\, a$. The current is assumed to be flowing parallel to the *x*-axis, so that current density is $\overline{J} = J_x\, \overline{\boldsymbol{u}}_x$ amperes per metre width (measured along *y*-axis). By symmetry we see that at all points not lying on the plane, only H_y component can exist, because the H_x component does not exist. Also from $z = a$ to $z = -\infty$, H_y is directed along the unit vector $+\,\overline{\boldsymbol{u}}_y$, and from $z = a$ to $z = +\infty$, along $-\,\overline{\boldsymbol{u}}_y$. For any path *ABCD* in a plane x = constant,

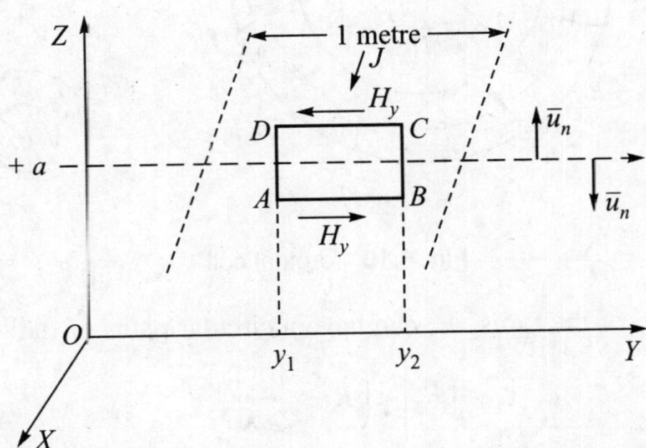

Fig. 8.9. Magnetic field intensity due to infinite plane sheet of current

with *CD* above the plane and *AB* below the plane, we get the line integral of $\overline{\boldsymbol{H}}$ as

$$\int_{ABCDA} \overline{\boldsymbol{H}} \cdot \overline{\boldsymbol{ds}} = \int_A^B \overline{\boldsymbol{H}}_y \cdot \overline{\boldsymbol{ds}} + \int_B^C \overline{0} \cdot \overline{\boldsymbol{ds}} + \int_C^D -\left(\overline{\boldsymbol{H}}_y\right) \cdot \overline{\boldsymbol{ds}} + \int_D^A \overline{0} \cdot \overline{\boldsymbol{ds}}$$

$$= 2H_y\,(y_2 - y_1)$$

where y_1 and y_2 are the *y* co-ordinates of *A*, *D* and *B*, *C*, respectively.

Also the current enclosed by $ABCDA = J_x\,(y_2 - y_1)$.

Equating these two

$$\bar{H}_y = \begin{cases} \dfrac{J_x}{2}\bar{u}_y & \text{for} \quad z < a \\ -\dfrac{J_x}{2}\bar{u}_y & \text{for} \quad z > a \end{cases}$$

or

$$\bar{H}_y = \frac{1}{2}\bar{J}_x \times \bar{u}_n \qquad \text{...(8.28)}$$

Compare this result with Eqn. (2.39) and observe their similarity.

8.13 TOROIDAL COIL

A torus (Fig. 8.10) of circular cross-section (as shown at *AA'*) has its centre at point *C*. The mean radius of the torus is R_4. It has a total of *N* turns of wire wound around it to form a toroid coil carrying a steady, direct current *I*.

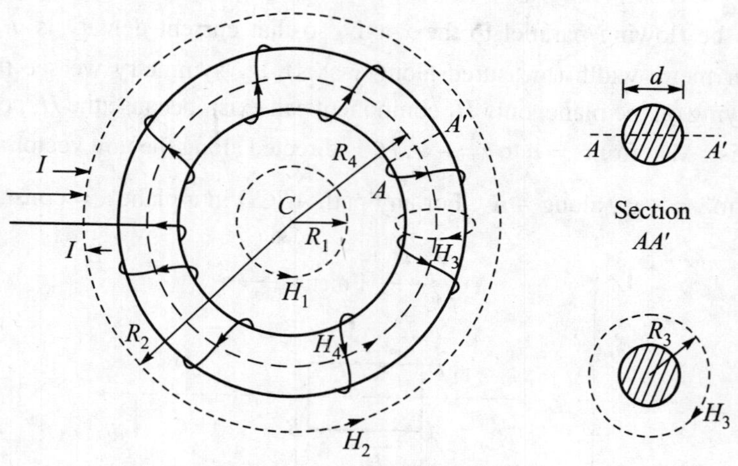

Fig. 8.10. Toroidal coil

In the plane of the torus, we can have a circular paths of radius

$$R_1 \left[< \left(R_4 - \frac{d}{2} \right) \right]$$

inside the torus, or of radius

$$R_1 \left[> \left(R_4 + \frac{d}{2} \right) \right]$$

outside the torus. Both these types of paths do not enclose any current and therefore for them, $H_1 = 0$ and $H_2 = 0$.

The third type of path is a circle in a plane through the centre and surrounding the torus. The coil wound around the torus crosses through this circle at *one* point only. Hence

$$H_3\,(2\pi\, R_3) = I$$

or $$H_3 = \frac{I}{2\pi\, R_3} \text{ A/m} \quad ...(8.29)$$

This result would have been obtained if there were only one conductor carrying the current I situated along the central circle of radius R_4 (the axis of the torque).

Finally, we take a path inside the torus along the central circle or radius R_4. Now

$$\mathcal{F} = \oint \overline{\boldsymbol{H}}_4 \cdot \overline{\boldsymbol{ds}} = H_4\,(2\pi\, R_4)$$

and current enclosed = $N\,I$

For these results we have assumed that $R_4 \gg d$ so that $R_4 - d/2 \approx R_4 + d/2 \approx R_4$. In other words, H_4 is assumed to be constant over the entire area of the cross-section of the torus. Hence

$$H_4 = \frac{NI}{2\pi\, R_4} = \left(\frac{N}{2\pi\, R_4}\right) I$$

$$= nI \quad \text{At/m} \quad ...(8.30)$$

The quantity within the bracket is denoted as n, the number of turns per metre length of coil measured along its circular axis. Comparing Eqns. (8.29) and (8.30) it is found that this number n is also nearly equal to the ratio of the magnetic field strength inside and outside the torus. Therefore, we can treat the field intensity outside the torus as negligibly small, or almost zero.

8.14 INFINITELY LONG SOLENOID

In the previous section, let us assume that the toroidal coil is made of thin wire closely wound on the torus. In other words we assume that the adjacent turns of the coil are touching each other all round the circumference of the coil. Next, in Eqn. (8.30) we let both N and R to increase but keeping their ratio always the same constant $\frac{N}{R_4} = 2\pi\, n$. Then when $N \to \infty$ and $R_4 \to \infty$, the toroidal coil becomes an infinitely long, closely wound, single-layer solenoid with n turns per metre length measured along the axis. But Eq. (8.30) still holds and

$$H_{\text{solenoid}} = nI \quad \text{At/m,} \quad ...(8.31)$$

the direction of $\overline{\boldsymbol{H}}$ being parallel to the axis of the solenoid. Also $\overline{\boldsymbol{H}}$ is constant everywhere inside the solenoid, while $\overline{\boldsymbol{H}}$ is zero outside the solenoid.

We could arrive at Eqn. (8.31) from another thinking process. As in the electrostatic case of two infinite parallel planes discussed in Section 2.16, the corresponding problem in magnetostatics of two infinite parallel planes with equal and opposite currents densities in them (see Ex. 8, Prob. 5), gives a uniform magnetic field intensity H_y between the planes where $H_y = J_x$, the surface current density in the planes. In the region outside the planes, the magnetic field intensity is zero.

Consider a small width dL of the solenoid (Fig. 8.11b). Since the solenoid is infinitely long, these two strips, of width dL at opposite ends of a diameter, are also of infinite length. Since the same current I flows through each turn of the solenoid, each strip is equivalent to a current sheet (nI) ampere per metre coming out of the paper [Fig. 8.11 (a)] at the top and a current sheet ($-nI$) ampere per metre going into the paper (as indicated by the minus sign) at the bottom. Hence this section resembles a section of width dL of the two infinite parallel planes, the strip being of infinite length. This means that in this region [Fig. 8.11(b)] the magnetic field intensity is nI

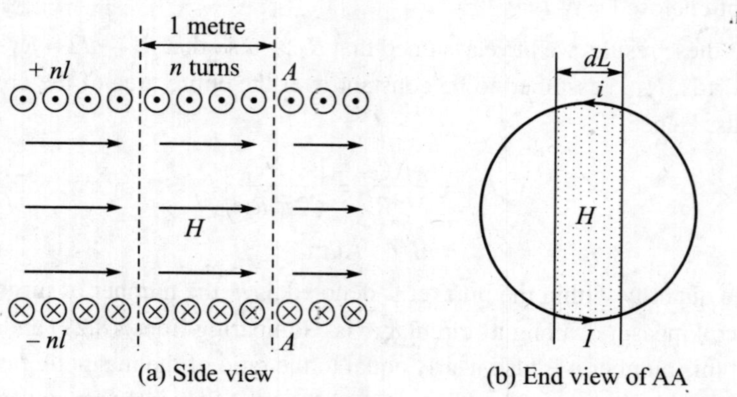

Fig. 8.11. Infinite solenoid

($= J_x$), directed as shown (to the right as at (a) and out of the paper as at (b)). This is true wherever dL is selected on the circumference of the solenoid. Hence the magnetic field intensity is (nI) At/m everywhere inside the solenoid and is zero outside the solenoid.

8.15 VECTOR MAGNETIC POTENTIAL

In electrostatics, to avoid the addition of vectors when applying Coulomb's law to get the electric field intensity at a point, we developed the concept of a potential $V = \dfrac{Q}{4\pi\,\varepsilon_0\, r}$ and obtained $\overline{E} = -\overline{\nabla} V$ by a process of differentiation. Since the electric charge Q which produces V is a scalar quantity, V is also a scalar quantity.

While looking for a similar potential function in magnetostatics, we first note a basic difference between the two types of fields. The magnetic field is produced by a current element ($I\,\overline{ds}$) which has magnitude and a direction and is therefore a vector quantity. This information about the direction of ($I\,\overline{ds}$) must be incorporated in the potential function and therefore we must now postulate a vector magnetic potential $\overline{A}$, which is such that an operation of differentiation on it will enable us to obtain the magnetic field intensity $\overline{H}$. Of the two vector differentiation operations possible, viz., $\overline{\nabla}\cdot$ and $\overline{\nabla}\times$, only the curl operation can produce a *vector* like $\overline{H}$. This leads us to define the vector magnetic potential as that function $\overline{A}$ which satisfies the relation

$$\overline{H} = \text{curl}\ \overline{A} = \overline{\nabla} \times \overline{A} \qquad ...(8.32)$$

Since $\overline{B} = \mu\overline{H}$, we could have absorbed the constant μ in the vector magnetic potential when defining $\overline{A}$, and written

$$\overline{B} = \text{curl}\ \overline{A} = \overline{\nabla} \times \overline{A} \qquad ...(8.33)$$

Logically both the forms of defining $\overline{A}$ are equally valid. So long as μ is constant and does not depend upon the magnitude of $\overline{H}$, we can use either Eqn. (8.32) or (8.33) to define the vector magnetic potential. It is found that only in the case of ferromagnetic materials does μ vary with $\overline{H}$. In that case, only Eqn. (8.33) is to be used, and this is the form which most standard reference books will be found to be using. However, as we shall not be dealing with ferromagnetic materials in this book, we shall use Eqn. (8.32) because it results in simpler equations that are symmetrical with the corresponding equations of electrostatics.

8.16 VECTOR MAGNETIC POTENTIAL OF A CURRENT ELEMENT $I\overline{ds}$

Having defined the vector magnetic potential $\overline{A}$, we proceed to fit it into the Biot-Savart Law, so that it may be consistent with other equations. Let $I\ \overline{ds}$ be an infinitesimally small current element. Choosing a cartesian system of co-ordinates with $I\ \overline{ds}$ located at the origin, we can write the magnetic field intensity at the point $P(x, y, z)$ in Fig. 8.12 as $\overline{dH}$ which has components dH_x, dH_y and dH_z along the three axes.

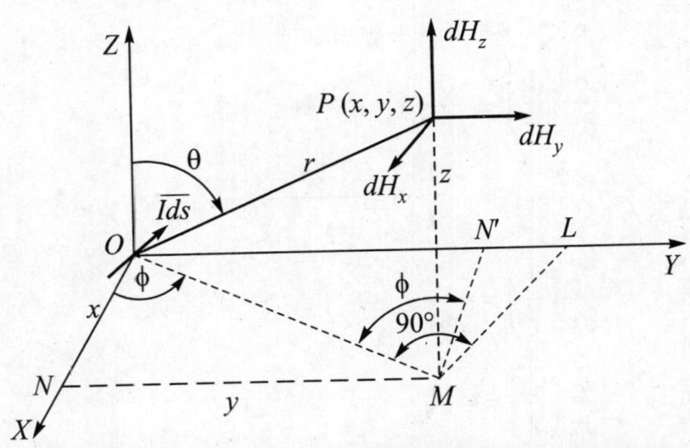

Fig. 8.12. Vector magnetic potential of a current element

Let $I\ dx$, $I\ dy$ and $I\ dz$ be the components of $I\ \overline{ds}$ parallel to the axes. We first find the contribution made by these current components to each of the field components like dH_x. Since $I\ dx\overline{u}_x \times \overline{u}_r$ will be perpendicular to the plane containing $\overline{u}_x$ and $\overline{u}_r$, it cannot have a component in the direction of $\overline{u}_x$. Hence $I\ dx$ contributes nothing to the term dH_x.

Now $I\,dz$ will give a magnetic field intensity $\dfrac{I\,dz \sin\theta}{4\pi r^2}$ perpendicular to the plane containing $\overline{dz}$ and $\bar{r}$, that is, parallel to $M L$, where $M L$ is perpendicular to OM (PM is the perpendicular from P on the x-y plane). If MN' is perpendicular to OY,

$$m\,\angle N'ML = m\,\angle OML - m\,\angle OMN'$$

$$= \left(\frac{\pi}{2} - \phi\right).$$

The component of $\left(\dfrac{I\,dz \sin\theta}{4\pi\, r^2} \text{ along } ML\right)$ resolved along MN' (parallel to x-axis) is

$$= -\frac{I\,dz \sin\theta}{4\pi\, r^2}\left[\cos\left(\frac{\pi}{2} - \phi\right)\right] = -\frac{I\,dz}{4\pi\, r^2} \sin\theta \sin\phi$$

$$= -\frac{I\,dz}{4\pi\, r^2} \sin\angle OPM \sin\angle NOM$$

$$= -\frac{I\,dz}{4\pi\, r^2}\,\frac{(OM)\;(NM)}{(OP)\;(OM)} = -\frac{I\,dz}{4\pi\, r^2}\,\frac{y}{r}$$

Finally, we write component contributed by $I\,dy$ to $d\,H_x$ as $+\dfrac{I\,dy}{4\pi\, r^2}\dfrac{z}{r}$, the plus sign occurring because the component is along $N'\,M$, the positive direction of the x-axis. The sum of these three results gives

$$dH_x = 0 + \frac{I\,dy}{4\pi}\left(\frac{z}{r^3}\right) - \frac{I\,dz}{4\pi}\left(\frac{y}{r^3}\right) \qquad \ldots(8.34)$$

Since $$r = (x^2 + y^2 + z^2)^{\frac{1}{2}}$$

$$\frac{\partial}{\partial x}\left(\frac{1}{r}\right) = \frac{\partial}{\partial x}(x^2 + y^2 + z^2)^{-\frac{1}{2}}$$

$$= -\frac{1}{2}\,\frac{2x}{(x^2 + y^2 + z^2)^{\frac{3}{2}}}$$

or $$\frac{\partial}{\partial x}\left(\frac{1}{r}\right) = -\frac{x}{r^3}$$

Similarly $$\frac{\partial}{\partial y}\left(\frac{1}{r}\right) = -\frac{y}{r^3}, \qquad \ldots(8.35)$$

and $$\frac{\partial}{\partial z}\left(\frac{1}{r}\right) = -\frac{z}{r^3}$$

Therefore using Eqn. (8.35), we can rewrite Eqn. (8.34) as shown below and, by symmetry, similar expressions for dH_y and dH_z :

$$\left.\begin{aligned} dH_x &= \frac{\partial}{\partial y}\left(\frac{I\,dz}{4\pi\,r}\right) - \frac{\partial}{\partial z}\left(\frac{I\,dy}{4\pi\,r}\right) \\ dH_y &= \frac{\partial}{\partial z}\left(\frac{I\,dx}{4\pi\,r}\right) - \frac{\partial}{\partial x}\left(\frac{I\,dz}{4\pi\,r}\right) \\ dH_z &= \frac{\partial}{\partial x}\left(\frac{I\,dy}{4\pi\,r}\right) - \frac{\partial}{\partial y}\left(\frac{I\,dx}{4\pi\,r}\right) \end{aligned}\right\} \qquad \text{...(8.36)}$$

which in vector calculus notation are abridged as

$$\begin{aligned} \overline{\boldsymbol{dH}} &= \text{curl}\left(\frac{I\,\overline{\boldsymbol{ds}}}{4\pi\,r}\right) \\ &= \overline{\nabla} \times \left(\frac{I\,\overline{\boldsymbol{ds}}}{4\pi\,r}\right) \end{aligned} \qquad \text{...(8.37)}$$

Comparing Eqns. (8.32) and (8.37) we conclude that the magnetic vector potential

$$\overline{\boldsymbol{dA}} \times \frac{I\,\overline{\boldsymbol{ds}}}{4\pi\,r} \qquad \text{...(8.38)}$$

This shows that the orientation of $\overline{\boldsymbol{dA}}$ in space is parallel to the current element ($I\,\overline{\boldsymbol{ds}}$). The magnetic vector potential is also inversely proportional to r, just like the electrostatic potential. To get the total vector magnetic potential at a point due to all current elements, we have to integrate Eqn. (8.38) and write

$$\overline{\boldsymbol{A}} = \int \frac{I\,\overline{\boldsymbol{ds}}}{4\pi\,r} \qquad \text{...(8.39)}$$

8.17 MAGNETIC FIELD INTENSITY AROUND INFINITELY LONG, STRAIGHT CONDUCTOR

In this case we first find the vector magnetic potential $\overline{\boldsymbol{A}}$ at a distance R from the conductor (Fig. 8.13). Since the current is only along the z-axis, dA_x and dA_y are equal to zero and only dA_z is to be taken at point P parallel to $I\,d_z$. Then

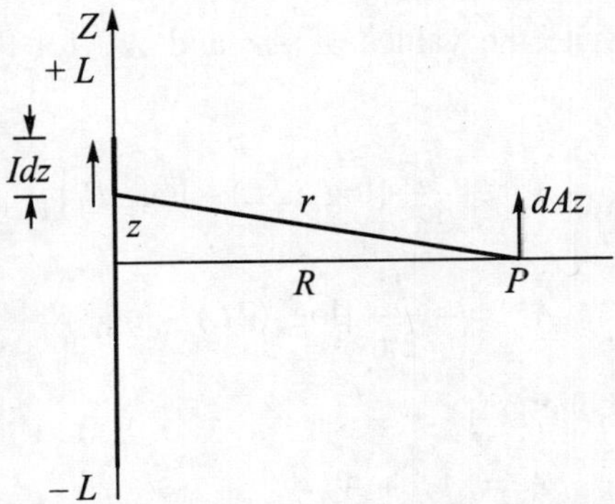

Fig. 8.13. Infinitely long straight conductor

$$A_z = \frac{1}{4\pi}\int_{-L}^{+L}\frac{I\,dz}{r}$$

$$= \frac{1}{4\pi}\int_{-L}^{+L}\frac{I\,dz}{\sqrt{z^2+R^2}} = \frac{2}{4\pi}\int_{0}^{L}\frac{I\,dz}{\sqrt{z^2+R^2}}$$

Using Eqn. (1.56), we get

$$A_z = \frac{I}{2\pi}\left[\log_e\left(z+\sqrt{z^2+R^2}\right)\right]_0^L$$

or

$$A_z = \frac{I}{2\pi}\left[\log_e\left(L+\sqrt{L^2+R^2}\right) - \log_e R\right] \quad ...(8.40)$$

This shows that A_z has the same magnitude for a constant value of R and it is independent of ϕ.

If we assume that $L \gg R$, then $L^2 + R^2 \approx L^2$ and

$$A_z \approx \frac{I}{2\pi}\left[\log_e(2L) - \log_e R\right] \quad ...(8.41)$$

Then using the expression for curl $\overline{A}$ in cylindrical co-ordinates [Eqn. (1.106)] and noting that $A_R = 0$ and $A_\phi = 0$, we get

$$\overline{\boldsymbol{H}}_\phi = \left[\frac{\partial A_R}{\partial z} - \frac{\partial A_z}{\partial R}\right]\overline{\boldsymbol{u}}_\phi = \frac{I}{2\pi R}\overline{\boldsymbol{u}}_\phi \text{ A/m (8.8)}$$

which is the same as Eqn. (8.8).

Equation (8.41) shows that as $L \to \infty$, $A_z \to \infty$, but still curl $\overline{A}$ is a finite quantity given by Eqn. (8.8). This must be carefully noted that $\overline{A} = 0$ or ∞ can still mean that curl $\overline{A}$ is finite.

8.18 INFINITE PARALLEL LINE

We choose the cartesian co-ordinate system with the origin O situated midway between the parallel lines which are spaced $2a$ apart (Fig. 8.14), the line being parallel to the z-axis. The conductors carry equal currents but in opposite directions. For a point P in the plane $z = 0$, we write the values of A_{z_1} and A_{z_2} for the two conductors using Eqn. (8.41) as

$$A_{z_1} = \frac{I_1}{2\pi}[\log_e(2L) - \log_e R_1]$$

$$A_{z_2} = -\frac{I_2}{2\pi}[\log_e(2L) - \log_e R_2]$$

With $I_1 = I_2 = I$, we get

$$A_z = A_{Z_1} + A_{Z_2}$$

$$= \frac{1}{2\pi}[\log_e(R_2) - \log_e R_1] \quad ...(8.42)$$

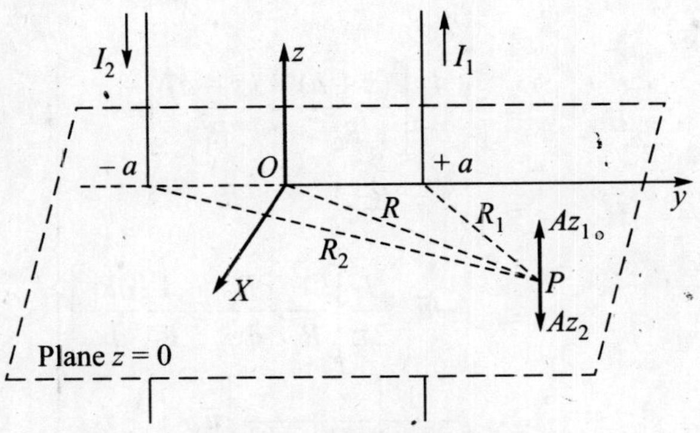

Fig. 8.14. Infinite parallel line

In Fig. (8.14) the co-ordinates of the point P being $(x, y, 0)$, we have $PN = x$, $ON = y$, and therefore

$$R_1^2 = x^2 + (y - a)^2,$$

$$R_2^2 = x^2 + (y + a)^2,$$

Hence differentiating

and
$$\left.\begin{aligned} 2R_1 \frac{\partial R_1}{\partial x} &= 2x, \\ 2R_2 \frac{\partial R_2}{\partial x} &= 2x, \\ 2R_1 \frac{\partial R_1}{\partial y} &= 2(y - a), \\ 2R_2 \frac{\partial R_2}{\partial y} &= 2(y + a), \end{aligned}\right\} \quad \text{...(8.43)}$$

Now
$$\overline{\boldsymbol{H}} = \text{curl } \overline{\boldsymbol{A}} = \begin{vmatrix} \overline{\boldsymbol{u}}_x & \overline{\boldsymbol{u}}_y & \overline{\boldsymbol{u}}_z \\ \dfrac{\partial}{\partial x} & \dfrac{\partial}{\partial y} & \dfrac{\partial}{\partial z} \\ A_x & A_y & A_z \end{vmatrix}$$

but in this case, $A_x = 0$ and $A_y = 0$ and A_z is given by Eqn. (8.42).

Hence using Eqn. (8.43), we get

$$\overline{\boldsymbol{H}}_x = \overline{\boldsymbol{u}}_x \frac{\partial A_z}{\partial y}$$

$$= \overline{\boldsymbol{u}}_x \frac{I}{2\pi} \left[\frac{I}{R_2} \frac{\partial R_2}{\partial y} - \frac{1}{R_1} \frac{\partial R_1}{\partial y} \right]$$

or

$$\left.\begin{aligned}\overline{\boldsymbol{H}}_x &= \frac{I}{2\pi}\left[\frac{(y+a)}{R_2^2}-\frac{(y-a)}{R_1^2}\right]\overline{\boldsymbol{u}}_x \\ \overline{\boldsymbol{H}}_y &= -\overline{\boldsymbol{u}}_y\frac{\partial A_z}{\partial x} \\ &= -\overline{\boldsymbol{u}}_y\frac{I}{2\pi}\left[\frac{1}{R_2}\frac{\partial R_2}{\partial x}-\frac{1}{R_1}\frac{\partial R_1}{\partial x}\right] \\ &= -\frac{1}{2\pi}\left[\frac{x}{R_2^2}-\frac{x}{R_1^2}\right]\overline{\boldsymbol{u}}_y \\ \overline{\boldsymbol{H}}_z &= 0\end{aligned}\right\} \qquad ...(8.44)$$

Again it should be noted that on the plane $y = 0$, $R_1 = R_2 = R$ and Eqn. (8.42) shows that $A_z = 0$, but Eqn. (8.44) shows that though $\overline{\boldsymbol{H}}_y = 0$, $\overline{\boldsymbol{H}}_x = \frac{I\,a}{\pi R^2}\overline{\boldsymbol{u}}_x$, a finite quantity.

8.19 MAGNETIC DIPOLE

Assume that a steady, direct current I is flowing in a conductor in the form of a circle of radius a in the plane $z = 0$, with the origin at the centre of the circle (Fig. 8.15).

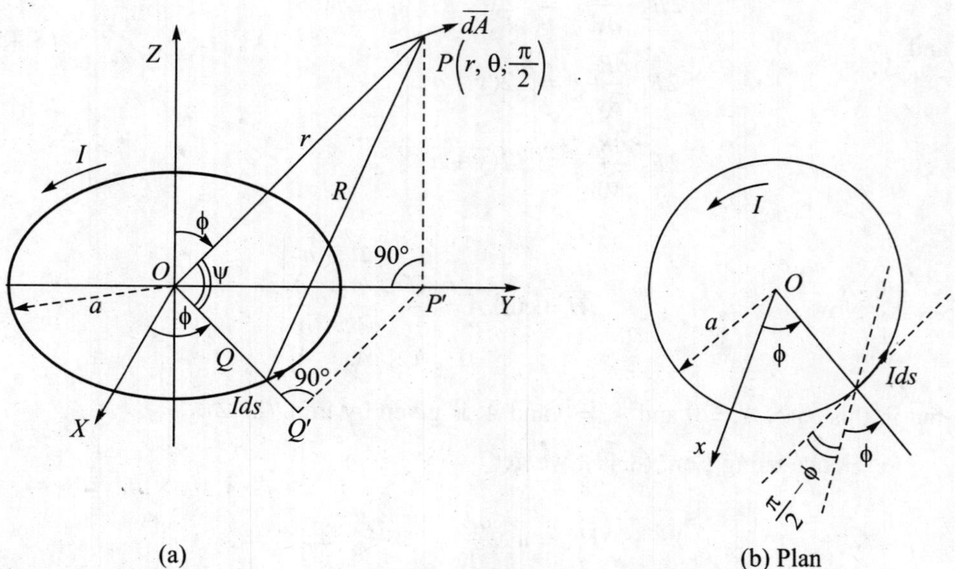

Fig. 8.15. The magnetic dipole

The infinitesimally small current element $I\,\overline{ds}$ at Q will give a vector magnetic potential at point P (r, θ, ϕ), where $\overline{dA}$ is parallel to $I\,\overline{ds}$. Since the current loop lies in the x-y plane, all elements $\overline{dA}$ will lie parallel to the x-y plane only, and there will be no component dA_z. Because of symmetry, the resolved parts of $\overline{dA}$ in other directions will cancel and only the components in the ϕ direction will exist. To simplify visualisation of the problem, we choose the x and y axes, such that the point P lies on the y-z plane $x = 0$, making its co-ordinates $(r, \theta, \pi/2)$. This makes $\overline{u}_\phi$ at P same as the vector $-\overline{u}_x$. and hence

$$\overline{A} = A_\phi\, \overline{u}_\phi = -\, A_x\, \overline{u}_x .$$

The vector magnetic potential $-dA_x\, \overline{u}_x$ contributed by the current element $I\, \overline{ds}$ is given by

$$dA_x = -\frac{I\, ds}{4\pi\, R} \cos\left(\frac{\pi}{2} - \phi\right)$$

$$= -\frac{I\, ds}{4\pi\, R} \sin\phi \qquad \text{...(8.45)}$$

since the angle between $I\, ds$ and the x-axis is $\left(\frac{\pi}{2} - \phi\right)$ [see Fig. 8.15(*b*)]. The total vector magnetic potential is obtained by integrating Eqn. (8.45) over the whole current loop. Thus noting that $ds = a\, d\phi$,

$$A_x = -\frac{Ia}{4\pi} \int_0^{2\pi} \frac{\sin\phi\, d\phi}{R} \qquad \text{...(8.46)}$$

Now the projection of $OP\ (= r)$ along the radial line OQ is $OQ' = r \cos\psi$ where ψ is the angle POQ. Let PP' be perpendicular to y-axis so that $OP' = r \sin\theta$. But OQ' will also be the projection of OP' on OQ, so that

$$OQ' = r \cos\psi$$

$$= OP' \cos \angle P'OQ'$$

$$= (r \sin\theta) \cos\left(\frac{\pi}{2} - \phi\right)$$

or

$$r \cos\psi = r \sin\theta \sin\phi \qquad \text{...(8.47)}$$

Applying the law of Cosines, Eqn. (1.31), to triangle POQ, and assuming that $r \gg a$, we get, using Binomial theorem

$$PQ^2 = R^2 = r^2 + a^2 - 2\, r\, a \cos\psi$$

or

$$R^2 \approx r^2 - 2\, r\, a \sin\theta \sin\phi$$

$$\approx r^2 \left(1 - \frac{2a}{r} \sin\theta \sin\phi\right)$$

Therefore

$$\frac{1}{R} = \frac{1}{r}\left[1 - \frac{2a}{r} \sin\theta \sin\phi\right]^{-\frac{1}{2}}$$

$$\approx \frac{1}{r}\left[1 - \left(-\frac{1}{2}\right)\frac{2a}{r} \sin\theta \sin\phi\right]$$

or $$\frac{1}{R} = \frac{1}{r}\left[1 + \frac{a}{r}\sin\theta\sin\phi\right]$$

Using this value in Eqn. (8.46)

$$A_x = -\frac{Ia}{4\pi r}\int_0^{2\pi}\left[\sin\phi + \frac{a}{r}\sin\theta\sin^2\phi\right]d\phi$$

$$= -\frac{Ia}{4\pi r}\left[\int_0^{2\pi}\sin\phi\, d\phi + \frac{a\sin\theta}{r}\int_0^{2\pi}\left(\frac{1-\cos 2\phi}{2}\right)d\phi\right]$$

$$= -\frac{Ia}{4\pi r}\frac{a\sin\theta}{r}\left[\frac{\phi}{2} - \frac{\sin 2\phi}{4}\right]_0^{2\pi}$$

$$= -\frac{Ia^2\sin\theta}{4\pi r^2}\times\pi = -\frac{(I\pi a^2)\sin\theta}{4\pi r^2}$$

Therefore, $A_\phi = -A_x$

$$= \frac{(I\pi a^2)\sin\theta}{4\pi r^2} \qquad \text{...(8.48)}$$

Using Eqn. (1.111) to express curl $\overline{A}$ in spherical co-ordinates, we get, noting that $A_r = 0$ and $A_\theta = 0$

$$\left.\begin{aligned}
\overline{H}_r &= \frac{1}{r\sin\theta}\left[\frac{\partial}{\partial\theta}(A_\phi\sin\theta) - \frac{\partial A_\theta}{\partial\phi}\right]\overline{u}_r \\
&= \frac{(I\pi a^2)\,2\cos\theta}{4\pi r^3}\overline{u}_r \\
\overline{H}_\theta &= \frac{1}{r}\left[\frac{1}{\sin\theta}\frac{\partial A_r}{\partial\phi} - \frac{\partial}{\partial r}(rA_\phi)\right]\overline{u}_\theta \\
&= \frac{(I\pi a^2)\sin\theta}{4\pi r^3}\overline{u}_\theta \\
\overline{H}_\phi &= \frac{1}{r}\left[\frac{\partial}{\partial r}(rA_\theta) - \frac{\partial A_r}{\partial\theta}\right]\overline{u}_\phi = 0
\end{aligned}\right\} \qquad \text{...(8.49)}$$

The reason why Eqn. (8.49) have been written in this particular form will become apparent, when they are compared with Eqn. (3.23) for the electric field intensity of an electric dipole. By analogy therefore, the circular current ring is called a magnetic dipole with dipole moment $\overline{m} = (I\pi a^2)$ directed, in this case, along the z-axis. This direction of the magnetic dipole moment is the direction in which a right-handed screw will travel, when it is rotated in the direction in which the current I is flowing in the circular loop. Alternatively, if we hold the current loop in the right hand with the thumb pointing in the direction in which the current I is flowing, then the fingers of the right hand will pierce the plane of the loop in the direction in which the magnetic moment is directed at the centre of the loop.

8.20 SELF-INDUCTANCE AND MUTUAL INDUCTANCE

In this and the next section, we shall consider time varying magnetic fields. It has been found in electric circuit theory that Faraday's Law of Induction in the form of

Eqn. (8.3) as $V = -d\Phi/dt$ is not very convenient to use as it involves the magnetic flux Φ, whereas in circuit theory, we wish to confine our attention only to voltages and currents in the circuit.

This law is therefore modified by associating time-varying magnetic flux with time-varying electric currents. Thus if we have three circuits 1, 2 and 3 (Fig. 8.16) carrying varying currents i_1, i_2 and i_3 with *assigned* positive directions as shown, then the voltages v_1, v_2 and v_3 in each circuit with *assigned* positive polarity as indicated, and the flux linking that circuit are related by the equations

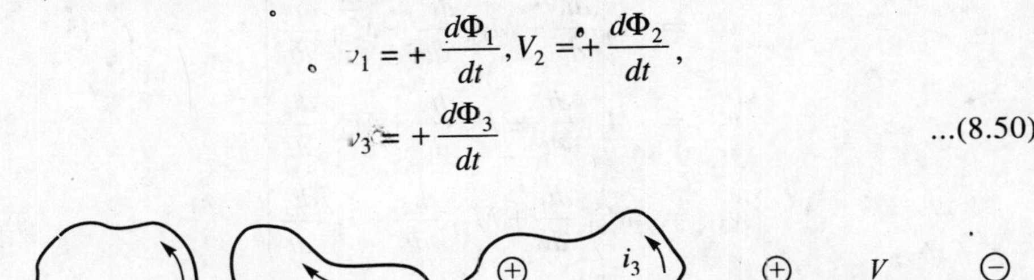

$$v_1 = +\frac{d\Phi_1}{dt}, V_2 = +\frac{d\Phi_2}{dt},$$

$$v_3 = +\frac{d\Phi_3}{dt} \qquad \text{...(8.50)}$$

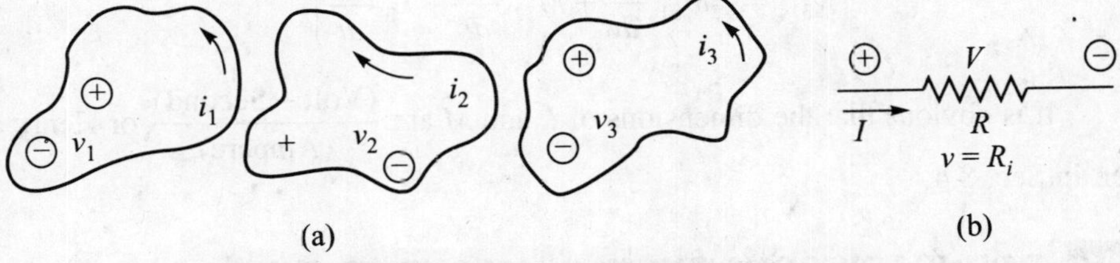

Fig. 8.16. Self-inductance and mutual inductance

The positive signs in Eqns. (8.50) are due to the fact that the assigned positive direction for a voltage v is opposed to the assigned positive direction for the current i in that circuit. In other words these are the voltage *drops* in that circuit associated with the current [see Fig. 8.16 (b)].

In any one circuit, the change of flux will be caused not only by the change of current with time in that circuit, but also by the time variations of the currents in *all the other circuits*. Therefore, we can rewrite Eqn. (8.50) as

$$\left.\begin{aligned} v_1 &= \frac{\partial\Phi_1}{\partial i_1}\frac{di_1}{dt} + \frac{\partial\Phi_1}{\partial i_2}\frac{di_2}{dt} + \frac{\partial\Phi_1}{\partial i_3}\frac{di_3}{dt} \\ v_2 &= \frac{\partial\Phi_2}{\partial i_1}\frac{di_1}{dt} + \frac{\partial\Phi_2}{\partial i_2}\frac{di_2}{dt} + \frac{\partial\Phi_2}{\partial i_3}\frac{di_3}{dt} \\ v_3 &= \frac{\partial\Phi_3}{\partial i_1}\frac{di_1}{dt} + \frac{\partial\Phi_3}{\partial i_2}\frac{di_2}{dt} + \frac{\partial\Phi_3}{\partial i_3}\frac{di_3}{dt} \end{aligned}\right\} \qquad \text{...(8.51)}$$

In these equations each Φ depends on all the three currents. Hence partial derivatives are written for this dependence. However, each current varying with time is written as a complete derivative.

A partial derivative of the form $\frac{\partial \Phi_k}{\partial i_k}$ where k = an integer (in this case 1 or 2 or 3) is called a coefficient of self-induction or simply self-inductance, and is denoted by the symbol L_k. A partial derivative of the form $\frac{\partial \Phi_k}{\partial i_m}$ where k and m are integers but $k \neq m$, is called a coefficient of mutual induction or simply mutual inductance and is denoted by the symbol M_{km}. The double subscript notation is used in the standard form, the first subscript being for the circuit in which the flux is being considered and the second subscript referring to the current causing it. This standard order (flux, current) or (effect, cause) must be strictly followed. With this standard notation, Eqn. (8.51) are usually written as

$$\left.\begin{aligned} v_1 &= L_1 \frac{di_1}{dt} + M_{12}\frac{di_2}{dt} + M_{13}\frac{di_3}{dt} \\ v_2 &= M_{21} \frac{di_1}{dt} + L_2\frac{di_2}{dt} + M_{23}\frac{di_3}{dt} \\ v_3 &= M_{31} \frac{di_1}{dt} + M_{32}\frac{di_2}{dt} + L_3\frac{di_3}{dt} \end{aligned}\right\} \quad \text{...(8.52)}$$

It is obvious that the dimensions of L and M are $\frac{\text{(Volt – Second)}}{\text{(Ampere)}}$ or Henry as seen in Sec. 8.4.

8.21 BOUNDARY CONDITIONS IN MAGNETIC FIELDS

First we start with a pill-box as in Fig. 7.6 (a) at the boundary between two magnetic materials, and by the same argument that was used in deriving Eqn. (7.12), we have, using Eqn. (8.20) that the net outward magnetic induction through the pill box [Fig. 8.17 (a)] is always zero, or

$$(B_{2n}\, a) - (B_{1n}\, a) = 0$$

where a is the area of the flat surface of the pill box. Hence

$$B_{2n} = B_{1n} \quad \text{...(8.53)}$$

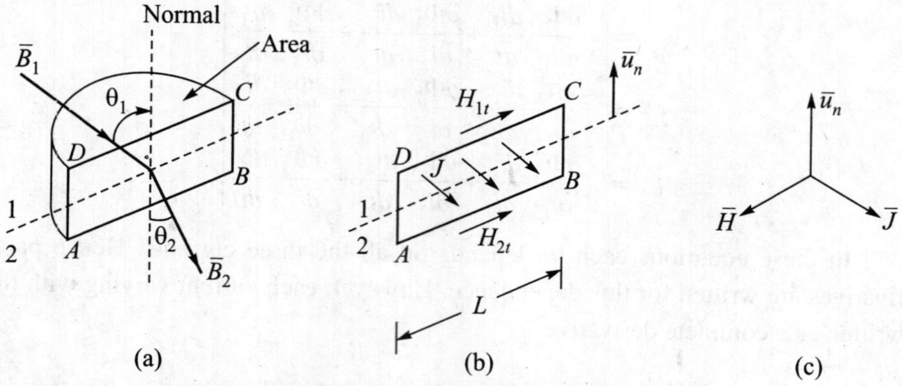

Fig. 8.17. Boundary conditions in magnetostatic fields

Again applying Ampere's circuital law, Eqn. (8.18) to the rectangular section *ABCD* of the pill box [Fig. 8.17(b) along the path *ABCDA*, we get

$$H_{2t}\,L + (-H_{1t})\,L = I_{\text{enclosed}} = J\,L$$

where J is the surface current density in A/m-width. Hence

$$H_{2t} - H_{1t} = J \qquad \text{...(8.54)}$$

If both media are perfect dielectrics with conductivity $\sigma = 0$, then $J = 0$, and in that case

$$H_{2t} = H_{1t} \qquad \text{...(8.55)}$$

If $\overline{u}_n$ is the outward normal unit vector perpendicular to the boundary plane and medium 1 is a perfect dielectric and medium 2 is a perfect conductor, then, since $\overline{E}$ is zero in a conductor, $\overline{H}$ must also be zero (for alternating field only) or $H = 0$ and Eqn. (8.4) changes to

$$H_{1t} = -J$$

or

$$\overline{J} = \overline{u}_n \times \overline{H} \qquad \text{...(8.56)}$$

as shown in Fig. 8.17 (c). Stated in words, this means that at the boundary between a perfect dielectric and a perfect conductor, the surface current density on the conductor surface is equal in magnitude to the magnetic field intensity in the dielectric just near the surface.

Also since the terms in Eqn. (8.53) can be written as

$$B_{1n} = \mu_1 H_1 \cos\theta_1,$$

$$B_{2n} = \mu_2 H_2 \cos\theta_2$$

and similarly Eqn. (8.55) can be written as

$$H_2 \sin\theta_2 = H_1 \sin\theta_1$$

by dividing, we get

$$\frac{\tan\theta_1}{\mu_1} = \frac{\tan\theta_2}{\mu_2} \qquad \text{...(8.57)}$$

8.22 SUMMARY AND COMMENTS

In this chapter we have developed the concept of the magnetic field as an effect produced by moving electric charges or electric current. It must have been noticed that we have carefully avoided any discussion of ferromagnetic materials or any mention of permanent magnets, since these subjects would have taken us into rather deep waters beyond the scope of this elementary textbook.

However, we have been able to observe the essential similarities between the electrostatic and the magnetostatic fields, particularly between the fields of linear electric charge and linear current, and also between the electric and the magnetic dipoles. The basic difference, that magnetic flux lines form loops whereas the electric flux lines begin and terminate on electric charges, will be useful when studying waveguides.

The great similarity between the equations for the electrostatic field and the magnetostatic field is very helpful in many problems. One way in which this analogy can be highlighted is to consider the relations as shown below:

Electrostatics	*Magnetostatics*
$\bar{D} = \dfrac{Q}{4\pi r^2}\bar{u}_r$	$d\bar{H} = \dfrac{I\,d\bar{s} \times \bar{u}_r}{4\pi r^2}$
$\bar{D} = \varepsilon \bar{E}$	$\bar{H} = \dfrac{1}{\mu}\bar{B}$

Just as $\bar{D}$ is dependent only on the source charge Q and is independent of the medium, similarly $d\bar{H}$ is directly related to the cause, the current element $I\,d\bar{s}$ and is independent of the medium. Also both obey the inverse square law. Thus the analogous quantities are as follows :

Electrical →	$\bar{D}$	ε	$\bar{E}$	dQ
Magnetic →	$\bar{H}$	$\dfrac{1}{\mu}$	$\bar{B}$	$I\,d\bar{s}$

But an objection can be raised against this analogy. While $\bar{D}$ and $\bar{E}$ are the electric flux density and the electric field intensity, the analogous quantities $\bar{H}$ and $\bar{B}$ are the magnetic field intensity and the magnetic induction (flux density), respectively. This difference in nomenclature probably arose because the early groups of workers in electric and magnetic subjects worked and thought more or less independently of each other, without any co-ordination in respect of nomenclature.

However, we could also draw an analogy on the basis of nomenclature in the two fields and the units of various quantities as follows:

Electrostatics	*Magnetostatics*
$\bar{D}$ = Flux Density, C/m^2	$\bar{B}$ = Induction (Flux density), Wb/m^2
$\bar{E}$ = Field Intensity, V/m	$\bar{H}$ = Field Intensity, A/m
ε = Permittivity, F/m	μ = Permeability, H/m
$\bar{D} = \varepsilon \bar{E}$	$\bar{B} = \mu \bar{H}$

The analogous quantities now become

Electrical →	$\bar{D}$	ε	$\bar{E}$
Magnetic →	$\bar{B}$	μ	$\bar{H}$

. This pairing of electrical and magnetic quantities has the advantage that it gives a symmetry to Maxwell's equations (see Chap. 10) and is the one we shall follow in that chapter. But it must be understood that both the analogies are equally correct, because, as in literature, a simile (or analogy) is used to bring out or high-light the *similarity on a few, but not all*, points.

In the next chapter we shall step aside a little to understand the motion of electric charges in electric and magnetic fields.

EXERCISE 8

1. A steady direct current I flows in a wire bent in the form of a square of side $2a$. Assuming that the z-axis passing through the centre of the square in normal to the plane of the square, find the magnetic field intensity $\overline{H}$ at any point on this z-axis. Find also the value of $\overline{H}$ at the centre of the square and verify it by using Eqn. (8.23).
2. A steady direct current I flows in a conductor forming a closed circuit of any shape. If this is placed in a uniform magnetic field of intensity $\overline{H} = H_u \overline{u}_x$, show that the total force on the conductor is zero.
3. Find the magnetic field intensity on the axis of a circular loop of radius a at distance z form the centre of the loop, if a steady direct current I flows around the loop. Verify that for $z = 0$, the value of $\overline{H}$ tallies with that given by Eqn. (8.25).
4. A coaxial cable, infinitely long, has an inner conductor of radius a, while the inner and the outer radii of the outer conductor are b and c, respectively. A steady direct current I flows through the inner conductor and returns via the outer conductor. Assume that the current densities J_i and J_o A/m^2 in the inner and the outer conductor, respectively are uniform over their cross-sections throughout each conductor. Find the variation of $\overline{H}$ as a function of R, as R varies from 0 to values greater than c, and plot a curve to show this variation.
5. Two infinite parallel conducting planes located at $z = +a$ and $z = -a$ carry steady direct currents of $J_x \overline{u}_x$ and $-J_x \overline{u}_x$ A/m-width, respectively. Find the variation of $\overline{H}$ as z varies from $-\infty$ to $+\infty$.

Chapter 9 Moving Charges in Electric and Magnetic Fields

9.1 INTRODUCTION

Till now we have been discussing the behaviour of point charges, regarding them as mathematical, abstract motions. Even when we had these charges in motion in the previous chapter, we never asked whether any mass was associated with them. In this chapter, however, we shall give a concrete, physical significance to the point charge and think of it as a material particle with a charge Q coulomb and a mass m kilogram. With this assumption of charge and mass, the particle will not only be subject to Coulomb's law, but when it moves, it will also be subject to Newton's laws of Motion. Since such movement of electric charge is called an electric current, it follows that if the charged particle moves in a magnetic field, it will experience a force as discussed in the previous chapter. These are the aspects which are considered in this chapter. However, the association of mass with the charged particle will be restricted to this chapter only.

9.2 MOVING CHARGE IN ELECTRIC FIELD

Because of the electric field intensity at every point in an electrostatic field, a particle having an electric charge Q coulomb placed anywhere in such a field will experience a force $\overline{F}_e$ given by

$$\begin{aligned}\overline{F}_e &= Q\,\overline{E} \\ &= Q\,E_x\,\overline{u}_x + Q\,E_y\,\overline{u}_y + Q\,E_z\,\overline{u}_z \qquad \text{...(9.1)}\end{aligned}$$

where E_x, E_y and E_z are the components of the electric field at the point $P(x, y, z)$.

If the mass of the particle is m kilogram and its velocity at the point P is $\overline{u}$ metre/second, it follows from Newton's second law (force is equal to the rate of change of momentum) that

$$Q\overline{E} = \frac{d}{dt}\,(m\,\overline{u}) \qquad \text{...(9.2)}$$

According to Albert Einstein's (1879–1955) Special Theory of Relativity (1905), the mass of a body depends on its velocity according to the equation

$$m = \frac{m_0}{\sqrt{1-\left(\frac{U}{c}\right)^2}} \qquad \text{...(9.3)}$$

where m_0 is called the rest mass, corresponding to $U = 0$, and c is the velocity of light in vacuum or free space.

Very high velocities are rarely encountered in ordinary work, except in nuclear physics. Hence assuming that $U \ll c$ we can regard the mass m of the particle to be constant, and rewrite Eqn. (9.2) as

$$Q\,\overline{\boldsymbol{E}} = m\frac{d}{dt}(\overline{\boldsymbol{u}})$$

$$= m\frac{d}{dt}(U_x\,\overline{\boldsymbol{u}}_x + U_y\,\overline{\boldsymbol{u}}_y + U_z\overline{\boldsymbol{u}}_z)$$

But $\quad U_x = \dfrac{dx}{dt},\quad U_y = \dfrac{dy}{dt},\quad U_z = \dfrac{dz}{dt}.$

Hence $\quad Q\overline{\boldsymbol{E}} = m\dfrac{d^2x}{dt^2}\,\overline{\boldsymbol{u}}_x + m\dfrac{d^2y}{dt^2}\,\overline{\boldsymbol{u}}_y + \dfrac{d^2z}{dt^2}\,\overline{\boldsymbol{u}}_z \qquad$ ...(9.4)

Equating the components of the force along the three axes as given in Eqns. (9.1) and (9.4), we get

$$\left.\begin{aligned} m\frac{d^2x}{dt^2} &= QE_x \\ m\frac{d^2y}{dt^2} &= QE_y \\ m\frac{d^2z}{dt^2} &= QE_z \end{aligned}\right\} \qquad \text{...(9.5)}$$

If the values of E_x, E_y and E_z in terms of the co-ordinates of the point P are substituted in Eqns. (9.5), then these equations become ordinary differential equations of mechanics, whose solution would give the equation to the curve described by the particle in the electric field. However, it must be remembered that the equation to the path will be in parametric form, the parameter being t.

9.3 MOVING ELECTRIC CHARGE IN UNIFORM MAGNETIC FIELD

In a region where the uniform magnetic flux density is $\overline{\boldsymbol{B}}$ tesla everywhere, if there is a current element $I\,\overline{\boldsymbol{dS}}$, then the force $\overline{\boldsymbol{dF}}_1$ acting on this element is given by Eqn. (8.13) as

$$\overline{\boldsymbol{dF}}_1 = I\,\overline{\boldsymbol{dS}} \times \overline{\boldsymbol{B}} \text{ newton} \qquad \text{...(8.13)}$$

Let us assume that the electric current in the results of the motion of charged particles, each carrying a charge $+Q$ coulomb, and all moving with a uniform velocity $\overline{\boldsymbol{U}}$ metre/second. Further, we assume that the element $\overline{\boldsymbol{ds}}$ has a cross-sectional area da square metre perpendicular to $\overline{U}$. Then in one second, all the charged particles contained in the volume *da U* will pass through the area *da*. If the charge density is ρ coulomb/m^3, then the current *I* is given by

$$I = \rho(da\ U) \qquad ...(9.6)$$

Since the volume of the current element is (*da ds*), the total charge in it is (ρ *da ds*) coulomb. If the number of charged particles in this current element is *n*, with each particle having charge + *Q* then the total charge in the current element can also be written as *nQ*. Thus

$$\rho\ da\ ds = n\ Q \qquad ...(9.7)$$

Multiplying Eqns. (9.6) and (9.7) and removing common factors

$$I\ ds = n\ Q\ U \qquad ...(9.8)$$

Using this value in Eqn. (8.13) given above, we get

$$\overline{\boldsymbol{dF}_1} = n\ Q\ \overline{\boldsymbol{U}} \times \overline{\boldsymbol{B}} \text{ newton}$$

and the mechanical force $\overline{\boldsymbol{F}}_m$ experienced by *each* charged particle becomes

$$\overline{\boldsymbol{F}}_m = \frac{\overline{\boldsymbol{dF}_1}}{n} = Q\,(\overline{\boldsymbol{u}} \times \overline{\boldsymbol{B}}) \text{ newton} \qquad ...(9.9)$$

This force on a moving charged particle in a magnetic field is known as the Lorentz Force in honour of the Dutch physicist Hendrik A. Lorentz (1853–1928), who in 1902 became the second person in the world to be awarded the Nobel Prize for Physics, the German inventor of X-rays, Wilhelm Konrad von Rontgen (1845–1923) being the first winner in 1901. It should be noted that the vector cross-product on the right hand side of Eqn. (9.9) shows that $\overline{\boldsymbol{F}}_m$ is perpendicular to the plane containing the velocity vector $\overline{\boldsymbol{u}}$ and the magnetic flux density $\overline{\boldsymbol{B}}$.

9.4 LORENTZ EQUATION

Newton stated an important property of forces, that they act independently of each other and their resultant effect is the sum of the effect of each. It follows that if in a region, both electric and magnetic fields are present simultaneously, then a particle carrying a charge + *Q* coulomb and moving with a velocity $\overline{\boldsymbol{U}}$ m/s will experience forces $\overline{\boldsymbol{F}}_e$ and $\overline{\boldsymbol{F}}_m$ due to the electric and magnetic fields respectively, and the resultant force will be obtained as the sum of Eqns. (9.1) and (9.9) as

$$\overline{\boldsymbol{F}} = \overline{\boldsymbol{F}}_e + \overline{\boldsymbol{F}}_m$$

$$= Q[\overline{\boldsymbol{E}} + (\overline{\boldsymbol{U}} \times \overline{\boldsymbol{B}})] \qquad ...(9.10)$$

This is the Lorentz equation. Two important points to be remembered about it are:

(*i*) The first term on the right hand side does not depend on $\overline{\boldsymbol{U}}$ and it is in the same direction as $\overline{\boldsymbol{E}}$;

(*ii*) The second term depends on $\overline{\boldsymbol{U}}$ and it is perpendicular to both $\overline{\boldsymbol{U}}$ and $\overline{\boldsymbol{B}}$.

Assume that only the magnetic field is present ($\overline{E} = 0$). Then if the mechanical forces are $\overline{F}_{m1}$ and $\overline{F}_{m2}$ corresponding to two different velocities $\overline{U}_1$ and $\overline{U}_2$, then

$$\overline{F}_{m1} = Q(\overline{U}_1 \times \overline{B})$$

and

$$\overline{F}_{m2} = Q(\overline{U}_2 \times \overline{B})$$

Using result of Exercise 1, Problem 17 (*b*), we get

$$\overline{F}_{m1} \times \overline{F}_{m2} = Q^2 [(\overline{U}_1 \times \overline{B}) \times (\overline{U}_2 \times \overline{B})]$$

$$= Q^2 [\{(\overline{U}_1 \times \overline{B}) \cdot \overline{B}\} \overline{U}_2 - \{(\overline{U}_1 \times \overline{B}) \cdot \overline{U}_2\} \overline{B}]$$

Now $(\overline{U}_1 \times \overline{B})$ is a vector perpendicular to $\overline{B}$. Hence its dot product with $\overline{B}$ is equal to zero. The equation thus reduces to

$$\overline{F}_{m1} \times \overline{F}_{m2} = -Q^2 [(\overline{U}_1 \times \overline{B}) \cdot \overline{U}_2] \overline{B}$$

$$= -Q [\overline{F}_{m1} \cdot \overline{U}_2] \overline{B}$$

Hence

$$\overline{B} = \frac{-(\overline{F}_{m1} \times \overline{F}_{m2})}{Q(\overline{F}_{m1} \cdot \overline{u}_2)}$$

$$= \frac{\overline{F}_{m2} \times \overline{F}_{m1}}{Q(\overline{F}_{m1} \cdot \overline{u}_2)} \quad ...(9.11)$$

Though Eqns. (9.10) and (9.11) are in a very compact form, in solving problems using them, it is simpler to put

$$\overline{E} = E_x \overline{u}_x + E_y \overline{u}_y + E_z \overline{u}_z,$$

$$\overline{B} = B_x \overline{u}_x + B_y \overline{u}_y + B_z \overline{u}_z,$$

$$\overline{U} = U_x \overline{u}_x + U_y \overline{u}_y + U_z \overline{u}_z,$$

$$\overline{F} = F_x \overline{u}_x + F_y \overline{u}_y + F_z \overline{u}_z,$$

9.5 ELECTROSTATIC DEFLECTION IN A CATHODE RAY TUBE

The electron gun in a cathode ray tube consists of an oxide-coated cathode *k* [Fig. 9.1 (*a*)] which emits electrons. These are accelerated by the positive anode *A* at a sufficiently high voltage V_{ak} (a few kilovolts). The anode *A* has a central hole through which the electrons pass into the region to the right with a constant velocity U_0. The cathode is enclosed in a cylinder *G* with a hole, the cylinder being maintained a few volts negative with respect to the cathode, This cylinder is called the control grid, because it controls the number of electrons going towards the anode *A*.

The electron beam passes between the deflecting plates [Fig. 9.1 (*b*)]. They are parallel plates of length L_p and separation *d* between them. The electron beam is assumed to be travelling along the *z*-axis at the middle of the plate system. The origin and the *y*-axis are as shown. The bottom plate being V_p volts positive with respect to the upper plate, the electric field intensity between the parallel plates is uniform and equal to $\bar{E} = \frac{V_p}{d}\bar{u}_y$ V/m. It is assumed that there is no field outside the plate system beyond O and P_0 and there is no fringing of the field at the edges.

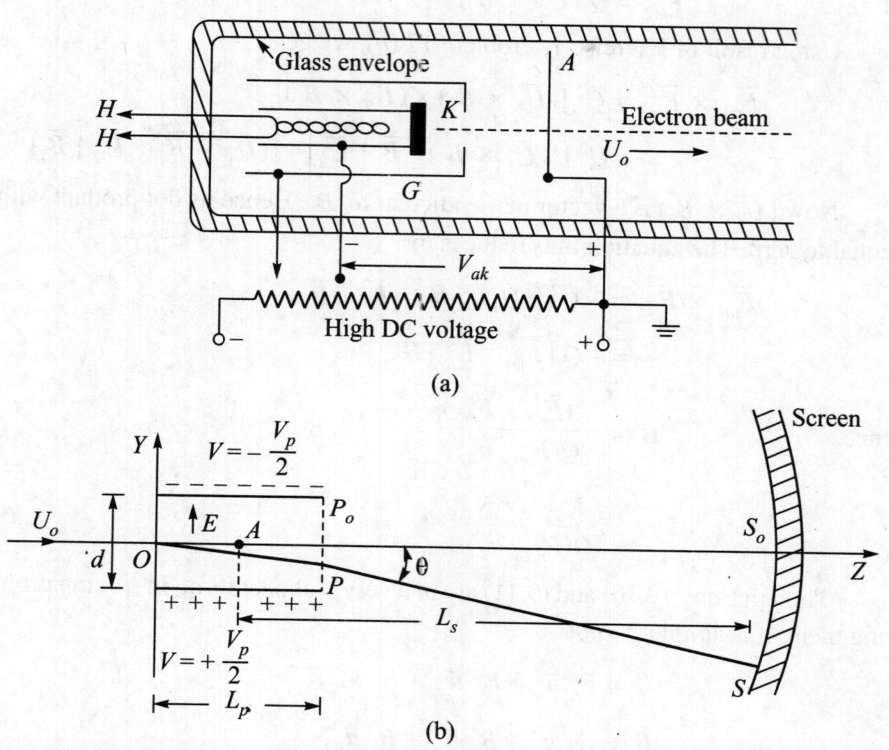

Fig. 9.1 (a) Electron gun and (b) Electrostatic deflection in cathode ray tube

Assuming as in a space charge limited current, that the electrons leave the cathode with zero velocity, we find that the velocity U_O of the electrons is given by Eqn. (6.3) as $U_0 = \sqrt{\frac{2eV_{ak}}{m}}$, e and m being the charge and the mass of the electron. Then the equations of motion in the plate system can be written using Eqn. (9.5) as

$$\left.\begin{aligned} m\frac{d^2x}{dt^2} &= 0, \\ m\frac{d^2y}{dt^2} = -e\quad E &= -\frac{eV_p}{d}, \\ m\frac{d^2z}{dt^2} &= 0, \end{aligned}\right\} \qquad \text{...(9.12)}$$

With the initial conditions that at $t = 0$,

$$x = y = z = 0$$

$$\left.\begin{aligned} \frac{dx}{dt} &= 0, \\ \frac{dy}{dt} &= 0, \\ \frac{dz}{dt} &= U_o\sqrt{\frac{2eV_{ak}}{m}} \end{aligned}\right\} \qquad \text{...(9.13)}$$

The first and the third equations in Eqn. (9.12) give the solutions

$$x = 0$$

and $$\frac{dz}{dt} = U_o \text{ always,}$$

or $$z = \sqrt{\frac{2eV_{ak}}{m}}\; t \qquad \text{...(9.14)}$$

Also the second equation $\dfrac{d^2y}{dt^2} = -\dfrac{eV_p}{md}$ gives

$$y = -\frac{eV_p}{md}\frac{t^2}{2} \qquad \text{...(9.15)}$$

If t is eliminated from Eqns. (9.14) and (9.15), the answer comes out to be a parabola having the equation

$$z^2 = \frac{2e\,V_{ak}}{m}\frac{2md\;y}{-e\,V_p}$$

$$= -\left(\frac{4d\,V_{ak}}{V_p}\right)y \qquad \text{...(9.16)}$$

At the point P where the electron leaves the plate system

$$x_p = 0,$$

$$z_p = L_p,$$

and $$y_p = -\frac{L^2 p\,V_p}{4d\,V_{ak}} \qquad \text{...(9.17)}$$

Beyond the point P, the electric field ends. Hence by Newton's first law, the electron continues to travel along the tangent to the parabola at P. This tangent intersects the z-axis at point A and the screen at point S. If the beam had not been deflected, it would have passed through the point S_o. Hence $S_o S$ is the deflection of the spot on the screen of the cathode ray tube.

To find the distance AP_o, where $OP_o = L_p$ we proceed as follows: From Eqn. (9.14), time t_o to reach P is given by

$$t_o = \frac{L_p}{\sqrt{\dfrac{2eV_{ak}}{m}}}$$

From Eqn. (9.15),

$$\frac{dy}{dt} = -\frac{eV_p t}{md}.$$

Hence

$$\left|\frac{dy}{dt}\right|_{\text{at } P} = \frac{eV_p}{md}\frac{L_p}{\sqrt{\frac{2eV_{ak}}{m}}} = \frac{eV_p L_p}{d\sqrt{2me\,V_{ak}}}$$

If θ is the angle $S_o AS$, then

$$\tan\theta = \left|\frac{dy}{dz}\right|_{\text{at } P} = \frac{\left|\frac{dy}{dt}\right|_{\text{at } P}}{\left|\frac{dz}{dt}\right|_{\text{at } P}}$$

$$= \frac{eV_p L_p}{d\sqrt{2me\,V_{ak}}}\sqrt{\frac{m}{2e\,V_{ak}}} = \frac{L_p V_p}{2d\,V_{ak}}$$

Therefore

$$AP_o = \frac{|y_p|}{\tan\theta}$$

$$= \frac{L_p^2 V_p}{4d\,V_{ak}}\frac{2d\,V_{ak}}{L_p V_p} = \frac{L_p}{2} \quad \text{[using Eqn. (9.17)]}$$

This shows that electrons reaching the screen appear to have travelled there along a straight line joining the centre A of the plate system to the point S. If the distance $AS_o = L_s$, then the deflection on the screen

$$= S_o S = L_s \tan\theta$$

$$= \left(\frac{L_p L_s}{2d\,V_{ak}}\right) V_p \qquad \text{...(9.18)}$$

Thus the deflection is directly proportional to the voltage difference between the two plates.

9.6 MAGNETOSTATIC DEFLECTION IN A CATHODE RAY TUBE

Instead of using electrostatic deflection plates in the cathode ray tube, we could put magnetising coils on either side of the neck of the cathode ray tube and thus use a magnetic field to produce the deflection. In television picture tubes where the angle of deflection θ has to be large, magnetic deflection is always used instead of electric deflection, because greater defocussing occurs at larger deflection angles in electrostatic deflection.

In Fig. 9.2, we have a uniform magnetic flux density $\overline{\boldsymbol{B}} = B_u\,\overline{\boldsymbol{u}}_x$ going into the plane of the paper, where it is assumed that this magnetic field is confined within the dotted rectangle of length L_m. An electron entering this field with the velocity $U_o\overline{\boldsymbol{u}}_z$ will experience a force

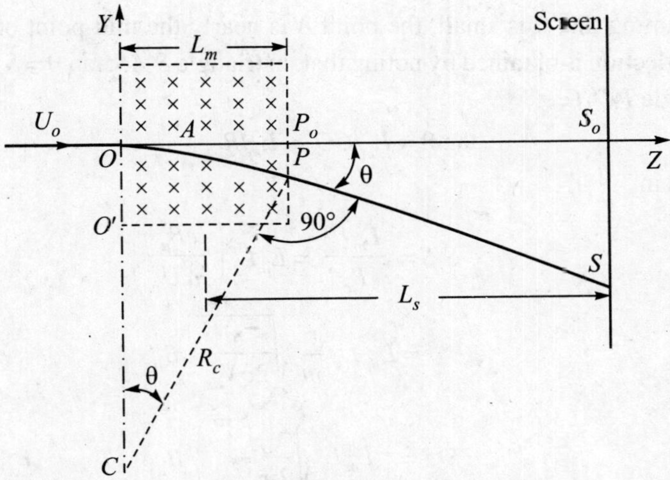

Fig. 9.2. Magnetostatic deflection in cathode ray tube

$$(-e)\,(U_o \bar{u}_z) \times (B_u \bar{u}_x) = -e\, U_o B_u\ \bar{u}_y \text{ newton.}$$

The negative sign shows that this force is directed along the negative direction of the y-axis. The path of the electron therefore curves downwards and it moves along a curve. At any point on the curve, the velocity $\bar{u}$ is directed along the tangent to the curve at that point, and the force ($\bar{U} \times \bar{B}$) is always normal to this tangent. Since the force is always perpendicular to the velocity, it cannot produce any change in the magnitude of the velocity, or U is always equal to U_o. Only the direction of $\bar{U}$ changes continuously along the curve. Thus at all points on the curve

$$F = e\, U_o B_u \qquad \text{...(9.19)}$$

If R_c is the radius of curvature at any point on the curve, then from kinematics, we know that this produces a centrifugal force $= \dfrac{mU_o^2}{R_c}$.

Equating this to the force on the moving electron due to the magnetic field

$$e\, U_o B_u = \frac{mU_o^2}{R_c}$$

whence

$$R_c = \frac{mU_o}{e\, B_u} \qquad \text{...(9.20)}$$

Since all the quantities on the right hand side of Eqn. (9.20) are constant, it follows that R_c = constant, or the path OP of the electron in the magnetic field is an arc of a circle with centre C. Beyond P, the electron moves along the tangent APS to this arc at P. Since the lines OC and S_oAO are perpendicular to each other, and so are lines CP and APS, it follows that

$$\angle PCO = \angle S_0AS = \theta.$$

Assuming that θ is small, the point A is nearly the mid-point of OP_o. On this basis the deflection is obtained by noting that, in triangle S_oAS, $\tan\theta = S_oS/AS = S_oS/L_s$, and in triangle PC'/C,

$$\tan\theta = L_m/CO' \approx L_m/R_c.$$

Equating them

Deflection
$$S_oS = \frac{L_m L_s}{R_c} = L_m L_s\left(\frac{eB_u}{m\,U_o}\right)$$

$$= L_m\,L_s\,\frac{e}{m}\sqrt{\frac{m}{2e\,V_{ak}}}\;B_u$$

or
$$S_oS = \left(L_m L_s\sqrt{\frac{e}{2m\,V_{ak}}}\right)B_u \qquad \text{...(9.21)}$$

The deflection is therefore proportional to the magnetic flux density $= B_u$.

9.7 PARALLEL PLATE MAGNETRON

We consider an idealized shape of the magnetron, as it will simplify the mathematics involved. The cathode and the anode are assumed to be infinite parallel planes separated by a distance d metres. If the anode is at potential V_p volt with respect to the cathode, the electric field intensity $\overline{E}$ is uniform [Fig. 9.3 (*a*)] and is given by

$$\overline{E} = E\,(-\,\overline{u}_y)$$

$$= \frac{V_p}{d}\,(-\,\overline{u}_y)\text{ volt/metre} \qquad \text{...(9.22)}$$

Between the anode and the cathode there is a uniform magnetic flux density perpendicular to the electric field given by

$$\overline{B} = B\,(-\,\overline{u}_z)\text{ tesla} \qquad \text{...(9.23)}$$

If an electron having a charge (– e) coulomb and mass m kg is released at the origin O, we wish to find its path when it moves under the influence of these crossed fields, as they are called.

The magnetic field has no influence on the stationary electron at O, but the positive cathode attracts it, *tending* to move it in the y- direction. The resulting velocity enables the magnetic field to produce a force exerted in the direction of the x-axis.

$$-\,e(\overline{U}_y)\times(-\,B\cdot\overline{u}_z) = e\,B\,U_y\,\overline{u}_x$$

Under the influence of the two fields, the electron leaves the cathode in the direction shown in Fig. 9.3 (*b*).

Using the Lorentz force equation, we can write the equations of motion as follows, where the velocity at any point is written as $\overline{U}$:

$$\overline{F} = -\,e\,[\overline{E} + (\overline{U}\times\overline{B})]$$

$$= -e\left[-E\,\bar{u}_y + \left(\frac{dx}{dt}\bar{u}_x + \frac{dy}{dt}\bar{u}_y + \frac{dz}{dt}\bar{u}_z\right)\times(-B\,\bar{u}_z)\right]$$

or $$m\frac{d^2x}{dt^2}\bar{u}_x + m\frac{d^2y}{dt^2}\bar{u}_y + m\frac{d^2z}{dt^2}\bar{u}_z = e\left[E\,\bar{u}_y - \frac{dx}{dt}B\bar{u}_y + \frac{dy}{dt}B\,\bar{u}_x + 0\right]$$

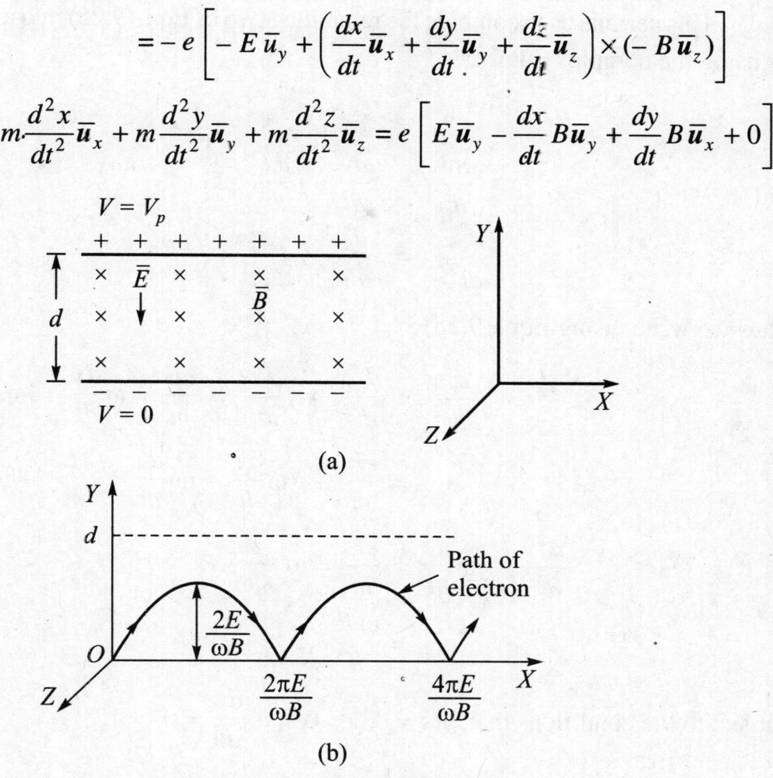

Fig. 9.3. Parallel-plate magnetron

Hence using the symbol

$$\omega = \frac{eB}{m} \qquad ...(9.24)$$

to make the equations more compact, we separate the components in the three directions and write

$$\left.\begin{aligned} \frac{d^2x}{dt^2} &= \frac{eB}{m}\frac{dy}{dt} = \omega\frac{dy}{dt} \\ \frac{d^2y}{dt^2} &= \frac{eE}{m} - \omega\frac{dx}{dt} \\ \frac{d^2z}{dt^2} &= 0 \end{aligned}\right\} \qquad ...(9.25)$$

These equations are to be solved subject to the initial conditions

at $t = 0$, $x = y = z = 0$,

$$\frac{dx}{dt} = \frac{dy}{dt} = \frac{dz}{dt} = 0 \qquad ...(9.26)$$

The third of Eqns. (9.25) gives the solution $z = 0$, indicating that the motion of the electron is confined to the x-y plane through the origin, or $z = 0$.

This permits us to combine the remaining two of Eqn. (9.25) into a single equation by using the complex number

$$\left.\begin{aligned} v &= y + jx \\ \text{and} \qquad \frac{dv}{dt} &= \frac{dy}{dt} + j\frac{dx}{dt}, \\ \frac{d^2v}{dt^2} &= \frac{d^2y}{dt^2} + j\frac{d^2x}{dt^2}, \end{aligned}\right\} \qquad \ldots(9.27)$$

Thus we write, using Eqn. (9.25)

$$\frac{d^2v}{dt^2} = \frac{d^2y}{dt^2} + j\frac{d^2x}{dt^2} = \frac{eE}{m} - \omega\frac{dx}{dt} + j\omega\frac{dy}{dt}$$

$$= \frac{eE}{m} + j^2\omega\frac{dx}{dt} + j\omega\frac{dy}{dt} = \frac{eE}{m} + j\omega\frac{dv}{dt}$$

or

$$\frac{d^2v}{dt^2} - j\omega\frac{dv}{dt} = \frac{eE}{m} = \omega\frac{e}{m\omega}E$$

$$= \frac{\omega E}{B} \qquad \ldots(9.28)$$

subject to the conditions that, at $t = 0$, $v = 0$ and $\frac{dv}{dt} = 0$.

This differential equation is now in the standard form $\frac{du}{dt} + Pu = Q$ and is to be solved by using the standard procedure of multiplying by the integrating factor $e^{\int P dt}$.

In this case $P = -j\omega$ and $\int (-j\omega)\ dt = -j\omega t$.

Therefore multiplying by $e^{-j\omega t}$, we get

or

$$e^{-j\omega t}\frac{d^2v}{dt} - j\omega\, e^{-j\omega t}\frac{dv}{dt} = \frac{\omega E}{B}e^{-j\omega t}$$

or

$$\frac{d}{dt}\left[e^{-j\omega t}\frac{dv}{dt}\right] = \frac{\omega E}{B}e^{-j\omega t}$$

Therefore

$$e^{-j\omega t}\frac{dv}{dt} = \frac{\omega E}{B}\int e^{-j\omega t}\, dt = \frac{\omega E}{B}\frac{e^{-j\omega t}}{-j\omega} + C_1$$

At $t = 0$, $\frac{dv}{dt} = 0$ Hence

$$C_1 = \frac{E}{jB} = \frac{-jE}{B}$$

Thus $$\frac{dv}{dt} = \frac{jE}{B} - \frac{jE}{B} e^{j\omega t}$$

Therefore $$v = \frac{jE}{B} t - \frac{jE}{B} \frac{e^{j\omega t}}{j\omega} + C_2$$

At $t = 0$, $v = 0$. Therefore

$$C_2 = \frac{E}{\omega B}$$

Thus, the solution is $$v = y + jx$$

$$= j\frac{E}{\omega B}(\omega t) - \frac{E}{\omega B}(\cos \omega t + j \sin \omega t) + \frac{E}{\omega B}$$

Separating the real and imaginary parts

$$\left.\begin{aligned} x &= \frac{E}{\omega B}(\omega t - \sin \omega t) \\ y &= \frac{E}{\omega B}(1 - \cos \omega t) \end{aligned}\right\} \qquad ...(9.29)$$

These are recognised as the parametric equations of a cycloid, the curve described by a point on a circle as the circle rolls, without slipping, on the x-axis.

As ωt varies from 0 to 2π, the y-co-ordinate of the electron's position varies from 0 to a value $\frac{2E}{\omega B}$ at $\omega t = \pi$. Thus the diameter of the rolling circle is $\frac{2E}{\omega B}$ metre. It is assumed that this is less than d, the separation between the anode and the cathode. For a critical value B_c of the magnetic flux density

$$\frac{2E}{\omega B_c} = d \quad \text{or} \quad B_c = \frac{2E}{\omega d} \qquad ...(9.30)$$

If $B < B_c$, the diameter of the rolling circle is greater than d and hence the electron reaches the anode and a plate current flows. As B is increased, the diameter of the rolling circle decreases, and the electron reaches some other point on the anode surface, but this does not affect the magnitude of the anode current which remains a constant. However, when $B > B_c$, the electron fails to reach the anode, but returns to the cathode, and the plate current abruptly drops to zero when $B \geq B_c$. Hence B_c is called the cut-off magnetic flux density (for a fixed d).

The electron leaving the origin again has its y-co-ordinate equal to zero when $\omega t = 2\pi$. For this value the

$$x\text{-co-ordinate} = \frac{E}{\omega B}(2\pi)$$

$$= \text{Circumference of rolling circle} \qquad ...(9.31)$$

In an actual magnetron, the anode is in the form of a hollow cylinder (instead of an infinite plane) surrounding a thin axial cathode.

9.8 MASS SPECTROGRAPH

The fact that in Eqn. (9.20), the radius R_c of the circular path is proportional to the mass m of the charged particle was used by F.W. Aston in 1919 in a device called a Mass Spectrograph, in which the spectrum lines obtained on a photographic plate did not correspond to the usual light waves of different wavelength, but they corresponded to charged particles of different mass.

In a vacuum, charged particles (ions) having different masses M but the same positive charge Q are first accelerated by an electrostatic field to acquire a velocity U_o given by

$$U_o = \sqrt{\frac{2QV_a}{M}} \quad \text{...(9.32)}$$

Then they enter a region where the magnetic flux density B is uniform and coming out of the paper (see Fig. 9.4). Hence each particle or ion travels in a circular path whose diameter is

$$2R_c = \frac{2MU_o}{QB} \quad \text{...(9.33)}$$

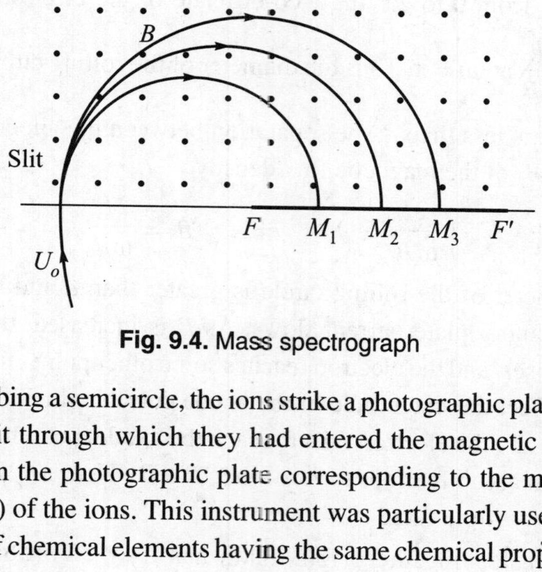

Fig. 9.4. Mass spectrograph

After describing a semicircle, the ions strike a photographic plate FF' and produce an image of the slit through which they had entered the magnetic field region. Thus lines are formed on the photographic plate corresponding to the masses M_1, M_2, M_3, etc. ($M_3 > M_2 > M_1$) of the ions. This instrument was particularly useful in the study of isotopes or atoms of chemical elements having the same chemical properties but different atomic weights.

9.9 SUMMARY AND COMMENTS

In this chapter we have discussed various devices based on the use of the Lorentz Force Equation. The practical uses of such devices based on Lorentz equation are numerous. Thus in 1899 Sir Joseph John Thomson (1856–1913) used Eqn. (9.30) in the form

$$\frac{e}{m} = \frac{2V_p}{B_c^2 d^2}$$

to determine the ratio e/m for the particles emitted in photoelectric emission and thus showed that electrons were emitted.

The angular velocity of the particle along the circular path in magnetic deflection is

$$\frac{d\theta}{dt} = \frac{U_o}{R_c} = \frac{BQ}{M}$$

from Eqn. (9.33). It is therefore *independent* of the velocity U_o. This means that though a particle with greater velocity U_o will travel along a circle of larger radius, the time taken by all particles to complete half a revolution is the same. This principle is used in the cyclotron to accelerate charged particles to very high velocities for nuclear physics.

The study of charged particles with mass ends here, and from the next chapter, we resume a study of abstract charges.

EXERCISE 9

1. In a uniform electrostatic field $\overline{E} = E_u\, \overline{u}_x$, a particle of mass m kg and charge + Q coulomb is placed at the origin at time $t = 0$, with its velocity at $t = 0$ being zero. Find the path of the particle, its position at any time t_1 and its velocity at the time t_1.
2. Two infinite parallel planes at $y = 0$ and $y = d$ are at potentials 0 and + V_p volts, respectively. In the plane $z = 0$, a particle of mass m kg and charge + Q C is released at time $t = 0$ at the origin with an initial velocity U m/s making an angle θ with the x-axis ($0 < \theta < 90°$).

   (*i*) Find the condition which must be satisfied in order that the particle may just fall to strike the plane $y = d$.

   (*ii*) If the particle does not strike the plane $y = d$, find

   (*a*) the equation to its path,

   (*b*) the maximum distance away from the x-axis attained by it,

   (*c*) the distance from the origin when it returns to the plane $y = 0$, and the magnitude and direction of its velocity at that point.
3. In a uniform magnetic field having flux density $\overline{B}$ tesla, an electron with charge ($-Q_e$) coulomb travelling along the x-axis with a velocity of one m/s experience a mechanical force

   $$\overline{F}_{m_1} = Q_e\,(\overline{u}_z - \overline{u}_x) \text{ newton.}$$

   But if the electron had been travelling with the same velocity along the y-axis, the force would have been

   $$\overline{F}_{m_2} = Q_e\,(\overline{u}_z - \overline{u}_x) \text{ newton.}$$

   Find $\overline{B}$.
4. In a region having uniform electric field intensity $\overline{E}$ V/m and uniform magnetic flux density $\overline{B}$ tesla, a particle having a charge + QC travels with a speed of one

metre per second. Depending on its direction of travel, the force acting on it is as follows:

Direction of travel	$\bar{u}_x$	$\bar{u}_y$	$\bar{u}_z$
Force in newton	$Q\bar{u}_x$	$Q(2\bar{u}_x + \bar{u}_y)$	$Q(\bar{u}_x + \bar{u}_y)$

Find the values of $\bar{E}$ and $\bar{B}$.

5. In a region having electric field intensity

$$\bar{E} = (\bar{u}_x + \bar{u}_y)$$

V/m and magnetic flux density $\bar{B} = \bar{u}_z$ tesla, what are the possible directions in which a charged particle may travel without experiencing any resultant force due to these two fields? What must be the magnitudes of the velocities in these directions?

6. In a uniform electrostatic field $\bar{E} = E_u \boldsymbol{u}_x$, a particle having a charge $+Q$ coulomb and a mass m kg is projected from the origin at time $t = 0$ with an initial velocity $U_{x_0}\ \bar{u}_x$ m/s. Find the position of the particle at time t, its velocity at that moment and its distance from the origin. Obtain these quantities for the special case where $U_{x_0} = 0$.

7. In a uniform region where the magnetostatic field $\bar{B} = B_u\ \bar{u}_z$ tesla, a particle having a charge $+Q$ C and a mass m kg is projected from the origin with an initial velocity $\bar{U}_0$. Describe its motion for the three special cases where

   (i) $\bar{U}_0 = 0,$

   (ii) $\bar{U}_0 = U_0\ \bar{u}_z$ and

   (iii) $\bar{U}_0 = U_0\ \bar{u}_y$.

8. In a region where the magnetic flux density $\bar{B} = B_z\ \bar{u}_z\ T$ (= tesla) a charged particle is projected at the origin at time $t = 0$ with an initial velocity

$$\bar{U}_0 = U_{x_0}\ \bar{u}_x + U_{y_0}\ \bar{u}_y + U_{z_0}\ \bar{u}_z.$$

Show that the path of the particle is a helix with its axis parallel to the z-axis. Find the equation to the curve formed by the projection of the helix on the x-y plane (z = constant) and also the pitch of the helix.

[This is the principle of the magnetic focussing in a television tube].

9. Find the magnitude of the velocity of the electron when it reaches the point S on the screen in the case of

   (i) electrostatic deflection [Fig. 9.1(*b*)] and

   (ii) magnetostatic deflection (Fig. 9.2).

10. In a cathode ray tube having both electrostatic deflection plates and magnetic deflection coils, it is found that when a small DC voltage is applied to the horizontal (x-) deflecting plates, the spot on the screen in seen to be displaced from the origin towards the right along the positive direction of the x-axis. What must be the direction of the magnetic field, which may now be applied to bring the spot back to the origin?

CHAPTER

10 MAXWELL'S EQUATIONS

10.1 AMPERE'S CIRCUITAL (WORK) LAW IN DIFFERENTIAL VECTOR FORM

In Section 4.19 we saw how Gauss's law in integral form applicable to a surface could be converted into a differential vector form to give the divergence of a vector at a point. In this section we proceed on a similar task and convert Ampere's circuital law from the integral form applicable round a closed loop to a differential vector form applicable at a point.

In Fig. 10.1 let H_x, H_y, H_z be the components of the magnetic field intensity at any point $A(x, y, z)$, where the components of the current density are J_x, J_y, J_z amperes/(metre)^2. In the plane passing through A and parallel to the y-z plane, choose a small rectangle $ABCD$ with sides Δy and Δz. Then the mmf around $ABCDA$ must be equal to the current enclosed, or $J_x \, \Delta y \, \Delta z$.

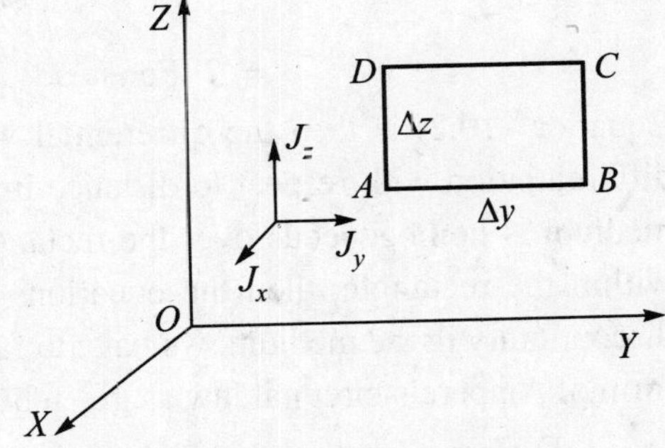

Fig. 10.1. To find curl $\overline{\mathbf{H}}$

To evaluate the mmf note that the values of the magnetic field intensity along the four sides of the rectangle are:

$$\text{Along } AB = H_y,$$

$$\text{Along } DC = H_y + \frac{\partial H_y}{\partial z} \Delta z$$

$$\text{Along } AD = H_z,$$

$$\text{Along } BC = H_z + \frac{\partial H_z}{\partial y} \Delta y$$

Hence

$$\oint_{ABCDA} \overline{\boldsymbol{H}} \cdot \overline{\boldsymbol{ds}} = \int_{AB} \overline{\boldsymbol{H}} \cdot \overline{\boldsymbol{ds}} + \int_{BC} \overline{\boldsymbol{H}} \cdot \overline{\boldsymbol{ds}} + \int_{CD} \overline{\boldsymbol{H}} \cdot \overline{\boldsymbol{ds}} + \int_{DA} \overline{\boldsymbol{H}} \cdot \overline{\boldsymbol{ds}}$$

$$= [H_y \, \Delta y] + \left[\left(H_z + \frac{\partial H_z}{\partial y} \Delta y\right) \Delta z\right]$$

$$+ \left[-\left(H_y + \frac{\partial H_y}{\partial z} \Delta z\right) \Delta y\right] + [-H_z \, \Delta z]$$

$$= \left(\frac{\partial H_z}{\partial y} - \frac{\partial H_y}{\partial z} \right) \Delta y \, \Delta z$$

Since this is equal to the enclosed current $J_x \, \Delta y \, \Delta z$, we get

$$\left. \begin{aligned} &\frac{\partial H_z}{\partial y} - \frac{\partial H_y}{\partial z} = J_x \\ \text{Similarly we can show that} \quad &\frac{\partial H_x}{\partial z} - \frac{\partial H_z}{\partial x} = J_y \\ \text{and} \quad &\frac{\partial H_y}{\partial x} - \frac{\partial H_x}{\partial y} = J_z \end{aligned} \right\} \qquad ...(10.1)$$

These equations are combined and written in a short-hand notation using Eqn. (1.95) as

$$\text{curl } \overline{\boldsymbol{H}} = \overline{\nabla} \times \overline{\boldsymbol{H}}$$

$$= \begin{vmatrix} \overline{\boldsymbol{u}}_x & \overline{\boldsymbol{u}}_y & \overline{\boldsymbol{u}}_z \\ \dfrac{\partial}{\partial x} & \dfrac{\partial}{\partial y} & \dfrac{\partial}{\partial z} \\ H_x & H_y & H_z \end{vmatrix}$$

$$= \overline{\boldsymbol{J}} \text{ (For steady DC only)} \qquad ...(10.2)$$

Equation (10.2) is then the differential vector form of Ampere's circuital law. The differentiation with respect to distance implied in Eqn. (10.1) is possible only if the medium is homogeneous over the rectangle *ABCD*. If there is a change of medium within the rectangle, then the equations will no longer be true. In case of such a discontinuity in the medium, we have to fall back upon the use of the original integral form of Ampere's circuital law as given by Eqn. (8.18).

Further it has to be noted that Eqn. (10.2) is valid only for steady, direct current. If the current is direct, but changing in magnitude with time, then Eqn. (10.2) is no longer valid. Such time-varying currents will be considered later and Eqn. (10.2) will then be suitably modified.

10.2 STOKE'S THEOREM

Equation (10.2) is applicable at a point only. It has been derived on the assumption that Δx, Δy and Δz are infinitesimally small changes in values of x, y and z at that point. Let us try to integrate Eqn. (10.2) over an area (Fig. 10.2). Then we notice that so far as the inner rectangles are concerned, each side of an inner rectangle appears in two integrations around the rectangles of which it is the common side, but the directions of integration in the two cases are opposite. Hence their magnitude being equal but signs opposite, they will cancel out.

Thus the only sides of the rectangles which do not cancel out are those that form the outer boundary of the area. This implies that the surface integral of curl $\overline{\boldsymbol{H}}$ over the area is equal to the line integral of $\overline{\boldsymbol{H}}$ along the closed boundary enclosing that area, or

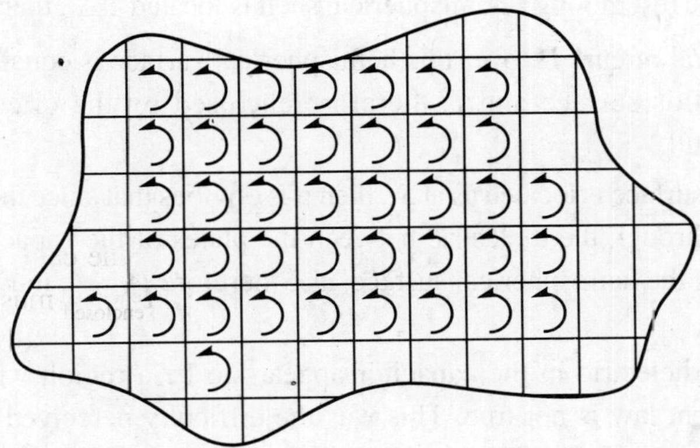

Fig. 10.2. Line integral and integral of curl

$$\int_{\text{surface}} (\overline{\nabla} \times \overline{H}) \cdot \overline{da} = \oint \overline{H} \cdot \overline{ds} \qquad \text{...(10.3)}$$

This result is called Stoke's theorem. It has a remarkable similarity with the Divergence Theorem given by Eqn. (4.49), but whereas the latter related a volume integral to a surface integral where the surface enclosed the volume, Stoke's theorem, which is true for *any* vector, relates a surface integral to a line integral along the boundary enclosing the surface.

By integrating Eqn. (10.2), we get

$$\int_{\text{surface}} (\overline{\nabla} \times \overline{H}) \cdot \overline{da} = \oint \overline{J} \cdot \overline{da} = I_{\text{enclosed}} \qquad \text{...(10.4)}$$

10.3 COMMENTS ON EQUATION (10.2)

Let us examine in detail the significance of the comment at the end of Sec. 10.1, that Eqn. (10.2) is only valid in the case of steady direct currents. Taking the divergence of both sides of Eqn. (10.2), we have

$$\overline{\nabla} \cdot (\overline{\nabla} \times \overline{H}) = \overline{\nabla} \cdot \overline{J}$$

But by Eqn. (1.117) the divergence of the curl of any vector is always zero. This makes the left-hand side always equal to zero. Hence

$$\overline{\nabla} \cdot \overline{J} = 0 \text{ (For steady DC only)} \qquad \text{...(10.5)}$$

But Eqns. (4.29) and (4.49) show that Eqn. (10.5) is only a statement of Kirchhoff's current law at a point. It implies that there is no net accumulation or loss of charge at any point. This is true in the case of steady, direct currents only, which are continuous throughout the circuit consisting of conductors.

However, with a capacitor (having a dielectric) in the circuit, and with varying direct current or an alternating current in the circuit, things become different. This happens because the capacitor is a device which has the ability to store charge. The region of the plates of the capacitor (Fig. 10.3) in a circuit with either varying *DC* as at (*a*) or *AC* as in (*b*) is a region where charge accumulates continuously or periodically, respectively.

If a surface resembling a hemispherical shell is located at S_1, then by Eqn. (10.4), the surface integral of curl $\overline{H}_o$ over this hemispherical surface is equal to the current I which *pierces* this surface and is therefore enclosed by the circular rim of the hemispherical shell.

But if the surface is located as at S_2, then it is obvious that since the hemispherical shell is passing through the dielectric between the plates of the capacitor, there is no current I piercing the hemispherical surface, and therefore I_{enclosed} must be considered to be zero.

Thus the dielectric in the capacitor appears to be a region where apparently Kirchhoff's current law is not true. This was the difficulty perceived in 1864 by the Scottish mathematician, James Clerk Maxwell (1831–1879) and to resolve it, he proposed an explanation. He based his study of time-varying electric and magnetic fields on this explanation and came up with astounding results which were verified experimentally 24 years later.

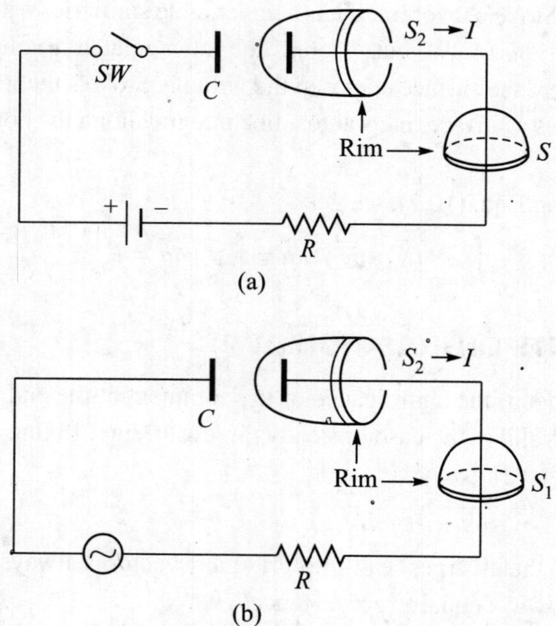

Fig. 10.3. Capacitor with (a) varying *DC* and (b) *AC*

10.4 EQUATION OF CONTINUITY

Since we find that divergence $\overline{J}$ is not equal to zero in time-varying electric fields, we first wish to find out what it is equal to in the general case. For this purpose we consider (Fig. 10.4) an infinitesimally small volume $dx\,dy\,dz$ surrounding a point $P(x, y, z)$, where the current density $\overline{J}$ (A/m²) is related to the volume charge density ρ (C/m³) and the velocity of the charge, $\overline{U}$ (m/s), by the relation

$$\overline{J} = \rho\overline{U} \quad (\text{A/m}^2) \qquad \text{...(10.6)}$$

First we consider only the x-component U_x of the velocity. Then the amount of charge entering the volume $dx\,dy\,dz$ through the left-hand face, because of velocity U_x is given by $\rho U_x\,dy\,dz$ coulomb per second. This is therefore the current entering the volume. The current leaving the volume is given by $[\rho U_x + \frac{\partial}{\partial x}(\rho U_x)\,dx]\,dy\,dz$. It flows out of the right-hand face.

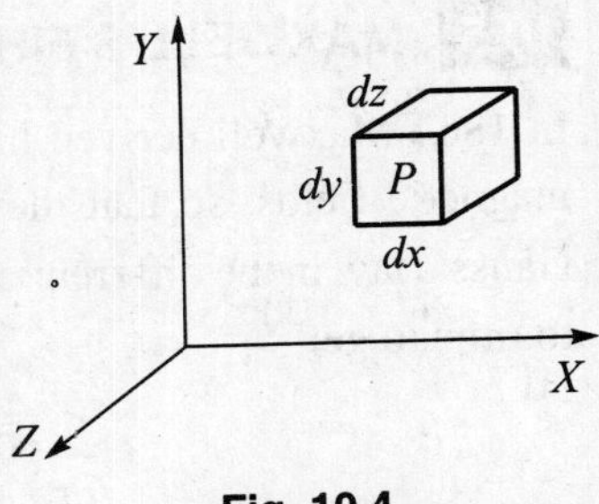

Fig. 10.4

If the current flowing in is greater than the current flowing out, then there will be an increase in the amount of charge in the volume, and this increase will be given by

$$(\rho U_x\,dy\,dz) - \left[\rho U_x + \frac{\partial}{\partial x}(\rho U_x)\,dx\right] dy\,dz$$

$$= -\frac{\partial}{\partial x}(\rho U_x)\,dx\,dy\,dz \text{ coulomb per second}$$

Similarly for the other components U_y and U_z, we will get

$$-\frac{\partial}{\partial y}(\rho U_y)\,dx\,dy\,dz$$

and

$$-\frac{\partial}{\partial z}(\rho U_z)\,dx\,dy\,dz \text{ respectively.}$$

Therefore,

(Total increase in charge per second per unit volume)

$$= \frac{\text{Sum of the above}}{\text{Volume } dx\,dy\,dz}$$

$$= -\left[\frac{\partial}{\partial x}(\rho U_x) + \frac{\partial}{\partial y}(\rho U_y) + \frac{\partial}{\partial z}(\rho U_z)\right]$$

$$= -\overline{\nabla}\cdot(\rho\overline{U}) \quad \text{by Eqn. (1.93)}$$

$$= -\overline{\nabla}\cdot\overline{J} \quad \text{by Eqn. (10.6)}$$

But the left-hand side can also be written as (Total increase in charge per unit volume per second), the mathematical expression for which is $\frac{\partial \rho}{\partial t}$. Hence the above equation becomes

$$\frac{\partial \rho}{\partial t} = -\overline{\nabla}\cdot\overline{J} \qquad \text{...(10.7)}$$

This is called the Equation of Continuity. This expression was derived first by Maxwell who, taking into consideration the negative sign on the right-hand side, defined what he called "convergence"–the negative of what we now call the divergence of a vector. In other words, his convergence $\overline{J}$ was the rate of flow of the vector *into* the volume $dx\,dy\,dz$, instead of divergence $\overline{J}$ which is the rate of flow *out of* the volume.

10.5 MAXWELL'S FIRST EQUATION

In 1864 Maxwell derived his famous four equations about time-varying electric and magnetic fields, so that they are also known as his "field equations". Starting with Gauss's law in the differential vector form as $\overline{\nabla} \cdot \overline{\boldsymbol{D}} = \rho$, he differentiated it with respect to time to get

$$\frac{\partial}{\partial t}(\overline{\nabla} \cdot \overline{\boldsymbol{D}}) = \frac{\partial \rho}{\partial t}$$

$$= -\,\overline{\nabla} \cdot \overline{\boldsymbol{J}} \quad \text{by Eqn. (10.7)}$$

Now divergence operation involves space differentiation. This is independent of the differentiation with respect to time, which the left-hand side of the equation requires to be performed next. Because of this independence, we can change the order of these two differentiations without affecting the result. Thus we write above equation as

$$\frac{\partial}{\partial t}(\overline{\nabla} \cdot \overline{\boldsymbol{D}}) = \overline{\nabla} \cdot \left(\frac{\partial \overline{\boldsymbol{D}}}{\partial t}\right)$$

$$= -\,\overline{\nabla} \cdot \overline{\boldsymbol{J}}$$

or $$\overline{\nabla} \cdot \left[\overline{\boldsymbol{J}} + \frac{\partial \overline{\boldsymbol{D}}}{\partial t}\right] = 0 \qquad \text{(Time-varying fields)} \qquad ...(10.8)$$

Equation (10.8) is thus the complete form of Kirchoff's current law, which is applicable to both *DC* as well as to *AC*. In the case of steady, direct currents, $\frac{\partial \overline{\boldsymbol{D}}}{\partial t} = 0$ and Eqn. (10.8) reduces to Eqn. (10.5). In other words, this shows that in the equations for magnetostatic field where we had used $\overline{\boldsymbol{J}}$, we should have used $\left(\overline{\boldsymbol{J}} + \frac{\partial \overline{\boldsymbol{D}}}{\partial t}\right)$ to make the equation suitable for time-varying fields. Maxwell did just this for the Ampere's critical law in differential vector form and he modified Eqn. (10.2) as

$$\overline{\nabla} \times \overline{\boldsymbol{H}} = \overline{\boldsymbol{J}} + \frac{\partial \overline{\boldsymbol{D}}}{\partial t} \quad \text{(Time-varying fields)} \qquad ...(10.9)$$

We may call this as Maxwell's first equation. Different authors present the four Maxwell's equations in different order, because there is no fixed or standard numbering scheme for them. To avoid all confusion, it is therefore best to refer to Eqn. (10.9) simply as "Maxwell's curl equation for $\overline{\boldsymbol{H}}$."

10.6 DISPLACEMENT CURRENT

We proceed to examine this new expression $\left(\overline{\boldsymbol{J}} + \frac{\partial \overline{\boldsymbol{D}}}{\partial t}\right)$ in greater detail. First we notice that the units for $\overline{\boldsymbol{J}}$ are amperes per square metre area (normal to the direction of vector $\bar{J}$). So far as $\bar{D}$ is concerned, its units are coulomb per square metre area (normal to the direction of the vector $\bar{D}$). Hence dimensions of $\frac{\partial \bar{D}}{\partial t}$ are (coulomb/second per square metre), or ampere/(metre)^2. Thus $\frac{\partial D}{\partial t}$ is also a current density. But where and how does this current arise?

In Section 4.12, we had considered briefly the flow of current in conductors as being caused by the simultaneous drift of charge carriers at all points in the circuit, and we had an analogy of children forming a closed loop and transferring lighted candles to their neighbours. Now we modify this analogy and imagine that each child does not transfer the lighted candle to his neighbour. Instead he shifts it from his outstretched left hand to his outstretched right hand in one second, and again from his right hand to his left hand in the next one second. Also he repeats this cycle over and over again. Then the distant observer will see candle lights travelling in one direction in one second, and in the opposite direction in the next second. This is like an alternating current flow in a circuit. But the difference now is that the candle does not drift from child to child, but is held by the same child all the time. And yet an alternating current of lighted candles is flowing in the chain formed by the children.

This is the mechanism by which this current $\frac{\partial \bar{D}}{\partial t}$ is able to flow in a dielectric. It is the turning of the polar molecules to and fro as in Fig. 7.2(*b*), that provides the movement or flow of charge, first in one direction and then in the reverse direction, but the charge never leaves the parent molecule. To indicate this displacement of charges to and fro in the polarised molecules, Maxwell gave the name "Displacement Current" to the term $\frac{\partial \bar{D}}{\partial t}$ to distinguish it from the conventional "Conduction Current" $\bar{J}$, in which charges actually drift about in the conductor. Otherwise, argued Maxwell, the two currents are indistinguishable from each other, and therefore, they will produce identical effects. Thus the displacement current will produce an I^2R heating effect in the medium. It will also produce a magnetic field around it.

This reciprocal effect, an electric charge displacement producing a magnetic field, and a varying magnetic field inducing an electromotive force, is the direct consequence of the assumption that a displacement current can flow where no conduction current is possible. Since Maxwell's work with his equation involved only paper and pen, he was able to postulate that a displacement current can also flow in a vacuum. It was only 24 years later in May 1888, that the German professor of Physics, Heinrich Rudolph Hertz (1857–1894) published his paper "On Electric Radiation" giving experimental proof of the possibility of the existence of electromagnetic waves produced by time-varying fields as predicted by Maxwell. Hertz was also able to show that these waves obeyed the laws of reflection, refraction and polarisation just like light waves.

Maxwell's assumption of a displacement current in a vacuum was a very daring one. We are accustomed to thinking about waves in a physical medium, but in a vacuum there is no such medium. But Maxwell argued that since the conductivity $\sigma = 0$ in a vacuum, there is no conduction current $\bar{J}$ possible, if we assume that the capacitor plates are situated in a vacuum and there is no other dielectric material between them. But the equation of continuity must be satisfied between the plates of a capacitor, and therefore there must be a displacement current even in a vacuum. In other words, the displacement current is necessary to bridge the gap between the plates, where the conduction current is zero.

Maxwell's final masterly touch was the assumption that in all materials, whether conductors or dielectrics or even vacuum, both the conduction and the displacement currents exist, but their relative magnitudes differ. Thus so far as a good conductor is concerned, it is impossible to measure its dielectric constant in the laboratory, because the conduction current dominates over the negligibly small displacement current. To get an idea of the order of magnitudes of these currents in a conductor, let us assume that $E = E_0 \sin (2\pi ft)$. Then

$$\text{the total current} \quad = J + \frac{\partial D}{\partial t}$$

$$= \sigma E + \frac{\partial(\varepsilon E)}{\partial t}$$

$$= \sigma E_0 \cos (2\pi ft) + 2\pi f \varepsilon E_0 \sin (2\pi ft)$$

$$= \sigma E_0 \sin (2\pi ft) + 2\pi f \varepsilon E_0 \sin \left(2\pi f t + \frac{\pi}{2}\right)$$

We notice that whereas the conduction current is in phase with the electric field intensity, the displacement current leads over it by the phase angle $\frac{\pi}{2}$.

Now as Tables 7.1 and 7.2 show, for most conductors $\sigma \approx (1 \text{ to } 6) \times 10^7$ S/m, and for most materials, $\varepsilon_r < 10$. For a good conductor therefore, we may arbitrarily put $\left(\frac{\varepsilon_r}{\sigma}\right) \approx 1.8 \times 10^{-7}$ and then we find that the ratio

$$\frac{\text{Amplitude of displacement current}}{\text{Amplitude of conduction current}}$$

$$= \frac{2\pi f \varepsilon_r \varepsilon_0}{\sigma} \approx 2\pi f (1.8 \times 10^{-7}) \frac{10^{-9}}{36\pi}$$

$$\approx f \times 10^{-17}$$

This shows that the frequency f has to be as high as 10^{17} Hz before these currents become equal in magnitude!! The yellow light from a sodium vapour lamp has a wavelength of about 600 nm. Hence its frequency is obtained as (velocity)/(wavelength) $= (3 \times 10^8)/(600 \times 10^{-9}) = 5 \times 10^{14}$ Hz. The required frequency of 10^{17} Hz is 200 times higher than this!

10.7 MAXWELL'S SECOND EQUATION

Just as Maxwell obtained his first equation by putting Ampere's circuital law in differential vector form, he obtained his second equation by putting Faraday's induction law, $V = -\dfrac{\partial \Phi}{\partial t}$, in differential vector form.

We have seen that in electrostatics, an electric field is conservative and $\oint \overline{E} \cdot \overline{ds}$ along any closed path is always equal to zero. But in Faraday's induction law, the fields are varying with time. For such a case, the line integral along a closed path is equal to the voltage V induced in the loop, and not equal to zero. Combining these we can write

$$\oint \overline{E} \cdot \overline{ds} = V = -\frac{\partial \Phi}{\partial t} \qquad ...(10.10)$$

We transform the left-hand side of this equation by applying Stoke's theorem, Eqn. (10.3), and write

$$\oint \overline{E} \cdot \overline{ds} = \int_{\text{surface}} (\overline{\nabla} \times \overline{E}) \cdot \overline{da} \qquad ...(10.11)$$

On the right-hand side of Eqn. (10.10), we use Eqn. (8.4) giving the value of Φ. Then we observe that a differentiation with respect to time is an independent operation, so far as integration with respect to an area is concerned. Therefore, the order of the differentiation and integration can be interchanged. Thus we write

$$-\frac{\partial \Phi}{\partial t} = -\frac{\partial}{\partial t}\left[\int_{\text{surface}} \overline{B} \cdot \overline{da}\right]$$

$$= \int_{\text{surface}} \left(-\frac{\partial \overline{B}}{\partial t}\right) \cdot \overline{da} \qquad ...(10.12)$$

Then equating Eqns. (10.11) and (10.12), we get

$$\int_{\text{surface}} (\overline{\nabla} \times \overline{E}) \cdot \overline{da} = \int_{\text{surface}} \left(-\frac{\partial \overline{B}}{\partial t}\right) \cdot \overline{da}$$

It follows that the terms within the brackets must be equal, or

$$\overline{\nabla} \times \overline{E} = -\frac{\partial \overline{B}}{\partial t} \qquad ...(10.13)$$

This is Maxwell's second equation or his curl equation for $\overline{E}$. This form of Faraday's law of induction gives us an insight into the curl or $\overline{\nabla}\,\times$ operator. For it tells us that if we have a tiny wire loop surrounding the point P where this curl is to be taken, then by connecting the ends of the loop to a voltmeter, we have what we may call a "curl meter" to measure the curl $\overline{E}$ at that point. This is so because the curl $\overline{E}$ equals the voltmeter reading. Naturally depending upon the orientation of the normal to the plane containing the loop, we shall get the component of the curl in the direction of the normal.

10.8 MAXWELL'S THIRD AND FOURTH EQUATIONS

Maxwell's third equation is nothing but the differential vector form of Gauss's law which we have already obtained as Eqn. (4.48), viz.

$$\overline{\nabla} \cdot \overline{D} = \rho \qquad ...(10.14)$$

Maxwell's Fourth Equation can be obtained either from the definition of the vector magnetic potential or from Eqn. (10.13) by remembering Eqn. (1.117), that the divergence of the curl of any vector is equal to zero. Thus if $\overline{A}$ is the vector magnetic potential, then

$$\begin{aligned} \text{div } \overline{B} &= \overline{\nabla} \cdot \overline{B} \\ &= \mu(\overline{\nabla} \cdot \overline{H}) = \mu\, \overline{\nabla} \cdot (\overline{\nabla} \times \overline{A}) \\ &= 0 \end{aligned}$$

or using Eqn. (10.13)

$$\overline{\nabla} \cdot (\overline{\nabla} \times \overline{E}) = \overline{\nabla} \cdot \left[-\frac{\partial \overline{B}}{\partial t} \right]$$

or

$$0 = -\frac{\partial}{\partial t}[\overline{\nabla} \cdot \overline{B}]$$

In either case,

$$\overline{\nabla} \cdot \overline{B} = 0 \qquad ...(10.15)$$

This differential vector form of Eqn. (8.20) is Maxwell's fourth equation.

The auxiliary equations $\overline{D} = \varepsilon \overline{E}$ and $\overline{B} = \mu \overline{H}$ combined with the four Maxwell's equations give the complete set of equations needed for studying time-varying fields.

10.9 SUMMARY AND COMMENTS

For ready reference, Maxwell's equations in different forms are collected together below:

10.9.1 Differential Vector Form

I. $$\text{curl } \overline{H} = \overline{\nabla} \times \overline{H} = \overline{J} + \frac{\partial \overline{D}}{\partial t} \qquad ...(10.9)$$

II. $$\text{curl } \overline{E} = \overline{\nabla} \times \overline{E} = -\frac{\partial \overline{B}}{\partial t} \qquad ...(10.13)$$

III. $$\text{div } \overline{D} = \overline{\nabla} \cdot \overline{D} = \rho \qquad ...(10.14)$$

IV. $$\text{div } \overline{B} = \overline{\nabla} \cdot \overline{B} = 0 \qquad ...(10.15)$$

Notice that in both the Eqns. (I) and (II), all electrical quantities are on one side of the equation and the magnetic quantities are on the other side. The advantage of the differential vector form of the equations is that they are compact, but more importantly, they are independent of any co-ordinate system.

10.9.2 Differential Scalar Form

(For rectangular Cartesian co-ordinates), for obtaining these equations, the curl equations are first written in the determinant form and then expanded out.

I.
$$\frac{\partial H_z}{\partial y} - \frac{\partial H_y}{\partial z} = J_x + \frac{\partial D_x}{\partial t} \quad \text{...(10.16a)}$$
$$\frac{\partial H_x}{\partial z} - \frac{\partial H_z}{\partial x} = J_y + \frac{\partial D_y}{\partial t} \quad \text{...(10.16b)}$$
$$\frac{\partial H_y}{\partial x} - \frac{\partial H_x}{\partial y} = J_z + \frac{\partial D_z}{\partial t} \quad \text{...(10.16c)}$$

II.
$$\frac{\partial E_z}{\partial y} - \frac{\partial E_y}{\partial z} = -\frac{\partial B_x}{\partial t} \quad \text{...(10.17a)}$$
$$\frac{\partial E_x}{\partial z} - \frac{\partial E_z}{\partial x} = -\frac{\partial B_y}{\partial t} \quad \text{...(10.17b)}$$
$$\frac{\partial E_y}{\partial x} - \frac{\partial E_x}{\partial y} = -\frac{\partial B_z}{\partial t} \quad \text{...(10.17c)}$$

III.
$$\frac{\partial D_x}{\partial x} + \frac{\partial D_y}{\partial y} + \frac{\partial D_z}{\partial z} = \rho \quad \text{...(10.18)}$$

IV.
$$\frac{\partial B_x}{\partial x} + \frac{\partial B_y}{\partial y} + \frac{\partial B_z}{\partial z} = 0 \quad \text{...(10.19)}$$

10.9.3 Integral Form

I.
$$\oint \overline{H} \cdot \overline{ds} = \int_{\text{surface}} \left(\overline{J} + \frac{\partial \overline{D}}{\partial t} \right) \cdot \overline{da} \quad \text{...(10.20)}$$

II.
$$\oint \overline{E} \cdot \overline{ds} = -\int_{\text{surface}} \frac{\partial \overline{B}}{\partial t} \cdot \overline{da} \quad \text{..(10.21)}$$

III.
$$\oint_{\text{closed surface}} \overline{D} \cdot \overline{da} = \int_{\text{volume}} \rho \, dv \quad \text{...(4.11)}$$

IV.
$$\oint_{\text{closed surface}} \overline{B} \cdot \overline{da} = 0 \quad \text{....(8.20)}$$

It should be remembered that Maxwell's curl equations assume that the region to which they are to be applied is source-free, *i.e.*, in that region there should be no voltage generator or current generator present. Also the medium is assumed to be homogeneous and isotropic, so that the space derivatives are continuous.

These equations form the most powerful tools developed by mathematicians for the study of time-varying electric and magnetic fields, which will engage our attention in the next chapter.

EXERCISE 10

1. In a dielectric in which $\rho = 0$ (charge-free) and $J = 0$ (current-free), show that for time-varying fields, Maxwell's divergence equations can be derived on the assumption that the curl equations are true and valid.
2. Can a time-varying electric field exist without a corresponding magnetic field, and vice-versa? Explain with reasons.
3. Show that inside a perfect conductor, the magnetic field intensity must be zero for time-varying fields.
4. In a copper rod of 1 mm × 1 mm cross-section, a conduction current of one ampere is flowing at a frequency of 50 Hz. Find the displacement current in the rod assuming that for copper,
$$\varepsilon = \varepsilon_0, \quad \mu = \mu_0$$
and $$\sigma = 5.800 \times 10^7 \text{ siemens per metre.}$$
5. For a charge-free region, write the four Maxwell's equations using only vectors $\overline{\boldsymbol{E}}$ and $\overline{\boldsymbol{B}}$.
6. Show that for a conductor obeying Ohm's law, placed in an electric field
$$E = E_0\, e^{jwt},\ \text{div}\ \overline{\boldsymbol{D}} = \overline{\nabla} \cdot \overline{\boldsymbol{D}} = 0$$
inside the conductor.
7. Show that for fields which do not vary with time, if
$$\overline{\nabla} \cdot \overline{\boldsymbol{D}} = \rho,$$
then $$\text{div (grad } D) = \text{grad (div } D)$$
or $$\nabla \cdot (\nabla \boldsymbol{D}) = \nabla(\nabla \cdot \boldsymbol{D})$$
[Hint: Expand the two sides in Cartesian co-ordinates]
8. Show that if at any point P inside a conductor obeying Ohm's law, a charge density ρ_0 C/m^3 is somehow created at time $t = 0$, it will be reduced exponentially to the value $\dfrac{\rho_0}{e}$ in a certain time T. Find the value of T.
[Hint: Use equation of continuity and Gauss's law]

CHAPTER

11 TIME-VARYING FIELDS AND ELECTROMAGNETIC WAVES

11.1 SOLUTION OF MAXWELL'S EQUATIONS FOR CHARGE-FREE UNBOUNDED REGION

We assume that we have a linear, homogeneous, isotropic, perfect dielectric, extending to infinity in all directions, in which there are no sources of currents and no free charges. The material being homogeneous, isotropic and source-free, we can apply Maxwell's curl equations to it. The perfect dielectric has $\rho = 0$ and hence $J = 0$. Also since there are no free charges, $\rho = 0$. Further we assume that the electric and magnetic fields that are present vary with time in some arbitrary manner. Then we can write Maxwell's equations and the auxiliary equations as follows:

$$\text{curl } \overline{\boldsymbol{H}} = \overline{\nabla} \times \overline{\boldsymbol{H}} = \frac{\partial \overline{\boldsymbol{D}}}{\partial t} \qquad ...(11.1)$$

$$\text{curl } \overline{\boldsymbol{E}} = \overline{\nabla} \times \overline{\boldsymbol{E}} = -\frac{\partial \overline{\boldsymbol{B}}}{\partial t} \qquad ...(11.2)$$

$$\text{div } \overline{\boldsymbol{D}} = \overline{\nabla} \cdot \overline{\boldsymbol{D}} = 0 \qquad ...(11.3)$$

$$\text{div } \overline{\boldsymbol{B}} = \overline{\nabla} \cdot \overline{\boldsymbol{B}} = 0 \qquad ...(11.4)$$

$$\overline{\boldsymbol{D}} = \varepsilon \overline{\boldsymbol{E}} \qquad ...(11.5)$$

$$\overline{\boldsymbol{B}} = \mu \overline{\boldsymbol{H}} \qquad ...(11.6)$$

The standard method for solving Maxwell's equations consists in differentiating one of the curl equations with respect to time t, and to take the curl of the other curl equation. First we differentiate Eqn. (11.1) with respect to time t, and use Eqn. (11.5) assuming that ε is independent of time. Also since the curl is a space differentiation, the order of differentiation with respect to time and space can be reversed without altering the result. Then

$$\frac{\partial}{\partial t}[\overline{\nabla} \times \overline{\boldsymbol{H}}] = \frac{\partial}{\partial t}\left[\frac{\partial \overline{\boldsymbol{D}}}{\partial t}\right] = \frac{\partial}{\partial t}\left[\varepsilon \frac{\partial \overline{\boldsymbol{E}}}{\partial t}\right]$$

or

$$\overline{\nabla} \times \left(\frac{\partial \overline{\boldsymbol{H}}}{\partial t}\right) = \varepsilon \frac{\partial^2 \overline{\boldsymbol{E}}}{\partial t^2} \qquad ...(11.7)$$

Next we take the curl of Eqn. (11.2) and use Eqn. (11.6) assuming that μ is independent of time t. Then

$$\overline{\nabla} \times (\overline{\nabla} \times \overline{\boldsymbol{E}}) = -\overline{\nabla} \times \frac{\partial}{\partial t}(\mu\, \overline{\boldsymbol{H}})$$

$$= -\mu \overline{\nabla} \times \left(\frac{\partial \overline{\boldsymbol{H}}}{\partial t} \right) \qquad ...(11.8)$$

Putting the value from Eqn. (11.7) in Eqn. (11.8), we get

$$\overline{\nabla} \times (\overline{\nabla} \times \overline{\boldsymbol{E}}) = -\mu\, \varepsilon \frac{\partial^2 \overline{\boldsymbol{E}}}{\partial t^2} \qquad ...(11.9)$$

But from Eqns. (11.9), (11.5) and (11.3), we have

$$\overline{\nabla} \times (\overline{\nabla} \times \overline{\boldsymbol{E}}) = \overline{\nabla}(\overline{\nabla} \cdot \overline{\boldsymbol{E}}) - \overline{\nabla}^2\, \overline{\boldsymbol{E}}$$

$$= \overline{\nabla}\left[\frac{1}{\varepsilon} (\overline{\nabla} \cdot \overline{\boldsymbol{D}}) \right] - \overline{\nabla}^2\, \overline{\boldsymbol{E}}$$

$$= -\overline{\nabla}^2\, \overline{\boldsymbol{E}} \qquad ...(11.10)$$

Combining Eqns. (11.9) and (11.10), we get

$$\overline{\nabla}^2\, \overline{\boldsymbol{E}} = \mu\, \varepsilon \frac{\partial^2 \overline{\boldsymbol{E}}}{\partial t^2} \qquad ...(11.11)$$

If we had taken the curl of Eqn. (11.1) and differentiated Eqn. (11.2) with respect to time, and followed the various steps taken above, we would have got

$$\overline{\nabla}^2\, \overline{\boldsymbol{H}} = \mu\, \varepsilon \frac{\partial^2 \overline{\boldsymbol{H}}}{\partial t^2} \qquad ...(11.12)$$

Equations (11.11) and (11.12) are together referred to as the wave equations, for reasons which will become obvious in the next section. The important points to be noticed about them are that they are identical indicating that *variations* of $\overline{\boldsymbol{E}}$ and $\overline{\boldsymbol{H}}$ are similar, but it does not mean that $\overline{\boldsymbol{E}}$ and $\overline{\boldsymbol{H}}$ are equal. In fact $\overline{\boldsymbol{E}} \neq \overline{\boldsymbol{H}}$ in any sense. Further it must be noted that $\overline{\nabla}^2\, \overline{\boldsymbol{E}}$ and $\overline{\nabla}^2\, \overline{\boldsymbol{H}}$ are shorthand notations, which mean something quite different from $\overline{\nabla}^2 V$ of Poisson's and Laplace's equations, because $\overline{\boldsymbol{E}}$ and $\overline{\boldsymbol{H}}$ are vectors, whereas V is a scalar quantity. This will become obvious by looking at their expanded forms. Thus

$$\overline{\nabla}^2 V = \frac{\partial^2 V}{\partial x^2} + \frac{\partial^2 V}{\partial y^2} + \frac{\partial^2 V}{\partial z^2}$$

but

$$\left.\begin{aligned} \overline{\nabla}^2 \overline{\boldsymbol{E}} = \; & \overline{\boldsymbol{u}}_x \left(\frac{\partial^2 E_x}{\partial x^2} + \frac{\partial^2 E_x}{\partial y^2} + \frac{\partial^2 E_x}{\partial z^2} \right) \\ & + \overline{\boldsymbol{u}}_y \left(\frac{\partial^2 E_y}{\partial x^2} + \frac{\partial^2 E_y}{\partial y^2} + \frac{\partial^2 E_y}{\partial z^2} \right) \\ & + \overline{\boldsymbol{u}}_z \left(\frac{\partial^2 E_z}{\partial x^2} + \frac{\partial^2 E_z}{\partial y^2} + \frac{\partial^2 E_z}{\partial z^2} \right) \end{aligned}\right\} \qquad ...(11.13)$$

$$= \bar{u}_x \nabla^2 E_x + \bar{u}_y \nabla^2 E_y + \bar{u}_z \nabla^2 E_z$$

Thus $\overline{\nabla}^2 V$ represents three terms, but $\overline{\nabla}^2 \overline{E}$ represents nine terms.

We have still not got the solution of Maxwell's equations, but we have got them converted into a form which yields solutions very easily for certain conditions as indicated in the next section.

11.2 UNIFORM WAVES

To solve Eqns. (11.11) and (11.12), we make some simplifying assumptions which, *without loss of generality*, will highlight the nature of the solution to be expected. For this purpose we assume that $\overline{E}$ and $\overline{H}$ do not vary in the x- and y-directions. This makes six of the terms in Eqn. (11.13) equal to zero, and we are left with Eqn. (11.11) in the form

$$\frac{\partial^2 E_x}{\partial z^2}\bar{u}_x + \frac{\partial^2 E_y}{\partial z^2}\bar{u}_y + \frac{\partial^2 E_z}{\partial z^2}\bar{u}_z$$

$$= \mu\,\varepsilon\left(\frac{\partial^2 E_x}{\partial t^2}\bar{u}_x + \frac{\partial^2 E_y}{\partial t^2}\bar{u}_y + \frac{\partial^2 E_z}{\partial t^2}\bar{u}_z\right) \qquad \text{...(11.14)}$$

Since we are interested in knowing only the *form* of the solution, we deal with only one component part of the solution, viz.

$$\frac{\partial^2 E_x}{\partial z^2} = \mu\,\varepsilon \frac{\partial^2 E_x}{\partial t^2}$$

$$= \left(\frac{1}{v^2}\right)\frac{\partial^2 E_x}{dt^2}, \quad \text{say} \qquad \text{...(11.15)}$$

Comparing Eqn. (11.15) with the original Eqn. (11.11), we are satisfied that we have been able to retain the *form* of the solution, because the two equations are very similar.

By actual differentiation, one can at once verify that any function of $(z - vt)$ like $f_1\,(z- vt)$ is a solution of Eqn. (11.15). This is an interesting function. Thus if for $z = z_1$ and $t = t_1$, this function has a value $f_1\,(z_1 - vt_1) = A$ (say), then for $z = z_2$, it will have the same value A at a later time t_2 which is given by

$$t_2 = t_1 + (z_2 - z_1)\,/v \qquad \text{...(11.16)}$$

This fact is shown in Fig. 11.1 (a). A phenomenon which displays this property defined by Eqn. (11.16) is called a wave. Stated in words, a wave is a physical phenomenon in which some modification of a medium or similar change occurs at one place at a given time, and an identical modification or change occurs at another place at a later time, in such a way that the time delay is proportional to the space separation between the two places. Thus the ratio of the space separation $(z_2 - z_1)$ to the time separation $(t_2 - t_1)$ is called the velocity v of wave propagation. If $v = (z_2 - z_1)\,/(t_2 - t_1)$ is positive, the wave is said to be a forward travelling wave, travelling in the positive direction of the z-axis. In such a wave, $(z_2 - z_1)$ and $(t_2 - t_1)$ are both positive quantities.

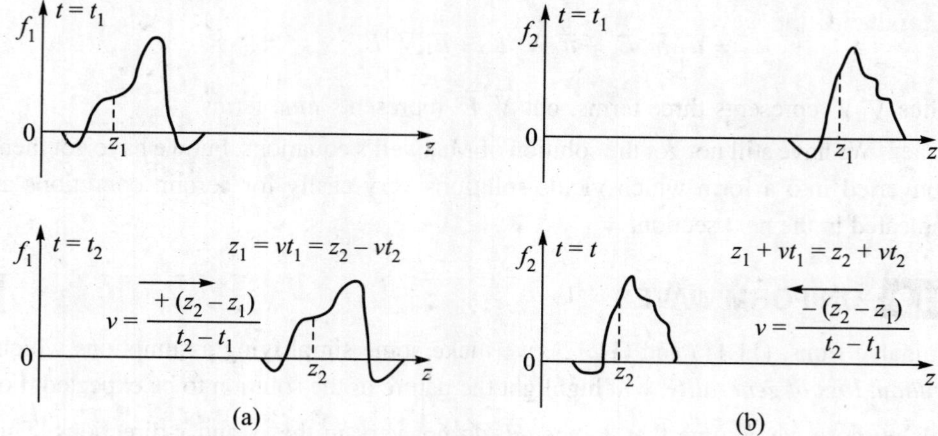

Fig. 11.1. Forward and reverse travelling waves

By actual differentiation, one can verify that if $f_1\,(z - vt)$ is a solution of Eqn. (11.15), then some other function of $(z + vt)$ of the form $f_2\,(z + vt)$ is also a solution. But now, if $(t_2 - t_1)$ is positive, $(z_2 - z_1)$ has to be negative, giving a negative velocity. This wave therefore travels in the opposite direction, that is, it travels in the negative direction of the z-axis, as shown in Fig. 11.1 (b). It is called a reverse travelling wave or a reflected wave. Thus in the forward-going or in the reverse-going wave, the magnitude of the velocity of the wave is obtained from Eqn. (11.15) as

$$v = \frac{1}{\sqrt{\mu\,\varepsilon}} \quad \text{m/s} \qquad \ldots(11.17a)$$

In the case of vacuum, or free space,

$$v_0 = \frac{1}{\sqrt{\mu_0\,\varepsilon_0}} \quad \text{m/s} \qquad \ldots(11.17b)$$

11.2.1 Sinusoidal Variations with Time

In the beginning of Sec. 11.1, it was assumed that $\overline{\boldsymbol{E}}$ and $\overline{\boldsymbol{H}}$ are varying in any arbitrary manner with time. Now Fourier had shown that non-sinusoidal time variations could be broken up into a sum of a series of purely sinusoidal variations. Hence it is useful to study the form of equations when sinusoidal time variations take place. These could be of the form cos ωt or sin ωt, but it is better to use the form $e^{j\omega t}$, because it is easier to handle and also because it includes both the forms, since

$$e^{j\omega t} = \cos\omega t + j\sin\omega t.$$

Then assuming that $E_x = E_o\,e^{j\omega t}$, Eqn. (11.15) becomes

$$\frac{\partial^2 E_x}{\partial z^2} = \mu\,\varepsilon\,\frac{\partial^2 (E_o e^{j\omega t})}{\partial t^2}$$

$$= \mu\,\varepsilon\,(j\omega)^2 E_o e^{j\omega t} = -\mu\,\varepsilon\,\omega^2 E_x$$

or

$$\frac{\partial^2 E_x}{\partial z^2} = -\left(\frac{\omega}{v}\right)^2 E_x \qquad \ldots(11.18)$$

Introducing the wavelength λ and the frequency f, we note that

$$\frac{\omega}{v} = \frac{2\pi f}{f\lambda} = \frac{2\pi}{\lambda} = \beta, \text{ say} \qquad \text{...(11.18}a\text{)}$$

Then Eqn. (11.18) becomes

$$\frac{\partial^2 E_x}{\partial z^2} = -\beta^2 E_x = (\pm j\beta)^2 E_x$$

Therefore, $E_x = Ae^{\pm j\beta z}$, where A = some constant not varying with z. Using $E_o\, e^{j\omega t}$ as this constant A which is not varying with z, we can write the solution of the equation as

$$E_x = \underset{\text{(Forward wave)}}{E_{o1}\, e^{j(\omega t - \beta z)}} + \underset{\text{(Reverse wave)}}{E_{o2}\, e^{j(\omega t + \beta z)}} \qquad \text{...(11.19)}$$

11.3 UNIFORM PLANE WAVES

In the previous section, we had selected only one term out of nine in Eqn. (11.13) to get the idea of the form of the solution, and we saw that in general, Eqn. (11.11) represents a wave travelling in space. Assume that we have a wave in which there is sinusoidal variation of the amplitude with time, but that this amplitude has the same value (no variation with distance) at all points on any infinite plane. Such a wave is therefore said to be a uniform wave. Since this condition is not always fulfilled, in general the wave will get attenuated in amplitude as it travels.

For making a beginning, however, a plane wave is the simplest to study. A *plane* wave is defined as one for which the phase ωt has the same value at all points on *an infinite plane*. Therefore, in the previous section, the simplifying assumption we made, that $\overline{\boldsymbol{E}}$ does not vary in the x and y directions, actually implied (though we did not expressly say so) that we are selecting a plane wave for study, for which the x-y plane (or z = constant) represented a wavefront (or a surface at every point of which ωt has the same value) and which satisfied Eqn. (11.14):

$$\frac{\partial^2 E_x}{\partial z^2}\overline{\boldsymbol{u}}_x + \frac{\partial^2 E_y}{\partial z^2}\overline{\boldsymbol{u}}_y + \frac{\partial^2 E_z}{\partial z^2}\overline{\boldsymbol{u}}_z = \mu\varepsilon\left(\frac{\partial^2 E_x}{\partial t^2}\overline{\boldsymbol{u}}_x + \frac{\partial^2 E_y}{\partial t^2}\overline{\boldsymbol{u}}_y + \frac{\partial^2 E_z}{\partial t^2}\overline{\boldsymbol{u}}_z\right) \qquad \text{...(11.14)}$$

A similar equation can be overwritten for H with E_x, E_y, E_z in above equation replaced by H_x, H_y and H_z respectively. Our problem now is to find what are the restrictions imposed on these six components of E and H, and to find the direction in which the wave travels.

We note first that we had assumed charge-free region which gave Eqn. (11.3) as

$$\overline{\nabla}\cdot\overline{\boldsymbol{D}} = 0$$

or

$$\varepsilon\,\overline{\nabla}\cdot\overline{\boldsymbol{E}} = 0$$

or, since ε is a constant

$$\nabla \cdot \overline{E} = \frac{\partial E_x}{\partial x} + \frac{\partial E_y}{\partial x} + \frac{\partial E_z}{\partial x} = 0$$

But since we have assumed that there is no variation in x- and y-directions, the first two terms are equal to zero, and

$$\frac{\partial E_z}{\partial z} = 0. \text{ Therefore } \frac{\partial^2 E_z}{\partial z^2} = 0.$$

Then from Eqn. (11.14) equating coefficients of $\overline{u}_z$ on both sides, we get

$$\frac{\partial^2 E_z}{\partial z^2} = \mu \varepsilon \frac{\partial^2 E_z}{\partial t^2}, \quad \text{or} \quad 0 = \frac{\partial^2 E_z}{\partial t^2}.$$

Therefore $\quad E_z = C_1 + C_2 t \quad$ ...(11.20)

Three possibilities arise:

(*i*) If $C_2 = 0$ but $C_1 \neq 0$, then $E_z = C_1$ means that E_z is a constant, independent of time. But we have seen that in a wave, the component must vary with time and distance, like $f_1(z - vt)$ or $f_2(z + vt)$. Hence $C_2 = 0$ does not give a plane wave which is what we want.

(*ii*) If $C_1 = 0$ but $C_2 \neq 0$, then $E_z = C_2 t$. Now E_z is varying with time but is increasing continuously. Also it does not vary with z, because $\frac{\partial E_z}{\partial z} = 0$. Hence this solution does not produce a wave.

(*iii*) This leaves us with the only possibility that both $C_1 = 0$ and $C_2 = 0$, or $E_z = 0$. Thus we see that for plane wave, the component E_z must be equal to zero everywhere, if the amplitude is to have no variation in the plane z = constant. A similar reasoning shows that H_z must also be equal to zero everywhere.

This means that in the plane wave Eqn. (11.14), only the components E_x, E_y, H_x, H_y exist while $E_z = H_z = 0$. In the previous section, we had analysed the equation for the x-component E_x and had seen that the equation represents a wave travelling in the z-direction. A similar analysis for E_y, H_x and H_y will also give similar solution of a wave travelling in the z-direction. Hence we conclude that for a plane wave having components E_x, E_y, H_x and H_y in the x-y plane, the direction for the propagation is the z-direction which is perpendicular to the x-y plane, and in this direction the $\overline{E}$ and $\overline{H}$ vectors do not have a component because, as shown above $E_z = 0$ and $H_z = 0$. This implies that the magnitudes of the electric and magnetic vectors in this case are given by

$$E = \sqrt{E_x^2 + E_y^2 + E_z^2} = \sqrt{E_x^2 + E_y^2}$$

and $$H = \sqrt{H_x^2 + H_y^2 + H_z^2} = \sqrt{H_x^2 + H_y^2} \quad \text{...(11.21)}$$

Both these vectors are transverse to the direction of propagation. This uniform plane wave is therefore described as a transverse electro-magnetic wave or TEM wave.

11.4 RELATION BETWEEN $\overline{E}$ AND $\overline{H}$ IN A UNIFORM PLANE WAVE

As seen in the previous section, for a plane wave travelling in the z-direction in a perfect and charge-free dielectric ($\sigma = 0$, $\rho = 0$), $E_z = 0$ and $H_z = 0$, and hence Maxwell's curl equation for $\overline{\boldsymbol{H}}$ becomes

$$\overline{\nabla} \times \overline{\boldsymbol{H}} = \begin{vmatrix} \overline{\boldsymbol{u}}_x & \overline{\boldsymbol{u}}_y & \overline{\boldsymbol{u}}_z \\ \dfrac{\partial}{\partial x} & \dfrac{\partial}{\partial y} & \dfrac{\partial}{\partial z} \\ H_x & H_y & 0 \end{vmatrix}$$

$$= \frac{\partial \overline{\boldsymbol{D}}}{\partial t} = \varepsilon \frac{\partial \overline{\boldsymbol{E}}}{\partial t} = \varepsilon \left[\frac{\partial E_x}{\partial t} \overline{\boldsymbol{u}}_x + \frac{\partial E_y}{\partial t} \overline{\boldsymbol{u}}_y \right]$$

Hence equating the components along three axes, we get

$$-\frac{\partial H_y}{\partial z} = \varepsilon \frac{\partial E_x}{\partial t},$$

$$+\frac{\partial H_x}{\partial z} = \varepsilon \frac{\partial E_y}{\partial t}, \qquad \text{...(11.22)}$$

$$\frac{\partial H_y}{\partial x} - \frac{\partial H_x}{\partial y} = 0$$

If we assume that $E_x = f_1\,(z - vt)$ and use the symbols to denote

$$f_1 = f_1\,(z - vt),\; f_1{}' = \frac{\partial f_1(z - vt)}{\partial (z - vt)}$$

then the first of Eqn. (11.22) becomes

$$\frac{\partial H_y}{\partial z} = -\,\varepsilon \frac{\partial f_1}{\partial t} = -\,\varepsilon \frac{\partial f_1(z - vt)}{\partial (z - vt)} \frac{\partial (z - vt)}{\partial t}$$

$$= -\,\varepsilon\, f_1{}'(-v)$$

Therefore $\quad H_y = \varepsilon\, v \int f_1{}'\, dz$

$$= \varepsilon\, v \int f_1{}'\, d(z - vt)$$

$$= \varepsilon v\, f_1\,(z - vt) + \text{constant} \qquad \text{...(11.23)}$$

The constant being independent of $(z - vt)$ will give a term which does not take part in wave phenomenon. Hence we may ignore it. Then making use of Eqn. (11.17a) we write Eqn. (11.23) as

$$H_y = \varepsilon v\, f_1\,(z - vt) = \varepsilon \frac{1}{\sqrt{\mu\, \varepsilon}} E_x$$

$$= \sqrt{\frac{\varepsilon}{\mu}}\, E_x$$

Therefore, $\quad \dfrac{E_y}{H_y} = \sqrt{\dfrac{\mu}{\varepsilon}} = \eta$ say $\qquad$ ...(11.24)

If we had started with the second equation in Eqn. (11.22), we would have obtained the relation

$$\frac{E_y}{H_x} = -\sqrt{\frac{\mu}{\varepsilon}} = -\eta \qquad \text{...(11.25)}$$

Therefore from Eqn. (11.21), we get

$$\begin{aligned} \frac{E}{H} &= \frac{\sqrt{E_x^2 + E_y^2}}{\sqrt{H_x^2 + H_y^2}} \\ &= \frac{\sqrt{(H_y\eta)^2 + (-H_x\eta)^2}}{\sqrt{H_x^2 + H_y^2}} \qquad \text{...(11.26)} \\ &= \eta \end{aligned}$$

For vacuum or free space,

$$\begin{aligned} \frac{E}{H} = \eta_o &= \sqrt{\frac{\mu_o}{\varepsilon_0}} \\ &\approx \sqrt{\frac{4\pi \times 10^{-7}}{10^{-9}/(36\,\pi)}} \\ &= 120\pi = 377 \text{ ohms} \qquad \text{...(11.27)} \end{aligned}$$

The dimensions of $\sqrt{\frac{\mu_o}{\varepsilon_0}}$ are $\sqrt{\frac{\text{Henry/meter}}{\text{Farad/meter}}}$ or $\sqrt{\frac{\text{Henry}}{\text{Farad}}}$. Now multiplying the dimensions of the equations $L\frac{di}{dt} = V$ and $Q = CV$, we get

$$\text{Henry} \times \frac{\text{Ampere}}{\text{second}} \times \text{Coulomb} = \text{Farad} \times (\text{Volts})^2$$

Therefore, $\sqrt{\frac{\text{Henry}}{\text{Farad}}} = \frac{\text{Volts}}{\text{Ampere}}$ = ohms. This is the reason why the units of η in Eqn. (11.27) are given as ohms.

The important point to be noted about η is that, though, because of the dimensions of their units, η and η_o are called the characteristic impedance or intrinsic impedance of the dielectric and free space respectively, it may be better to use the term *characteristic wave impedance* to remind us that it is the ratio of the components E and H of a wave. Eqn. (11.26) and (11.27) are *not* Ohm's law relationships, even though the ratio of E/H has dimensions of ohms. This is because H is not an electrical quantity (current) but a magnetic quantity. The difference will become at once apparent, if one asks the question : "377 ohms is the resistance between which points or surfaces and for what spacing between the two?", and find that answer to this is absurd.

The second thing to be noted about it is that in a wave, $\bar{\boldsymbol{E}}$ and $\bar{\boldsymbol{H}}$ are at right angles to each other, and the wave is propagated in the direction $\bar{\boldsymbol{E}} \times \bar{\boldsymbol{H}}$. When we consider the ratios of their perpendicular components as in Eqns. (11.24) and (11.25), the ratio is positive if the 90° rotation from E-component to H-component is clockwise when looking in direction of propagation, and is negative if this rotation is counter-clockwise.

Finally, it must remembered that Eqns. (11.24) and (11.25) are for the forward travelling wave because for deriving them, we had assumed $E_x = f_1\ (z - vt)$. For a reflected wave we would have assumed $E_x = f_2\ (z + vt)$ and this change of sign of v would have given an additional negative sign in Eqns. (11.24) and (11.25). Thus, we write

For forward wave,

$$\frac{E_{x+}}{H_{y+}} = +\eta$$

$$\frac{E_{y+}}{H_{x+}} = -\eta$$

For reflected wave,

$$\frac{E_{x-}}{H_{y-}} = -\eta \qquad \text{...(11.28)}$$

$$\frac{E_{y-}}{H_{x-}} = +\eta$$

However, this characteristic wave impedance plays the same significant role in field theory as the characteristic impedance Z_c of a transmission line plays in the study of reflection in a line terminated in a load impedance $Z_l \neq Z_c$.

11.5 POLARIZATION

For discussing polarization in optics, physicists talk of a plane of polarization. In electromagnetic field theory, we only talk of a *direction* of polarization, the wave being said to be *polarized in the direction of the* $\overline{\boldsymbol{E}}$ *vector*. For instance, for a plane wave travelling in the z-direction, $E_z = 0$, and we have

$$\overline{\boldsymbol{E}} = E_x\,\overline{\boldsymbol{u}}_x + E_y\,\overline{\boldsymbol{u}}_y$$

and the wave is said to be polarized in the direction of this $\overline{\boldsymbol{E}}$ vector [Fig.11.2 (a)]. If $E_x = E_{x0} \sin \omega t$ and $E_y = E_{y0} \sin \omega t$ at any point on a plane z = constant, then since $E = \sqrt{E_x^2 + E_y^2}$, the length of the vector $\overline{\boldsymbol{E}}$ varies from 0 to $E_{\max}$ and back to zero in one direction, and then from 0 to $-E_{\max}$ and back to zero in the reverse direction (shown dotted). As the vector $\overline{\boldsymbol{E}}$ always lies along the straight line $P\,P'$ shown, this wave is said to be *linearly* polarized.

It must be noted, however, that the components E_x and E_y need not always be in time phase, even in a plane wave. Thus if at any point on the plane z = constant,

$$x = E_x = E_{x0} \sin \omega t$$

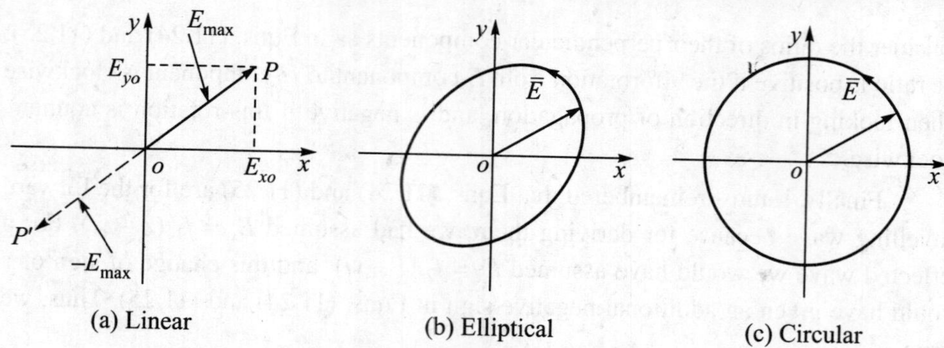

(a) Linear (b) Elliptical (c) Circular

Fig. 11.2. Polarization of waves

and
$$y = E_y = E_{y_o} \sin(\omega t + \phi),$$
then by eliminating ωt from these equations, we get an equation representing the locus of the tip of the vector $\overline{E}$. It is found to be an equation to an ellipse with its axis inclined to the x and y axes. This means that the vector $\overline{E}$ constantly changes both its *magnitude* and *direction*, and hence the tip of the vector $\overline{E}$ describes the ellipse shown in Fig. 11.2 (b). The magnitude attains its maximum value E_{max} in the direction of the major axis of the ellipse, and its minimum value E_{min} in the direction of the minor axis. The rate at which the vector $\overline{E}$ rotates around the origin is ω radians per second. Such a wave is said to be an elliptically polarized plane wave.

In the very special case of elliptical polarization in which $E_{x_o} = E_{y_o} = E_o$ and the phase difference $\phi = \frac{\pi}{2}$, we get
$$x = E_x = E_o \sin \omega t$$
and
$$y = E_y = E_o \sin\left(\omega t + \frac{\pi}{2}\right) = E_o \cos \omega t,$$
and eliminating ωt, we get
$$x^2 + y^2 = E_o^2,$$
the equation to a circle of constant radius E_o described by the tip of the vector $\overline{E}$ rotating at the ratio of ω radians per second as shown in Fig. 11.2 (c). This wave is therefore said to be a circularly polarized plane wave.

11.6 WAVE EQUATIONS FOR A CHARGE-FREE MEDIUM WITH FINITE CONDUCTIVITY

Having obtained an idea of how waves are possible in a perfect dielectric ($\sigma = 0$), in which there are no sources and no charges, and how they are propagated, we now take up a more practical case with a medium for which $0 < \sigma < \infty$, and in which there are no charges ($\rho = 0$). However, as before, the medium is assumed to be linear, homogeneous and isotropic, and extending to infinity in all directions. Further, the electric and magnetic fields are assumed to be varying sinusoidally with time, or

$$E = E_1 \, e^{j\omega t}$$

and $$H = H_1 e^{j\omega t}$$

Moreover Ohm's law gives $\bar{J} = \sigma \bar{E}$.

Then Maxwell's curl equations become

$$\bar{\nabla} \times \bar{H} = \bar{J} + \frac{\partial \bar{D}}{\partial t}$$

$$= \sigma \bar{E} + \varepsilon \frac{\partial \bar{E}}{\partial t}$$

$$= (\sigma + j\omega \, \varepsilon) \, \bar{E} \qquad \text{...(11.29)}$$

and $$\bar{\nabla} \times \bar{E} = -\frac{\partial \bar{B}}{\partial t}$$

$$= -\mu \frac{\partial \bar{H}}{\partial t} \qquad \text{...(11.30)}$$

Following the standard method of solution, we first differentiate Eqn. (11.29) with respect to t. Then changing the order of time and space differentiation, we get

$$\frac{\partial}{\partial t}(\bar{\nabla} \times \bar{H}) = \frac{\partial}{\partial t}[(\sigma + j\omega\varepsilon)\, \bar{E}]$$

or $$\bar{\nabla} \times \left(\frac{\partial \bar{H}}{\partial t}\right) j\omega\,(\sigma + j\omega\varepsilon)\, \bar{E} \qquad \text{...(11.31)}$$

Then taking the curl of Eqn. (11.30) and using Eqn. (11.31), we get

$$\bar{\nabla} \times (\bar{\nabla} \times \bar{E}) = -\mu \left[\bar{\nabla} \times \left(\frac{\partial \bar{H}}{\partial t}\right)\right]$$

$$= -j\omega\mu\,(\sigma + j\omega\varepsilon)\, \bar{E}$$

The left hand side is changed using Eqns. (11.29), (11.5) and (11.3) as

$$\bar{\nabla} \times (\bar{\nabla} \times \bar{E}) = \bar{\nabla}\,(\bar{\nabla} \cdot \bar{E}) - \bar{\nabla}^2 \bar{E}$$

$$= \bar{\nabla}\left[\frac{1}{\varepsilon}(\bar{\nabla} \cdot \bar{D})\right] - \bar{\nabla}^2 \bar{E}$$

$$= -\,\bar{\nabla}^2 \bar{E}$$

Hence $$\bar{\nabla}^2 \bar{E} = j\omega\mu\,(\sigma + j\omega\varepsilon)\, \bar{E}$$

$$= \gamma^2 \, \bar{E} \ \ \text{(say)} \qquad \text{...(11.32)}$$

If we had taken the curl of Eqn. (11.29) and differentiated Eqn. (11.30) with respect to t, we would have got the relation

$$\bar{\nabla}^2 \bar{H} = j\omega\mu\,(\sigma + j\omega\varepsilon)\, \bar{H}$$

$$= \gamma^2 \, \bar{H} \qquad \text{...(11.33)}$$

where, as in Eqn. (11.32)

$$\gamma^2 = j\omega\mu\,(\sigma + j\omega\varepsilon) \qquad ...(11.34)$$

Equations (11.32) and (11.33) are now the form that the wave equations (11.11) and (11.12) have taken.

11.7 CHARACTERISTIC WAVE IMPEDANCE AND PROPAGATION CONSTANT

As before if we assume that the wave is being propagated along the z-axis, the $E_z = 0$ and $H_z = 0$, and the solution of the wave equations (11.32) and (11.33) is of the form

$$\overline{\boldsymbol{E}} = \overline{\boldsymbol{E}}_0 e^{(j\omega t - \gamma z)} \qquad ...(11.35)$$

$$\overline{\boldsymbol{H}} = \overline{\boldsymbol{H}}_0 e^{(j\omega t - \gamma z)} \qquad ...(11.36)$$

Then Eqn. (11.29) can be written as

$$\overline{\nabla} \times \overline{\boldsymbol{H}} = \begin{vmatrix} \overline{\boldsymbol{u}}_x & \overline{\boldsymbol{u}}_y & \overline{\boldsymbol{u}}_z \\ \dfrac{\partial}{\partial x} & \dfrac{\partial}{\partial y} & \dfrac{\partial}{\partial z} \\ H_x & H_y & 0 \end{vmatrix}$$

$$= (\sigma + j\omega\varepsilon)\,[E_{0_x}\,\overline{\boldsymbol{u}}_x + E_{0_y}\,\overline{\boldsymbol{u}}_y + E_{0_z}\,\overline{\boldsymbol{u}}_z]\,e^{(j\omega t - \gamma z)}$$

Equating the x-components on the two sides

$$-\frac{\partial H_y}{\partial z} = (\sigma + j\omega\varepsilon)\,E_{0_x}\,e^{(j\omega t - \gamma z)}$$

Therefore

$$H_y = (\sigma + j\omega\varepsilon)\,E_{0_x}\,e^{(j\omega t)} \int (-e^{-\gamma z})\,dz$$

$$= (\sigma + j\omega\varepsilon)\,E_{0_x}\,e^{(j\omega t)}\,\frac{e^{-\gamma z}}{\gamma}$$

$$= \frac{(\sigma + j\omega\varepsilon)\,E_x}{\sqrt{j\omega\mu\,(\sigma + j\omega\varepsilon)}}, \text{ using Eqn. (11.34).}$$

Hence

$$= \frac{E_x}{H_y} = \sqrt{\frac{j\omega\mu}{\sigma + j\omega\varepsilon}} = \eta \qquad ...(11.37)$$

Then this is the complete expression for the characteristic wave impedance which reduces to the form $\sqrt{\dfrac{\mu}{\varepsilon}}$ as given in Eqn. (11.24) only, if $\sigma = 0$ (for a *perfect* dielectric), In Eqn. (11.28) also, the value of η is to be taken as that given in Eqn. (11.37).

The quantity γ which determines the values of $\overline{\boldsymbol{E}}$ and $\overline{\boldsymbol{H}}$ for different values of z as the wave is propagated, is called the propagation constant. Usually it is a complex quantity, so we write

$$\gamma = \alpha + j\beta \qquad ...(11.38)$$

Using this in Eqn. (11.34), we proceed to determine the values of α and β. Thus

$$\gamma^2 = (\alpha + j\beta)^2$$

$$= (\alpha^2 - \beta^2) + j2\alpha\,\beta$$
$$= j\omega\,\mu\,\sigma - \omega^2\mu\,\varepsilon$$

Separating real and imaginary parts

$$\left.\begin{aligned}(\alpha^2 - \beta^2) &= -\omega^2\mu\varepsilon\\ 2\alpha\beta &= \omega\mu\sigma\end{aligned}\right\} \qquad ...(11.39)$$

Therefor $\alpha^2 + \beta^2 = \sqrt{(\alpha^2 - \beta^2)^2 + 4\alpha^2\beta^2}$

$$= \sqrt{(\omega^2\mu\,\varepsilon)^2 + (\omega\mu\,\sigma)^2}$$

or $$\alpha^2 + \beta^2 = \omega^2\,\mu\,\varepsilon\sqrt{1 + \left(\frac{\sigma}{\omega\varepsilon}\right)^2} \qquad ...(11.40)$$

From Eqns. (11.39) and (11.40), we get by addition and subtraction

$$\alpha = \left[\frac{\omega^2\mu\varepsilon}{2}\left\{\left(1 + \left(\frac{\sigma}{\omega\varepsilon}\right)^2\right)^{\frac{1}{2}} - 1\right\}\right]^{\frac{1}{2}} \qquad ...(11.41)$$

$$\beta = \left[\frac{\omega^2\mu\varepsilon}{2}\left\{\left(1 + \left(\frac{\sigma}{\omega\varepsilon}\right)^2\right)^{\frac{1}{2}} + 1\right\}\right]^{\frac{1}{2}} \qquad ...(11.42)$$

The physical significance of α and β can be understood by writing Eqns. (11.35) and (11.36) using them instead of γ. Thus we write Eqn. (11.35) as

$$E = E_0\, e^{[j\,\omega\,t - (\alpha + j\,\beta)\,z]}$$
$$= (E_0\, e^{-\alpha\,z})\, e^{j\,(\omega t - \beta z)} \qquad ...(11.43)$$

The term in brackets decreases continuously as z increases. This means that the amplitude of the $\overline{E}$ vector gets attenuated as z increases, because of the term $(-\alpha\,z)$ in the exponent. Therefore α is called the attenuation constant. An equation similar to Eqn. (11.43) can also be written for the vector $\overline{H}$.

The factor $\quad e^{j\,(\omega t - \beta z)} = \cos(\omega t - \beta z) + j\sin(\omega t - \beta z)$

shows that βz introduces a phase change as z increases. Hence β is called the phase-shift constant.

In Eqn. (11.43) since e is raised to the power $(-\alpha\,z)$, and such power must be a number (with no dimensions), it follows that αz being a number, the units for α must be some number per metre. If the amplitudes of E in Eqn. (11.43) for E_1 and E_2 at two points separated by one metre, then amplitudes are

$$E_1 = E_0\, e^{-\alpha\,z},$$
$$E_2 = E_0\, e^{-\alpha\,(z+1)} = E_0\, e^{-\alpha\,z}\, e^{-\alpha}$$

Hence $\quad \log_e\left(\dfrac{E_1}{E_2}\right) = \log_e(e^{+\alpha}) = \alpha$ neper/metre.

Neper is the name given to this unit in honour of John Napier (1550–1617), the inventor of logarithms. The attenuation constant is thus a logarithmic unit, an attenuation of one neper per metre will make $E_2 = E_1/e$ in the above example. However, the commonly used unit now-a-days is decibels (dB) which is a power ratio, and according to its definition

$$\text{Attenuation in dB} = N_{dB}$$
$$= 20 \log_{10} (E_1/E_2).$$

Using Eqns. (1.8) and (1.10) with $E_2 = \frac{E_1}{e}$, we have

$$\text{Attenuation in nepers} = N_{neper} = \log_e (E_1/E_2)$$
$$= \log_{10} (E_1/E_2) \log_e 10$$
$$= \frac{2.30}{20} [20 \log_{10}(E_1/E_2)]$$

or
$$N_{neper} = 0.115\, N_{dB}$$

This shows that *dB* is a unit of smaller size than neper, because the number of dB has to be multiplied by 0.115, a number less than one. Successively putting N_{dB} and N_{neper} equal to unity in the above equation, we get

$$1 \text{ dB} = 0.115 \text{ neper}$$

and
$$1 \text{ neper} = \frac{1}{0.115} = 8.686 \text{ dB} \qquad ...(11.44)$$

The unit for β, the phase constant, is found from Eqn. (11.43) by noting that $(\omega t - \beta z)$ must be number of radians. Just as ω is in radians/second, and t is in seconds and therefore ωt is in radians, similarly βz must be in radians, or units for β must be radians/metre.

11.8 CLASSIFICATION OF MEDIUM AS DIELECTRIC OR CONDUCTOR

From Table 7.1 we conclude that for most conductors, σ has value of the order of 10^7 siemens/metre, while for most dielectrics, it is of the order of 10^{-3} to 10^{-15} siemens/metre. From Table 7.2 we see that for most materials, ε_r lies between 1 and 100. Hence it may appear that there need not be any difficulty about classifying a medium, in which an electromagnetic wave is being propagated as a conductor or as a dielectric on the basis of its conductivity only.

However, in Eqns. (11.41) and (11.42), the significant quantity is found to be not σ alone but the ratio $\sigma/(\omega\, \varepsilon)$, which suggests that the influence of σ can be materially altered by the frequency dependent term in the denominator.

Now the Eqn. (11.29) shows that the ratio

$$\frac{\text{Amplitude of conduction current}}{\text{Amplitude of displacement current}} = \frac{\sigma}{\omega \varepsilon} \qquad ...(11.45)$$

Therefore, the question of whether to treat a medium *at a given frequency* as a conductor or as a dielectric, actually means asking whether the conduction current is greater or the displacement current is greater. It also means that a medium may act as a fairly good conductor below a certain frequency and as a fairly good dielectric above that frequency. For the sake of simplicity, we assume that if $\sigma < \omega\varepsilon$, it is a dielectric, and if $\sigma > \omega\varepsilon$, it is a conductor. In practical cases, for a good dielectric, $\sigma << \omega\,\varepsilon$, or $\frac{\sigma}{\omega\varepsilon} < \frac{1}{100}$ and for a good conductor, $\sigma >> \omega\,\varepsilon$, or $\frac{\sigma}{\omega\varepsilon} > 100$.

11.9 WAVE PROPAGATION IN A GOOD DIELECTRIC

Here we assume that $\sigma << \omega\,\varepsilon$ (or $\sigma / \omega\,\varepsilon < 1/100$). As a first approximation, therefore, we may assume that the wave Eqns. (11.11) and (11.12) are valid, (or in Eqns. (11.32) and (11.33) , σ is put equal to zero). The velocity of propagation of the wave is

$$v' = \frac{1}{\sqrt{\mu\,\varepsilon}}$$

$$= \frac{1}{\sqrt{\mu_0\,\varepsilon_0}}\,\frac{1}{\sqrt{\mu_r\,\varepsilon_r}}$$

$$= \frac{v_0}{\sqrt{\mu_r\,\varepsilon_r}} \qquad ...(11.46)$$

This shows that in any medium, the velocity of wave propagation v' is less than v_0, the velocity of propagation in free space or in a vacuum. For all non-magnetic materials, $\mu_r = 1$. Hence $v' = v_0\sqrt{\varepsilon_r}$. But we know from optics (and it will also be proved for electromagnetic waves as in Eqn. (12.14) in next chapter), that the refractive index of the medium is the ratio of the phase velocity in vacuum to the phase velocity in the medium. In other form,

$$\text{Refractive index} = \frac{v_0}{v'} = \sqrt{\mu_r\,\varepsilon_r} = \sqrt{\varepsilon_r} \qquad ...(11.47)$$

A comparison of the refractive index and $\sqrt{\varepsilon_r}$ for some materials is shown in Table 11.1.

Table 11.1 Refractive index and dielectric constant

Material	*Dielectric constant* (ε_r)	$\sqrt{\varepsilon_r}$	*Optical refractive Index*
Air	1.000590	1.000295	1.000294
CO_2	1.000946	1.000473	1.000449
Plate Glass,			1.500
Crown Glass,	4.84	2.2	1.530
Flint Glass			1.635
Ethyl Alcohol	25.0	5.0	1.360
Water	81.0	9.0	1.333

It is seen that except in the case of gases, Eqn. (11.47) does not appear to be true. However in optics it is found that the refractive index varies with the frequency (or wavelength) of light, giving rise to phenomenon of optical dispersion and the production of a spectrum by a prism. The formula (11.47) disregards this aspect which arises because of the presence of dipoles in the dielectric. The measurement of dielectric is done at the radio frequencies below 10 GHz, whereas optical refractive index is measured at 500,000 GHz or so, as seen at the end of Sec. 10.6. This accounts for the apparent discrepancy between Eqn. (11.47) and Table 11.1.

The fact that $\sigma/\omega\,\varepsilon << 1$ enables us to apply binomial theorem [Eqn. (1.1)] taking only the first two terms, and to write Eqns. (11.41) and (11.42) as follows:

$$\left[1+\frac{\sigma^2}{\omega^2\varepsilon^2}\right]^{\frac{1}{2}} \pm 1 = \left[1+\frac{\sigma^2}{2\omega^2\varepsilon^2}\right] \pm 1$$

Hence $$\alpha = \left[\frac{\omega^2\mu\varepsilon}{2}\,\frac{\sigma^2}{2\omega^2\varepsilon^2}\right]^{\frac{1}{2}} = \frac{\sigma}{2}\sqrt{\frac{\mu}{\varepsilon}} \qquad \text{...(11.48)}$$

and $$\beta = \left[\frac{\omega^2\mu\varepsilon}{2}\left(\frac{\sigma^2}{2\omega^2\varepsilon^2}+2\right)\right]^{\frac{1}{2}} = \omega\sqrt{\mu\varepsilon}\left[1+\frac{\sigma^2}{4\omega^2\varepsilon^2}\right]^{\frac{1}{2}}$$

or $$\beta = \omega\sqrt{\mu\varepsilon}\left(1+\frac{\sigma^2}{8\omega^2\varepsilon^2}\right) \qquad \text{...(11.49)}$$

Now $\omega\sqrt{\mu\varepsilon} = \frac{\omega}{v} = \frac{2\pi f}{f\lambda} = \frac{2\pi}{\lambda} = k$, where λ is the wavelength in the medium, and k is called the wave number, or the number of radians phase change per metre length in the direction of wave propagation. This is the phase-shift constant for a perfect dielectric having $\sigma = 0$ and in which the wave would travel with the velocity v'. However in a medium with a finite value of σ, the correction term $\frac{\sigma^2}{8\omega^2\varepsilon^2}$ indicates that the actual phase shift is slightly greater. Also the velocity of wave propagation with $\sigma \neq 0$ is, by Eqn. (1.2),

$$v = \frac{\omega}{\beta} = \frac{\omega}{\omega\sqrt{\mu\varepsilon}\left(1+\frac{\sigma^2}{8\omega^2\varepsilon^2}\right)} = v'\left[1-\frac{\sigma^2}{8\omega^2\varepsilon^2}\right] \qquad \text{...(11.50)}$$

Thus from Eqns. (11.46) and (11.50), we see that the velocity of propagation v' in a perfect dielectric is less than the velocity v_0 in free space, because $\varepsilon_r > 1$. This velocity is still further reduced, because of the power loss that occurs on account of the finite conductivity.

Similarly, from Eqn. (11.37) we find that for a perfect dielectric with $\sigma = 0$.

$$\eta' = \sqrt{\frac{\mu}{\varepsilon}} = \sqrt{\frac{\mu_r}{\varepsilon_r}}\sqrt{\frac{\mu_0}{\varepsilon_0}} = \sqrt{\frac{\mu_r}{\varepsilon_r}}\,\eta_0 \qquad \text{...(11.51)}$$

is less than η_0 or 377 ohms, because $\varepsilon_r > 1$ and $\mu_r = 1$. This is further altered for a dielectric having a finite value of σ as follows:

$$\eta = \sqrt{\frac{j\omega\mu}{\sigma + j\omega\varepsilon}} = \sqrt{\frac{j\omega\mu}{j\omega\varepsilon}\frac{1}{1+\dfrac{\sigma}{j\omega\varepsilon}}}$$

$$= \eta'\left[1 - \frac{\sigma}{j\omega\varepsilon}\right]^{\frac{1}{2}}$$

$$= \eta'\left(1 - \frac{\sigma}{2j\omega\varepsilon}\right)$$

$$= \eta'\left(1 + j\frac{\sigma}{2\omega\varepsilon}\right) \quad \text{...(11.52)}$$

Thus a small reactive (inductive) component gets added to the characteristic wave impedance, because of the losses in the dielectric.

11.10 WAVE PROPAGATION IN A GOOD CONDUCTOR

In this case $\sigma >> \omega\,\varepsilon$ or $\frac{\sigma}{\omega\varepsilon} > 100$ (say). Therefore Eqn. (11.34) can be simplified using Eqn. (1.26) as

$$\gamma = \sqrt{j\omega\mu(\sigma + j\omega\varepsilon)}$$

$$= \sqrt{j\omega\mu\sigma\left(1 + j\frac{\omega\varepsilon}{\sigma}\right)}$$

$$\approx \sqrt{\omega\mu\sigma}\,\sqrt{j} = \sqrt{\omega\mu\sigma}\left(\frac{1}{\sqrt{2}} + j\frac{1}{\sqrt{2}}\right)$$

$$= \sqrt{\omega\mu\sigma}\;\underline{/\,45^\circ} \quad \text{...(11.53a)}$$

Therefore $$\sigma = \beta = \sqrt{\frac{\omega\mu\sigma}{2}} \quad \text{...(11.53b)}$$

The first thing we note about Eqn. (11.53) is that, since α is very large, σ is very large. In other words, the wave gets attenuated very rapidly inside a good conductor. The second thing is to be noted that β is also very large, which implies that

$v = \omega/\beta$ is very small at lower values of ω (radio frequencies). In fact v is of the order of the velocity of sound in air (about 331.45 m/s), the value being

$$v = \frac{\omega}{\beta} = \omega\sqrt{\frac{2}{\omega\mu\sigma}}$$

$$= \sqrt{\frac{2\omega}{\mu\sigma}} = \sqrt{\frac{4\pi f}{\mu\sigma}} \quad \text{...(11.54)}$$

So far as the characteristic wave impedance is concerned, we have

$$\eta = \sqrt{\frac{j\omega\mu}{\sigma + j\omega\varepsilon}} \approx \sqrt{\frac{j\omega\mu}{\sigma}}$$

$$= \sqrt{\frac{\omega\mu}{\sigma}}\left(\frac{1}{\sqrt{2}} + j\frac{1}{\sqrt{2}}\right) \qquad ...(11.55)$$

This shows that η is very small in magnitude (less than one ohm) at radio frequencies. Also at all these frequencies, its resistive and reactive components are equal, that is, its phase angle is 45°. In other words, the electric field intensity E leads the magnetic field intensity H by an angle of 45° at all frequencies in all good conductors.

11.11 DEPTH OF PENETRATION IN A GOOD CONDUCTOR

A very useful parameter can be thought of to give an idea of the rate at which the wave gets attenuated in a good conductor as per value of α given by Eqn. (11.53*b*). Assume that the plane $z = 0$ is the surface of the good conductor and that the wave is travelling into the conductor perpendicular to the surface. Then if E_0 is the amplitude of the electric field intensity at the surface $z = 0$, the amplitude will have a value $E_0 e^{-\alpha z}$ at a distance z inside the surface. Hence we define the parameter d, which is the value of z at which the amplitude of the electric field intensity becomes $1/e$ times its amplitude at the surface. This means that

$$e^{-\alpha\,\delta} = e^{-1} \text{ or } \alpha\,\delta = 1$$

or

$$\delta = \frac{1}{\alpha} = \sqrt{\frac{2}{2\pi f \mu \sigma}}$$

$$= \sqrt{\frac{1}{\pi f \mu \sigma}} \text{ metre} \qquad ...(11.56)$$

This distance δ is called the depth of penetration. It follows from Table 1.1 that at a depth of 5δ, the amplitude may be regarded as being almost zero. This is the reason why the parameter δ gives a better mental picture of the situation than the rate of attenuation α per metre does.

For the case of copper having $\mu = \mu_0$ and $\sigma = 5.800 \times 10^7$ S/m, the depth of penetration of an electromagnetic wave can be written as

$$\delta = \frac{1}{\sqrt{f}\,\sqrt{\pi(4\pi \times 10^{-7})(5.8 \times 10^7)}}$$

or

$$\delta = \frac{0.667}{\sqrt{f}} \text{ metre} \quad (f \text{ in Hertz}) \qquad ...(11.57)$$

At 100 Hz, it is 6.67 mm but at 100 MHz, it is only 0.00667 mm. This is the reason why the coils used in the radio frequency circuits of the high power stages in a transmitter are made of hollow copper tubes, because the inner part of the conductor is not used for conducting current at all, and hence removing it does not affect the working of the circuit, but saves copper cost.

11.12 SURFACE IMPEDANCE OF GOOD CONDUCTOR TO SINUSOIDAL CURRENTS

As before, we assume the definition of a good conductor that $\frac{\sigma}{\omega\varepsilon} > 100$ with charge density inside the conductor being zero. Also the electric field is assumed to very as $e^{j\omega t}$. Since the displacement current is negligibly small, Eqns. (11.32) and (11.33) take the form

$$\overline{\nabla}^2 \overline{E} = j\omega\mu\sigma \overline{E} = \gamma^2 \overline{E} \qquad \text{...(11.58)}$$

and

$$\overline{\nabla}^2 \overline{H} = j\omega\mu\sigma \overline{H} = \gamma^2 \overline{H} \qquad \text{...(11.59)}$$

where

$$\gamma^2 = j\,\omega\mu\sigma \qquad \text{...(11.60)}$$

These differential equations must be satisfied at every point in the conducting medium. To get a picture of what this implies, we assume, to simplify the calculations and to highlight some aspects. that the conductor is semi-infinite in extent, extending from $z = 0$ to $z = \infty$, but in the x- and y-directions, it extends from $-\infty$ to $+\infty$. How such an enormous size of conductor actually simplifies the calculations will be seen later.

Next we assume that the electric vector has a component in the x-direction only, so that everywhere $E_y = 0$ and $E_z = 0$. In other words, it has the same value at all points on a plane z = constant. With these assumptions, Eqn. (11.58) reduces to

$$\frac{d^2 E_x}{dz^2} = \gamma^2 E_x$$

which gives the solution

$$E_x = C_1 e^{+\gamma z} + C_2 e^{-\gamma z}$$

Obviously $C_1 = 0$, otherwise E_x will become infinite when $z \to \infty$. Also using Eqns. (11.26) and (11.56), we find that

$$-\gamma = -\sqrt{j\omega\mu\sigma} = -\left[\frac{1}{\sqrt{2}} + j\frac{1}{\sqrt{2}}\right]\sqrt{2\pi f\mu\sigma}$$

$$= \pm\,(1 + j1)\,\sqrt{\pi f\mu\sigma} = \pm\left(\frac{1}{\delta} + j\frac{1}{\delta}\right) \qquad \text{...(11.60}a\text{)}$$

Taking the negative sign only, the complete solution is

$$E_x = \left(C_2\, e^{-\frac{z}{\delta}}\right) e^{j(\omega t - \frac{z}{\delta})}$$

If we assume that at time $t = 0$, on the *surface* of the conductor at $z = 0$, the value of $E_x = E_0$, then $C_2 = E_0$ and the solution becomes

$$E_x = \left(E_0\, e^{-\frac{z}{\delta}}\right) e^{j(\omega t - \frac{z}{\delta})}$$

$$= \left(E_0\, e^{-\frac{z}{\delta}}\right) \cos\left(\omega t - \frac{z}{\delta}\right) \qquad \text{...(11.61)}$$

But this is the same as Eqn. (11.43) with

$$\alpha = \beta = \frac{1}{\delta} \qquad \text{...(11.62)}$$

Hence this represents a wave travelling in the positive direction of the z-axis. However, as the wave advances inside the conductor, the electric field amplitude gets attenuated by the factor $e^{-\frac{z}{\delta}}$ and its phase gets retarded with respect to the field at the surface, by an angle z/δ radians at a depth z from the surface.

If instead of the negative sign, we had used the positive sign in Eqn. (11.60*a*), we would have obtained a reflected wave travelling in the negative direction of the z-axis. By taking a conductor extending to infinity in the z-direction, we have placed the other surface of the conductor at infinity. The forward wave travelling in the positive direction of the z-axis will never reach the surface at infinity (or stated differently, it will take an infinite time to reach that surface) and hence there will be no reflected wave. This is the simplification resulting from the assumption of a conductor of infinite extent. It has enabled us to put $C_1 = 0$.

The final picture that has emerged is that with only E_x present everywhere in the conductor, a current flows in the conductor only in the x-direction. This current is alternating because E_x varies as $e^{j\omega t}$. However, there is also a wave travelling in the z-direction, which gets attenuated because of the factor $e^{-\frac{z}{\delta}}$ in the Eqn. (11.62). The implication of this is that, though the value of E_x (and current) is constant for any plane z = constant, this value goes on diminishing exponentially, as the value of z is increases, and we go deeper and deeper into the conductor from its surface at $z = 0$. This is shown in Fig. 11.3 at a time which makes $\omega t = 2n\pi$, the value of E_x at $z = 0$ being E_0.

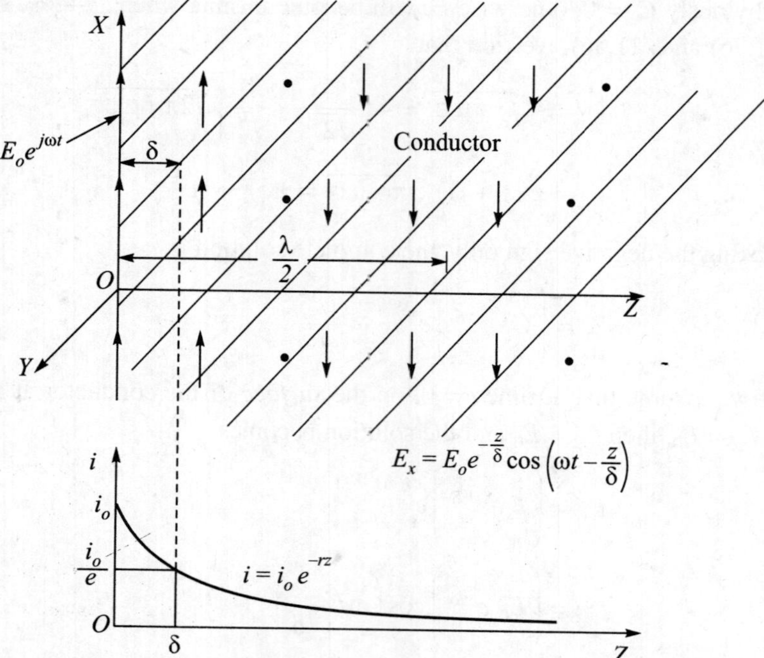

Fig. 11.3. Surface impedance versus voltage and current in a semi-infinite conductor

We are now in a position to define the surface impedance of the conductor, denoted by $Z_{surface}$, as the ratio of the tangential component E_t at the *surface* of the conductor to the *total current density J* per metre width (measured in this example in the *y*-direction) that flows in the conductor under the influence of this time-varying E_t. In equation form, we can write this as

$$Z_{surface} = \frac{E_t}{J} \quad ...(11.63)$$

It is important to notice that J in this definition is the *total* current that flows in the *x*-direction across a plane x = constant, in the entire slab of one metre width lying between the planes $y = y_0$ and $y = y_0 + 1$, and extending from $z = 0$ to $z = \infty$.

Let the current at the surface of the slab of one metre width in plane $z = 0$ be denoted as i_0. Then by Ohm's law

$$i_0 = \sigma E_t \quad ...(11.64)$$

where $E_t = E_0$ in this case.

Since E decreases inside the conductor to the value $E_o e^{-\gamma z}$, it follows that the current i in any plane z = constant will be related to i_0 by the equation

$$i = i_0 e^{-\gamma z} \quad ...(11.65)$$

Then the current in the entire slab of one metre width is given by

$$J = \int_0^\infty i dz = i_0 \int_0^\infty e^{-\gamma z} dz$$

$$= -\frac{i_0}{\gamma}\left[e^{-\gamma z}\right]_0^\infty = \frac{i_0}{\gamma} \quad ...(11.66)$$

Combining this with Eqn. (11.64)

$$J\gamma = i_0 = \sigma E_t$$

Hence by Eqns. (11.63) and (11.60*a*)

$$Z_{surface} = \frac{E_t}{J} = \frac{\gamma}{\sigma} = \frac{\frac{1}{\delta} + j\frac{1}{\delta}}{\sigma}$$

or

$$R_{surface} + jX_{surface} = \frac{1}{\delta\sigma} + j\frac{1}{\delta\sigma}$$

$$= \frac{\sqrt{2}}{\delta\sigma} \angle 45° \quad ...(11.67)$$

Thus the phase angle of the surface impedance is always equal to 45°. Moreover, the magnitude of the resistive component is $1/\delta\sigma$. Stated in words, this means that the *surface resistance* of a conductor at any frequency is equal to the *resistance of a thickness of δ metres* of the conductor at the surface. Because of its location near the surface, δ is referred to as the *skin depth* and the phenomenon is called the *skin effect*.

This is the basis of an interesting exhibit called a Tesla coil. Tesla had originally made it using an induction coil, but its modern version uses a high frequency (say 1 MHz) power oscillator with its coil coupled to a secondary coil having a large number of turns. The ends of the secondary coil terminate in small spheres situated a few centimetres apart. Because of the step-up transformer action, a high voltage is produced between the spheres. The intensity of the electric field in their neighbourhood is so high that a fluorescent tube brought near the spheres begins to glow! Yet a person can touch the sphere without getting electrocuted!! The reason for this is that it requires a current of about 50 mA to flow through the heart of a person to kill him. Because of the high frequency used in the Tesla coil, the current flows only through the skin of the person, and not only does the current not go through the heart at all, but it does not even penetrate the skin surface sufficiently to reach the nerve ends which are below the skin. Hence the person does not even get an electric shock. Truly an example justifying the name "skin effect"!

Again noting that using Eqn. (11.60), we can write

$$Z_{\text{surface}} = \frac{\gamma}{\sigma} = \frac{\sqrt{j\omega\mu\sigma}}{\sigma} = \sqrt{\frac{j\omega\mu}{\sigma}} \quad \text{...(11.68)}$$

and comparing it with Eqn. (11.37) for the case where $\sigma >> \omega\,\varepsilon$, we find that $Z_{surface} = \eta$, the characteristic wave impedance of the conducting medium. From transmission line theory, we know that the *input impedance* of a two-wire line or a co-axial cable of infinite length is called the characteristic impedance of the line. In the present case, $Z_{surface}$ is equal to the input impedance of the one-metre-wide infinitely long strip, and therefore the name characteristic wave impedance given to η is justified.

As we have seen in Sec. 11.11, the conductor need not be infinitely long. The result of this section will hold if the dimension of the conductor is >> 5δ. This is a very useful concept from the practical point of view because conductors of even a few millimetres thickness can be regarded as being infinitely long at radio frequencies, or radii of curvature of curved conductors can be treated as infinite, if they are greater than a few millimetres.

11.13 SUMMARY AND COMMENTS

It is found that any type of time variation of the electric and magnetic fields gives the solution of Maxwell's equations in the form of an electromagnetic wave, which is propagated at a velocity that is independent of the frequency. The ratio $\dfrac{\overline{E}}{\overline{H}}$ for the wave gives the characteristic wave impedance of the medium. In this chapter, we have studied the behaviour of waves in unbounded media, whether dielectrics or conductors. In the next chapter, we shall deal with wave phenomenon at boundary between media.

EXERCISE 11

1. Show that for sinusoidal time variations of the electric and magnetic fields of the form $e^{j(\omega t - \beta z)}$, the wave Eqns. (11.11) and (11.12) can be written as

$$\left(\overline{\nabla}^2 + k^2\right)\overline{E} = 0$$

and $$\left(\overline{\nabla}^2 + k^2\right)\overline{H} = 0$$

where k is the wave number, $k = \dfrac{2\pi}{\lambda}$.

2. In Eqn. (11.19) it is stated that

$$E_1 = E_0\, e^{j(\omega t - \beta z)}$$

represents a wave travelling in the positive direction of the z-axis. Find what

$$E_2 = E_0\, e^{(\omega t - \beta z)}$$

represents. Assuming that $\beta = 2\pi$ radians per metre, sketch the two functions over the range $z = 0$ to $z = 2$ for values of

(i) $\omega t_1 = 0$,

(ii) $\omega t_2 = \dfrac{\pi}{2}$ and

(iii) $\omega t_3 = \pi$.

3. Following the concept of Sec. 11.10, let us assume that the Gangetic plain can be considered to be a good conductor if $\dfrac{\sigma}{\omega \varepsilon} > 100$. Assuming the values $\sigma = 8 \times 10^{-3}$ S/m and $\varepsilon_r = 15$ for the soil there, find the frequency below which the plain may be treated as a good conductor.

4. The inner surfaces of brass waveguides for microwaves are usually silvered to reduce the losses, silver being the best conductor. Assuming that the thickness of this silver coating must be at least 5δ where δ is the skin depth, find the minimum thickness needed at 10 GHz (3-cm waveguide). For silver, assume that $\mu = \mu_0$ and $\sigma = 6.1 \times 10^7$ S/m.

5. Find the displacement current in a copper wire in which the conduction current at 10 MHz is one milliampere. Assume that for copper, $\varepsilon_r = 1$ and $\sigma = 5.800 \times 10^7$ S/m.

6. Prove that the depth of penetration δ is always very much shorter than one wavelength λ.

7. Find the skin depth at 1 MHz and at 10 GHz for the following materials:

Material	*σ in S/m*	*Relative permittivity μ_r*
Silver	6.15×10^7	1.0
Copper	5.80×10^7	1.0
Nickel	1.30×10^7	100.0
Magnetic iron	1.00×10^7	1200.0
Graphite	1.00×10^5	1.0

CHAPTER

12

WAVES AT BOUNDARY BETWEEN TWO MEDIA

12.1 INTRODUCTION

Throughout this chapter, unless otherwise stated, the wave will be assumed to be travelling in medium 1 and meeting the boundary between this and the medium 2 at $z = 0$, where both media are assumed to be semi-infinite or extending to infinity on either side of the boundary. This assumption ensures that the only phenomenon at the boundary will be those due to the incident wave only and not due to any waves reflected from any other boundaries. The quantities ε_1, μ_1 and η_1 will refer to medium 1 and ε_2, μ_2, and η_2 to medium 2.

Similarly the electric and magnetic field intensities in the incident wave and its direction of propagation (or direction of Poynting's vector) will be denoted as E_i, H_i, P_i in the reflected wave as E_r, H_r, P_r and in the transmitted wave as E_t, H_t, P_t.

12.2 WAVE INCIDENT NORMALLY ON BOUNDARY BETWEEN PERFECT DIELECTRICS

Let $z = 0$ be the plane separating the two semi-infinite perfect dielectrics 1 and 2 (Fig. 12.1). The condition that the incident wave is incident normally on the boundary means that the Poynting vector P_i is perpendicular to the plane $z = 0$. We choose the x-axis in the direction of the electric vector E_i. Since the wave travels in the direction $\overline{\boldsymbol{E}} \times \overline{\boldsymbol{H}}$, this implies that Hi must be in the direction of the y-axis.

At the plane $z = 0$, a part of the energy is reflected back in medium 1 along the normal, giving a reflected wave with P_r directed opposite to P_i, and the remaining energy is transmitted as a wave along the normal in medium 2. Thus P_t is in the same direction as P_i.

Next we apply Eqn. (7.17) to the boundary and equate the tangential components of the wave on the two sides to write

$$\left.\begin{array}{l}\text{Tangential component of electric}\\ \text{field in medium 1}\end{array}\right\} = \left\{\begin{array}{l}\text{Tangential component of electric}\\ \text{field in medium 2}\end{array}\right.$$

or

$$E_i + E_r = E_t \qquad ...(12.1)$$

Similarly, applying the boundary condition for magnetic fields as in Eqn. (8.55) and noting that $\sigma = 0$ for perfect dielectrics, we get

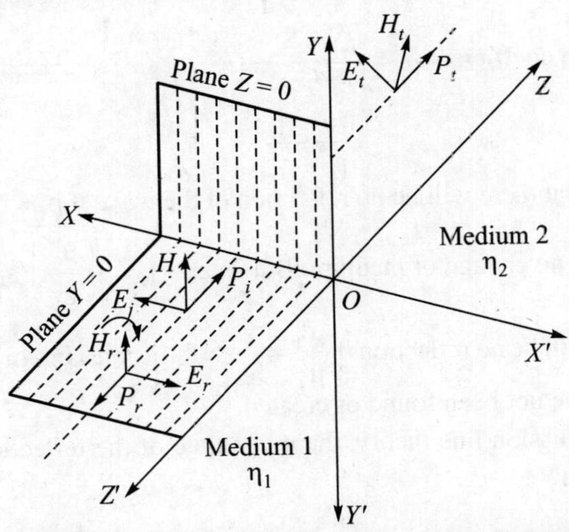

Fig. 12.1. Wave incident normally on perfect dielectric

$$H_i + H_r = H_t \quad \text{...(12.2)}$$

Also by Eqn. (11.28)

$$E_i = \eta_1 H_i,$$

$$E_r = -\eta_1 H_r$$

and

$$E_t = \eta_2 H_t \quad \text{...(12.3)}$$

Substituting these values in Eqn. (12.2), and combining with Eqn. (12.1)

$$\frac{1}{\eta_1}(E_i - E_r) = \frac{E_t}{\eta_2} = \frac{1}{\eta_2}(E_i + E_r)$$

so that

$$\frac{E_i - E_r}{E_i + E_r} = \frac{\eta_1}{\eta_2}$$

Therefore,

$$\frac{E_r}{E_i} = \frac{\eta_2 - \eta_1}{\eta_2 + \eta_1} \quad \text{...(12.4)}$$

Also from Eqn. (12.1)

$$\frac{E_t}{E_i} = 1 + \frac{E_r}{E_i}$$

$$= 1 + \frac{\eta_2 - \eta_1}{\eta_2 + \eta_1} = \frac{2\eta_2}{\eta_2 + \eta_1} \quad \text{...(12.5)}$$

Again from Eqn. (12.3)

$$\frac{H_r}{H_i} = -\frac{E_r}{E_i} = \frac{\eta_1 - \eta_2}{\eta_1 + \eta_2} \quad \text{...(12.6)}$$

and

$$\frac{H_t}{H_i} = \frac{\eta_1}{\eta_2}\frac{E_t}{E_i} = \frac{2\eta_1}{\eta_1 + \eta_2} \quad \text{...(12.7)}$$

Thus the values of E_r, H_r, E_t and H_t are known in terms of the values of E_i and H_i of the incident wave and the constants η_1 and η_2 for the two media.

Borrowing the idea from transmission line theory, the ratio $\frac{E_r}{E_i}$ is called the reflection coefficient ρ_r, and the ratio $\frac{E_t}{E_i}$ is called the transmission coefficient. Thus

$$\text{Reflection coefficient } \rho_r = \frac{E_r}{E_i}$$

$$= \frac{\eta_2 - \eta_1}{\eta_2 + \eta_1} \qquad ...(12.8)$$

It follows that there will be no reflection of the wave if $\rho_r = 0$. This implies that $\eta_2 = \eta_1$, or there is no change of medium. But since $\eta_1 = \sqrt{\left(\frac{\mu_1}{\varepsilon_1}\right)}$ and $\eta_2 = \sqrt{\left(\frac{\mu_2}{\varepsilon_2}\right)}$, it follows that there will be no reflection if $\frac{\mu_2}{\mu_1} = \frac{\varepsilon_2}{\varepsilon_1} = k_0$, a constant. In practice, however, such dielectrics have not been found or created.

As in transmission line theory, the magnitude of the reflection coefficient $|\rho_r|$ lies between 0 and 1.

12.3 WAVE INCIDENT OBLIQUELY ON BOUNDARY BETWEEN PERFECT DIELECTRICS

In Fig. 12.2 an incident beam meets the boundary over the region AB and gets partly reflected and partly transmitted. The incident ray FA, the reflected ray AD and the transmitted ray AE all lie in one plane, and θ_i, θ_r and θ_t are the angles made respectively

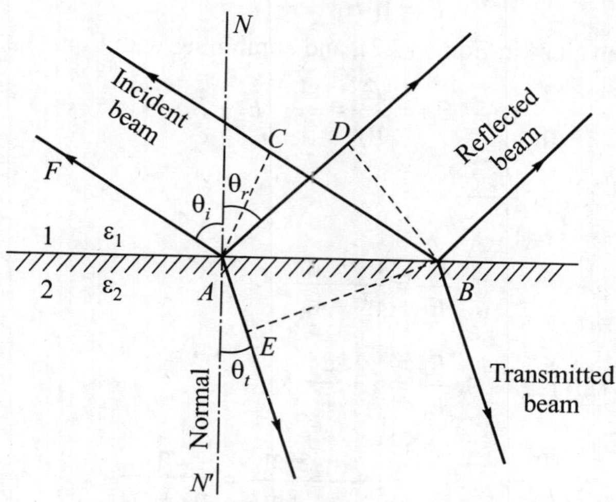

Fig. 12.2. Oblique incidence

by them with the normal NAN' to the boundary at the point A. Draw AC perpendicular to the incident ray, BD perpendicular to the reflected ray and BE perpendicular to the transmitted ray.

Let ε_1 and ε_2 be the permittivities of medium 1 and medium 2 and μ_0, the permeability of both media, since the permeability of all known dielectrics is the same as for free space. Then the velocities of propagation of waves in the two media will be

$$v_1 = \frac{1}{\sqrt{\mu_0 \varepsilon_1}}$$

and
$$v_2 = \frac{1}{\sqrt{\mu_0 \varepsilon_2}} \qquad ...(12.9)$$

If we assume that the incident beam reaches the point A at time t_1 and the point B at time t_2, then

$$CB = v_1(t_2 - t_1),$$
$$AD = v_1(t_2 - t_1),$$
$$AE = v_2(t_2 - t_1) \qquad ...(12.10)$$

From the geometry of the figure, using the concept that the angle between two lines is equal to the angle between the perpendiculars to these lines, we get

$$\angle NAF = \theta_i = \angle BAC,$$
$$\angle NAD = \theta_r = \angle ABD,$$
$$\angle N'AE = \theta_t = \angle ABE, \qquad ...(12.11)$$

Hence from Eqn. (12.10), using Eqn. (12.11), we get

$$CB = v_1(t_2 - t_1) = AD,$$

or
$$AB \sin \theta_i = AB \sin \theta_r$$

Therefore
$$\theta_i = \theta_r \qquad ...(12.12)$$

This is Snell's law of optics showing that the angle of incidence is equal to the angle of reflection.

Also from Eqns. (12.10) and (12.11)

$$\frac{CB}{AE} = \frac{v_1}{v_2}$$

or
$$\frac{AB \sin \theta_i}{AB \sin \theta_t} = \sqrt{\frac{\mu_0 \varepsilon_2}{\mu_0 \varepsilon_1}}$$

or
$$\frac{\sin \theta_i}{\sin \theta_t} = \frac{v_1}{v_2} = \sqrt{\frac{\varepsilon_2}{\varepsilon_1}}$$

or
$$\sqrt{\varepsilon_1} \sin \theta_i = \sqrt{\varepsilon_2} \sin \theta_t \qquad ...(12.13)$$

This is Snell's law of sines in optics and it shows that the

$$\text{Refractive index of the medium} = \frac{v_1}{v_2}$$
$$= \frac{\sqrt{\varepsilon_2}}{\sqrt{\varepsilon_1}} \qquad ...(12.14)$$

It will be shown in the next chapter in Sec. 13.3, that power flows in the direction $\overline{E} \times \overline{H}$ and is equal to $\overline{E} \times \overline{H}$. Since in the wave $\overline{E}$ and $\overline{H}$ are at right angles and $\frac{E}{H} = \eta$, it follows that the power that flows is equal to

$$EH \sin 90° = E\left(\frac{E}{\eta}\right) = \frac{E^2}{\eta}.$$

Therefore from Fig. 12.2, considering the component of power flowing in the direction of the normal *NAN′*, we get, by law of conservation of energy

$$\frac{E_i^2}{\eta_1}\cos\theta_i = \frac{E_r^2}{\eta_1}\cos\theta_i + \frac{E_t^2}{\eta_2}\cos\theta_t$$

Dividing throughout by $\frac{E_i^2}{\eta_1}\cos\theta_i$ and transposing

$$1-\frac{E_r^2}{E_i^2} = \frac{\eta_1}{\eta_2}\frac{\cos\theta_t}{\cos\theta_i}\frac{E_t^2}{E_i^2} = \frac{\sqrt{\varepsilon_2}\cos\theta_t}{\sqrt{\varepsilon_1}\cos\theta_i}\left(\frac{E_t}{E_i}\right)^2 \qquad \text{...(12.15)}$$

This important relation will be used in the two special cases discussed in the next two sections.

12.4 WAVE POLARIZED PERPENDICULAR TO THE PLANE OF INCIDENCE

In practice all antennas from which the electromagnetic waves originate are located near the ground and we are interested in the reflection of these waves from the ground. If the antenna conductor is horizontal, the electric vector of the electromagnetic wave launched by it will also be horizontal, and the wave is said to be horizontally polarized. If the antenna conductor is in the vertical plane, the electric vector need not be vertical; it will always lie in the vertical plane, but it need not itself be vertical. The term vertical polarization is often applied to this case, but strictly speaking it is not quite correct. It only implies that the antenna is in the vertical plane. To avoid this confusion caused by the terms horizontal and vertical, it is better to use the expression "wave perpendicular (or parallel) to the plane of incidence" instead of horizontal (or vertical) polarization.

In Fig. 12.3 let us assume that the electric vector $\overline{\boldsymbol{E}}$ of the incident waves is perpendicular to the plane of incidence (plane of the paper) coming out of the paper in the *y*-direction. Hence it is shown in the figure as a dot inside a tiny circle. For the reflected and transmitted waves, we *assign the same direction to be the positive direction* for E_r and E_t. Since Eqn. (12.1) holds at the boundary between the two diclectrics, it follows that

$$\frac{E_t}{E_i} = 1 + \frac{E_r}{E_i} \qquad \text{...(12.16)}$$

Substituting this value in Eqn. (12.15) we get, applying componendo and dividendo

$$\left(1-\frac{E_r^2}{E_i^2}\right) = \left(1-\frac{E_r}{E_i}\right)\left(1+\frac{E_r}{E_i}\right)$$

$$= \frac{\sqrt{\varepsilon_2}\cos\theta_t}{\sqrt{\varepsilon_1}\cos\theta_i}\left(1+\frac{E_r}{E_i}\right)^2$$

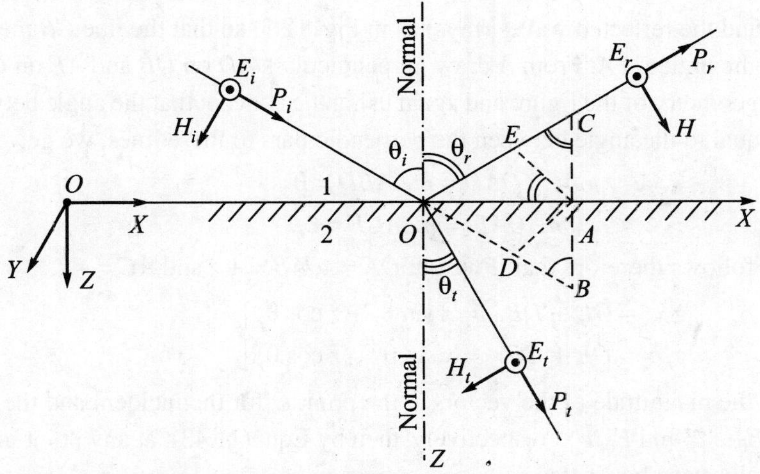

Fig. 12.3. Wave polarized perpendicular to plane of incidence

or
$$\frac{1-\dfrac{E_r}{E_i}}{1+\dfrac{E_r}{E_i}} = \frac{\sqrt{\varepsilon_2}\cos\theta_t}{\sqrt{\varepsilon_1}\cos\theta_i}$$

or
$$\rho_r = \frac{E_r}{E_i}$$
$$= \frac{\sqrt{\varepsilon_1}\cos\theta_i - \sqrt{\varepsilon_2}\cos\theta_t}{\sqrt{\varepsilon_1}\cos\theta_i + \sqrt{\varepsilon_2}\cos\theta_t} \quad ...(12.17)$$

where ρ_r is the reflection coefficient.

Thus we have obtained an expression giving the value of E_r in terms of E_i, the constants ε_1 and ε_2 of the media, and the angles θ_i and θ_t. It is obvious that E_r is less than E_i, or the coefficient of reflection lies between 0 and 1 as seen at end of Sec. 12.2.

Obviously there will be no reflected ray, or $E_r = 0$, if

$$\sqrt{\varepsilon_1}\cos\theta_i = \sqrt{\varepsilon_2}\cos\theta_t$$

Squaring and using Snell's law of sines [Eqn. (12.13)]

$$\varepsilon_1(1-\sin^2\theta_i) = \varepsilon_2(1-\sin^2\theta_t)$$
$$= \varepsilon_2\left(1-\frac{\varepsilon_1}{\varepsilon_2}\sin^2\theta_i\right)$$

Therefore $\varepsilon_1 = \varepsilon_2$

as before indicating that there will be no reflection, only if there is no change of medium. With electric vector perpendicular to plane of incidence, there must be a reflected ray in all practical cases if the medium changes.

With an incident and a reflected wave simultaneously present in medium 1, at any point, the values of E and H for the two waves have to be added. Let $OB = +Z_i$ and $OC = +Z_r$ be the two vectors equal in magnitude and *in the directions of travel* of the incident and the reflected waves (rays) as in Fig. 12.3, so that the line CB meets the x-axis at right angles at A. From A draw perpendiculars AD on OB and AE on OC. Then from the geometry of the figure and again using the concept that the angle between two lines is equal to the angle between the perpendiculars to these lines, we get

$$m\,\angle OAD = m\angle ABD = \theta_i$$

and
$$m\angle OAE = m\angle ACE = \theta_r.$$

It follows therefore that if we put $OA = x$, $AB = +z$ and $AC = -z$,

$$\left.\begin{aligned} z_i &= OD + DB &&= x \sin\theta_i + z\cos\theta_i \\ z_r &= OE + EC &&= x\sin\theta_r - z\cos\theta_r \end{aligned}\right\} \quad \text{...(12.18)}$$

If the magnitude of the vectors at the point O for the incident and the reflected rays are $E_{0i}\, e^{j\omega t}$ and $E_{0r}\, e^{j\omega t}$ respectively, then by Eqn. (11.43), at any point in medium 1, their values can be written as

$$E_i = E_{0i} e^{j(\omega t - \beta_1 z_i)}$$

and
$$E_r = E_{0r} e^{j(\omega t - \beta_1 z_r)} \quad \text{...(12.19)}$$

where $\beta_1 = \dfrac{2\pi}{\lambda_1} = k_1$, the wave number for medium 1, and λ_1 is the wavelength in medium 1. Notice that the expressions for both E_i and E_r can be written in the *same form*, because Z_i and Z_r are both measured *in the direction* in which the incident and the reflected waves are travelling.

Then using the reflection coefficient ρ_r given by Eqn. (12.17), we can write with the help of Eqns. (12.19) and (12.18), that the total electric vector in medium 1 is

$$\begin{aligned} E_y &= E_i + E_r \\ &= E_i + \rho_r E_i \\ &= E_0\, i\, e^{j\omega t}\,[e^{-j\beta_1 z_i} + \rho_r e^{-j\beta_1 z_r}] \\ &= E_{0i}\, e^{j\omega t}\,[e^{-j\beta_1 (x\sin\theta_i + z\cos\theta_i)} + \rho_r e^{-j\beta_1 (x\sin\theta_r - \cos\theta_r)}] \\ &= E_{0i}\, e^{j\omega t}\, e^{-j(\beta_1 \sin\theta_i)x}\,[e^{-j(\beta_1\cos\theta_i)z} + \rho_r e^{+j(\beta_1\cos\theta_i)z}] \end{aligned}$$

or
$$E_y = E_{0i}\, e^{j\omega t}\, e^{-j\beta_x x}\,[e^{-j\beta_z z} + \rho_r e^{+j\beta_z z}] \quad \text{...(12.20)}$$

where $\theta_r = \theta_i$,

$$\left.\begin{aligned} \beta_x &= \beta_1 \sin\theta_i &&= \frac{2\pi}{\lambda_1}\sin\theta_i &&= k_1 \sin\theta_i \\ \beta_z &= \beta_1\cos\theta_i &&= \frac{2\pi}{\lambda_1}\cos\theta_i &&= k_1\cos\theta_i \end{aligned}\right\} \quad \text{...(12.21)}$$

From these relations, we get

$$\begin{aligned} H &= H_i + H_r \\ &= \frac{E_{0i}}{\eta_1} + \frac{E_{0r}}{\eta_1} \end{aligned}$$

Hence noting the directions of H_i and H_r in Fig. 12.3, we have

$$H_x = \frac{E_{0i} e^{j\omega t} e^{-j\beta_x x}}{\eta_1}\,[(-\cos\theta_i)\, e^{-j\beta_z z} + \rho_r \cos\theta_i\, e^{+j\beta_z z}] \quad \text{...(12.22)}$$

and $$H_z = \frac{E_{0i} e^{j\omega t} e^{-j\beta_x x}}{\eta_1} [(\sin\theta_i)\, e^{-j\beta_z z} + \rho_r \sin\theta_i\, e^{+j\beta_z z}]$$ ...(12.23)

Equations (12.20), (12.22) and (12.23) define the values of the E and H vectors for the resultant wave in medium 1. The factor $e^{j(\omega t - \beta_x x)}$ in these equations indicates that the resultant is a wave travelling in the x-direction, that is, parallel to the boundary. A change in the angle of incidence does not affect the E vector [see Eqn. (12.20) which is independent of θ_i] but the components of the H vector only change with the angle of incidence. In the z-direction in medium 1, there is a standing wave which is discussed in detail in Sec. 12.6.

12.5 WAVE POLARIZED PARALLEL TO PLANE OF INCIDENCE

In this case since the electric vector E_i is in the plane of incidence, it follows that the magnetic vector H_i must be perpendicular to the plane of incidence. Hence we choose the magnetic vector as the reference vector and assign the positive direction for all of them (H_i, H_r, H_t) as being in the y-direction coming out of the paper (Fig. 12.4).

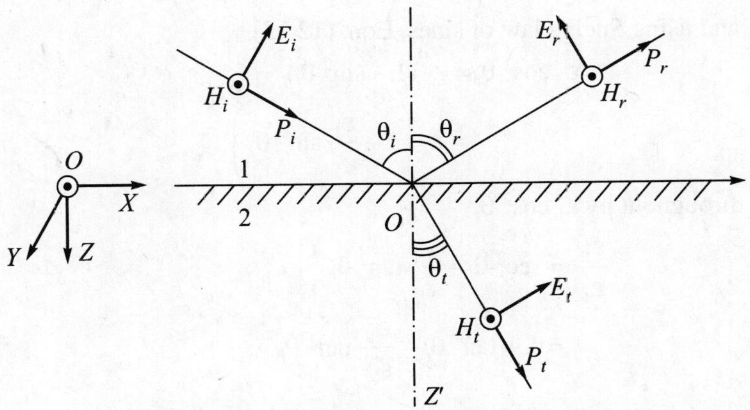

Fig. 12.4. Wave polarized parallel to plane of incidence

This is indicated by the dots within the three small circles. From these known direction of P and H in the three rays, the direction for E_i, E_r and E_t can be deduced to be as shown in the figure, since $\overline{P} = \overline{\boldsymbol{E}} \times \overline{\boldsymbol{H}}$ in direction. Then applying the boundary condition that the tangential component of E must have the same value on both sides of the boundary between medium 1 and medium 2, we write , since $\theta_r = \theta_i$

$$E_i \cos\theta_i - E_r \cos\theta_i = E_t \cos\theta_t$$ ....(12.24*a*)

Dividing throughout by $E_i \cos\theta_i$

$$\frac{E_t}{E_i} = \left(1 - \frac{E_r}{E_i}\right) \frac{\cos\theta_i}{\cos\theta_t}$$ ...(12.24)

Substituting this value in Eqn. (12.15), we get, after applying componendo and dividendo

$$\left(1-\frac{E_r^2}{E_i^2}\right)=\left(1-\frac{E_r}{E_i}\right)\left(1+\frac{E_r}{E_i}\right)$$

$$=\frac{\sqrt{\varepsilon_2}\cos\theta_t}{\sqrt{\varepsilon_1}\cos\theta_i}\left(1-\frac{E_r}{E_i}\right)^2\left(\frac{\cos\theta_i}{\cos\theta_t}\right)^2$$

or

$$\frac{1+\dfrac{E_r}{E_i}}{1-\dfrac{E_r}{E_i}}=\frac{\sqrt{\varepsilon_2}\cos\theta_i}{\sqrt{\varepsilon_1}\cos\theta_t}$$

Therefore

$$\rho_r'=\frac{E_r}{E_i}=\frac{\sqrt{\varepsilon_2}\cos\theta_i-\sqrt{\varepsilon_1}\cos\theta_t}{\sqrt{\varepsilon_1}\cos\theta_i+\sqrt{\varepsilon_1}\cos\theta_t} \quad \text{...(12.25)}$$

where ρ_r' is the reflection coefficient in this case. Since ρ_r' is less than unity, E_r is less than E_i. Also the minus sign in the numerator of Eqn. (12.25) indicates the possibility of E_r being equal to zero for a certain condition, viz.

$$\sqrt{\varepsilon_2}\cos\theta_i=\sqrt{\varepsilon_1}\cos\theta_t$$

Squaring and using Snell's law of sines, Eqn. (12.13)

$$\varepsilon_2\cos^2\theta_i=\varepsilon_1(1-\sin^2\theta_t)$$

$$=\varepsilon_1\left(1-\frac{\varepsilon_1}{\varepsilon_2}\sin^2\theta_i\right)$$

Dividing throughout by $\varepsilon_1\cos^2\theta_i$

$$\frac{\varepsilon_2}{\varepsilon_1}=\sec^2\theta_i-\frac{\varepsilon_1}{\varepsilon_2}\tan^2\theta_i$$

$$=1+\tan^2\theta_i-\frac{\varepsilon_1}{\varepsilon_2}\tan^2\theta_i$$

or

$$\frac{\varepsilon_2-\varepsilon_1}{\varepsilon_1}=\left(\frac{\varepsilon_2-\varepsilon_1}{\varepsilon_2}\right)\tan^2\theta_i$$

Therefore

$$\tan^2\theta_i=\frac{\varepsilon_2}{\varepsilon_1}$$

or

$$\theta_i=\tan^{-1}\left(\sqrt{\frac{\varepsilon_2}{\varepsilon_1}}\right) \quad \text{...(12.26)}$$

This angle is called the Brewster angle in honour of Sir David Brewster (1781–1868). For this angle of incidence in this case, there is no reflected ray and the entire energy of the incident ray goes into the transmitted ray. At any other angle of incidence, in the medium 1, there will be both the incident and the reflected rays present.

Following the same method of setting up Z_i and Z_r as in Sec. 12.4, we write Eqn. (12.18) and get for the magnetic components an equation similar to Eqn. (12.20), noting that

$$\frac{E_{ix}}{H_{iy}} = -\frac{E_{rx}}{H_{ry}} = \eta_1$$

$$H_y = H_i + H_r$$

$$= H_i - \rho'_r H_i$$

or $$H_y = H_{oi} e^{j\omega t} e^{-j\beta_x x} [e^{-j\beta_z z} - \rho'_r e^{+j\beta_z z}] \quad \text{...(12.27)}$$

Then $$E_x = \eta_1 H_{oi} e^{j\omega t} e^{-j\beta_x x} [(\cos\theta_i) e^{-j\beta_z z} - \rho'_r (-\cos\theta_i) e^{+j\beta_z z}] \quad \text{...(12.28)}$$

and $$E_z = \eta_1 H_{oi} e^{j\omega t} e^{-j\beta_x x} [(-\sin\theta_i) e^{-j\beta_z z} - \rho'_r (-\sin\theta_i) e^{+j\beta_z z}] \quad \text{...(12.29)}$$

where Eqn. (12.21) define the various quantities.

As before we have a travelling wave propagated in the x-direction in medium 1, that is parallel to the boundary, and also a standing wave in the z-direction. In Sections 12.4 and 12.5, since the coefficient of reflection is less than unity, the maximum value of amplitude in the standing wave is less than twice the value for the incident wave, and also the minimum value is not zero, because amplitude cancellation of the incident and the reflected wave is not complete.

12.6 WAVE INCIDENT NORMALLY ON PERFECT CONDUCTOR

We assume that the wave is a plane wave travelling in the z-direction, so that the incident wave can be represented as

$$E_i = E_{oi} e^{j(\omega t - \beta z)}$$

Let the conductor be situated with its surface on the plane $z = 0$ and extending from $-\infty < x < +\infty$, $-\infty < y < +\infty$ and $0 < z <$ any value. Since the conductor is perfect ($\sigma = \infty$), energy cannot enter into it, because as Eqn. (11.56) shows, when $\sigma = \infty$, the depth of penetration $\delta = 0$. Thus the wave undergoes total reflection at the surface of the perfect conductor and the reflected wave is given by

$$E_r = E_{or} e^{j(\omega t + \beta z)}$$

In these two equations, E_{oi} and E_{or} are the amplitudes of the two waves at time $t = 0$ and at the surface of the conductor. Let the direction of E be chosen as the axis of x.

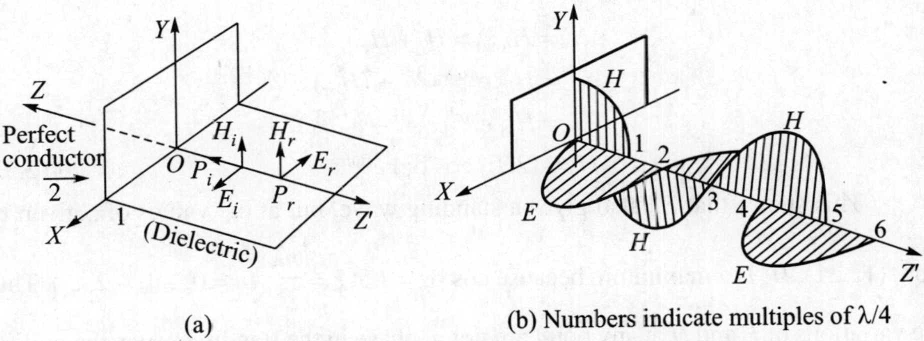

(a) (b) Numbers indicate multiples of $\lambda/4$

Fig. 12.5. Plane wave incident normally on a perfect conductor

Now in the plane wave incident *normally* to the surface of the conductor, both E and H are parallel to the conductor surface in the incident as well as in the reflected wave [Fig. 12.5(a)]. Just inside the conductor the *tangential component of E must be zero at all times*. Hence just outside the conductor, the total tangential component of E must be zero at all times. This implies that $E_{or} = -E_{oi}$. With the known directions of $\overline{P}_i$ and $\overline{E}_i$, and $\overline{P}_r$ and $\overline{E}_r$, the directions of $\overline{H}_i$ and $\overline{H}_r$ are found to be as shown in Fig. 12.5(a).

Hence in the dielectric where both the incident and the reflected waves are present, at any point,

$$E_x = E_{\text{total}} = E_i + E_r$$

$$= E_{oi}\, e^{j(\omega t - \beta z)} + (-E_{oi})\, e^{j(\omega t + \beta z)}$$

$$= E_{oi}\, e^{j\omega t}\left(\frac{e^{-j\beta z} - e^{+j\beta z}}{-2j}\right)(-2j)$$

or $$E_x = E_{\text{total}} = -j(2\,E_{oi} \sin \beta z)\, e^{j\omega t} \qquad \text{...(12.30)}$$

An examination of Eqn. (12.30) shows that it does not contain any term like

$$e^{j(\omega t - \beta z)} \quad \text{or} \quad e^{j(\omega t + \beta z)}.$$

This shows that it does not represent a wave travelling either in the z or in the $(-z)$-direction. But the term $e^{j\omega t}$ shows that at any constant value of z, E_{total} varies sinusoidally with time as it does in any wave. Hence this is referred to as a stationary wave, to distinguish it from a wave that travels.

The term in brackets in Eqn. (12.30) shows that the amplitude of the stationary wave is a function of z. The electric field vector E_{total} is zero at $z = 0$ on the surface of the conductor *at all times* or for all values of ωt as required by the boundary condition for a perfect conductor. But it is also zero at points for which $\sin \beta z = 0$ or

$$\beta z = \frac{2\pi}{\lambda} z = n\,\pi$$

Therefore, $$z = \frac{n\lambda}{2} \quad (n = 0, -1, -2, \ldots) \qquad \text{...(12.31)}$$

So far as H is concerned, both H_i and H_r are along the y-axis and $H_{or} = H_{oi}$, so that

$$H_y = H_{\text{total}} = H_i + H_r$$

$$= H_{oi}\, e^{j(\omega t - \beta z)} + (H_{or})\, e^{j(\omega t + \beta z)}$$

$$= H_{oi}\, e^{j\omega t}\,(e^{-j\beta z} - e^{+j\beta z})$$

or $$H_y = (2H_{oi} \cos \beta z)\, e^{j\omega t} \qquad \text{...(12.32)}$$

This shows that H_y also gives a standing wave, but at the values of z given by Eqn. (12.31), H_y is a maximum, because $\cos \beta z = 1$ at $z = \frac{n\lambda}{2}$ $(n = 0, -1, -2, \ldots)$. Thus the variations of E and H at any point are not in phase in the standing wave, the nulls of E corresponding to the maxima of H, and vice versa.

In particular, we notice that at the surface of the conductor at $z = 0$, E_{total} is equal to zero but H_{total} has a maximum value equal to the surface current density J ampere/metre width (measured in the y-direction).

Comparison of Eqns. (12.30) and (12.32) shows why no energy or power flows into the conductor. The coefficient $(-j)$ in Eqn. (12.30) can be written as

$$-j = \cos\left(\frac{-\pi}{2}\right) + j\sin\left(\frac{-\pi}{2}\right)$$

$$= e^{j\left(-\frac{\pi}{2}\right)} \qquad \text{...(12.33)}$$

which means that E lags behind H by a quarter cycle or 90°. This is shown in Fig. 12.5 (*b*). This means that in Eqn. (12.32), H_y is varying as $\cos \omega t$ but in Eqn. (12.30), E_x varies as $\cos\left(\omega t - \frac{\pi}{2}\right)$ or $\sin \omega t$. Hence power, which is shown in the next chapter to be given by

$$\overline{\boldsymbol{E}} \times \overline{\boldsymbol{H}} = E_x H_y \sin 90°$$

becomes proportional to $(\sin \omega t \cos \omega t)$. The value of this over one cycle is found to be zero showing that no power is transmitted in a standing wave.

12.7 USE OF DIRECTION COSINES

Next we have to consider the cases of the uniform plane wave incident at an oblique angle as in Secs. 12.4 and 12.5, but instead of following the method of attack of those sections based on Eqn. (12.18), another method is first developed in this section.

We have put the expression for the uniform plane wave travelling in the z-direction in a perfect dielectric in the form

$$E = E_0\, e^{j(\omega t - \beta z)}$$

This can be put in a slightly different form using Eqn. (11.54), stating that $v = \dfrac{\omega}{\beta}$ as follows:

$$E = E_0\, e^{j\omega\left(t - \frac{\beta}{\omega} z\right)} = E_0\, e^{j\omega\left(t - \frac{z}{v}\right)} \qquad \text{...(12.34)}$$

In this wave, all points in the plane z = constant have the same phase for E and H. This leads us to the idea that for a wave travelling in any other direction, we have only to find a function $f(x, y, z)$ of the coordinates x, y, z, which is linear (power of x, y, z is unity), and which when put equal to a constant describes the equiphase plane perpendicular to the direction of propagation of the wave. Then we replace z in Eqn. (12.34) by this $f(x, y, z)$ and thus get the desired expression for the wave.

In our search for this function $f(x, y, z)$, we note that if $P(x, y, z)$ is *any* point on the equiphase surface, then its distance from the origin O is given by

$$\overline{OP} = \overline{r} = x\overline{\boldsymbol{u}}_x + y\overline{\boldsymbol{u}}_y + z\overline{\boldsymbol{u}}_z \qquad \text{...(12.35)}$$

Let ON be the perpendicular from the origin O to the plane and let the direction of travel of the wave be ON (and not NO). Then if $\bar{u}_n$ be the unit vector in the direction ON, and if l, m and n are the components of this unit vector in the directions of the three axes, it follows that

$$l = \cos \angle NOX,$$

$$m = \cos \angle NOY,$$

$$n = \cos \angle NOZ, \qquad \text{...(12.36)}$$

Because of the relation given by Eqn. (12.36), l, m and n are called the direction cosines of the straight line ON.

Next we consider the dot product $\bar{u}_n \cdot \bar{r}$. We can write $\bar{u}_n$ in terms of the direction cosines as

$$\bar{u}_n = l\bar{u}_x + m\bar{u}_y + n\bar{u}_z \qquad \text{...(12.37)}$$

and hence combining Eqns. (12.35) and (12.37)

$$\bar{u}_n \cdot \bar{r} = (l\bar{u}_x + m\bar{u}_y + n\bar{u}_z) \cdot (x\bar{u}_x + y\bar{u}_y + z\bar{u}_z)$$

$$= lx + my + nz \qquad \text{...(12.38)}$$

But $\bar{u}_n \cdot \bar{r}$ is also the projection of the vector $\overline{OP} = \bar{r}$ in the direction of $\bar{u}_n$, hence

$$\bar{u}_n \cdot \bar{r} = ON = \text{Constant.} \qquad \text{...(12.39)}$$

Equation (12.39) involves the co-ordinates (x, y, z) of *any* point on the equiphase plane perpendicular to the direction of propagation of the wave. The expression is linear with powers of x, y and z equal to unity. The expression is also equal to a constant. Hence, it is the function $f(x, y, z)$ we have been looking for to define the equiphase surface and hence, replacing z with this $f(x, y, z)$ in Eqn. (12.34), the uniform plane wave travelling perpendicular to the plane defined by Eqn. (12.39) is given by

$$\left.\begin{aligned} E &= E_0\, e^{j\omega\left(t - \frac{\bar{u}_n \cdot \bar{r}}{v}\right)} \\ \text{or} \quad E &= E_0\, e^{j\omega\left(t - \frac{lx + my + nz}{v}\right)} \end{aligned}\right\} \qquad \text{...(12.40)}$$

12.8 PLANE WAVE INCIDENT OBLIQUELY ON PERFECT CONDUCTOR WITH ELECTRIC VECTOR PERPENDICULAR TO PLANE OF INCIDENCE

Since the wave does not enter the perfect conductor, Fig. 12.6 for this case is similar to Fig. 12.3, but without a transmitted wave.

Using Eqn. (12.40) we represent the uniform plane wave incident on the conductor 2 at point O in the form

$$E_i = E_{oi}\, e^{j\omega\left[t - \frac{lx + my + nz}{v}\right]}$$

where, from the geometry of Fig. 12.6, we have, since P_i is the direction of $\bar{u}_n$,

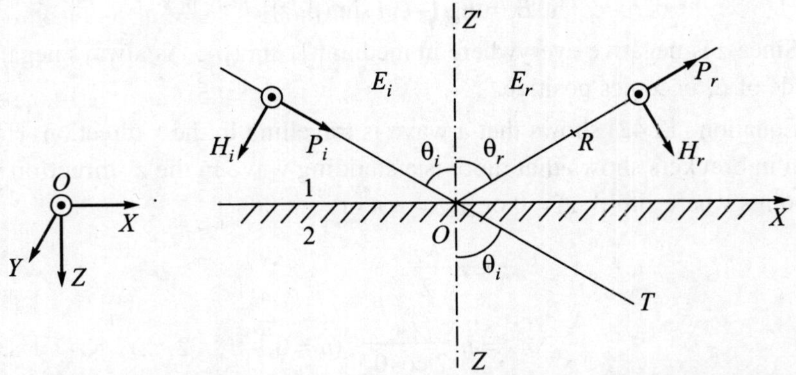

Fig. 12.6. Oblique incidence on conductor with E perpendicular to plane of incidence

$$l = \cos \angle XOT$$

$$= \cos\left(\frac{\pi}{2} - \theta_i\right) = \sin \theta_i,$$

$$m = 0$$

and $$n = \cos \angle ZOT = \cos \theta_i.$$

Thus $$E_i = E_{oi}\, e^{j\omega\left[t - \frac{x \sin \theta_i + z \cos \theta_i}{v}\right]}$$

$$= E_{oi}\, e^{j[\omega t - \beta_1 (x \sin \theta_i + z \cos \theta_i)]}$$

as we have had from Eqns. (12.18) and (12.19).

Also the reflected ray would have

$$E_r = E_{or}\, e^{j\omega\left[t - \frac{l' x + m' y + n' z}{v}\right]}$$

where, since P_r is the direction of $\bar{u}_{n_r}$, and $\theta_r = \theta_i$,

$$l' = \cos \angle XOR$$

$$= \cos\left(\frac{\pi}{2} - \theta_r\right) = \sin \theta_i,$$

$$m' = 0$$

and $$n' = \cos \angle ZOR$$

$$= \cos (\pi - \theta_r) = -\cos \theta_i.$$

Thus $$E_r = E_{or}\, e^{j\omega\left[t - \frac{x \sin \theta_i - z \cos \theta_i}{v}\right]}$$

$$= E_{or}\, e^{j[\omega t - \beta_1 (x \sin \theta_i - z \cos \theta_i)]}$$

Also using the boundary condition that on the surface of a perfect conductor $E_{or} = -E_{oi}$, we get the total electric vector at any point in medium 1 as

$$E_{\text{total}} = E_y$$

$$= E_i + E_r$$

$$= E_{oi}\, e^{j\omega t}\, e^{-j(\beta_1 \sin \theta_i)x}\, [e^{-j(\beta_1 \cos \theta_i)z} - e^{+j(\beta_1 \cos \theta_i)z}] \qquad ...(12.41)$$

or $$E_y = E_{oi}\,[-2\,j \sin(\beta_z z)]e^{j(\omega t - \beta_x x)} \qquad ...(12.42)$$

Since z is negative everywhere in medium 1, $\sin(\beta_z z)$ is always negative and amplitude of E_y becomes positive.

Equation (12.42) shows that a wave is travelling in the x-direction. However, the term in brackets shows that there is a standing wave in the z'-direction with $E_y = 0$ at points where $\sin(\beta_z z) = 0$ or $\beta_z z = n\,\pi$ or

$$z = n\,\frac{\lambda_z}{2}$$

$$= n\,\frac{\lambda_1}{2\cos\theta_i},\ (n = 0, -1, -2, ...) \qquad ...(12.43)$$

Comparing Eqns. (12.20) and (12.42), we see that they are identical if we assume $\rho_r = -1$, as indicated above by the relation $E_{or} = -E_{oi}$. Using this idea, the following equations are written from Eqns. (12.22) and (12.23) as

$$H_x = -\frac{E_{oi}\,(2\cos\beta_z z)}{\eta_1}\cos\theta_i\, e^{j(\omega t - \beta_x x)} \qquad ...(12.44)$$

$$H_z = -\frac{E_{oi}\,(2\cos\beta_z z)}{\eta_1}\sin\theta_i\, e^{j(\omega t - \beta_x x)} \qquad ...(12.45)$$

12.9 PLANE WAVE INCIDENT OBLIQUELY ON PERFECT CONDUCTOR WITH ELECTRIC VECTOR PARALLEL TO PLANE OF INCIDENCE

Figure 12.7, which is drawn similar to Fig. 12.4, shows the geometry of the situation with the assigned positive directions for the magnetic vectors H_i and H_r in the y-direction, perpendicular to the plane of the paper.

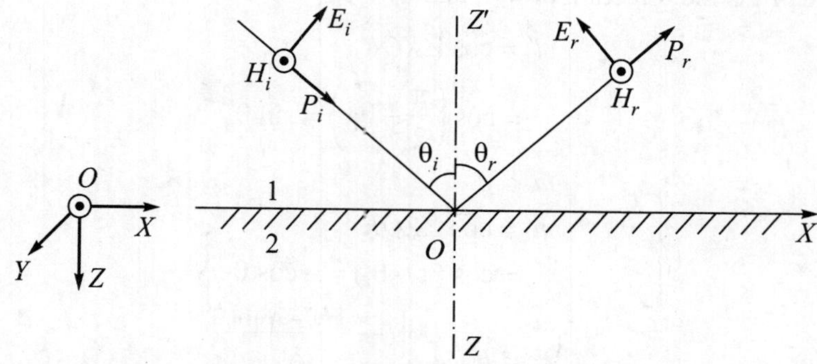

Fig. 12.7. Oblique incidence on conductor with E parallel to plane of incidence

Reflection of the wave at the surface of the conductor causes the *tangential component* of the electric vector to reverse its direction. Hence it follows that the magnetic vector will get reflected *without* any phase reversal, so as to give propagation in the reverse direction.

Using the reasoning of the previous section, we can write

$$H_i = H_{oi}\, e^{j[\omega t - \beta_1 (x \sin \theta_i + z \cos \theta_i)]}$$

and

$$H_r = H_{or}\, e^{j[\omega t - \beta_1 (x \sin \theta_i - z \cos \theta_i)]}$$

Since $H_{or} = H_{oi}$, we get in medium 1,

$$H_{total} = H_y = H_i + H_r$$

or

$$H_y = H_{oi}\, e^{j[\omega t - (\beta_1 \sin \theta_i)x]}\, [e^{-j(\beta_1 \cos \theta_i)z} + e^{+j(\beta_1 \cos \theta_i)z}]$$

$$= H_{oi}\,(2 \cos \beta_z z)\, e^{j(\omega t - \beta_x x)} \quad ...(12.46)$$

$$E_x = \eta_1 H_{oi} \cos \theta_i [-2j \sin \beta_z z]\, e^{j(\omega t - \beta_x x)} \quad ...(12.47)$$

and

$$E_z = \eta_1 H_{oi} (-\sin \theta_i)\, [2 \cos \beta_z z]\, e^{j(\omega t - \beta_x x)} \quad ...(12.48)$$

These equations follow from Eqns. (12.27), (12.28) and (12.29), if ρ'_r is assumed to be equal to – 1.

12.10 SUMMARY AND COMMENTS

In this chapter, we have seen that the electromagnetic waves obey the laws of reflection and refraction as in optics. The phenomenon of standing waves in medium 1 caused by fields of the incident and the reflected ray overlapping each other, is similar to what is shown in laboratory for sound waves. It is presented in optics under the name interference phenomenon. Before going on to a study of electromagnetic waves guided by metal boundaries, we shall devote the next chapter to a study of power flow and energy in a wave.

EXERCISE 12

1. In a plane electromagnetic wave in air, the rms value of the electric field vector is 1 V at 1 MHz. It is incident normally on water for which the air-to-water refractive index is 9 at this frequency. Find the rms values of E and H for the incident, reflected and transmitted waves.

2. Show that if for any two dielectric media, the refractive indices are n_1 and n_2, and $\mu_1 = \mu_2 = \mu_0$, then Eqns. (12.4) and (12.5) for normal incidence can be written as

$$\frac{E_r}{E_i} = \frac{\sqrt{\varepsilon_1} - \sqrt{\varepsilon_2}}{\sqrt{\varepsilon_1} + \sqrt{\varepsilon_2}} = \frac{n_1 - n_2}{n_1 + n_2}$$

and

$$\frac{E_t}{E_i} = \frac{2\sqrt{\varepsilon_1}}{\sqrt{\varepsilon_1} + \sqrt{\varepsilon_2}} = \frac{2n_1}{n_1 + n_2}$$

3. Assuming Eqn. (13.11), that $\overline{\boldsymbol{E}} \times \overline{\boldsymbol{H}}$ gives in magnitude and direction the rate of energy flow per unit area at any point, show that for a uniform plane wave normally incident on the boundary between two media with $\mu_1 = \mu_2 = \mu_0$.

P_i = Average incident power flow per unit area

$$= \frac{1}{2}\left(\frac{\varepsilon_1}{\mu_0}\right)^{\frac{1}{2}} E_{oi}^2$$

P_r = Average reflected power flow per unit area

$$= \frac{1}{2}\left(\frac{\varepsilon_1}{\mu_0}\right)^{\frac{1}{2}} E_{or}^2$$

P_t = Average transmitted power flow per unit area

$$= \frac{1}{2}\left(\frac{\varepsilon_2}{\mu_0}\right)^{\frac{1}{2}} E_{ot}^2$$

4. In Ex. 12, Prob. 3 above for normal incidence between two dielectrics, if R_n and T_n are the coefficients of reflection and transmission for average power defined as $P_r = R_n P_i$ and $P_t = T_n P_i$, then show that

$$R_n = \left(\frac{n_1 - n_2}{n_1 + n_2}\right)^2 \text{ x}$$

and
$$T_n = \frac{4n_1 n_2}{(n_1 + n_2)^2}$$

5. A plane electromagnetic wave is normally incident on the boundary between two dielectrics. What must be the ratio of the refractive indices of the two media, in order that the reflected and the transmitted waves may have average power of equal magnitude?

6. Two hypothetical dielectrics have $\varepsilon_1 = \varepsilon_2$ but $\mu_1 \neq \mu_2$. As in Secs. 12.4 and 12.5, find the condition in which there will be no reflected wave, and the corresponding value of the angle of incidence θ_i.

CHAPTER

13

THE POYNTING VECTOR

13.1 ENERGY STORAGE IN ELECTRIC FIELD

We have already seen that a capacitor has the ability or capacity to store electric charges on its plates. When a capacitor discharges, an electric current flows in an external circuit and does work like producing heat due to I^2R loss in a resistor. Obviously, the energy spent in doing this work had come from the capacitor, where it was stored in the electric field in the dielectric between the plates.

Consider a parallel-plate capacitor in which each side of each plate has an area of A square metres, and the separation d between the plates is small as compared to the dimensions of the plates, so that over the whole area A, a uniform electric field intensity E may be assumed to exist. In other words, the effect produced by the fringing of the electric field at the edges of the capacitor plates is neglected.

Let $+q$ coulomb be the total charge on one plate (and therefore automatically, $(-q)$ is the total charge on the other plate) and v volts the potential difference between the plates. Then the capacitance parameter C is defined by the relation.

$$q = C\,v$$

or
$$v = \frac{q}{C} \quad \text{...(13.1)}$$

where C is in farad. If an additional charge $(+dq)$ is placed on the positive plate, then the work done in placing this charge is, by definition of the potential, equal to $v\,dq$. Let us assume that initially both q and v are zero, and finally they acquire the values $+Q$ and V. Then the work done in this process is given by

$$\text{Total work done} = \int_0^{+Q} v\,dq$$

$$= \int_0^{+Q} \frac{q}{C}\,dq = \frac{Q^2}{2C} \quad \text{...(13.2)}$$

If σ coulomb/(metre)^2 is the uniform surface charge density on the plate, $Q = \sigma A$. Also from Sec. 2.16, $E = \dfrac{\sigma}{\varepsilon}$ in the dielectric between the plates, and capacitance $C = \dfrac{\varepsilon A}{d}$ by Eqn. (4.34). Using these results and noting that the volume of the dielectric between the plates is $A\,d$ cubic metres.

Energy stored *per unit volume* in dielectric

$$= \frac{\frac{Q^2}{2C}}{Ad} = \frac{(\sigma A)^2}{2C\,Ad} = \frac{1}{2}\frac{(\sigma^2 A^2)d}{(\varepsilon A)(Ad)}$$

$$= \frac{1}{2}\varepsilon E^2 \qquad \text{...(13.3)}$$

Because we are discussing electric fields, we have used the expression "energy stored in the dielectric". In books on circuit theory, the energy is associated with the charges $+Q$ and $-Q$ on the two plates. Any discussion, as to which of these statements is correct (or even more correct) is meaningless. It is like trying to define in which part of the body does a person's soul reside. In this chapter, we shall have more instances of differences between the circuit and the field points of view. The important thing to remember is that; both these views are really in our minds, and are different ways of interpreting the observed results. As we shall always find, both views arrive at the same *final* result!!

In this case of the capacitor, $\frac{1}{2}\varepsilon E^2$ joule/metre³ is the energy density in the electric field in the dielectric.

13.2 ENERGY STORAGE IN MAGNETIC FIELD

To determine this, we use the toroidal coil of Sec. 8.13, with cross-sectional area A (metre)², and mean length L metre. The coil has N turns. If i ampere is the instantaneous value of the current flowing through the coil, the magnetic field intensity in the core is given by

$$H_c = \frac{Ni}{L} \qquad \text{...(13.4)}$$

If this current magnitude changes, a back emf will be induced in the coil as given by Eqn. (8.3) and

$$v = -\frac{d}{dt}(N\phi) = -\frac{d}{dt}(NAB)$$

$$= -NA\mu\frac{dH_c}{dt} \qquad \text{...(13.5)}$$

Let the current i be zero at time $t = 0$, and let the current be established to have a final value $i = I$ in time t_1. The work done in changing current i to $(i + di)$ in time dt is $(vi)\,dt$ joule. Hence the total work done to establish the current I in time t_1 second is :

$$\text{Total work done} = -\int_0^{t_1}(vi)\,dt$$

$$= -\int_0^{t_1}(-NA\mu)\left(\frac{dH_c}{dt}\right)\left(\frac{H_c L}{N}\right)dt$$

$$= \mu(LA)\int_0^{H} H_c\,dH_c = \frac{1}{2}\mu H^2(LA)$$

or

$$\left.\begin{array}{l}\text{Work done } \textit{per unit volume} \text{ and} \\ \text{stored in the magnetic field}\end{array}\right\} = \frac{\frac{1}{2}\mu H^2 (LA)}{LA}$$

$$= \frac{1}{2}\mu H^2 \frac{\text{joule}}{\text{metre}^3} \qquad ...(13.6)$$

where LA is the volume of the toroid, and H is the final value of H_c when current is equal to I.

Note the similarity in the expressions for the energy storage in the electric and magnetic fields.

13.3 POYNTING VECTOR FOR A PLANE WAVE IN A DIELECTRIC

Consider a plane wave being propagated in a linear, homogeneous, isotropic dielectric of infinite extent. These assumptions ensure that all derivatives are continuous, and also that there are no reflected waves. If we choose the x-axis in the direction of the $\boldsymbol{E}$-vector (so that $E_y = 0$, $E_z = 0$) and the y-axis in the direction of the $\boldsymbol{H}$-vector (so that $H_x = 0$, $H_z = 0$), then the z-axis will give the direction of propagation and the wave equations (11.11) and (11.12) reduce to

$$\frac{d^2 E_x}{dz^2} = \mu\varepsilon \frac{d^2 E_x}{dt^2},$$

$$\frac{d^2 H_y}{dz^2} = \mu\varepsilon \frac{d^2 H_y}{dt^2} \qquad ...(13.7)$$

If sinusoidal variations with respect to time are assumed to E_x and H_y, the solutions of Eqn. (13.7) are

$$\overline{\boldsymbol{E}}_x = \left[E_{0x} e^{j(\omega t - kz)}\right]\overline{\boldsymbol{u}}_x,$$

$$\overline{\boldsymbol{H}}_y = \left[H_{0y} e^{j(\omega t - kz)}\right]\overline{\boldsymbol{u}}_y, \qquad ...(13.8)$$

where the wave number $k = \dfrac{2\pi}{\lambda}$, and

$$\frac{E_x}{H_y} = \sqrt{\frac{\mu}{\varepsilon}} = \frac{E_{ox}}{H_{oy}} \qquad ...(13.9)$$

By the results established in Secs. 13.1 and 13.2, the energy density in the electric and in the magnetic fields is given by $\frac{1}{2}\varepsilon E_x^2$ and $\frac{1}{2}\mu E_y^2$, respectively. Using Eqn. (13.9), we can write

$$\frac{1}{2}\mu H_y^2 = \frac{1}{2}\mu\left[\frac{\varepsilon}{\mu} E_x\right]^2$$

$$= \frac{1}{2}\varepsilon E_x^2 \qquad ...(13.10)$$

This shows that half of the total instantaneous energy density of (εE_x^2) or (μH_y^2) joule/metre^3 is stored in the electric field and the other half in the magnetic field.

The wave travels in the direction of the z-axis with velocity v metre/second. If we consider a volume of length v metre in the direction of the z-axis and having cross-sectional area 1 metre^2, then energy stored in this value is

$$\left(\frac{1}{2}\varepsilon E_z^2 + \frac{1}{2}\mu H_y^2\right)v.$$

Hence using Eqn. (13.9) and noting that $v = \dfrac{1}{\sqrt{\mu\varepsilon}}$

Rate of flow of energy per square metre cross-sectional area

$$= \left(\frac{1}{2}\varepsilon E_x^2 + \frac{1}{2}\mu H_y^2\right)v\bar{u}_z$$

$$= \left[\frac{1}{2}\varepsilon E_x\left(\sqrt{\frac{\mu}{\varepsilon}}H_y\right) + \frac{1}{2}\mu H_y\left(\sqrt{\frac{\varepsilon}{\mu}}E_x\right)\right]\frac{1}{\sqrt{\mu\varepsilon}}\bar{u}_z$$

$$= (E_x H_y)\,\bar{u}_z = \bar{E} \times \bar{H}$$

$$= \bar{P} \qquad \text{...(13.11)}$$

The vector $\bar{P}$ is called the Poynting vector. It is equal to $\bar{E} \times \bar{H}$ which gives the magnitude as well as the direction, in which power flows in time-varying electromagnetic fields. This is true for a wave in a conductor also.

13.4 POYNTING'S THEOREM

Premultiplying Maxwell's curl equation for $\bar{H}$ by $\bar{E}$, we get

$$\bar{E} \cdot (\bar{\nabla} \times \bar{H}) = \bar{E} \cdot \bar{J} + \bar{E} \cdot \frac{\partial \bar{D}}{\partial t}$$

$$= \bar{E} \cdot \bar{J} + \varepsilon \bar{E} \cdot \frac{\partial \bar{E}}{\partial t} \qquad \text{...(13.12)}$$

Using the result of Ex. 1, Prob. 16, we can write

$$\bar{\nabla} \cdot (\bar{E} \times \bar{H}) = \bar{H} \cdot (\bar{\nabla} \times \bar{E}) - \bar{E} \cdot (\bar{\nabla} \times \bar{H}) \qquad \text{...(13.13)}$$

Combining Eqns. (13.12) and (13.13), and using Maxwell's curl equation for $\bar{E}$, we get

$$-\bar{\nabla} \cdot (\bar{E} \times \bar{H}) = \bar{E} \cdot \bar{J} + \varepsilon \bar{E} \cdot \frac{\partial \bar{E}}{\partial t} - \bar{H} \cdot (\bar{\nabla} \times \bar{E})$$

$$= \bar{E} \cdot \bar{J} + \varepsilon \bar{E} \cdot \frac{\partial \bar{E}}{\partial t} + \mu \bar{H} \cdot \frac{\partial \bar{H}}{\partial t}$$

Since $\quad \dfrac{\partial}{\partial t}(E^2) = 2E\dfrac{\partial E}{\partial t}$

and $\quad \dfrac{\partial}{\partial t}(H^2) = 2H\dfrac{\partial H}{\partial t}.$

We can write the above equation as

$$-\overline{\nabla}\cdot(\overline{\boldsymbol{E}}\times\overline{\boldsymbol{H}})=\overline{\boldsymbol{E}}\cdot\overline{\boldsymbol{J}}+\frac{1}{2}\varepsilon\frac{\partial}{\partial t}(E^2)+\frac{1}{2}\mu\frac{\partial}{\partial t}(H^2)$$

Integrating this over a volume v, we get

$$-\int_{\text{vol}}\overline{\nabla}\cdot(\overline{\boldsymbol{E}}\times\overline{\boldsymbol{H}})\,dv=\int_{\text{vol}}(\overline{\boldsymbol{E}}\cdot\overline{\boldsymbol{J}})\,dv+\int_{\text{vol}}\left[\frac{1}{2}\varepsilon\frac{\partial}{\partial t}(E^2)+\frac{1}{2}\mu\frac{\partial}{\partial t}(H^2)\right]dv$$

Since time-derivation and volume-integration are independent operations, their *order* can be interchanged. Then using Divergence Theorem Eqn. (4.49), we change the volume integral on left-hand side to a surface integral over area of enclosing surface of volume v, and get

$$-\oint_{\text{surface}}(\overline{\boldsymbol{E}}\cdot\overline{\boldsymbol{H}})\,\overline{\boldsymbol{da}}=\int_{\text{vol}}(\overline{\boldsymbol{E}}\cdot\overline{\boldsymbol{J}})\,dv+\frac{\partial}{\partial t}\left[\int_{\text{vol}}\left(\frac{1}{2}\varepsilon E^2+\frac{1}{2}\mu H^2\right)\right]dv \quad ...(13.14)$$

This result is called Poynting's theorem. Notice the *negative* sign of the left-hand side, which indicates that it represents the instantaneous power flowing *into* the volume v, over the enclosing surface of which the Poynting vector is integrated. The integral with a *plus sign* would give the rate of energy flow *outwards* from inside the volume through the bounding surface.

The dimensions of Eqn. (13.14) are easily found. Since terms in Maxwell's curl equation for $\overline{\boldsymbol{H}}$ have dimensions of (amp/metre^2) and it is multiplied by $\overline{\boldsymbol{E}}$ which has dimensions of (volt/metre), Eqn. (13.12) has terms with dimensions of (volt-amp)/metre^3 or watt/metre^3. This makes terms in Eqn. (13.14) to have dimensions of watt.

Next we consider the first term on the right-hand side of Eqn. (13.14). Here $\overline{\boldsymbol{E}}$ is in volt/metre and $\overline{\boldsymbol{J}}$ is the volume current density in (amp/metre^2) of area perpendicular to direction of current flow. Hence $\overline{\boldsymbol{E}}\cdot\overline{\boldsymbol{J}}$ integrated over the volume gives the power in watts being dissipated in the volume, or entering from outside the bounding surface into the volume.

It is interesting to examine how the product $\overline{\boldsymbol{E}}\cdot\overline{\boldsymbol{J}}$ gives different interpretations in various situations. If as shown in Fig. 13.1 (*a*), $\overline{\boldsymbol{E}}$ is the electric field intensity which produces the current density $\overline{\boldsymbol{J}}$ as per Ohm's law, and $\overline{\boldsymbol{E}}$ and $\overline{\boldsymbol{J}}$ act in the resistor in the same direction, $\overline{\boldsymbol{E}}\cdot\overline{\boldsymbol{J}}=EJ$ gives the power dissipated as heat in the resistor per metre^3. It must be remembered, however, that Ohm's law does not required that $\overline{\boldsymbol{E}}$ and $\overline{\boldsymbol{J}}$ be in the same direction. They can be in two different directions as shown in Fig. 13.1(*b*), the direction of the current density $\overline{\boldsymbol{J}}$ being *forced* to be in a particular direction, because of the boundary of the conductor. In such a case, $\overline{\boldsymbol{E}}\cdot\overline{\boldsymbol{J}}=EJ\cos\theta$, because only the component $E\cos\theta$ is able to contribute to the current density $\overline{\boldsymbol{J}}$ and the component $E\sin\theta$ is forced to be ineffective by the boundary of the conductor.

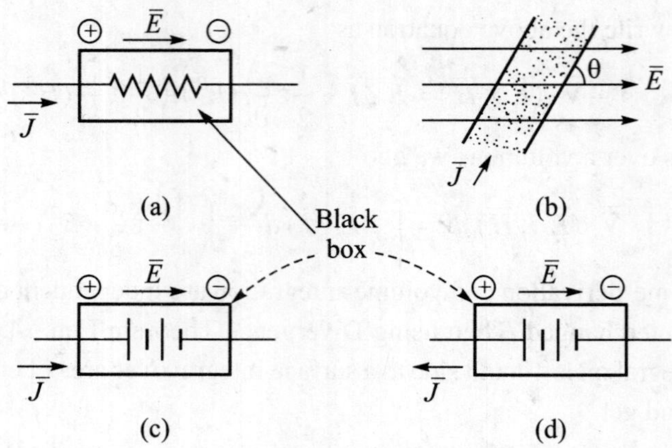

Fig. 13.1. The product $\overline{E} \cdot \overline{J}$

From outward appearances, Fig. 13.1 (*a*) and (*c*) are identical, and in both cases, the product $\overline{\boldsymbol{E}} \cdot \overline{\boldsymbol{J}}$ is the power per unit volume going into the black box. But in (*a*) this power is dissipated as heat, while in (*c*), it gets converted into chemical form and the energy is stored as chemical energy, as the battery gets charged. Finally in Fig. 13.1(*d*), $\overline{\boldsymbol{E}}$ and $\overline{\boldsymbol{J}}$ are in opposite directions, and $\overline{\boldsymbol{E}} \cdot \overline{\boldsymbol{J}} = -EJ$, the negative sign indicating that power is coming out of the black box, and the battery is discharging, or converting chemical energy into electrical form.

In the last term of Eqn. (13.14), the sum of the two terms inside the brackets gives the total energy stored per unit volume in the electric and magnetic fields. When integrated over the whole volume, it gives the total energy stored in these fields, and $\frac{\partial}{\partial t}$ of this, gives the rate at which this stored energy is varying. A positive value for this rate indicates that the stored energy is increasing, which is possible only if power flows into the volume from outside. A negative rate would mean power being delivered by the fields inside the volume to the region outside the volume.

In the next few sections, some examples of the application of Poynting's theorem are taken, not only to illustrate it but also to highlight some basic differences in concepts of field theory from those of circuit theory.

13.5 FLOW OF DIRECT CURRENT IN CYLINDRICAL RESISTOR

In Fig. 13.2 is a cylindrical resistor of radius *a* with its axis coinciding with the *z*-axis.

A direct and constant current of I_z amperes flows through it. If R_0 ohm is the resistance per metre length (in *z*-direction) of the resistor, the electric field intensity everywhere in the resistor is

$$\overline{\boldsymbol{E}}_z = I_z R_0 \overline{\boldsymbol{u}}_z \text{ volt/metre}$$

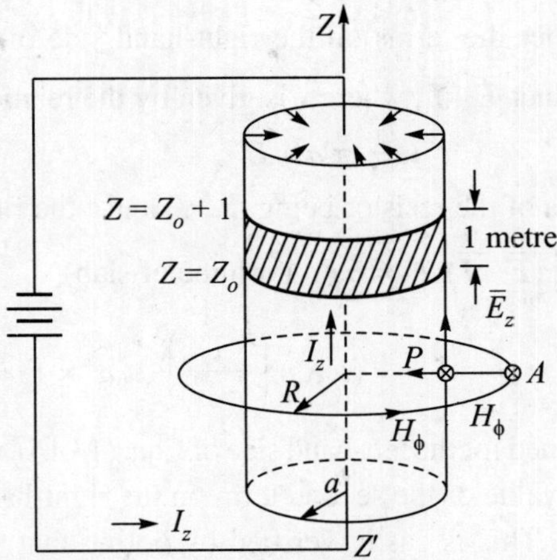

Fig. 13.2. Poynting's theorem applied to direct current in resistor

in the direction of the z-axis. In a plane perpendicular to the z-axis, along a circular path of radius R metre measured from the z-axis, the magnetic field intensity H_ϕ is everywhere tangential to the circular path and is directed into the paper at the point A. Its magnitude everywhere is

$$\overline{\boldsymbol{H}}_\phi = \frac{I_z}{2\pi R}\overline{\boldsymbol{u}}_\phi \text{ ampere/metre}$$

Obviously, on the curved surface of the resistor where $R = a$, the Poynting vector $\overline{\boldsymbol{P}}$ is everywhere radial and it is directed towards the z-axis. This is depicted by the short arrows shown on the top surface of the resistor. Thus we have

$$\overline{\boldsymbol{P}} = \overline{\boldsymbol{E}} \times \overline{\boldsymbol{H}} = (I_z R_o \overline{\boldsymbol{u}}_z) \times \left(\frac{I_z}{2\pi a}\overline{\boldsymbol{u}}_\phi\right)$$

or
$$\overline{\boldsymbol{P}} = \frac{I_z^2 R_o}{2\pi a}(-\overline{\boldsymbol{u}}_r) \qquad \text{...(13.15)}$$

Consider a slab of the resistor between two planes perpendicular to the z-axis and separated by one metre, that is, at $z = z_o$ and $z = z_o + 1$. Applying Poynting's theorem to it, we find that power flows into this volume normal to the curved surface (shown shaded in Fig. 13.2), but there is no power flow into this volume across the two flat end faces, because the Poynting vector is parallel to these flat ends. Then the left-hand side of Eqn. (13.14) can be written, using Eqn. (13.15), as

$$\text{Total power going into slab} = -\int_{\text{surface}} (\overline{\boldsymbol{E}}_z \times \overline{\boldsymbol{H}}_\phi) \cdot \overline{\boldsymbol{da}}$$

$$= -\left(-\frac{I_z^2 R_o}{2\pi a}\right)(2\pi\, a \times 1) = I_z^2 R_o \qquad \text{...(13.16)}$$

But this is the answer that the circuit theory gives for the power dissipated in one metre length of the resistor, which has a resistance of R_o ohm. Thus the Poynting vector does enable us to arrive at the same result from the field point of view that we are accustomed to deriving from the circuit theory.

Next we consider the terms on the right-hand side of Eqn. (13.14). The first term involves the product $\bar{E} \cdot \bar{J}$, where J is given by the relation

$$I_x = (\pi a^2)\, J,$$

the cross-sectional area of the resistor being πa^2. Hence the first term becomes

$$\int_{\text{volume}} (\bar{E} \cdot \bar{J})\, dv = EJ \quad \text{(volume of slab)}$$

$$= (I_z R_o)\left(\frac{1}{\pi a^2}\right)(\pi a^2 \times 1) = I_z^2 R_o$$

which is the value obtained for the left-hand side of Eqn. (13.14) as given by Eqn. (13.16). This suggests that the value of the second term on the right-hand side of Eqn. (13.14) must be equal to zero. This is easily verified by noting that since a *constant*, direct current is flowing, both E_z and H_ϕ are constant at all times, and hence the time derivative $\frac{d}{dt}\left(\frac{1}{2}\varepsilon E_z^2 + \frac{1}{2}\mu H_\phi^2\right)$ must be zero.

This last sentence may arouse another doubt in one's mind, leading one to argue that since E_z and H_ϕ do not vary with time, there is no electromagnetic wave. Then why does power flow in the direction $\bar{E} \times \bar{H}$, which is the direction of propagation of an electromagnetic wave? The reason for this is that whereas $\bar{E}$ can be produced by static charges which process does not involve any work, for *producing and maintaining H* at a constant value, all the time the charges have to be in motion to give the current I_z. This maintains the flow of energy radially towards the z-axis, even when there is no electromagnetic wave.

13.6 FLOW OF DIRECT CURRENT IN CO-AXIAL CABLE

Assume that the inner and outer conductors of the co-axial cable are perfect conductors ($\sigma = \infty$, or resistance = 0). A constant, direct current I_z flows through the inner conductor [Fig. 13.3(*a*)] to the load resistance R_L, and returns via the outer conductor, making the voltage of the inner conductor to be $V_i = V_L$ and of the outer (grounded) conductor to be $V_0 = 0$. The outer diameter of the inner conductor is $2a$ metre and the inner diameter of the outer conductor is $2b$ metre.

If we take a section of the coaxial cable by the plane $z = z_0$, [Fig. 13.3 (*b*)], it is found that in the dielectric between the conductors, the magnetic field is everywhere circular with centre on z-axis and along any circular path, its magnitude H_ϕ is a constant, given by

$$\bar{H}_\phi = \frac{I_z}{2\pi R}\bar{u}_\phi \text{ ampere/metre}$$

If the length of the co-axial cable is large as compared to its diameter $2b$, the formulas for infinite line can be applied to it, with negligible error, to determine the electric field intensity, which is everywhere directed outwards along the radial line.

Then using formulas (2.27) and (3.38), we can write, if q_l coulomb is the charge on the central conductor per unit length

$$\overline{\boldsymbol{E}}_R = \frac{q_l}{2\pi\varepsilon R}\overline{\boldsymbol{u}}_R$$

and
$$V_L = \frac{q_l}{2\pi\varepsilon}\log_e\frac{b}{a}$$

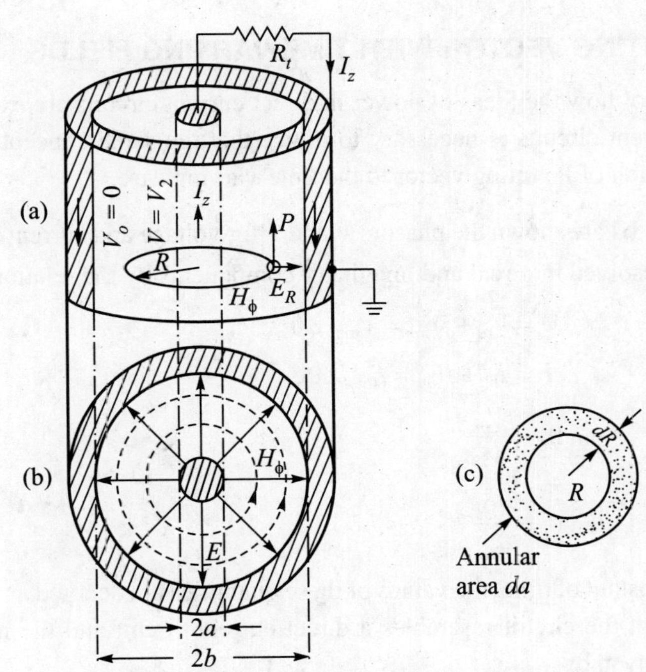

Fig. 13.3. Flow of direct current in co-axial cable

Therefore
$$\overline{\boldsymbol{E}}_R = \frac{V_L}{R\log_e\frac{b}{a}}\overline{\boldsymbol{u}}_R$$

Hence Poynting vector

$$\overline{\boldsymbol{E}}_R \times \overline{\boldsymbol{H}}_\phi = E_R H_\phi(\overline{\boldsymbol{u}}_R \times \overline{\boldsymbol{u}}_\phi) = E_R H_\phi \overline{\boldsymbol{u}}_z$$

at all points where $a < R < b$. In words, it means that power is flowing parallel to the z-axis at every point in the dielectric. The total power flowing through the section of the dielectric in Fig. 13.3 (b), is obtained by noting that over the annular area da shown in Fig. 13.3 (c) E_R and H_ϕ are constant as R is constant for the annular ring. Hence using the area of the annular ring as $(2\pi R)$ (dR), and integrating from $R = a$ to $R = b$.

$$\text{Total power} = \int_{\text{surface}} (\overline{\boldsymbol{E}}_R \times \overline{\boldsymbol{H}}_\phi)\cdot\overline{\boldsymbol{da}}$$

$$= \int_a^b \left(\frac{V_L}{R\log_e\frac{b}{a}}\right)\left(\frac{I_z}{2\pi R}\right)(2\pi R)\,dR$$

$$= \frac{V_L I_z}{\log_e \dfrac{b}{a}} \int_a^b \frac{dR}{R} = \frac{V_L I_z}{\log_e \dfrac{b}{a}} \left[\log_\varepsilon R\right]_a^b = V_L I_z$$

This is the power being supplied to the resistor R_L according to circuit theory, and thus we find that the field approach also gives the same final answer which is obtained by a much shorter method by circuit theory.

13.7 POYNTING VECTOR WITH TIME-VARYING FIELDS

A brief review of how the ideas of power in direct current circuits are extended to the alternating current circuits is necessary to notice that the same concepts are used to extend application of Poynting vector to the time-varying case.

In Fig. 13.4 are shown the phasors V and I for voltage and current in ac circuits. These can be resolved into real and imaginary components by the relations

$$V = V_{re} + jV_{im} = V_{max}\angle\theta_v,$$

$$I = I_{re} + jI_{im} = I_{max}\angle\theta_i,$$

where $$\tan\theta_v = \frac{V_{im}}{V_{re}}$$

and $$\tan\theta_i = \frac{I_{im}}{I_{re}} \quad ...[13.16(a)]$$

At any instant of time, the values of these quantities are denoted as V_{inst} and I_{inst}. For that moment the circuit resembles a direct current circuit and the instantaneous power W_{inst} is given by

$$W_{inst} = V_{inst} I_{inst} \quad ...(13.17)$$

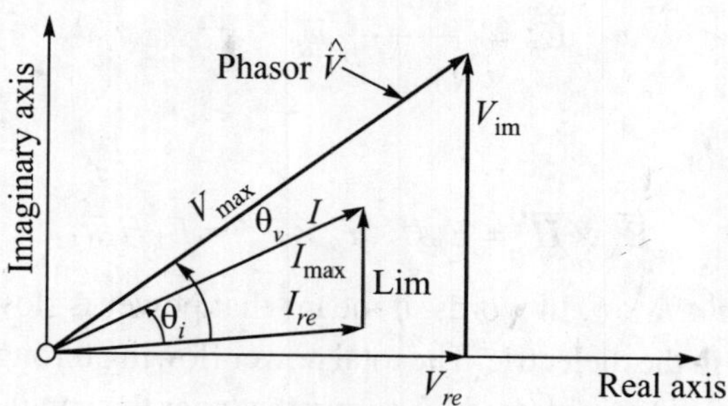

Fig. 13.4. Voltage and current phasors in ac circuit

The active (or real) power is the *average of this instantaneous power over one complete cycle.* In terms of rms or effective values

$$W_{active} = V_{eff} I_{eff} \cos(\theta_v - \theta_i) \quad ...(13.18)$$

where $$V_{inst} = V_{max}\cos(\omega t + \theta_v),$$

$$I_{inst} = I_{max}\cos(\omega t + \theta_i), \quad ...(13.19)$$

and
$$V_{\text{eff}} = \frac{V_{\max}}{\sqrt{2}},$$
$$I_{\text{eff}} = \frac{I_{\max}}{\sqrt{2}}, \qquad \text{...(13.20)}$$

Then for the instant $\omega t = 0$ in Eqn. (13.19), for which Fig. 13.4 is drawn and Eqns. (13.16) written, Eqn. (13.18) can be modified as

$$W_{\text{active}} = \frac{V_{\max}}{\sqrt{2}}\frac{I_{\max}}{\sqrt{2}}(\cos\theta_v \cos\theta_i + \sin\theta_v \sin\theta_i)$$
$$= \frac{V_{\max} I_{\max}}{2}\left(\frac{V_{re}\, I_{re}}{V_{\max}\, I_{\max}} + \frac{V_{im}\, I_{im}}{V_{\max}\, I_{\max}}\right)$$

or
$$W_{\text{active}} = \frac{1}{2}[V_{re} I_{re} + V_{im} I_{im}] \qquad \text{...(13.21)}$$

Also the reactive power (which does no useful work) is given by a similar expression as

$$W_{\text{reactive}} = V_{\text{eff}}\, I_{\text{eff}} \sin(\theta_v - \theta_i)$$
$$= V_{\text{eff}}\, I_{\text{eff}} (\sin\theta_v \cos\theta_i - \cos\theta_v \sin\theta_i)$$

or
$$W_{\text{reactive}} = \frac{1}{2}[V_{im} I_{re} - V_{re} I_{im}] \qquad \text{...(13.22)}$$

The Eqns. (13.21) and (13.22) can be put into a neat form like Eqn. (13.17) by introducing the conjugate of the complex number I as defined in the last paragraph of Sec. 1.6, thus:

$$I = I_{re} + j\, I_{im}$$

Therefore
$$I^* = I_{re} - j\, I_{im}$$

Now
$$V_{\text{eff}}\; I^*{}_{\text{eff}} = \frac{V}{\sqrt{2}}\frac{I^*}{\sqrt{2}}$$
$$= \frac{1}{2}(V_{re} + jV_{im})\,(I_{re} - jI_{im})$$
$$= \frac{1}{2}[\{V_{re} I_{re} + V_{im} I_{im}\} + j\{V_{im} I_{re} - V_{re} I_{im}\}]$$

Hence Eqns. (13.21) and (13.22) can be written as

$$\text{Real power} = W_{\text{active}}$$
$$= \text{Re}\,(V_{\text{eff}}\; I^*{}_{\text{eff}}) \qquad \text{...(13.23)}$$
$$\text{Apparent power} = W_{\text{reactive}}$$
$$= \text{Im}\,(V_{\text{eff}}\; I^*{}_{\text{eff}}) \qquad \text{...(13.24)}$$

where Re and Im stand for Real part of and Imaginary part of, respectively.

In all the above relations for power, V is in volts and I in amperes. For electromagnetic field theory, power is given by the Poynting vector $\overline{\boldsymbol{E}} \times \overline{\boldsymbol{H}}$ where $\overline{\boldsymbol{E}}$ is in volt/metre and $\overline{H}$ in ampere/metre, while $\overline{P}$ is in watt/(metre)^2. Hence similar equations are true. Thus,

$$\overline{P}_{\text{inst}} = \overline{E}_{\text{inst}} \times \overline{H}_{\text{inst}} \quad ...(13.25)$$

$$\overline{P}_{\text{real}} = \text{Re}\left[\overline{E}_{\text{eff}} \times \overline{H}^{*}{}_{\text{eff}}\right] \quad ...(13.26)$$

$$\overline{P}_{\text{apparent}} = \text{Im}\left[\overline{E}_{\text{eff}} \times \overline{H}^{*}{}_{\text{eff}}\right] \quad ....(13.27)$$

the only difference is that since vector cross-products are involved in these equations, the direction of $\overline{P}$ is at right angles to the plane containing $\overline{E}$ and $\overline{H}$ vectors.

13.8 SUMMARY AND COMMENTS

The examples discussed in Secs. 13.5 and 13.6 have shown that the Poynting vector does give a *final* answer about the power dissipation in a region and the answer tallies with what we learnt in circuit theory. It was Faraday who has shifted the attention of the scientific world from the circuit components to the region or field surrounding these components. Therefore, the fact that the field approach gives the same answer for power, as the circuit point of view can be considered to be an added success for the field theory.

But whether the field point of view is more satisfying than the circuit view is a doubtful proposition. For instance in Sec. 13.5 we have seen that the power $I_z^2 R_o$ is entering the slab section of the resistor *from the surrounding field*! In circuit theory we never bother about the region outside the respective slab!! Can we obtain that one point of view is more reasonable than the others?

The queerness of the situation is more clearly highlighted in Sec. 13.6. Here we have assumed that the current is flowing through the perfect inner and outer conductors, and the circuit theory wants us to believe that it is through these conductors that the power $V_L I_z$ is being delivered to the load resistor R_L. Yet the integration of the Poynting vector $\overline{E}_R \times \overline{H}_\phi$ is done between the limits $R = a$ and $R = b$, that is, in the region excluding these inner and outer conductors, and we are asked to believe that the power $V_L I_Z$ flows to the load resistor through the dielectric only, of the co-axial cable!!

There is no need to bother about answering the question as to which of these two view points is more correct. Actually both the points of view are only creations of our imagination in an effort to explain the observed phenomenon of delivery of power to the load. In this sense therefore, both are equally valid, but it is only the field point of view that enables us to arrive at the idea that time-varying $\overline{E}$ and $\overline{I}$ in a conducting wire can permit a certain amount of power to detach itself from the conducting wire and be propagated away (never to return to wire) in the surrounding space as an electromagnetic wave—an answer to which the circuit theory does not lead us. This aspect will be the subject of study in Chapter 15. However in the next chapter we shall discuss wave propagation in hollow tubes.

Finally, it must be remembered that the Poynting vector in Eqn. (13.14) gives the power correctly if, and only if, the integration is done over a *completely closed* surface, but it does not mean that the value of $\overline{\boldsymbol{E}} \times \overline{\boldsymbol{H}}$ at every point of the surface is correctly given by the expression on the right-hand side of the equation.

EXERCISE 13

1. In a cylindrical co-ordinate system, there is an infinitely long, cylindrical conductor of radius a metre, and with its axis coinciding with the z-axis. The resistance of the conductor is R_o ohm/metre length (measured in the z-direction). Assume that there is a uniform electric field intensity $\overline{\boldsymbol{E}}_z = E_u \overline{\boldsymbol{u}}_z$ V/m everywhere, including the region occupied by the conductor. For a cylindrical surface of radius R metre, with its axis coinciding with the z-axis, and lying between the planes $z = z_0$ and $z = z_0 + 1$, determine the total power entering the surface
   (*i*) for $R > a$ and
   (*ii*) for $R < a$, and thus verify Eqn. (13.14).
2. Prove that the following relations are not true:
$$W_{\text{active}} = \text{Re}\,(V_{\text{eff}}\, I_{\text{eff}})$$
$$W_{\text{reactive}} = \text{Im}\,(V_{\text{eff}}\, I_{\text{eff}})$$
3. Verify the following relations by plotting V_{inst}, I_{inst} and W_{inst} over the range $\omega t = 0$ to $\omega t = 2\pi$:
$$W_{\text{active}} = V_{\text{eff}}\, I_{\text{eff}} \cos(\theta_v - \theta_i)$$
$$= \frac{1}{2} V_{\text{max}}\, I_{\text{max}} \cos(\theta_v - \theta_i)$$
$$W_{\text{reactive}} = V_{\text{eff}}\, I_{\text{eff}} \sin(\theta_v - \theta_i)$$
$$= \frac{1}{2} V_{\text{max}}\, I_{\text{max}} \sin(\theta_v - \theta_i)$$
4. Verify whether the following relations are true or false:
$$W_{\text{active}} = \text{Re}\,(V^*{}_{\text{eff}}\, I_{\text{eff}})$$
$$W_{\text{reactive}} = \text{Im}\,(V^*{}_{\text{eff}}\, I_{\text{eff}})$$
5. Verify whether it makes any difference to Eqn. (13.14), if on the left-hand side $(\overline{\boldsymbol{E}} \times \overline{\boldsymbol{H}})$ is replaced by $[(\overline{\boldsymbol{E}} \times \overline{\boldsymbol{H}}) + \overline{\nabla} \times \overline{\boldsymbol{F}}]$, where vector $\overline{\boldsymbol{F}}$ is an arbitrary vector having dimensions of watts/metre.
   [**Hint:** Use divergence theorem. The result proves the correctness of the statement made in the last sentence of Sec. 13.8]

CHAPTER

14 GUIDED WAVES AND WAVE-GUIDES

14.1 ESSENTIAL CONDITIONS FOR GUIDED WAVES

To get a simplified picture of the solution of Maxwell's equations, we have reduced the problem to that of one-dimensional variation and obtained Eqn. (11.19) representing two plane waves travelling in opposite directions. If we have a medium of *finite* size, it implies that at a finite distance from the origin where we are considering the wave phenomenon, there is a boundary, that is, the linear, uniform, homogeneous, isotropic medium ends at that surface, and another medium exists beyond that surface. Such a change or discontinuity causes a reflected wave to be produced. Thus the infinite extent of the medium enabled us to confine our attention only on the forward wave of Eqn. (11.19), because the forward wave never reached the boundary of the medium in any finite time, however, large in magnitude that time may be, or in other and more accurate words, the wave would take an infinite time to reach the boundary, and thus there would be no reflected wave till then!

In practice, however, we are interested in a solution to the problem of sending power or information via an electromagnetic wave to a point a finite distance away. Thus, we send power using electric conductors (open wire lines or underground cables) in much the same way as water is supplied through pipes to people living in a city. This process ensures two things:

(*i*) the electric power (or water) is duly conveyed to the desired points, and

(*ii*) the electric power (or water) is prevented from going anywhere else.

In a radio broadcasting station, the first condition is met but not the second, because everybody around the transmitting station can receive the broadcast by only tuning the radio set to the correct frequency. But in other information carrying systems, such as telephony for instance, the second condition is equally important, for the conversation should not be heard by a third party. Metal conductors are therefore used to guide the electromagnetic waves from one point to another in such point-to-point communication systems. These conductors can be in one of the following forms:

(*i*) Two-wire transmission lines,

(*ii*) Co-axial cables,

(*iii*) Two parallel conducting planes,

(*iv*) Hollow tubes (called wave-guides) of rectangular or circular cross-section or (*v*) Single conductor in the form of a strip, the wave clinging to the surface of the strip.

For the purpose of analysis, the dielectric in which the wave is being propagated is assumed to be perfect ($\sigma = 0$) and the conductor guiding the wave is also assumed to be perfect ($\sigma = \infty$). Further, it is assumed that the dielectric is charge-free ($\rho = 0$). In the region of the dielectric, the electric and magnetic fields must satisfy Maxwell's equations. The physical picture of the electric and magnetic field lines that emerges when these conditions are imposed is summarised below, and these conditions must be satisfied and *be true always:*

Electric field lines must either

(*i*) begin from and end on electric charges located on the conducting boundary, or

(*ii*) form continuous loops or closed paths, and such loops must surround a time-varying magnetic field.

Magnetic field lines must

(*iii*) never begin or end at a point (because there is nothing magnetic corresponding to an electric charge), but

(*iv*) they must form continuous loops or closed paths surrounding either (*a*) a conduction current or (*b*) a displacement current produced by a time-varying electric field.

So far as the conducting boundary is concerned, the boundary conditions which must be satisfied there are:

(*v*) the electric field must be normal to the conductor surface (otherwise, with $\sigma = \infty$ any tangential component E_t will produce an infinite surface current, which is physically not realizable in practice), and

(*vi*) the magnetic field cannot have any component normal to the conductor surface (because this will imply that the magnetic field terminates at a point on the conductor surface, which is contrary to condition (*iii*) above).

These six conditions enable us to form a picture of the wave that can be guided by the different conducting structures mentioned above. However, in actual practice, no conductor is perfect, and σ, though large, has a finite value. In such a case, it becomes possible for a very small component $E_{\text{tangential}}$ to exist, giving rise to a penetration of the wave into the conductor as discussed in Secs. 11.11 and 11.12. At frequencies above 50 MHz, the skin depth is exceedingly small, and hence even moderately thick conductors do not give rise to any reflected wave from their outer surfaces.

14.2 TEM WAVES IN CO-AXIAL CABLES AND TWO WIRE TRANSMISSION LINES

In this chapter also, as in the previous ones, we shall assume that in the dielectric, the wave is being propagated in the z-direction. The simplest wave that was studied in

Sec. 11.3 was the plane transverse electromagnetic (TEM) wave, in which there was no component of the electric and magnetic fields in the direction of propagation of the wave, that is, $E_z = 0$ and $H_Z = 0$. We shall now proceed to see whether this type of wave can be guided by a co-axial cable or a two-wire transmission line. Both the electric and the magnetic fields will be assumed to be varying sinusoidally with time.

In a co-axial cable, the inner and the outer conductors are at different potentials and the electric field is everywhere symmetrical and radial in the dielectric (Fig. 14.1), and thus, if the axis of the cable is chosen as the *z*-axis, everywhere $E_z = 0$. The electric field lines therefore start and end on electric charges located on the inner and the outer conductors. In a wave travelling along the *z*-axis, the phase varies continuously with *z*, points separated by a distance λ along the *z*-axis being in the same phase, those separated by $\frac{\lambda}{2}$ being 180° out of phase, and so on [Fig. 14.1 (*a*)].

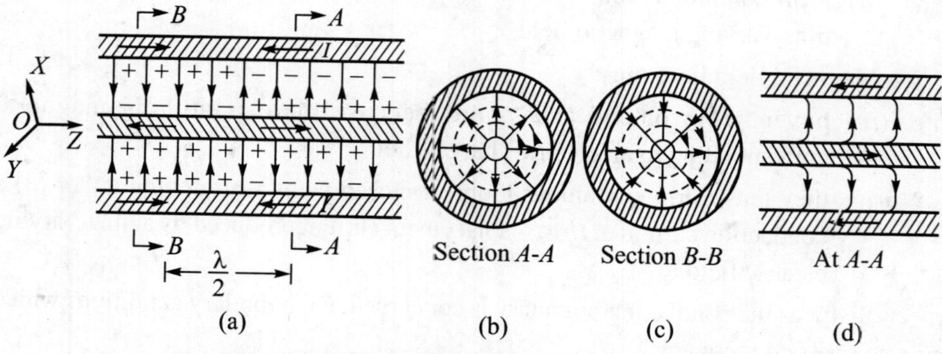

Fig. 14.1. TEM-mode guided by co-axial cable

The sinusoidally varying radial electric field produces radial displacement currents in the dielectric between the inner and the outer conductors. The magnetic field cannot surround this displacement current because to do so in Fig. 14.1(a), the magnetic field loop will have some part of it parallel to the *z*-axis, which is contrary to our assumption that $H_z = 0$. By condition (*iv*) above, we are thus forced to the conclusion that the magnetic field must surround the conduction current in the inner conductor as shown at (b) and (c) in Fig. 14.1. Thus both the electric and magnetic field lines are confined to a plane z = constant, and this ensures that $E_z = 0$ and $H_z = 0$ everywhere in the dielectric.

In an actual co-axial cable with finite conductivity for the inner and outer conductors, the picture is slightly modified. Now the entire length of the electric field lines does not lie in the plane *z*-constant, but these lines are slightly bent near the two conductor surfaces. This is because the resistance of the conductors causes the field lines to drag along the conductor surface, when the field sweeps past the plane. [In Fig. 14.1(d)] the curvature is very much exaggerated for the sake of clarity.] The $E_{\text{tangential}}$ component thus created by this drag gives the currents in the inner and the outer conductor surfaces shown in Fig. 14.1(a). An identical drag is also produced at the plane *B-B*, but since the direction of the electric field is reversed at *B-B* from that at *A-A*, the currents in the conductors are also reversed. This point should be carefully noted, because it is contrary to the impression brought from high-school days that the current is in the same direction at *all* points in a conductor forming a circuit. In this case the closed circuit between the inner and outer conductors is completed by the displacement currents in the dielectric. Finally, it is emphasised again that $E_z = 0$ and $H_z = 0$ are still true everywhere in the dielectric, except in the immediate neighbourhood of the conducting surfaces.

In the end we examine whether a TEM wave is the only possible form of wave in a co-axial cable. By the symmetry and geometry of the situation, we have already established that everywhere $\overline{E}$ must be radial only, and therefore E_z must always be zero everywhere. However, when we concluded that the magnetic field loop must enclose a conduction current, we chose the current in the inner conductor to be so enclosed and hence found that TEM mode with $H_z = 0$ also equal to zero could be possible. Why did we not select the current in the outer conductor as the one to be enclosed? The answer lies in the fact that if on a plane z = constant, we choose a circular path with radius R greater than the radius of the outer conductor, so that the path lies entirely outside the co-axial cable, then we find that the *total* current enclosed by this path is always zero, since the inner and outer conductors carry equal and opposite currents for every value of z. This implies that $\overline{H}$ is zero everywhere outside the cable, and we must have the magnetic loop only within the outer conductor. The only conduction current in this region is in the inner conductor, which we enclosed as at (*b*) and (*c*) in Fig. 14.1. Finally, the possibility of a wave with magnetic field lines enclosing the radial electric field or displacement current and with H_z not equal to zero has to be ruled out, because it would required a field $+ H_z$ on one side of the displacement current and $- H_z$ on the other side. But since the electric field has symmetry in the ϕ direction, the magnetic field also should have such a symmetry and this means that H_z cannot change sign as required above. Thus, we conclude that TEM mode is the only possible mode of transmission in a co-axial cable.

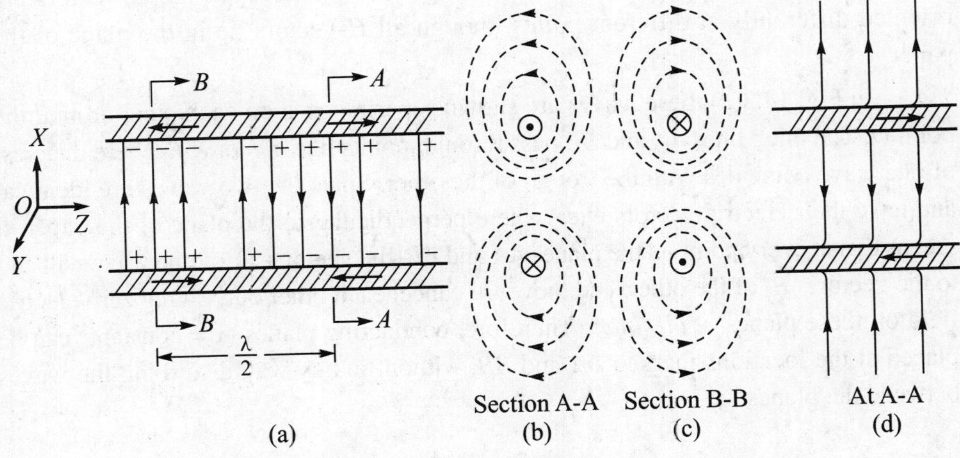

Fig. 14.2. TEM wave on two wire transmission line

Similarly on a two-wire transmission line also, only a TEM wave can exist. The only difference from the co-axial cable is that, now both the conductors have magnetic field line loops surrounding them in the plane z = constant (Fig 14.2).

In this case also the finite conductivity of the conductors causes the electric field lines to drag along the surface of the conductors, but away from the immediate neighbourhood of the conductor everywhere, $E_z = 0$ and $H_z = 0$. The component $E_{tangential}$ at the surface produces current in the conductor, which generated the magnetic field surrounding the conductor.

14.3 TE WAVES GUIDED BETWEEN PARALLEL CONDUCTING PLANES

Assume that we have in free space a TEM plane wave travelling along a straight line which makes an angle $(-\theta)$ with the z-axis as shown in Fig. 14.3(a). The *electric* vector everywhere is assumed to be perpendicular to the plane of the paper as E_y. The position of the planes corresponding to $+E_{y_{max}}$ and $-E_{y_{max}}$ are shown by straight lines perpendicular to the direction of propagation; they are continuous lines for $+E_{y_{max}}$ (positive maximum or crest of the wave) and they are dotted lines for $-E_{y_{max}}$ (negative maximum or trough of the wave). These crests and troughs occur alternately, and their separation in the direction of wave propagation is $\frac{\lambda_0}{2}$, where λ_0 is the wavelength in free space. The small arrows on these lines indicate the direction of propagation of the wave.

Further we assume that in the same region, there is another TEM plane wave present *identical* with the above wave, but travelling at an angle $+\theta$ with respect to the z-axis as shown in Fig. 14.3 (b). [Note that though the two waves are in the same region, two separate figures are drawn initially for the purpose of clarity. The waves are superimposed in Fig. 14.3 (c)]. It is obvious that in both waves, the $\boldsymbol{E}$ vector is always perpendicular to the plane of the paper everywhere, but the $\boldsymbol{H}$-vectors are oriented differently at different points, though all $\boldsymbol{H}$-vectors lie in the plane of the paper.

In Fig. 14.3(c) these waves are shown superimposed on each other around the point O. The thick lines AA and BB pass through points like a, b or e, f, where the crest of one wave coincides with the trough of the other. Since the two waves are identical and have their electric vectors everywhere perpendicular to the plane of the paper, it means that at every point on the planes AA and BB, the vector $+E_y$ of one wave will add to the vector $-E_y$ of the other, and they will cancel each other out, giving zero electric field on these planes *at all times*. Therefore, conducting planes (x = constant) can be placed at the locations marked AA and BB, without in any way disturbing the waves between the planes.

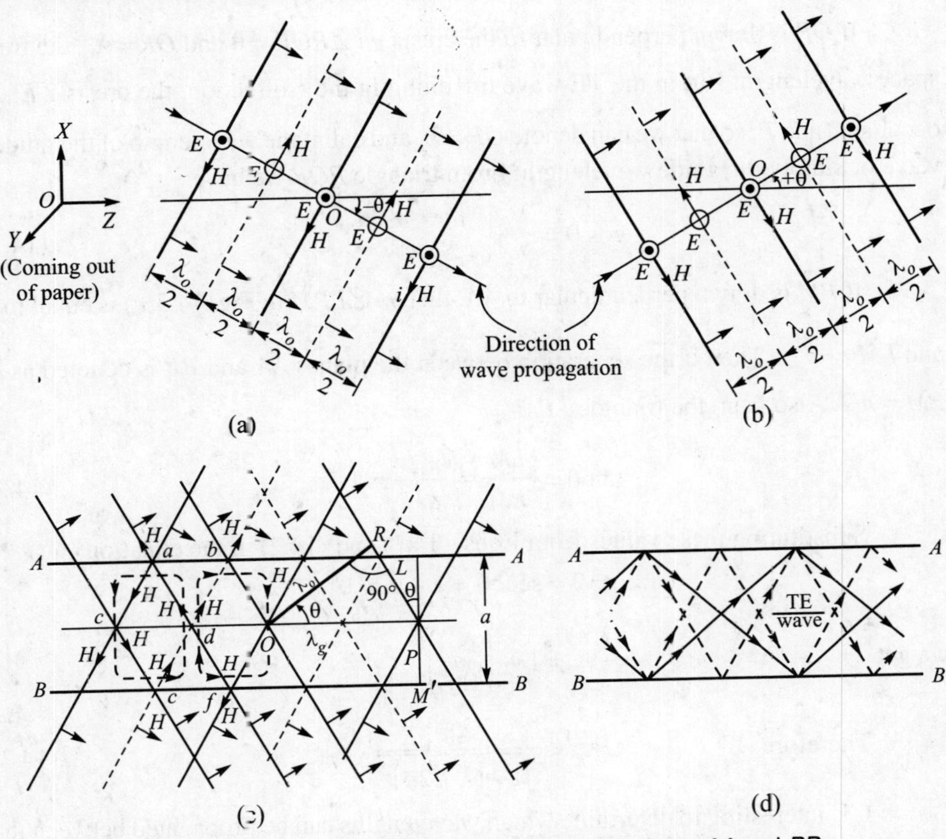

Fig. 14.3. TE waves guided between parallel planes *AA* and *BB*

The magnetic vectors $\overline{\boldsymbol{H}}$ at a, b and e, f indicate that their x-components cancel, but their z-components add showing that there is a resultant component H_z parallel to *AA* and *BB* near these planes, but there is no component perpendicular to these planes. Thus conditions (*v*) and (*vi*) are satisfied at these conducting planes.

At points like c, d, O, P, we have two crests or two troughs crossing each other. Hence the electric vectors of the two waves add giving the values $2\,E_{y_{max}}$ at c, O and P and $-\,2\,E_{y_{max}}$ at d. It is seen that the magnetic vectors at these points are so oriented that their resultant has no H_z component, but only the H_x component is present. Since at all points the $\overline{\boldsymbol{H}}$ = vectors lie in the plane of the paper, it follows that the loops for the magnetic field lines also lie in the plane of the paper, and similar loops will be obtained for all planes, like y = constant.

These E_y and H_x components give rise to a wave travelling in the z-direction. Since this wave has $E_z = 0$ everywhere, but H_z is not zero everywhere, it is called a Transverse Electric (TE) wave. The time-varying electric field of the wave gives rise to displacement currents in the y-direction and these currents are surrounded by loops of magnetic field as shown in Fig. 14.3 (c). The two original TEM waves with which we started undergo reflections at the planes *AA* and *BB* [Fig. 14.3 (d)] involving reversal of ***E***-vector as in Sec. 12.6, and their resultant is the TE wave travelling in the z-direction.

If OR is drawn perpendicular to the crests, $m\ \angle ROP = \theta$ and $OR = \lambda_o$, the free-space wavelength. But in the TE wave travelling in the z-direction, the crests $2\ E_{y_{max}}$ occur at O and P, so that we can denote $OP = \lambda_g$ and call it the wavelength of the guided wave or simply the guide wavelength. From triangle POR we have

$$\cos\theta = \frac{\lambda_o}{\lambda_g} \quad \text{...(14.1)}$$

If PM is drawn perpendicular to AA, then $m\ \angle LPM$ in Fig. 14.3(c) is equal to θ and $LM = \dfrac{\lambda_g}{4}$. Then if the separation between the planes AA and BB is denoted as a, $PM = a/2$. Also from the triangle LPM

$$\tan\theta = \frac{LM}{PM} = \frac{\lambda_g/4}{a/2} = \frac{\lambda_g}{2a} \quad \text{...(14.2)}$$

Substituting these values from Eqns. (14.1) and (14.2) in the equation

$$1/\cos^2\theta = \sec^2\theta = 1 + \tan^2\theta$$

we get

$$\frac{\lambda_g^2}{\lambda_o^2} = 1 + \left(\frac{\lambda_g}{2a}\right)^2$$

Therefore

$$\lambda_g = \frac{\lambda_o}{\sqrt{1-(\lambda_o/2a)^2}} \quad \text{...(14.3)}$$

It is interesting to determine which wavelengths can be propagated between the conducting planes AA and BB, if the distance a between them is held constant. The denominator of Eqn. (14.3) shows that if the wave is propagated, then λ_g must be real and therefore $\lambda_o < 2a$, otherwise the denominator becomes imaginary. Further if λ_o is increased, the denominator decreases and λ_g increases. Then Eqn. (14.2) shows that keeping $2a$ constant, if λ_g is increased, the angle θ must increase, as shown by dotted curve in Fig. 14.3 (d). In the limit, when $\lambda_o = 2a$, θ becomes 90° and λ_g becomes infinite. In this way, the system acts as a high-pass filter, cutting off all frequencies less than that corresponding to $\lambda_o = 2a = \lambda_c$, the cut-off wavelength.

Another way of looking at the propagation of two TEM waves depicted in Fig. 14.3(d) is to consider each TEM wave as composed of two components, of which one travels along the z-axis and the other along the x-axis, as in Sec. 12.8. Then the z-axis components add to give the travelling TE wave in the z-direction. But the x-axis components of the two waves are travelling in opposite directions and therefore, they produce a standing wave pattern in the x-direction with adjacent minima separated by a distance a. This distance a is a constant irrespective of the frequency of the TEM waves. This is so, because the angle θ changes with the frequency to achieve this result, as mentioned in the preceding paragraph.

Further, it is clear that the separation between the planes AA and BB could have been changed to $2a$, $3a$, ... without any effect on wave propagation between them, except to have 2, 3, ... maxima for E_y in the x-direction instead of only one. Finally, we note that on all planes y = constant, the same electrical and magnetic field patterns occur. The electric fields are everywhere normal to the plane y = constant, while the magnetic fields are parallel to this plane. It follows therefore, that on any two planes $y = C_1$ and $y = C_2$ we can place conducting planes without disturbing the wave propagation. The electric field lines will now terminate on positive and negative charges located on the planes $y = C_1$ and $y = C_2$ while the magnetic field lines will form loops in planes parallel to these planes.

We have thus created a system consisting of four conducting planes parallel to the x and y axes, forming a pipe with rectangular cross-section, through which a TE wave is being propagated in the z-direction. This system, called a rectangular wave-guide, is examined mathematically in the next section.

14.4 RECTANGULAR WAVE-GUIDE

The notation to be used to describe wave propagation in rectangular and circular wave-guides has been standardized. For a rectangular wave-guide with section as shown in Fig. 14.4, the axes are chosen as shown in the figure. The x-axis is along the *longer* side of the rectangle. The origin is deliberately placed at the bottom right-hand corner of the rectangle, with the positive direction of the x-axis towards the left and the y-axis upwards, so that by the right-handed screw rule, the positive direction of the z-axis will go into the paper away from the observer.

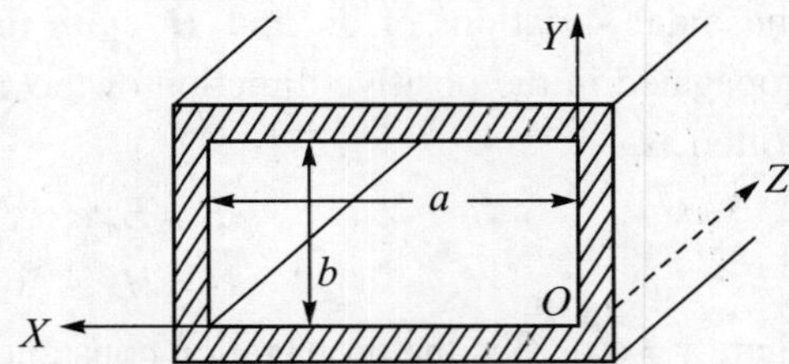

Fig. 14.4. Rectangular wave-guide—standard notation

The walls of the wave-guide are assumed to be perfect conductors, while for the dielectric inside the waveguide, $\sigma = 0$. Also as there are to be no free charges, $\rho = 0$. The wave is to travel down the length of the wave-guide in the positive direction of the z-axis. As in Sec. 14.3, we first try to find physically whether a TEM wave can be propagated in the guide.

As in the previous section, we assume that the $\boldsymbol{E}$ vector is everywhere parallel to the y-axis and therefore $E_z = 0$. Now if a TEM wave is to be propagated, we have to find a suitable magnetic field, which satisfies the condition that $H_z = 0$. The condition (*iv*) of Sec. 14.1 stipulates that such a magnetic field must be in the form of closed loops enclosing either a time-varying displacement current or a time-varying conduction current. If we think of a magnetic field loop enclosing the $\overline{E}$ vectors that are parallel to the y-axis, we find that such a closed loop cannot be formed without H_z components, in opposite directions, on either side of the $\overline{E}$ vector. This means that a TEM wave is not possible with magnetic field enclosing time-varying displacement current. Therefore, we must search for a magnetic field loop that can enclose a conduction current.

Since there is no inner conductor in the wave-guide, like the inner conductor in a co-axial cable, the field must enclose a conduction current in the outer rectangular conductor. But condition (*vi*) of Sec. 14.1 requires that the magnetic field near the surface of a conductor must be parallel to that surface. Also if H_z is to be zero to get a TEM wave, the loop must lie entirely in the plane z = constant. These two conditions will give a magnetic field loop enclosed by the rectangular conducting boundary instead of vice versa. This shows that there is no way in which we can have both E_z and H_z equal to zero. Or in other words, a TEM wave cannot be propagated inside a hollow wave-guide. In the next section this result is derived from Maxwell's equations.

14.5 ELECTRIC AND MAGNETIC FIELDS INSIDE RECTANGULAR WAVE-GUIDE

It is assumed that the wave-guide is made of a perfect conductor ($\sigma = \infty$) the dielectric is perfect ($\sigma = 0$) (it is usually free space or air), and it is charge free ($\rho = 0$). Further sinusoidal variations of $\overline{\boldsymbol{E}}$ and $\overline{\boldsymbol{H}}$ with time are assumed to cause the wave to be propagated in the positive direction of the z-axis (Fig. 14.4) so that the fields can be written as

$$E = E_o\, e^{(j\omega t - \gamma z)},$$
$$H = H_o\, e^{(j\omega t - \gamma z)} \qquad \text{...(14.4)}$$

where $\gamma = \alpha + j\beta$ is the propagation constant. Using these expressions, Maxwell's curl equations can be written as

$$\overline{\nabla} \times \overline{\boldsymbol{H}} = 0 + \frac{\partial \overline{D}}{\partial t} = \varepsilon \frac{\partial \overline{\boldsymbol{E}}}{\partial t} = j\omega\varepsilon\, \overline{\boldsymbol{E}},$$

$$\overline{\nabla} \times \overline{\boldsymbol{E}} = -\frac{\partial \overline{B}}{\partial t} = -\mu \frac{\partial \overline{\boldsymbol{H}}}{\partial t} = -j\omega\mu\, \overline{\boldsymbol{H}},$$

or resolving the curl into its components along the three axes (using determinant form) and noting that

$$\frac{\partial E}{\partial z} = -\gamma E \quad \text{and} \quad \frac{\partial H}{\partial z} = -\gamma H,$$

we get

$$\left.\begin{aligned}
&\frac{\partial H_z}{\partial z} - (-\gamma H_y) = j\omega\varepsilon E_x && (i)\\
&\frac{\partial E_z}{\partial y} - (-\gamma E_y) = -j\omega\mu H_x && (ii)\\
&(-\gamma H_x) - \frac{\partial H_z}{\partial x} = j\omega\varepsilon E_y && (iii)\\
&(-\gamma E_x) - \frac{\partial E_z}{\partial x} = -j\omega\mu H_y && (iv)\\
&\frac{\partial H_y}{\partial x} - \frac{\partial H_x}{\partial y} = j\omega\varepsilon E_z && (v)\\
&\frac{\partial E_y}{\partial x} - \frac{\partial E_x}{\partial y} = -j\omega\mu H_z && (vi)
\end{aligned}\right\} \qquad \text{...(14.5)}$$

From these six equations, the values of E_x, E_y, H_x and H_y can be obtained in terms of E_z and H_z. Thus to get the value of E_x, we eliminate H_y from Eqns. (14.5) (*i*) and (*iv*) by multiplying (*i*) by $j\omega\mu$ and (*iv*) by γ and adding. Similarly, for other quantities. Thus we have

$$\left.\begin{aligned}
E_x &= \frac{-1}{\gamma^2+\omega^2\mu\in}\left[\gamma\frac{\partial E_z}{\partial x}+j\omega\mu\frac{\partial H_z}{\partial y}\right] && (i)\\
E_y &= \frac{1}{\gamma^2+\omega^2\mu\in}\left[-\gamma\frac{\partial E_z}{\partial y}+j\omega\mu\frac{\partial H_z}{\partial x}\right] && (ii)\\
H_x &= \frac{1}{\gamma^2+\omega^2\mu\in}\left[j\omega\varepsilon\frac{\partial E_z}{\partial y}-y\frac{\partial H_z}{\partial x}\right] && (iii)\\
H_y &= \frac{-1}{\gamma^2+\omega^2\mu\in}\left[j\varepsilon\frac{\partial E_z}{\partial x}+\gamma\frac{\partial H_z}{\partial y}\right] && (iv)
\end{aligned}\right\} \quad ...(14.6)$$

These equations reveal why a TEM wave is not possible in a hollow wave-guide. For if we put $E_z = 0$ and also $H_z = 0$ in Eqn. (14.6), all the terms in the brackets become zero and thus all fields become zero.

Though E_z and H_z cannot both be zero, we can make them zero one at a time. Thus when $E_z = 0$ but $H_z \neq 0$, the electric field is always transverse to the direction of propagation (*z*-axis) and hence such a wave is called a transverse electric (TE) or *H*-wave. Similarly, when $H_z = 0$ but $E_z \neq 0$, the magnetic field is always transverse and we get a transverse magnetic (TM) or *E*-wave.

14.6 SOLUTION OF WAVE EQUATIONS FOR RECTANGULAR WAVE-GUIDE

Since Eqn. 14.4 have been assumed, the following equations result:

$$\left.\begin{aligned}
\frac{\partial E}{\partial z} &= -\gamma E, & \frac{\partial^2 E}{\partial z^2} &= +\gamma^2 E\\
\frac{\partial H}{\partial z} &= -\gamma H, & \frac{\partial^2 H}{\partial z^2} &= +\gamma^2 H\\
\frac{\partial E}{\partial t} &= j\omega E, & \frac{\partial^2 E}{\partial t^2} &= (j\omega)^2 E = -\omega^2 E\\
\frac{\partial H}{\partial t} &= j\omega H, & \frac{\partial^2 H}{\partial t^2} &= (j\omega)^2 H = -\omega^2 H
\end{aligned}\right\} \quad ...(14.7)$$

Using Eqn. (11.13), the wave equations (11.11) and (11.12) for the *z*-components only can be written with the help of Eqn. (14.7) as

$$\overline{\nabla}^2 H_z = \mu\varepsilon\frac{\partial^2 H_z}{\partial t^2},$$

$$\overline{\nabla}^2 E_z = \mu\varepsilon\frac{\partial^2 E_z}{\partial t^2},$$

or
$$\frac{\partial^2 H_z}{\partial x^2}+\frac{\partial^2 H_z}{\partial y^2}+\frac{\partial^2 H_z}{\partial z^2}=\mu\,\varepsilon\,\frac{\partial^2 H_z}{\partial t^2},$$

$$\frac{\partial^2 E_z}{\partial x^2}+\frac{\partial^2 E_z}{\partial y^2}+\frac{\partial^2 E_z}{\partial z^2}=\mu\,\varepsilon\,\frac{\partial^2 E_z}{\partial t^2},$$

or
$$\left.\begin{aligned}\frac{\partial^2 H_z}{\partial x^2}+\frac{\partial^2 H_z}{\partial y^2}&=-(\gamma^2+\omega^2\mu\varepsilon)\,H_z \quad (a)\\ \frac{\partial^2 E_z}{\partial x^2}+\frac{\partial^2 E_z}{\partial y^2}&=-(\gamma^2+\omega^2\mu\varepsilon)\,E_z \quad (b)\end{aligned}\right\} \qquad ...(14.8)$$

For TE mode, $E_z = 0$ and therefore the wave equation (14.8*a*) is to be solved. For TM mode, $H_z = 0$ and in this case, Eqn. (14.8*b*) has to be solved. The two equations being similar will give similar solutions. Their solutions are easily found by the method given in Sec. 6.4 by assuming a product solution.

Thus, to solve Eqn. (14.8*a*), we assume that $H_z = XY$ and obtain the solutions as

$$\left.\begin{aligned}X &= A'\cosh k_x' x + B'\sinh k_x' x\\ Y &= C'\cosh k_y' y + D'\sinh k_y' y\end{aligned}\right\} \qquad ...(14.9)$$

where
$$k_x'^2 + k_y'^2 = -(\gamma^2 + \omega^2\mu\varepsilon)$$

The constants $k_x'^2$ and $k_y'^2$ give rise to three possible types of solutions:

(*i*) $k_x'^2$ and $k_y'^2$ are negative.

(*ii*) Both are positive, and

(*iii*) One is positive while the other is negative.

As discussed in Sec. 6.4, a positive value for $k_x'^2$ and $k_y'^2$ gives a solution involving hyperbolic sines and cosines, which (see Fig. 1.5) increase continuously. But since in the wave-guide we have identical conducting planes at two places in the *x*- and in the *y*-directions, the fields must have repeating values in both *x*- and *y*-directions. We note that if we put $k_x'^2 = -k_x^2$ and $k_y'^2 = -k_y^2$ we have $k_x' = jk_x$ and $k_y' = jk_y$, and the terms in Eqn. (14.9) change to sines and cosines by Eqn. (1.39) and these undergo the cyclic variations that we desire in the *x*- and *y*-directions.

The constants in Eqn. (14.9) have to be determined from the boundary conditions. These are, that there are perfectly conducting planes at $x = 0$, $x = a$, $y = 0$ and $y = b$. At these boundaries, $E_{\text{tangential}}$ and H_{normal} must be equal to zero. This can be stated in the form of the following boundary conditions:

$$\left.\begin{aligned}&\text{At } x=0 \text{ and at } x=a \;: E_y=0,\, E_z=0,\, H_x=0\\ &\text{At } y=0 \text{ and at } y=b \;: E_x=0,\, E_z=0,\, H_y=0\end{aligned}\right\} \qquad ...(14.10)$$

But these conditions cannot be applied directly, because they involve E_x, E_y, H_x, H_y whereas the wave equations (14.8) involve E_z and H_z, only. Therefore, Eqn. (14.10) is first applied to Eqn. (14.6) to give the boundary conditions in the form:

$$\left.\begin{aligned} &\text{At } x=0 \text{ and at } x=a \quad :[\text{From } (ii) \text{ and } (iii)] \quad \frac{\partial H_z}{\partial x}=0 \\ &\text{At } y=0 \text{ and at } y=b \quad :[\text{From } (i) \text{ and } (iv)] \quad \frac{\partial H_z}{\partial y}=0 \end{aligned}\right\} \quad \text{...(14.11)}$$

Since $k_x'^2$ and $k_y'^2$ are both negative (to avoid hyperbolic sines and cosines in the solution), the Eqn. (14.9) are rewritten as

$$\left.\begin{aligned} H_z &= XY \\ &= (A\cos k_x x + B\sin k_x\ x)(C\cos k_y\ y + D\sin k_y\ y) \\ \text{where} \quad & k_x^2 + k_y^2 = \gamma^2 + \omega^2\mu\varepsilon \end{aligned}\right\} \quad \text{...(14.12)}$$

The constants are now determined as follows:

$$\text{At } x=0,\ \frac{\partial H_z}{\partial x}=0, \text{ therefore } B=0.$$

$$\text{At } y=0,\ \frac{\partial H_z}{\partial y}=0, \text{ therefore } D=0.$$

$$\text{At } x=a,\ \frac{\partial H_z}{\partial x}=0, \text{ therefore } \frac{\partial}{\partial x}[AC\cos k_x x\cos k_y y]=0, \text{ or } k_x=\frac{m\pi}{a}.$$

$$\text{At } y=b,\ \frac{\partial H_z}{\partial y}=0, \text{ therefore } \frac{\partial}{\partial y}[AC\cos k_x x\cos k_y y]=0, \text{ or } k_y=\frac{n\pi}{b}.$$

where m and n are integers or zero. The final solution of Eqn. (14.8a) can now be written as

$$H_z = C_1\cos\left(\frac{m\,\pi\,x}{a}\right)\cos\left(\frac{n\,\pi\,y}{b}\right)e^{(j\,\omega\,t-\gamma\,z)} \quad \text{...(14.13)}$$

Next we take up the solution for Eqn. (14.8b), whose product solution can be written as

$$E_z = (A_0\cos k_x x + B_0\sin k_x x)(C_0\cos k_y y + D_0\sin k_y y)$$

Using the boundary conditions of Eqn. (14.10), we find the constants.

At $x = 0$, $E_z = 0$, therefore $A_0 = 0$.

At $y = 0$ also, $E_z = 0$, therefore $C_0 = 0$.

At $x = a$, $E_z = 0$, therefore $k_x = \dfrac{m\pi}{a}$.

At $y = b$, $E_z = 0$, therefore $k_y = \dfrac{n\pi}{b}$.

The final solution of Eqn. (14.8b) then is written as

$$E_z = C_2\sin\left(\frac{m\pi x}{a}\right)\sin\left(\frac{n\pi y}{b}\right)e^{(j\,\omega\,t-\gamma\,z)} \quad \text{...(14.14)}$$

For both the solutions given by Eqns. (14.13) and (14.14)

$$\gamma^2 = k_x^{\,2} + k_y^{\,2} - \omega^2\,\mu\,\varepsilon$$

or
$$\gamma = \sqrt{\left(\frac{m\pi}{a}\right)^2 + \left(\frac{n\pi}{b}\right)^2 - \omega^2 \mu \varepsilon} \qquad \text{...(14.15)}$$

Equations (14.13) and (14.14) are thus the complete solutions of the wave equation (14.8). These equations along with Eqns. (14.6) and (14.15) completely define all the waves that can be propagated in the wave-guide and which are defined by the factor $e^{(j\omega t - \gamma z)}$. All these equations satisfy the boundary conditions defined by Eqns. (14.10) and (14.11), when the wave-guide is made of a material which is a perfect conductor. However, notice that Eqns. (14.13) and (14.14) still involve two constants C_1 and C_2. These constants are to be determined from the information about the wave at some point in the dielectric (that is, not situated on the conducting walls of the wave-guide). This is done in the next two sections.

14.7 TRANSVERSE ELECTRIC WAVE IN RECTANGULAR WAVE-GUIDE

In a TE (or *H*-) wave in a wave-guide, *at every point* the electric field has no E_z component. This is possible only if $C_2 = 0$ in Eqn. (14.14). Then using Eqns. (14.13) and (14.6), we get these equations defining a TE wave :

$$\left.\begin{aligned}
E_z &= 0,\ H_z = C_1 \cos\left(\frac{m\pi x}{a}\right)\cos\left(\frac{n\pi y}{b}\right) e^{(j\omega t - \gamma z)}\\
E_x &= \frac{-j\omega\mu C_1}{\gamma^2 + \omega^2\mu\varepsilon}\left(\frac{-n\pi}{b}\right)\cos\left(\frac{m\pi x}{a}\right)\sin\left(\frac{n\pi y}{b}\right) e^{(j\omega t - \gamma z)}\\
E_y &= \frac{-j\omega\mu C_1}{\gamma^2 + \omega^2\mu\varepsilon}\left(\frac{-m\pi}{a}\right)\sin\left(\frac{m\pi x}{a}\right)\cos\left(\frac{n\pi y}{b}\right) e^{(j\omega t - \gamma z)}\\
H_x &= \frac{-\gamma C_1}{\gamma^2 + \omega^2\mu\varepsilon}\left(\frac{-m\pi}{a}\right)\sin\left(\frac{m\pi x}{a}\right)\cos\left(\frac{n\pi y}{b}\right) e^{(j\omega t - \gamma z)}\\
H_y &= \frac{-\gamma C_1}{\gamma^2 + \omega^2\mu\varepsilon}\left(\frac{-n\pi}{b}\right)\cos\left(\frac{m\pi x}{a}\right)\sin\left(\frac{n\pi y}{b}\right) e^{(j\omega t - \gamma z)}
\end{aligned}\right\} \qquad \text{...(14.16)}$$

where
$$\gamma = \sqrt{\left(\frac{m\pi}{a}\right)^2 + \left(\frac{n\pi}{b}\right)^2 - \omega^2 \mu \varepsilon} \qquad \text{...(14.15)}$$

Very often in books, in Eqn. (14.16), the factor $e^{(j\omega t - \gamma z)}$ is understood to be included on the right-hand side, though it is not written out explicitly as above. This should be remembered when consulting other books.

Since the integers *m* and *n* can have any integral values, it is obvious that there are an infinite varieties of waves that can be propagated. These are called TE_{mn} modes of wave propagation. From the way in which these integers *m* and *n* were introduced in Eqn. (14.13), it is seen that *m* and *n* represent the numbers of half-cycles of sinusoidal field variations along the x- and y-axes, counting from the origin O. This implies that the subscripts m and n in TE_{mn} can have only positive values including zero. The only value that is not permissible is for both $m = 0$ and $n = 0$, for this makes all components of $\overline{E}$ and $\overline{H}$ to become zero, except $H_z = C_1\, e^{(j\omega t - \gamma z)}$ and thus there is no electromagnetic wave propagation.

The most useful mode of transverse electric wave propagation is the one with $m = 1$ and $n = 0$, called the TE_{10} mode which is discussed in detail in a later section.

14.8 TRANSVERSE MAGNETIC WAVE IN RECTANGULAR WAVE-GUIDE

In a TM (or E-) wave in a wave-guide, *at every point* the magnetic field has no H_z component. For this to be true everywhere, $C_1 = 0$ in Eqn. (14.13). Then using Eqns. (14.14) and (14.6), we have the following equations defining the TM wave propagation:

$$\left.\begin{aligned}
H_z &= 0,\ E_z = C_2 \sin\left(\frac{m\pi x}{a}\right)\sin\left(\frac{n\pi y}{b}\right)e^{(j\omega t-\gamma z)}\\
E_x &= \frac{-\gamma C_2}{\gamma^2+\omega^2\mu\varepsilon}\left(\frac{m\pi}{a}\right)\cos\left(\frac{m\pi x}{a}\right)\sin\left(\frac{n\pi y}{b}\right)e^{(j\omega t-\gamma z)}\\
E_y &= \frac{-\gamma C_2}{\gamma^2+\omega^2\mu\varepsilon}\left(\frac{n\pi}{b}\right)\sin\left(\frac{m\pi x}{a}\right)\cos\left(\frac{n\pi y}{b}\right)e^{(j\omega t-\gamma z)}\\
H_x &= \frac{j\omega\varepsilon C_2}{\gamma^2+\omega^2\mu\varepsilon}\left(\frac{n\pi}{b}\right)\sin\left(\frac{m\pi x}{a}\right)\cos\left(\frac{n\pi y}{b}\right)e^{(j\omega t-\gamma z)}\\
H_y &= \frac{j\omega\varepsilon C_2}{\gamma^2+\omega^2\mu\varepsilon}\left(\frac{m\pi}{a}\right)\cos\left(\frac{m\pi x}{a}\right)\sin\left(\frac{n\pi y}{b}\right)e^{(j\omega t-\gamma z)}
\end{aligned}\right\} \quad \text{...(14.17)}$$

where
$$\gamma = \sqrt{\left(\frac{m\pi}{a}\right)^2 + \left(\frac{n\pi}{b}\right)^2 - \omega^2\mu\varepsilon} \quad \text{...(14.15)}$$

As with Eqn. (14.16), in most books the factor $e^{(j\omega t - \gamma z)}$ is not explicitly written as above, and only the amplitudes of the components are given.

It is obvious that we can have an infinite number of modes of wave propagation designated as TM_{mn} modes, as the integers m and n take different positive values. The only value that is not permitted is $m = 0$ or $n = 0$, so that modes like TM_{on} and TM_{mo} are not permissible, because these make all field components equal to zero.

The most useful mode of this type is the TM_{11} mode, with $m = 1$ and $n = 1$. It is discussed in a later section.

14.9 CUT-OFF FREQUENCY AND WAVELENGTH

Equation (14.15) applies to both TE and TM modes of wave propagation. If we want that the wave should be propagated through the wave-guide, then we must have the attenuation constant $\alpha = 0$ or $\gamma = \alpha + j\beta = 0 + j\beta$. Then Eqn. (14.15) can be written as

$$j\beta = \sqrt{\left(\frac{m\pi}{a}\right)^2 + \left(\frac{n\pi}{b}\right)^2 - \omega^2\mu\varepsilon}$$

If the square-root term on the right-hand side is to yield an imaginary quantity, it follows that

$$\omega^2\mu\varepsilon > \left(\frac{m\pi}{a}\right)^2 + \left(\frac{n\pi}{b}\right)^2$$

Assuming that we have free space inside the waveguide, the left-hand side becomes $\omega^2 \mu_0 \varepsilon_0$ and using Eqns. (11.17*b*) and (11.18*a*), this gives

$$\omega^2 \mu_0 \varepsilon_0 = \frac{\omega^2}{v_0^2} = \left(\frac{2\pi f}{f\lambda_0}\right)^2 = \left(\frac{2\pi}{\lambda_0}\right)^2$$

Using this value, the above equation becomes

$$\left(\frac{2\pi}{\lambda_0}\right)^2 > \left(\frac{m\pi}{a}\right)^2 + \left(\frac{n\pi}{b}\right)^2$$

Since $v_0 = f\lambda_0$ is a constant, it follows that as f is decreased, λ_0 increases and the left-hand side of the above equation decreases. At frequency f_c, we get a wavelength λ_c, such that

$$v_0 = f_c\,\lambda_c$$

and
$$\left(\frac{2\pi}{\lambda_c}\right)^2 = \left(\frac{m\pi}{a}\right)^2 + \left(\frac{n\pi}{b}\right)^2$$

or
$$[\lambda_c]_{mn} = \frac{2}{\sqrt{\left(\frac{m}{a}\right)^2 + \left(\frac{n}{b}\right)^2}} \qquad \text{...(14.18)}$$

Frequencies less than f_c or wavelengths larger than λ_c are not transmitted in the waveguide. Hence these are called cut-off frequency and cut-off wavelength, respectively. It must be remembered that Eqn. (14.18) is equally applicable to both TE and TM modes of propagation, or in other words, both modes of propagation have the same cut-off frequency and wavelength.

14.10 DOMINANT TE_{10} MODE

Since the values $m = 0$ and $n = 0$ do not give a wave propagation in TE mode, the next mode to be considered is one, in which only one of the constants m or n is made equal to one, while the other is zero. In rectangular wave-guides, $a > b$ (because a is always chosen, by standard convention, as the longer side). Therefore, we conclude from Eqn. (14.18) that

$$[\lambda_c]_{10} > [\lambda_c]_{01} \quad \text{or} \quad [f_c]_{10} < [f_c]_{01}.$$

The cut-off frequency for TE_{10} mode being less than that for TE_{01} mode, the TE_{10} mode is called the *dominant mode*. It corresponds to the *largest range* of frequencies transmitted.

The equations for the field components of the TE_{10} mode are obtained directly from Eqn. (14.16), by putting $m = 1$ and $n = 0$. They are

$$\left.\begin{aligned} &E_z = 0, H_z = C_1 \cos\left(\frac{\pi x}{a}\right) e^{(j\omega t - \gamma z)} \\ &E_x = 0, H_y = 0 \\ &E_y = \frac{j\omega\mu}{\gamma^2 + \omega^2\mu_0\varepsilon_0}\left(\frac{-C_1\pi}{a}\right) \sin\left(\frac{\pi x}{a}\right) e^{(j\omega t - \gamma z)} \\ &H_x = \frac{1}{\gamma^2 + \omega^2\mu_0\varepsilon_0}\left(\frac{\gamma C_1\pi}{a}\right) \sin\left(\frac{\pi x}{a}\right) e^{(j\omega t - \gamma z)} \end{aligned}\right\} \quad ...(14.19)$$

where $\gamma^2 = \left(\frac{\pi}{a}\right)^2 - \omega^2 \mu_0 \varepsilon_0, \lambda_c = 2a$

The remarkable thing about Eqn. (14.19) is the fact that everywhere, the electric field vector is parallel to the y-axis, ($E_x = 0, E_z = 0$). If the maximum value of the electric field intensity (or the amplitude of the E_y component) is written as E_0, we get the relation between C_1 and E_0 as

$$\frac{j\omega\mu}{\gamma^2 + \omega^2\mu_0\varepsilon_0}\left(\frac{C_1\pi}{a}\right) = E_0$$

Therefore $$C_1 = \frac{\gamma^2 + \omega^2\mu_0\varepsilon_0}{j\omega\mu}\left(\frac{-a}{\pi}\right) E_0$$

$$= \frac{\left(\frac{\pi}{a}\right)^2 \lambda_0}{j(2\pi f\lambda_0)\mu}\left(\frac{j^2 a}{\pi}\right) E_0$$

Putting $$f\lambda_0 = v_0 = \frac{1}{\sqrt{\mu_0\varepsilon_0}} \quad \text{and} \quad \eta_0 = \sqrt{\frac{\mu_0}{\varepsilon_0}}$$

$$C_1 = \frac{jE_0}{\eta_0}\left(\frac{\lambda_0}{2a}\right) \quad ...(14.20)$$

Then Eqn. (14.19) for the TE_{10} mode becomes

$$\left.\begin{aligned}
&E_z = 0, E_x = 0, H_y = 0 \\
&E_y = E_0 \sin\left(\frac{\pi x}{a}\right) e^{(j\omega t - \gamma z)} \\
&H_x = \frac{-E_0\gamma}{j\omega\mu_0} \sin\left(\frac{\pi}{a}\right) e^{(j\omega t - \gamma z)} \\
&H_z = \frac{jE_0}{\eta_0}\left(\frac{\lambda_0}{2a}\right) \cos\left(\frac{\pi x}{a}\right) e^{(j\omega t - \gamma z)} \\
&\gamma^2 = \left(\frac{\pi}{a}\right)^2 - \omega^2 \mu_0 \varepsilon_0, \lambda_c = 2a
\end{aligned}\right\} \quad ...(14.21)$$

The electric and magnetic fields are shown in Fig. 14.5 at a time when $\omega t = 2n'\pi$, n' being any integer. The end view at plane $z = 0$ shows that electric field lines (continuous lines) start from positive charges located on the bottom conductor and end on negative charges located on the top conductor. The side view on plane $x = \frac{a}{2}$ shows that at a distance $z = \frac{\lambda_g}{2}$ inside the guide, the situation gets reversed from that at the end face. On every plane z = constant, the electric field intensity has a half-sinusoidal variation from the edge $x = 0$ to the edge $x = a$. The magnetic field lines are shown in all views as dotted lines. In the end view, they are parallel to the x-axis because $H_y = 0$. In the top view for the plane $y = \frac{b}{2}$, they form loops surrounding the electric field lines, and for all planes y = constant, they have the same shape as shown in the top view. At every point, the direction of $\bar{P}$, the Poynting vector, is always in the positive direction of the z-axis.

The cut-off wavelength $\lambda_c = 2a$ shows that the smaller dimension b of the wave-guide is irrelevant for determining λ_c. However, if TE_{10} is the only mode that is to exist in the wave-guide and *if all other modes are to be suppressed*, then the dimensions of the guide have to be carefully determined. Thus to prevent the TE_{20} mode being propagated, the dimension a must be less than λ_0. Further, to prevent the TE_{01} mode being propagated, the dimension b must be less than $\frac{\lambda_0}{2}$. For use at nominal 3 cm wavelength or 10 GHz frequency, the RCSC (Radio Components Standardization Committee) Type WG16 or American EIA (Electronics Industries Association) Type WR90 wave-guide is used. Its outside dimensions are 1 inch × 0.5 inch, with a wall thickness of 0.050 inch. This makes its inner dimension as a = 0.900 inch (22.9 mm) and b = 0.400 inch (10.2 mm). The operating frequency range is 8.20 to 12.40 GHz. The power rating is 250 kW and attenuation is 3.24 dB per 100 feet.

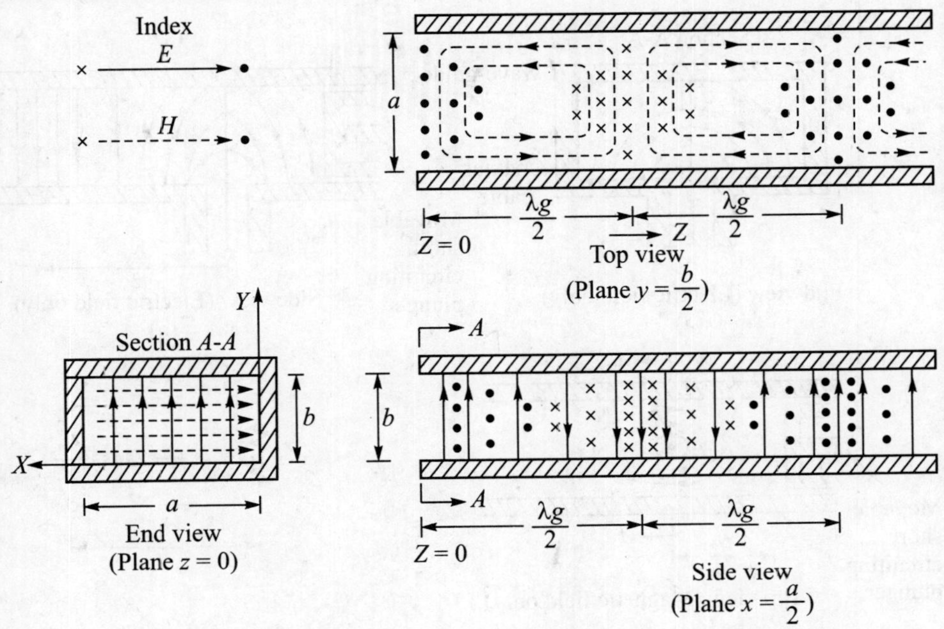

Fig. 14.5. Dominant TE_{10} mode in rectangular wave-guide

The reason why a single mode of wave propagation is desirable in a wave-guide is that, it eliminates the losses and field distortions that would result, if another mode gets reflected from a discontinuity in the guide or from the other end of the guide. Even if the walls of the guide are not perfect conductors, the power loss is the least for the TE_{10} mode.

The fact that the electric field is everywhere polarized in the y-direction is an advantage, because it simplifies the way in which the TE_{10} mode may be launched inside the wave-guide from a co-axial cable, as shown in Fig. 14.6. The function of the moveable plunger in Fig. 14.6 (b) and (c) is to vary the position of the short circuiting plane, to make its distance from the central conductor of the co-axial cable equal to $\frac{\lambda_g}{4}$. By the impedance-inversion property of a quarter-wave line shown in Fig. 14.6 (d), input impedance of a shorted $\frac{\lambda}{4}$ line becomes infinite.

The surface currents that flow along the four *inner* faces of the wave-guide are shown in Fig. 14.7. These current densities are equal to the H_x and H_z components of the magnetic field intensity just outside the conducting surface, as indicated by Eqn. (8.56). Thus the current density J_y along the y-axis equals the magnetic field intensity $[-H_z]$ ampere/metre. On the top plane, the current density $-J_z$ equals $[+H_x]$ ampere/metre.

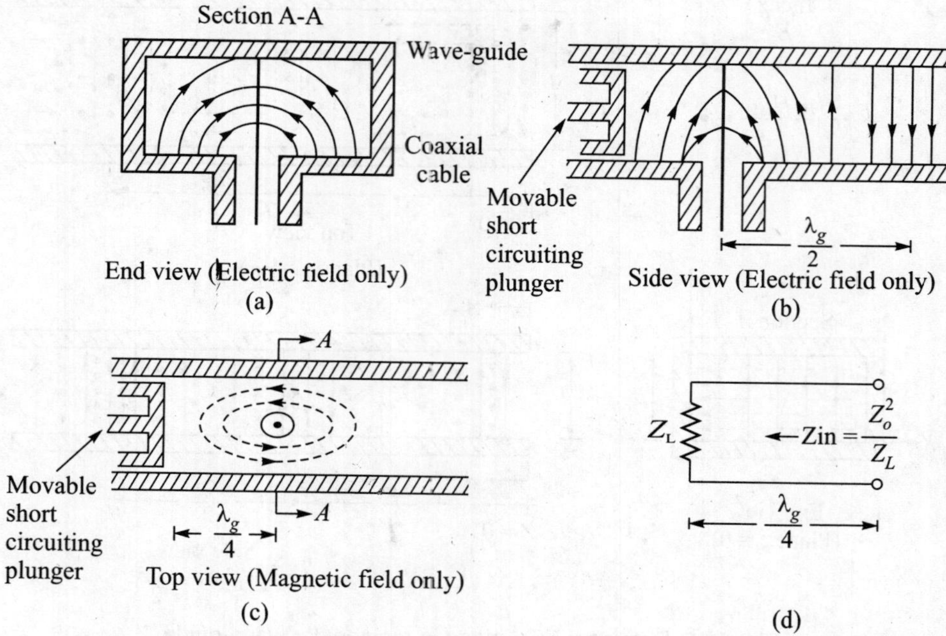

Fig. 14.6. Exciting the TE_{10} mode in a wave-guide from a co-axial cable

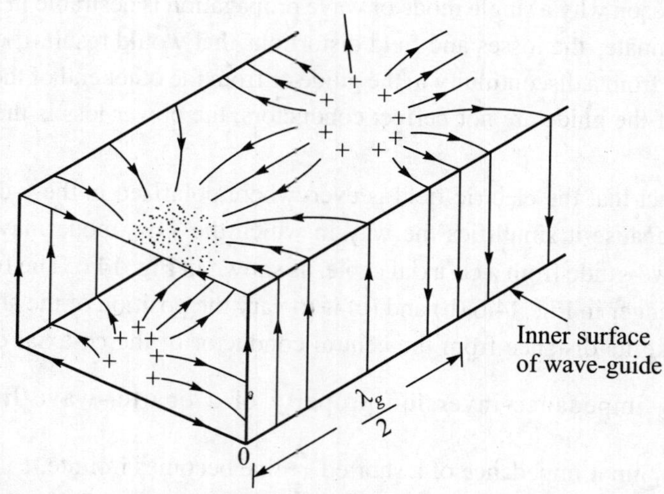

Fig. 14.7. Surface currents along inner walls of wave-guide for TE_{10} mode

14.11 LOWEST FREQUENCY TRANSVERSE MAGNETIC TM_{11} MODE

Equation (14.17) shows that if either $m = 0$ or $n = 0$, then all field components disappear (become zero) and therefore, there cannot be any waves in the modes TM_{oo}, TM_{on} or TM_{mo}. The lowest frequency transverse magnetic mode corresponds to TM_{11} mode for which Eqn. (14.17) become (assuming $\alpha = 0$ and $\gamma = j\beta$) as follows:

$$\left.\begin{aligned}
&H_z = 0, E_z = C_2 && \sin\left(\frac{\pi x}{a}\right) \sin\left(\frac{\pi y}{b}\right) e^{j(\omega t - \beta z)} \\
&E_x = \frac{-j\beta C_2}{\gamma^2 + \omega^2 \mu \varepsilon}\left(\frac{\pi}{a}\right) && \cos\left(\frac{\pi x}{a}\right) \sin\left(\frac{\pi y}{b}\right) e^{j(\omega t - \beta z)} \\
&E_y = \frac{-j\beta C_2}{\gamma^2 + \omega^2 \mu \varepsilon}\left(\frac{\pi}{b}\right) && \sin\left(\frac{\pi x}{a}\right) \cos\left(\frac{\pi y}{b}\right) e^{j(\omega t - \beta z)} \\
&H_x = \frac{j\omega \varepsilon C_2}{\gamma^2 + \omega^2 \mu \varepsilon}\left(\frac{\pi}{b}\right) && \sin\left(\frac{\pi x}{a}\right) \cos\left(\frac{\pi y}{b}\right) e^{j(\omega t - \beta z)} \\
&H_y = \frac{j\omega \varepsilon C_2}{\gamma^2 + \omega^2 \mu \varepsilon}\left(\frac{\pi}{a}\right) && \cos\left(\frac{\pi x}{a}\right) \sin\left(\frac{\pi y}{b}\right) e^{j(\omega t - \beta z)}
\end{aligned}\right\} \qquad ...(14.22)$$

where $\gamma^2 + \omega^2 \mu \varepsilon = \left(\frac{\pi}{a}\right)^2 + \left(\frac{\pi}{b}\right)^2, \lambda_c = \frac{2ab}{\sqrt{a^2 + b^2}}$

The electric and the magnetic fields for the TM_{11} mode are shown in Fig. 14.8. It should be carefully noted that the side view shown in this case is at a plane $x = \frac{a}{4}$, and not $\frac{a}{2}$ as in Fig. 14.6.

14.12 CIRCULAR WAVE-GUIDES

Hollow conducting tubes of circular cross-section can also be used for transmitting electromagnetic waves from one point to another. However, their mathematical analysis is not undertaken here, because it leads to solutions involving Bessel's functions. Unlike the case of rectangular wave-guides, the circular wave-guide do not have a unique orientation, because it is perfectly symmetrical all round the axis. As a result, the electric field vector may undergo a change of direction as the wave travels down the wave-guide, causing a change in the plane of polarization of the wave. This is a drawback. The circular wave-guides are therefore only used where their circular symmetry is required to overcome some problem.

As with rectangular wave-guides, we have TE waves with $E_z = 0$ and TM waves with $H_z = 0$. But instead of the suffixes *m* and *n*, we use letters related to the cylindrical system of co-ordinates (Fig. 14.9) as *p* and *r* (*p* for periphery or phi-direction, and *r* for radial or *R*-direction). Thus in TE_{pr} and TM_{pr} modes we confine our attention to the electric and the magnetic field component respectively, in the plane z = constant, and then *p* refers to the number of *full sinusoidal* variations of this field along the periphery of a circle in the phi-direction, as shown in Fig. 14.9 and *r* represents the number of *half-sinusoidal* variations of this field measured from the centre of the circle to the inner surface of the conducting cylinder along the radial or *R*-direction.

The three modes of wave propagation in cylindrical circular wave-guides that are important are TE_{11}, TM_{01} and TE_{01} (Fig. 14.10).

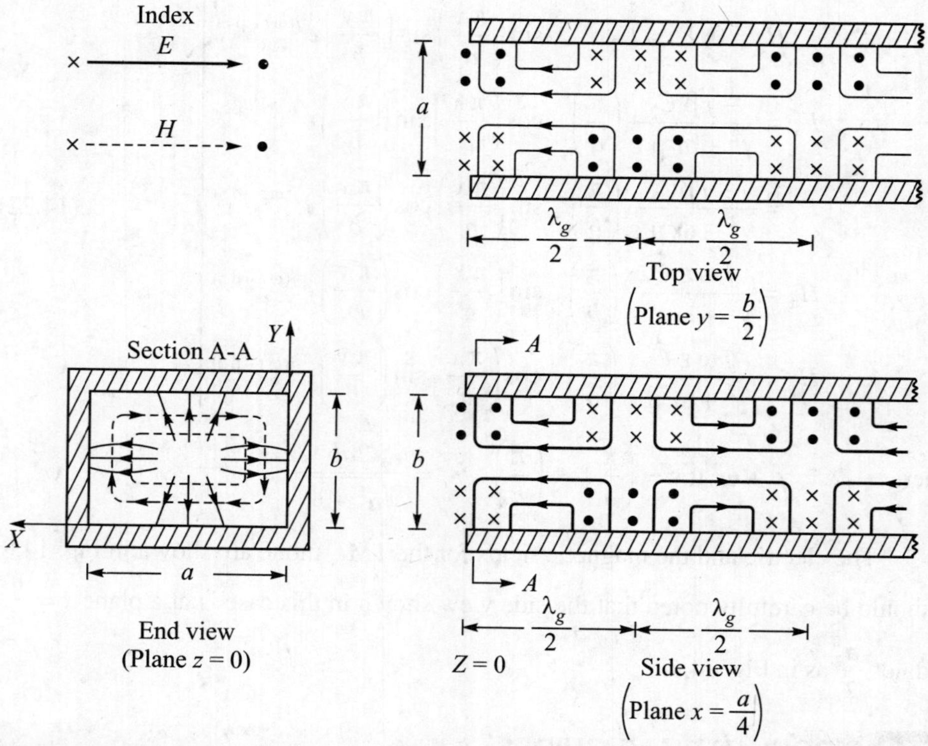

Fig. 14.8. Lowest frequency transverse magnetic TM_{11} mode in rectangular wave-guide

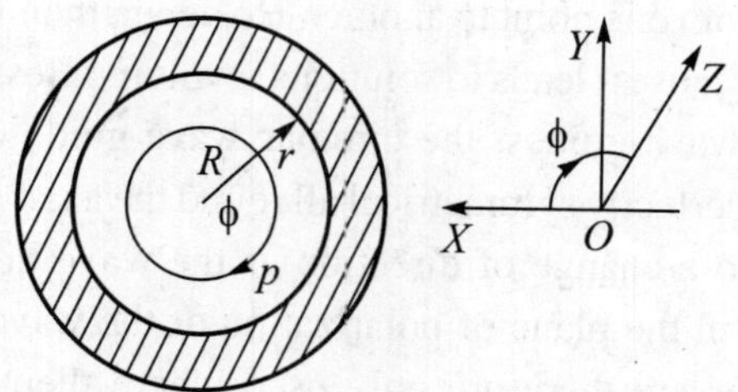

Fig. 14.9. Circular wave-guide mode notation

The TE_{11} mode has the lowest cut-off frequency and therefore, it is the dominant mode in the circular wave-guide. However, the frequency difference between the cut-off frequencies of the TE_{11} and the next higher TM_{01} mode is very small, their ratio being only (1.71/1.31) = 1.3, as indicated in Fig. 14.10. Because of the difficulty of eliminating this higher mode, circular wave-guides are not generally used, unless there is a very important consideration dictating its use. The circular symmetry is a further disadvantage, as there is no inherent structural factor prohibiting the rotation of the plane of polarization of the wave as it propagates.

The TM_{01} and TE_{01} fields have symmetrically distributed fields in the ϕ-direction and the electric field in one has the same configuration as the magnetic field in the other. The radial electric field in the TM_{01} mode permits a circular wave-guide to be used as a rotating joint between two rectangular wave-guides, having TE_{10} mode

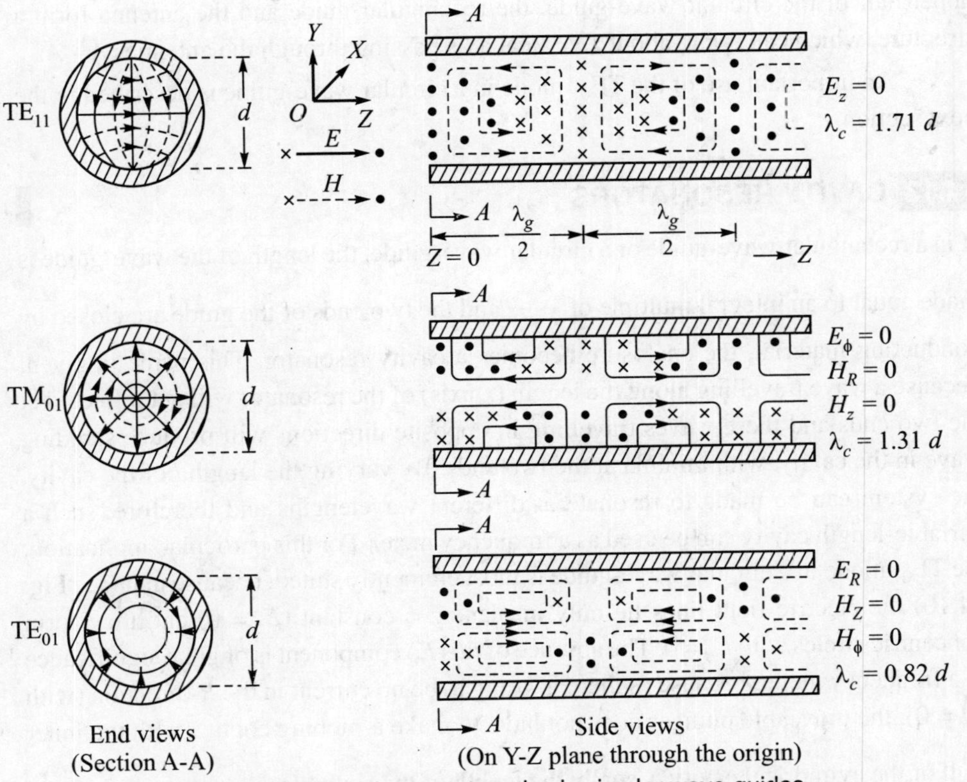

TE$_{11}$ — $E_z = 0$, $\lambda_c = 1.71\,d$

TM$_{01}$ — $E_\phi = 0$, $H_R = 0$, $H_z = 0$, $\lambda_c = 1.31\,d$

TE$_{01}$ — $E_R = 0$, $H_Z = 0$, $H_\phi = 0$, $\lambda_c = 0.82\,d$

End views (Section A-A) — Side views (On Y-Z plane through the origin)

Fig. 14.10. Three important modes in circular wave-guides

of propagation. This is a requirement in a radar unit, where the power from the magnetron received along a *fixed* rectangular wave-guide has to be fed to a *rotating* antenna system (Fig. 14.11). In the rectangular wave-guide (RW) the exciting antenna A is parallel to the narrow side and at the centre of the broad side, while for the circular wave-guide, it is along the axis. Maximum current is therefore produced in the antenna in the rectangular wave-guide by its TE$_{10}$ mode, while it launches a TM$_{10}$ mode in the circular wave-guide (CW) with electric field lines radiating from A towards the inner wall of the circular wave-guide. The magnetic field lines surround the conduction current in the antenna. The reverse phenomenon takes place at the other end for transfer of wave from circular to rectangular wave-guide. The black section of Fig. 14.11 remains fixed in space, but the lightly hatched section consisting of the

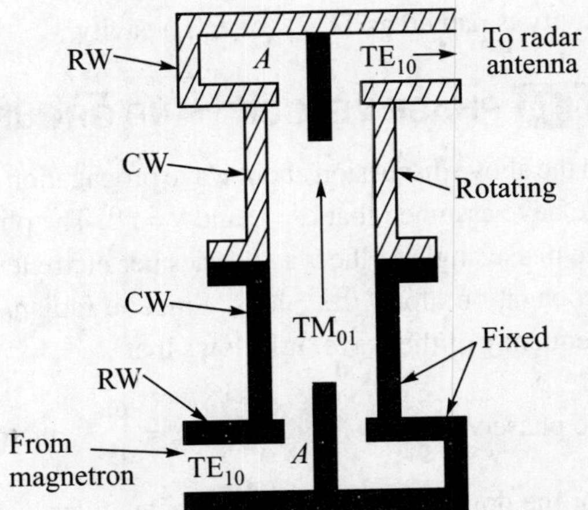

Fig. 14.11. Rotating joint using TM$_{01}$ mode in circular wave-guide

upper half of the circular wave-guide, the rectangular guide and the antenna form a structure, which can rotate about a vertical axis passing through the antennas $A - A$.

An important use of the TE_{01} mode in a circular wave-gui le is described in the next section.

14.13 CAVITY RESONATORS

If in a rectangular wave-guide or a circular wave-guide, the length of the wave-guide is made equal to an integral multiple of $\frac{\lambda_g}{2}$, and the two ends of the guide are closed by conducting material, the enclosure becomes a cavity resonator. This name is given, because a wave travelling along the length (z-axis) of the resonator will get reflected at the two ends and these waves travelling in opposite directions will produce standing wave in the cavity, with minima at the two ends. By varying the length l of the cavity, the system can be made to resonate at different wavelengths and therefore, such a variable-length cavity can be used as a frequency meter. For this particular application, the TE_{01} mode in a circular wave-guide is most eminently suited, because in it (see Fig. 14.10) the electric field lines lie only in plane z = constant ($E_z = 0$) and they form concentric circles with $E_R = 0$. The absence of this E_R component is of great significance in a frequency meter, because as there is going to be no current in the R-direction (with $E_R = 0$), the moveable plunger does not have to make a rubbing contact with the inner wall of the cylindrical cavity. Usually the length of the cavity is made equal to $l = \frac{\lambda_g}{2}$ over the entire frequency range, for which the frequency meter is designed. Hence this cavity is named as TE_{011} (or H_{011}) cavity.

14.14 PHASE VELOCITY AND GROUP VELOCITY

In the above discussion about wave propagation in rectangular and circular wave-guides, we have assumed that $\alpha = 0$ and $\gamma = j\beta$. The phase constant β defines the rate at which the phase angle of the wave varies per metre length, measured in the direction of wave propagation along the z-axis. Since ω radian/second is equal to $2\pi f$, where f is the frequency of the wave, it follows that

the phase velocity $$v_p = \frac{\omega}{\beta} \quad \text{metre/second} \qquad \text{...(14.23)}$$

For the dominant TE_{10} mode in rectangular wave-guide, we have $\lambda_c = 2a$ and for free space,

$$v_0 = \frac{1}{\sqrt{\mu_0 \varepsilon_0}} = f\lambda_0$$

so that $$\gamma = j\beta$$

$$= \sqrt{\left(\frac{\pi}{a}\right)^2 - \omega^2\mu_0\varepsilon_0} = j\sqrt{\left(\frac{\omega}{v_0}\right)^2 - \left(\frac{\pi}{a}\right)^2}$$

Therefore $$\beta = \sqrt{\left(\frac{2\pi f}{f\lambda_0}\right)^2 - \left(\frac{2\pi}{2a}\right)^2} = \frac{2\pi}{\lambda_0}\sqrt{1 - \left(\frac{\lambda_0}{\lambda_c}\right)^2}$$

However, in the guide a length λ_g in the *z*-direction indicates a phase change of 2π radians, so that $\beta = (2\pi)/\lambda_g$. Putting this value in the above equation and replacing $2a$ by λ_c, we get

$$\frac{2\pi}{\lambda_g} = \frac{2\pi}{\lambda_0}\sqrt{1 - \left(\frac{\lambda_0}{\lambda_c}\right)^2}$$

Squaring and transposing terms

$$\frac{1}{\lambda_0^2} = \frac{1}{\lambda_g^2} + \frac{1}{\lambda_c^2} \quad \text{...(14.24)}$$

It is seen that $\lambda_g > \lambda_0$.

Hence phase velocity is given by

$$v_p = \frac{\omega}{\beta} = \frac{2\pi f}{2\pi/\lambda_g} \times \frac{\lambda_0}{\lambda_0} = v_0\,\frac{\lambda_g}{\lambda_0}$$

$$= \frac{v_0}{\sqrt{1 - \left(\frac{\lambda_o}{\lambda_c}\right)^2}} \quad \text{...(14.25)}$$

This shows that phase velocity v_p in the wave-guide is greater than the free-space velocity v_0 of wave propagation.

An objection may be raised here that this last sentence violates the basic principle of the Special Theory of Relativity enunciated by Einstein in 1905, on the basis of which Eqn. (9.3) has been derived. The principle states that no measured velocity in the universe can exceed the velocity of propagation of light (or electro-magnetic) waves in a vacuum. The key word is "measured". To measure the velocity of propagation of a wave, we must measure the time that a particular cycle of a wave takes to cover a known distance. But a difficulty arises here. All cycles of a sinusoidally varying electro-magnetic wave are exactly alike, and therefore we cannot know when the particular cycle being used for measurement has reached the point a known distance away, unless we put some distinguishing mark on that particular cycle of the wave. This process of putting a distinguishing mark on a wave is called, modulation of a wave by a signal. This modification implies that we are no longer concerned with the particular cycle on which the distinguishing mark was put originally, because we now measure the time this mark has taken to cover the known distance. In fact we cannot be sure that the mark has not moved on to some other cycle of the wave, instead of remaining on the original particular cycle. In short we can only measure the velocity of propagation of the mark. Though the phase velocity v_p in the wave-guide is greater than the free-space velocity v_0, the group velocity v_g with which the mark or signal is propagated along the wave-guide (or energy is transmitted) is less than v_0 as demonstrated below.

Going back to the physical picture of wave propagation between two parallel planes discussed in Sec. 14.3, we combine Eqn. (14.1) with Eqn. (14.25) and get

$$v_p = \frac{v_0}{\lambda_0/\lambda_g} = \frac{v_0}{\cos\theta} \quad ...(14.26)$$

where $\pm\,\theta$ is the angle between the z-axis and the direction in which the two plane electromagnetic waves are travelling with velocity v_0. The wave components travel with velocity $v_0 \sin\theta$ at right angles to the z-axis in opposite directions, and being reflected by conducting planes at $x = a/2$ and $x = -a/2$, produce a *standing* wave which does not involve any propagation of energy. However, the wave components travelling with velocity $v_0 \cos\theta$ along the z-axis give a travelling wave which results in transmission of energy with this velocity along the z-axis. This means that the group velocity

$$v_g = v_0 \cos\theta \quad ...(14.27)$$

Combining Eqns. (14.26) and (14.27)

$$v_p\, v_g = v_0^2 \quad ...(14.28)$$

Since $\cos\theta$ lies between 1 and 0, Eqn. (14.27) shows that v_g is less than v_0 as required by the theory of relativity.

The reason why v_g is called the group velocity is now explained. First by transforming Eqn. (14.25), we get

$$v_p = \frac{v_0}{\sqrt{1-\left(\dfrac{\lambda_0}{\lambda_c}\right)^2}} = \frac{v_0}{\sqrt{1-\left(\dfrac{f_c}{f}\right)^2}} \quad ...(14.29)$$

This shows that the phase velocity in the wave-guide is a function of the frequency f. An *increase* of frequency f causes the phase velocity v_p to *decrease*.

A wave travelling along the z-axis is represented as $E_0\, e^{j(\omega t - \beta z)}$ or $E_0 \cos(\omega t - \beta z)$, the velocity of phase propagation being defined as $v_p = \dfrac{\omega}{\beta}$. If this carrier wave is amplitude modulated by a signal of (modulating) frequency, $\Delta\omega$, then additional waves of upper and lower side frequencies are created of amplitude, say E_1 and frequencies $\omega + \Delta\omega$ and $\omega - \Delta\omega$, respectively. Denoting the corresponding phase constants as $\beta + \Delta\beta$ and $\beta - \Delta\beta$, these waves can be represented as

Upper side-frequency wave

$$= E_1 \cos\{(\omega + \Delta\omega)\, t - (\beta + \Delta\beta)\, z\}$$

Lower side-frequency wave

$$= E_1 \cos\{(\omega - \Delta\omega)\, t - (\beta - \Delta\beta)\, z\}$$

These waves carry the signal or information, and together they combine to produce a recognizable mark. The addition of the waves gives a resultant wave as

$$\text{Resultant wave} = E_1 [\cos \{(\omega + \Delta\omega)\, t - (\beta + \Delta\beta)\, z\} + \cos \{(\omega - \Delta\omega)\, t - (\beta - \Delta\beta)\, z\}]$$

$$= E_1 \cos \{(\omega t - \beta z) + (\Delta\omega \,.\, t - \Delta\beta \,.\, z\} + \cos \{(\omega t - \beta z) - (\Delta \omega \cdot t - \Delta\beta \cdot z)\}]$$

$$= [2E_1 \cos (\Delta\omega \cdot t - \Delta\beta \cdot z)] \cos (\omega t - \beta z) \qquad ...(14.30)$$

Equation (14.30) shows that the resultant is a wave of frequency ω but its amplitude is not a constant, but a function of both t and z, as shown by the term is square brackets. This represents the envelope of the modulated wave which contains the signal.

In Fig. 14.12 the two waves of frequencies

$$(\omega + \Delta\omega) \quad \text{and} \quad (\omega - \Delta\omega)$$

are shown. The wave of frequency $(\omega + \Delta\omega)$ is shown by a continuous line and its crests are numbered (in the order in which they reach any point on z-axis) as 1+, 2+, 3+, ... The wave of frequency $(\omega - \Delta\omega)$ is shown by dotted lines and its crests are numbered as 1–, 2–, 3–, ... At time $t = t_1$, the crests (3+ and 2–) of the two waves are coincident at C_1 where $z = z_1$. However by Eqn. (14.29), the phase velocity of propagation $\beta + \Delta\beta$ for frequency $(\omega + \Delta\omega)$ is less than the phase velocity $\beta - \Delta\beta$ for frequency $(\omega - \Delta\omega)$. Hence the waveform for frequency $(\omega + \Delta\omega)$ slides backwards with respect to the waveform for frequency $(\omega - \Delta\omega)$. As a result, at a time $t = t_1 + \Delta t$ the crests (2+ and 2–) of the two waveforms coincides at C_2 where $z = z_1 + \Delta z$.

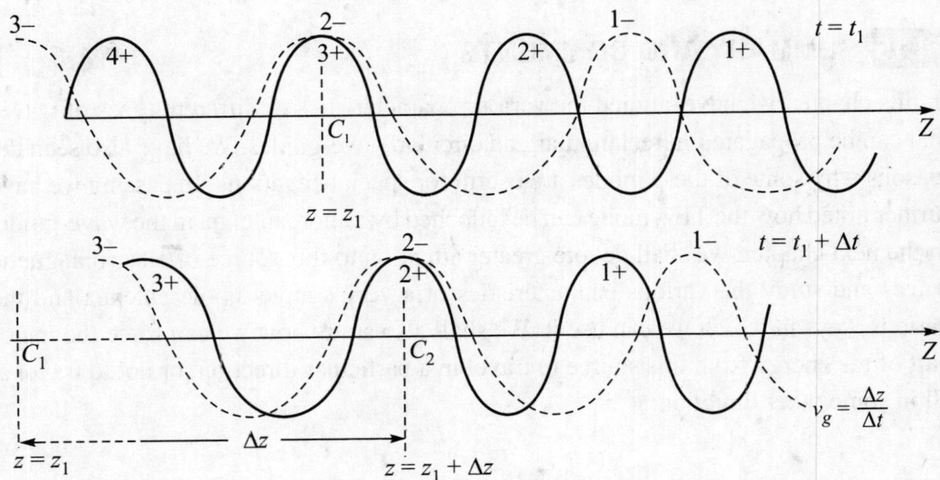

Fig. 14.12. Illustrating group velocity at which a signal travels.

If we regard the coincidence of the crests of the two waves as being the signal or mark, that we wish to utilize for velocity measurement, then it is obvious that the velocity with which this signal or mark is being propagated is $v_g = \dfrac{\Delta z}{\Delta t}$. We may also regard this coincidence as a bunching or grouping together of the two waves and hence call V_g as the group velocity, or the velocity at which this group of two crests is propagated. This group is really the point of maxima of the envelope of the wave of Eqn. (14.30). Just as the term cos $(\omega t - \beta z)$ gave us the phase velocity in Eqn. (14.23) as $v_p = \dfrac{\omega}{\beta}$, similarly from the term cos $(\Delta\omega \cdot t - \Delta\beta \cdot z)$ in the square brackets in Eqn. (14.30), we can write

$$\text{Group velocity } v_g = \frac{\Delta\omega}{\Delta\beta} = \frac{d\omega}{d\beta} \qquad ...(14.31)$$

By squaring Eqn. (14.29), we get

$$v_p^2 = \frac{\omega^2}{\beta^2} = \frac{v_0^2}{1 - \dfrac{f_c^2}{f^2}} = \frac{v_0^2}{1 - \dfrac{\omega_c^2}{\omega^2}}$$

or $$\omega^2 \left(1 - \frac{\omega_c^2}{\omega^2}\right) = \beta^2 v_0^2$$

or $$\omega^2 - \omega_c^2 = \beta^2 v_0^2$$

Hence differentiating this and noting that ω_c and v_0 are constants,

$$2\omega\, d\omega = 2\beta\, d\beta\, v_0^2$$

or $$\frac{\omega}{\beta} \cdot \frac{d\omega}{d\beta} = v_p\, v_g = v_0^2$$

as we had derived in another manner in Eqn. (14.26).

14.15 SUMMARY AND COMMENTS

In this chapter, we have studied the various characteristics of different types of waves that can be propagated in rectangular and circular wave-guides. We have also seen the reasons why some of these modes are useful for special situations. In passing we have further noted how the TE_{10} mode can be launched by a short antenna in the wave-guide. In the next chapter, we shall devote greater attention to this source of electromagnetic waves and study the various characteristics of a very simple dipole antenna and the various ways in which we can use it. We shall also study how we can force the major part of the energy from this source to travel in a particular direction, or not to travel at all in some other direction.

EXERCISE 14

1. (*a*) The plane $y = 0$ forms the boundary between a perfect dielectric ($\sigma = 0$) and a perfect conductor ($\sigma = \infty$). A plane electromagnetic wave travelling in the *z*-direction in the dielectric has its electric field defined by $\overline{E} = E_y\, e^{j(\omega t - \beta z)}\, \overline{u}_y$. Find the electric charge density q_s coulomb/(metre)^2 on the surface of the conductor, and also the surface current density as functions of z and t. Verify that at any given instant, the maximum charge density and the maximum current density occur for the same value of z.

(*b*) If the conductor is not perfect but σ is very large, we can assume that $\overline{H}$ remains the same as in (*a*) above. Then show that a tangential component of E will exist at the conductor surface, such that

$$E_z = -H_x \sqrt{\frac{(\omega \mu_0)}{\sigma}} \angle 45° \, e^{j(\omega t - \beta z)}$$

2. Find the direction of the Poynting vector $\overline{E} \times \overline{H}$ at different points in Figs. 14.1, 14.2, 14.5, 14.8 and 14.10.

3. Show that Maxwell's equation $\overline{\nabla} \cdot \overline{B} = 0$ implies that the magnetic field lines H must form *closed* loops.

4. Assume that a wave is travelling in the z-direction in a perfect dielectric ($\sigma = 0$) of infinite extent, situated between two perfect conducting planes ($\sigma = \infty$) at $x = 0$ and $x = a$. Also assume that the electric and magnetic fields do not vary in the y-direction ($\frac{\partial E}{\partial y}$ and $\frac{\partial H}{\partial y}$ are zero). Show that the electric and magnetic fields for the transverse electric and the transverse magnetic modes of propagation obtained as solutions of the wave equations in E_y or H_y will be given as follows:

TE Mode:

$$E_z = 0, \ \gamma^2 + \omega^2 \mu_0 \varepsilon_0 = \left(\frac{m\pi}{a}\right)^2$$

$$H_z = \frac{-m\pi C_1}{j\omega\mu_0 a} \cos\left(\frac{m\pi x}{a}\right) e^{(j\omega t - \gamma z)}$$

$$E_x = 0, \ E_y = C_1 \sin\left(\frac{m\pi x}{a}\right) e^{(j\omega t - \gamma z)}$$

$$H_y = 0, \ H_x = \frac{-\gamma C_1}{j\omega\mu_0} \sin\left(\frac{m\pi x}{a}\right) e^{(j\omega t - \gamma z)}$$

TM Mode:

$$H_z = 0, \ \gamma^2 + \omega^2 \mu_0 \varepsilon_0 = \left(\frac{m\pi}{a}\right)^2$$

$$E_z = \frac{-m\pi C_2}{j\omega\varepsilon_0 a} \sin\left(\frac{m\pi x}{a}\right) e^{(j\omega t - \gamma z)}$$

$$E_y = 0, \ E_x = \frac{\beta C_2}{\omega\varepsilon_0} \cos\left(\frac{m\pi x}{a}\right) e^{(j\omega t - \gamma z)}$$

$$H_x = 0, \ H_y = C_2 \cos\left(\frac{m\pi x}{a}\right) e^{(j\omega t - \gamma z)}$$

5. (*a*) From the equations of Ex. 14, Prob. 4, show that for a TE wave, $m = 0$ does not give a wave that is propagated, so that $m = 1$ gives the dominant TE wave.
(*b*) From the equations of Ex. 14, Prob. 4, show that for TM wave, $m = 0$ permits a TM wave to be propagated with components E_x and H_y. (This TEM wave is often referred to as the Principal Wave).

6. Given that the attenuation constant for the TM mode between two parallel conducting planes made of good conductor (σ_2 very large but not infinite) is given by

$$\alpha = \frac{2}{\eta a}\sqrt{\frac{(\pi f \mu_2)/\sigma_2}{1-\left(\frac{f_c}{f}\right)^2}} \text{ neper/metre,}$$

show that the attenuation will be a maximum at a frequency $f = \sqrt{3}\, f_c$.

7. For the RCSC Type WG-16 or EIA Type WR-90 wave-guide, determine the cut-off wavelength when the dielectric inside the wave-guide is
(*i*) air with $\varepsilon_r = 1$ and
(*ii*) water with $\varepsilon_r = 81$.

CHAPTER

15 ANTENNA AND ANTENNA ARRAYS

15.1 RETARDED MAGNETIC VECTOR POTENTIAL

In Chap. 3, we had found that it is easier to derive the electric field intensity $\overline{E}$ in an electromagnetic field by first finding the electric potential at that point. Similarly in Chap. 8, we had found the vector magnetic potential first, and from it, obtained the magnetic field intensity $\overline{H}$ in a magnetostatic field. In an attempt to relate the electromagnetic wave to the *time-varying current* in a conductor, we extend the method to this case, and first find the vector magnetic potential $\overline{A}$ due to a time-varying current element $(I \cos \omega t)\ \overline{dl}$ and thus derive $\overline{H}$ at a distant point P as $\overline{H} = \text{curl } \overline{A}$. Then $\overline{E}$ at the point P is obtained from Maxwell's curl equation for $\overline{H}$ by noting that $\overline{J}$, the conduction current density is zero in free space. Then

$$\overline{\nabla} \times \overline{H} = \varepsilon_0 \frac{\partial \overline{E}}{\partial t}$$

and hence

$$\overline{E} = \frac{1}{\varepsilon_0} \int (\overline{\nabla} \times \overline{H})\ dt \qquad \text{...(15.1)}$$

The procedure outlined above is initiated by assuming a current element in a conductor of length $\overline{dl}$, of small but finite area of cross-section a, situated at the origin O of spherical co-ordinates with dl along the z-axis [Fig. 15.1 (*a*)]. An infinitesimally thin current element $i\ \overline{dl}$ passing through O gives a vector magnetic potential $\overline{dA}$ as in Eqn. (8.38):

$$\overline{dA} = \frac{i\ \overline{dl}}{4 \pi r} \qquad \text{...(8.38)}$$

where i is a constant direct current and $\overline{dA}$ is parallel to $\overline{dl}$.

Next we let the current i vary sinusoidally with time, so that i can be replaced by $(i_{max} \cos \omega t)$. But in Chap 11, it was found that time-varying $\overline{E}$ and $\overline{H}$ give rise to an electromagnetic wave that travels with a finite velocity $v_0 = \dfrac{1}{\sqrt{\mu_0 \varepsilon_0}}$. This statement can be re-worded as follows: The information that the "current at O is varying" is conveyed to a point P at a distance r from O, but it does not reach there instantaneously. It reaches there at a slightly later time, because of the *finite* velocity v_0, and this time delay = $\frac{r}{v_0}$ second. Hence for the time-varying current (i_{max} cos ω t) in element $\overline{dl}$, Eqn. (8.38) above is modified by replacing t by t', where

$$t' = t - \frac{r}{v_0} \quad ...(15.2)$$

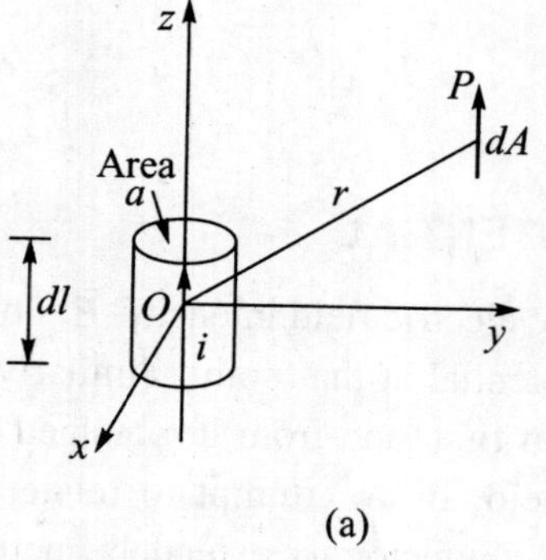

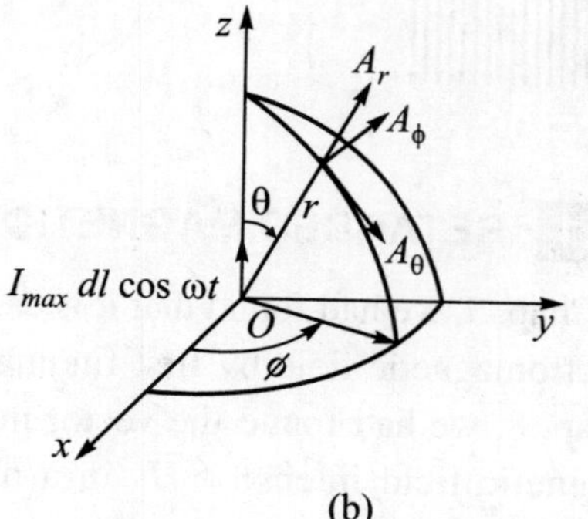

Fig. 15.1. (*a*) Retarded vector magnetic potential (*b*) Resolved parts of $\overline{A}$

to get

$$\overline{dA} = \frac{(i_{max} \cos \omega t')\, \overline{dl}}{4\pi r}$$

$$= \frac{\left[i_{max} \cos \omega \left(t - \frac{r}{v_0}\right)\right] \overline{dl}}{4\pi r} \quad ...(15.2a)$$

Let the total current in the finite-sized conductor be denoted as I_{max} cos ω t, where I_{max} is obtained by integrating i over the cross-section a of the conductor. Then total vector magnetic potential due to the finite sized conductor is obtained, integrating Eqn. (15.2*a*) over the whole cross-section of the conductor:

$$\overline{A} = \frac{(\cos \omega t')\overline{dl}}{4\pi} \int \frac{i_{max}}{r}\, da$$

For this integration, if $r \gg$ diameter of current carrying conductor, then r can be regarded as being constant and thus

$$\overline{A} = \frac{(\cos \omega t')\overline{dl}}{4\pi r} \int \frac{i_{max}}{r}\, da$$

$$= \frac{I_{max}\, \overline{dl} \cos \omega \left(t - \frac{r}{v_0}\right)}{4\pi r} \quad ...(15.3)$$

This vector magnetic potential is in the same direction as the current element (I_{max} cos ωt) $\overline{dl}$. However, the sinusoidal variation of $\overline{A}$ lags behind (in time phases) with respect to the variation of current I_{max} cos ω t. Therefore this $\overline{A}$ is referred to as retarded vector magnetic potential at point P, the retardation or phase-lag being caused by the finite velocity v_0 of wave propagation.

This $\overline{A}$ which is in the same direction as $\overline{dl}$ (that is, parallel to the z-axis) can be resolved into the three components in spherical co-ordinates [Fig. 5.1 (b)] as

$$A_r = A \cos\theta$$
$$A_\theta = -A \sin\theta$$
$$A_\phi = 0 \qquad ...(15.4)$$

15.2 MAGNETIC AND ELECTRIC FIELDS OF OSCILLATING HERTZIAN DIPOLE

The short element $\overline{dl}$ of previous section carrying time-varying current can be thought of as consisting of equal and opposite charges along the z-axis situated symmetrically with respect to the origin O. The magnitude of the charge at every point along $\overline{dl}$ varies sinusoidally with time. Such a system is therefore described as an oscillating Hertzian dipole (in honour of Heinrich Rudolf Hertz (1857–94), who was the first person to use it in his experiments) or a doublet antenna. To find the electric and magnetic field intensities in the region surrounding the dipole antenna, we make use of the retarded vector magnetic potential given by Eqns. (15.3) and (15.4).

We first write

$$\overline{H} = H_r \overline{u}_r + H_\theta \overline{u}_\theta + H_\phi \overline{u}_\phi$$
$$= \overline{\nabla} \times \overline{A}$$

where using Eqn. (1.111) for the curl in spherical co-ordinates and the components of $\overline{A}$ given in Eqn. (15.4), we get

$$H_r = \frac{1}{r\sin\theta}\left[\frac{\partial}{\partial\theta}(\sin\theta\, A_\phi) - \frac{\partial A_\theta}{\partial\phi}\right] = 0 \qquad ...(15.4a)$$

$$H_\theta = \frac{1}{r}\left[\frac{1}{\sin\theta}\frac{\partial A_r}{\partial\phi} - \frac{\partial}{\partial r}(rA_\phi)\right] = 0 \qquad ...(15.4b)$$

$$H_\phi = \frac{1}{r}\left[\frac{\partial}{\partial r}(r\, A_\theta) - \frac{\partial A_r}{\partial\theta}\right]$$

$$= \frac{I_{max}\, dl}{4\pi}\left[\frac{1}{r}\frac{\partial}{\partial r}\left\{-\sin\theta\cos\omega\left(t - \frac{r}{v_0}\right)\right\} - \frac{1}{r}\frac{\partial}{\partial\theta}\left\{\frac{\cos\theta}{r}\cos\omega\left(t - \frac{r}{v_0}\right)\right\}\right]$$

or $$H_\phi = \frac{I_{max}\, dl \sin\theta}{4\pi}\left[\frac{-\omega}{v_0 r}\sin\omega t' + \frac{1}{r}\frac{\cos\omega t'}{r}\right] \qquad ...(15.4c)$$

where $$t' = t - \frac{r}{v_0} \text{ as per Eqn. (15.2).}$$

It is interesting to find that H_r and H_θ components are zero, but the H_ϕ component contains two parts, the first term varying inversely with r and the second inversely with

r^2. Naturally the second term decreases more rapidly with distance from O and at great distances, it becomes negligibly small, so that for large values of r, only the first term remains. The first term is called the radiation or distant field, and the second term, the induction or near field. The reason for these names will become apparent later. At a certain critical distance r_c, the two terms will have equal *amplitudes* when

$$\frac{\omega}{v_0 r_c} = \frac{1}{r_c^2}$$

or

$$r_c = \frac{v_0}{\omega} = \frac{f\,\lambda_0}{2\pi f} = \frac{\lambda_0}{2\pi} \approx \frac{\lambda_0}{6} \qquad \text{...(15.5)}$$

This shows that for distances greater than $r_c = \dfrac{\lambda_0}{6}$, the amplitude of the radiation term is very much larger than that of the induction term, and hence for such distances, only the radiation term need be considered.

Next we proceed to find the E_r, E_θ and E_ϕ components of the electric field by using Eqn. (15.1) using the r, θ and ϕ components respectively, of the curl. Then with the values given in Eqns. (15.4 a, b, c)

$$E_r = \frac{1}{\varepsilon_0} \int (\overline{\nabla} \times \overline{\boldsymbol{H}})_r \; dt$$

$$= \frac{1}{\varepsilon_0} \int \left[\frac{1}{r \sin\theta} \frac{\partial}{\partial\theta} (\sin\theta \, H_\phi) - \frac{\partial H_\theta}{\partial\phi} \right] dt$$

$$= \frac{I_{\max}\, dl}{4\pi\,\varepsilon_0} \int \left[\frac{1}{r \sin\theta} \frac{\partial}{\partial\theta} \left\{ \sin^2\theta \left(\frac{-\omega}{v_0 r} \sin\omega t' \right) + \sin^2\theta \, \frac{\cos\omega t'}{r^2} \right\} \right] dt$$

$$= \frac{I_{\max}\, dl}{4\pi\,\varepsilon_0} \int \left[\frac{2\sin\theta\cos\theta}{r\sin\theta} \left(\frac{-\omega}{v_0 r} \sin\omega t' \right) + \frac{2\sin\theta\cos\theta}{r\sin\theta} \left(\frac{\cos\omega t'}{r^2} \right) \right] dt$$

$$= \frac{I_{\max}\, dl\, 2\cos\theta}{4\pi\,\varepsilon_0} \left[\left(\frac{-\omega}{v_0 r^2} \right) \left(\frac{-\cos\omega t'}{\omega} \right) + \frac{1}{r^3} \left(\frac{\sin\omega t'}{\omega} \right) \right]$$

or

$$E_r = \frac{I_{\max}\, dl\, 2\cos\theta}{4\pi\,\varepsilon_0} \left[\frac{\cos\omega t'}{v_0\, r^2} + \frac{\sin\omega t'}{\omega\, r^3} \right] \qquad \text{...(15.6}a\text{)}$$

Similarly

$$E_\theta = \frac{1}{\varepsilon_0} \int (\overline{\nabla} \times \overline{\boldsymbol{H}})_\theta \; dt$$

$$= \frac{1}{\varepsilon_0} \int \left[\frac{1}{r\sin\theta} \frac{\partial H_r}{\partial\phi} - \frac{1}{r} \frac{\partial}{\partial r} (r H_\phi) \right] dt$$

$$= \frac{I_{max}\, dl \sin\theta}{4\pi\,\varepsilon_0} \int \left[-\frac{1}{r}\frac{\partial}{\partial r}\left\{ \frac{-\omega}{v_0} \sin\omega\left(t - \frac{r}{v_0}\right) + \frac{1}{r}\cos\omega\left(t - \frac{r}{v_0}\right) \right\} \right] dt$$

$$= \frac{I_{max}\, dl \sin\theta}{4\pi\,\varepsilon_0} \int \left[-\frac{1}{r}\left\{ \frac{-\omega}{v_0}\cos\omega t' \left(\frac{-\omega}{v_0}\right) \right.\right.$$

$$\left.\left. + \frac{1}{r}(-\sin\omega t')\left(\frac{-\omega}{v_0}\right) + \left(\frac{-1}{r^2}\right)\cos\omega t' \right\} \right] dt$$

$$= \frac{I_{max}\, dl \sin\theta}{4\pi\,\varepsilon_0} \left[\frac{-\omega^2}{r v_0^2}\left(\frac{\sin\omega t'}{\omega}\right) - \frac{\omega^2}{r^2 v_0}\left(\frac{-\cos\omega t'}{\omega}\right) + \frac{1}{r^3}\left(\frac{\sin\omega t'}{\omega}\right) \right]$$

or $$E_\theta = \frac{I_{max}\, ui \sin\theta}{4\pi\,\varepsilon_0} \left[\frac{-\omega\sin\omega t'}{r v_0^2} + \frac{\cos\omega t'}{r^2 v_0} + \frac{\sin\omega t'}{r^3\omega} \right] \qquad \text{...(15.6b)}$$

and $$E_\phi = \frac{1}{\varepsilon_0}\int (\overline{\nabla} \times \overline{\boldsymbol{H}})_\phi \; dt$$

$$= \frac{1}{\varepsilon_0}\int \left[\frac{1}{r}\left\{ \frac{\partial}{\partial r}(rH_\theta) - \frac{\partial H_r}{\partial\theta} \right\} \right] dt = 0$$

The above expression for E_θ shows the same radiation field term varying as $\frac{1}{r}$ and the induction field term varying as $\frac{1}{r^2}$, and as before the amplitudes of these terms become equal when $r_c = \frac{\lambda_0}{2\pi}$. The expression for E_r however does not have term varying as $\frac{1}{r}$, thus giving a hint that E_r component does not contribute to the radiation field at all !! However, the induction field term varying as $\frac{1}{r^2}$ is present in E_r.

Moreover, both E_r and E_θ have an additional term varying as $\frac{1}{r^3}$, which reminds us of field components of the electrostatic fields of a dipole fields as found by Eqns. (3.23). Hence these are referred to as the electrostatic field terms. Because of the inverse cubic factor, these terms decrease very fast with distance, and are therefore negligible everywhere except very close to the origin.

15.3 RADIATION FIELD OF HERTZIAN DIPOLE ANTENNA

In this chapter we shall be concerned only with electromagnetic effects produced by Hertzian dipole antenna at a *great* distance from the antenna, when the antenna is carrying a current $I_{max} \cos \omega t$. Hence we confine our attention only to the radiation field components, which vary with $\frac{1}{r}$ in the previous section, and write the equations as follows, using the relations

$$v_o = \frac{1}{\sqrt{\mu_o \varepsilon_0}}, \quad \frac{\omega}{v_o} = \frac{2\pi}{\lambda_o}$$

and
$$\eta_o = \sqrt{\mu_o \varepsilon_0} = 120\,\pi$$

$$E_\theta = \frac{I_{max}\, dl \sin\theta}{4\pi\, \varepsilon_o v_o r} \frac{\omega}{v_o}\left[-\sin\omega\left(t - \frac{r}{v_o}\right)\right]$$

$$= \frac{I_{max}\, dl \sin\theta}{4\pi\, (\varepsilon_o/\sqrt{\mu_o\varepsilon_o})r} \frac{2\pi}{\lambda_o}(-\sin\omega\, t')$$

$$= \frac{I_{max}\, dl \sin\theta}{2\lambda_o r}\eta_o\, (-\sin\omega\, t')$$

or
$$E_\theta = \frac{60\pi\, I_{max}\, dl \sin\theta}{\lambda_o r}\left[-\sin\omega\left(t - \frac{r}{v_o}\right)\right] \quad \text{...(15.7a)}$$

$$H_\phi = \frac{E_\theta}{\eta_o} = \frac{I_{max}\, dl \sin\theta}{2\lambda_o r}\left[-\sin\omega\left(t - \frac{r}{v_o}\right)\right] \quad \text{...(15.7b)}$$

$$E_r = 0,\ E_\phi = 0,\ H_r = 0,\ H_\theta = 0. \quad \text{...(15.7c)}$$

15.4 POYNTING VECTOR AND RADIATION RESISTANCE OF SHORT DIPOLE ANTENNA

In spherical co-ordinates the Poynting vector has component P_r, P_θ, P_ϕ. Of these $P_\phi = E_r H_\theta - E_\theta H_r$ is always equal to zero since $H_r = 0$, $H_\theta = 0$ by Eqns. (15.4*a*) and (15.4*b*). This shows that there is no transfer of energy in the ϕ-direction anywhere, because that is the direction of the magnetic field everywhere. Similarly in P_θ and P_r, terms involving E_ϕ are equal to zero by Eqn. (15.7*c*).

The component $P_\theta = -E_r H_\phi$. Using Eqns. (15.6*a*) and (15.4*c*), the product of the terms is square brackets will give terms like $(\cos^2 \omega t' - \sin^2 \omega t')/v_0 r^4$ and $2 \sin \omega t'$ $\cos \omega t' \frac{1}{2}\left(\frac{-\omega}{v_0^2 r^3} + \frac{1}{\omega r^5}\right)$ which are proportional to $\cos 2\omega t'$ and $\sin 2\omega t'$, respectively.

This shows that the power moves back and forth in a cyclic manner in the θ-direction. But since the positive and negative parts of the curves for $\cos 2\omega t'$ and $\sin 2\omega t'$ have exactly equal areas, the *average* power over one complete cycle is zero. In other words, there is no net power flow in the θ-direction. This power is associated with the inductive and electrostatic field components of E_r.

Finally, we consider component $P_r = E_\theta H_\phi$. Using Eqns. (15.6*b*) and (15.4*c*), the product of the terms in square brackets will give terms involving cos $2\omega t'$ and sin $2\omega t'$ as above, corresponding to the inductive and electrostatic field components of E_θ. These terms give zero energy flow over a complete cycle in the *r*-direction.

However, the radiation terms of E_θ and H_ϕ give the terms

$$\left(\frac{I_{max}\, dl \sin\theta}{4\pi}\right)^2 \frac{\omega^2}{\varepsilon_0 v_0^3 r^2} \sin^2 \omega t' = \left(\frac{I_{max}\, dl \sin\theta}{4\pi}\right)^2 \frac{\omega^2}{\varepsilon_0 v_0^3 r^2}\left[\frac{1-\cos 2\omega t'}{2}\right]$$

Though the part cos $2\omega t'$ will contribute zero energy flow over a complete cycle, the constant term $\frac{1}{2}$ within the square brackets will give a *constant flow of energy all the time.* Hence there is a constant *outward* flow of energy in the *r*-direction from the centre of the Hertzian dipole and its magnitude is, using the relation $v_o \varepsilon_o = \frac{1}{\eta_o}$,

$$[P_r]_{average} = \left(\frac{I_{max}\, dl \sin\theta}{4\pi}\right)^2 \frac{\omega^2}{(\varepsilon_0 v_0) v_0^2 r^2} \times \frac{1}{2}$$

$$= \frac{\eta_0}{2}\left(\frac{\omega\, I_{max}\, dl \sin\theta}{4\pi v_0 r}\right)^2 \text{ W/m}^2$$

The most important thing to notice about this average power contributed by the radiation $\left(\frac{1}{r}\right)$ terms of E_θ and H_ϕ is that, the power is all the time travelling outwards, never to return to the antenna !! To get the total power thus radiated by the antenna in all directions, we must integrate this $[P_r]_{average}$ over the surface of a sphere of radius *r*. For this purpose we divide the surface of the sphere into infinitesimally narrow strips lying between two cones with semi-vertical angles θ and θ + *d*θ (Fig. 15.2). The radius of the strip is *NP* = *r* sin θ, so that the length of the strip is = 2π (*r* sin θ). The width of the strip is *r d*θ, and area = *da* = 2π (*r* sin θ) *r d*θ.

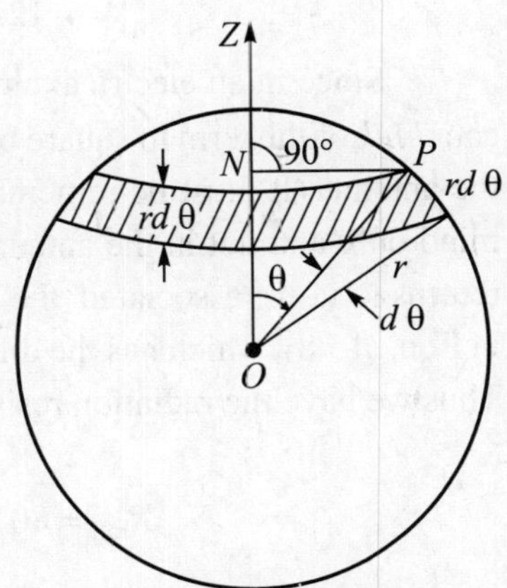

Fig. 15.2. To find the total radiated power

Thus, the total power radiated by the antenna is

$$P_{rad} = \int [\overline{P}_r]_{average} \cdot \overline{\boldsymbol{da}}$$

$$= \frac{\eta_0}{2}\left[\frac{\omega\, I_{max}\, dl}{4\pi v_0 r}\right]^2 \int_0^\pi \sin^2\theta\,(2\pi r \sin\theta\; r\, d\theta)$$

$$= \frac{\eta_0}{\pi}\left(\frac{\omega I_{max}\, dl}{4v_o}\right)^2 \int_0^{\pi} \sin^3\theta\, d\theta$$

Since $\sin 3\theta = 3\sin\theta - 4\sin^3\theta$,

$$\sin^3\theta = \frac{1}{4}(3\sin\theta - \sin 3\theta),$$

$$\int_0^{\pi} \sin^3\theta\, d\theta = \int_0^{\pi}\left(\frac{3}{4}\sin\theta - \frac{1}{4}\sin 3\theta\right) d\theta$$

$$= \left[\frac{3}{4}(-\cos\theta) - \frac{1}{4}\frac{(-\cos 3\theta)}{3}\right]_0^{\pi}$$

$$= \frac{3}{4}(1+1) + \frac{1}{12}(-1-1) = \frac{16}{12}$$

Therefore $$P_{rad} = \frac{\eta_0\omega^2 I_{max}^2\, dl^2}{12\pi\, v_0^2}$$

Using the relations $I_{max} = \sqrt{2}\, I_{rms}$, $\eta_0 = 120\pi$, $\dfrac{\omega}{v_o} = \dfrac{2\pi}{\lambda_0}$, this can be written as

$$P_{rad} = \frac{120\pi}{12\pi}\left(\frac{2\pi}{\lambda_0}\right)^2 2\, I_{rms}^2\, dl^2 = \left[80\pi^2\left(\frac{dl^2}{\lambda_0}\right)\right] I_{rms}^2 \quad \text{...(15.8)}$$

Since in an electrical circuit, the power dissipated in a resistor R is $I_{rms}^2\, R$, we can *think* of the term in square brackets in Eqn. (15.8) as the radiation resistance R_{rad} of the antenna. It must be remembered that this is a fictitious resistance existing in our mind only and not in the antenna. Since this power P_{rad} leaves the antenna and never returns to it, we associated the term resistance with the quantity in the square backets in Eqn. (15.8), which has the dimensions of ohm, because we had put $\eta_0 = 120\pi$ ohms. Thus we have the radiation resistance of a short Hertzian dipole as

$$R_{rad} = 80\,\pi^2\left(\frac{dl}{\lambda_0}\right)^2 \text{ ohm} \quad \text{...(15.9)}$$

15.5 QUARTER-WAVE MONOPOLE AND HALF-WAVE DIPOLE

The results obtained in the previous four sections for an infinitesimally small Hertzian dipole of length *dl* are now to be extended to a practical dipole antenna of finite legnth *L* metres. This length is always measured from one end of the antenna to the other even when, as in Fig. 15.3(a), the antenna is centre-fed. In this case, power is fed by a two-wire transmission line to the centre of the antenna. In fact, we may think of the antenna as a continuation of the two-wire line, in whcih the line opens out, or its wires are bent at right angles to form the antenna.

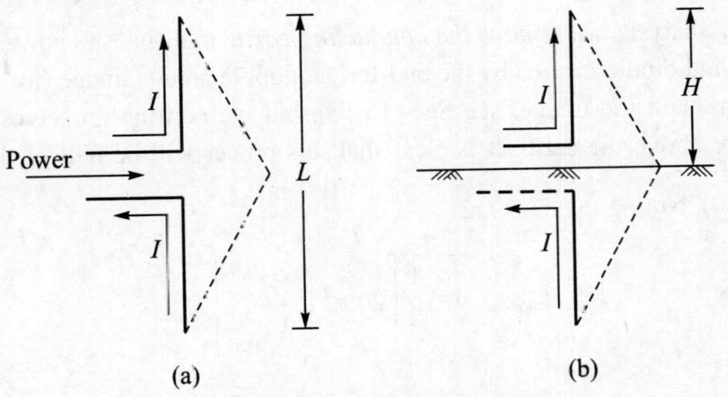

Fig. 15.3. (*a*) Dipole and (*b*) Monopole

If we assume that the length L of the antenna is very much less than λ_0, then a simplified analysis is immediately possible. We have seen in Sec. 14.2 that the current in a transmission line varies along its length. Also in the case of the antenna, the current must be zero at the two ends of the antenna and maximum at the centre. The power radiated from different parts of the antenna therefore varies. Since $L < \lambda_0$ we may assume that the current varies *linearly* from maximum value at centre to zero value at the ends. In that case the average value is

$$\frac{I_{max} + 0}{2} = \frac{I_{max}}{2},$$

and we may say that this $\frac{I_{rms}}{2}$ is the current throughout the *entire length* of the antenna. In other words, we may extend Eqn. (15.8) to this antenna, if we replace $I_{rms}\, dl$ by $\left(\frac{I_{rms}}{2}\right) L$. The power radiated by a centre-fed dipole of length $L \ll \lambda_0$ is then given by

$$\underset{L \;\; \lambda_0}{[P_{rad}]_{dipole}} = \left[80\,\pi^2 \left(\frac{L}{\lambda_0}\right)^2\right]\left(\frac{I_{rms}}{2}\right)^2$$

$$= \left[20\,\pi^2 \left(\frac{L}{\lambda_0}\right)^2\right] I_{rms}^2 \qquad \text{...(15.10)}$$

so that

$$\underset{L \;\; \lambda_0}{[R_{rad}]_{dipole}} = \left[20\,\pi^2 \left(\frac{L}{\lambda_0}\right)^2\right] \approx 200 \left(\frac{L}{\lambda_0}\right)^2 \text{ ohm} \qquad \text{...(15.11)}$$

since $\pi^2 = 9.8696 \ldots \approx 10$.

If at the frequency at which power is being fed to the antenna, we can regard the earth as a good conductor, we can take advantage of the method of images and use only the upper conductor of the transmission line and the upper half of the dipole antenna, the lower conductor being grounded (connected to earth). Then an image of the end fed monopole of length H will appear in the earth as shown dotted in Fig. 15.3(*b*). For the purpose of analysis, *we remove the conducting earth*, and find the power radiated by the centre-fed dipole formed by the end-fed monopole and its image (for the direction of the current in the image, see Sec. 15.16), but we confine ourselves only to the hemisphere above the earth. It is clear that this power will be half of that given by Eqn. (15.10). Noting that $H = L/2$

$$[P_{\text{rad}}]_{\substack{\text{monopole}\\ H << \frac{\lambda_0}{2}}} = \frac{1}{2}\left[20\,\pi^2\left(\frac{2H}{\lambda_0}\right)^2\right] I_{\text{rms}}^2$$

$$= 40\,\pi^2\left(\frac{H}{\lambda_0}\right)^2 I_{\text{rms}}^2 \qquad \text{...(15.12)}$$

so that

$$[P_{\text{rad}}]_{\substack{\text{monopole}\\ H << \frac{\lambda_0}{2}}} = 40\,\pi^2\left(\frac{H}{\lambda_0}\right)^2 \approx 400\left(\frac{H}{\lambda_0}\right)^2 \text{ ohm} \qquad \text{...(15.13)}$$

The assumption of linear variation of current in the antenna, on the basis of which Eqns. (15.10) to (15.13) have been derived, is justified by the fact that these equations are found to be valid in actual antennas for L up to $\frac{\lambda_0}{4}$ and H upto $\frac{\lambda_0}{8}$. However, for longer lengths we have to think of a better approximation. Most of the mathematical work in antenna theory is based on intelligent guess-work about the nature of the variation of the antenna current in different parts of the antenna. That guess work is acceptable, which gives answers that tally with actual measured values.

Since we have assumed that the dipole is an opened-up transmission line, and on lines we had assumed sinusoidal variation of current with distance, it seems logical to assume that on longer dipole antennas, the current varies sinusoidally with distance, as well as with time.

Equation (15.7*a*) was derived on the assumption that the current was varying with time as $I_{\max} \cos \omega t$. Now that the dipole antenna extends from $z = -H$ to $z = +H$ [see Fig. 15.4 (*a*)], we assume the current in it to be of the form

$$I = \begin{cases} I_{\max} \sin \beta\,(H - z)\, e^{j\omega t} & (z > 0) \\ I_{\max} \sin \beta\,(H + z)\, e^{j\omega t} & (z < 0) \end{cases} \qquad \text{...(15.14)}$$

where z is the distance of the current element $I\,dz$ from the origin O. The sine terms in Eqn. (15.14) are written in a form to ensure that $I = 0$ at the two ends of the dipole. At a point P, at a distance r_0 from O, the total electric field in θ-direction will be obtained by adding together the E_θ components contributed by the different parts of the antenna, as the integration is done with position of current element Idz shifted from $z = -H$ to $z = +H$. Though the distance of the point P from the different parts of the antenna varies [Fig. 15.4(*b*)], for the purpose of the above integration, we assume that the small variations in distance will not materially affect the amplitude of the E_θ components, but will only cause a variation in their phases. As seen from Fig. 15.4 (*b*) for *Idz* at + *z*, the phase will be advanced while for *I dz* at – *z*, it will be retarded as indicated by the equations:

$$r = \begin{Bmatrix} r_0 - OB = r_0 - z\cos\theta & (z > 0) \\ r_0 + OC = r_0 + z\cos\theta & (z < 0) \end{Bmatrix} \qquad ...(15.15)$$

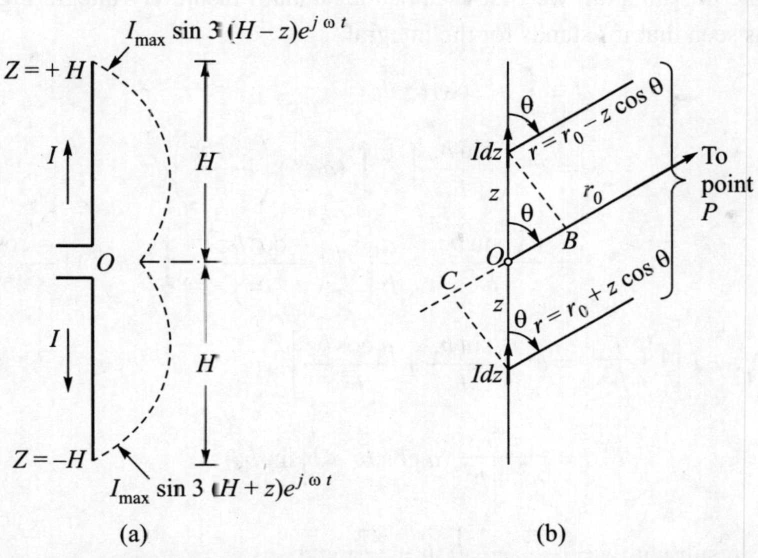

Fig. 15.4. Centre-fed dipole antenna with assumed sinusoidal current distribution

Since for power calculations we need only the amplitude of E_θ we omit the time-variation term of Eqn. 15.7 (*a*), and get the component E_θ for the centre-fed dipole as

$$[E_\theta]_{\substack{\text{dipole} \\ L=2H}} = \frac{60\,\pi\, I_{max}\,\sin\theta}{\lambda_0\, r}\left[\int_{-H}^{0} e^{j\frac{2\pi}{\lambda_0}z\cos\theta}\,\sin\frac{2\pi}{\lambda_0}(L+H)\,dz + \int_{0}^{+H} e^{j\frac{2\pi}{\lambda_0}z\cos\theta}\,\sin\frac{2\pi}{\lambda_0}(L-H)\,dz\right] \qquad ...(15.16)$$

In the first integral, since $z = -|z|$, $e^{j\frac{2\pi}{\lambda_0}|z|\cos\theta}$ gives a phase retardation. This general expression gets considerably simplified for the special case of a centre-fed half-wave dipole, for which $H = \frac{\lambda_0}{4}$. For this case

$$\sin\frac{2\pi}{\lambda_0}\left(\frac{\lambda_0}{4} + z\right) = \sin\left(\frac{\pi}{2} + \frac{2\pi}{\lambda_0}z\right) = \cos\frac{2\pi}{\lambda_0}z$$

$$\sin\frac{2\pi}{\lambda_0}\left(\frac{\lambda_0}{4} - z\right) = \sin\left(\frac{\pi}{2} - \frac{2\pi}{\lambda_0}z\right) = \cos\frac{2\pi}{\lambda_0}z$$

With these values, we have

$$[E_\theta]_{\substack{\text{dipole}\\ L=\frac{\lambda_0}{2}}} = \frac{60\pi\, I_{max}\, \sin\theta}{\lambda_0 r} \int_{-\frac{\lambda_0}{4}}^{-\frac{\lambda_0}{4}} e^{j\frac{2\pi}{\lambda_0} z\cos\theta} \cos\frac{2\pi}{\lambda_0} z\, dz$$

To evaluate this integral, we first establish a standard result. By integrating by parts twice, it is seen that if I stands for the integral.

$$I = \int e^{az} \cos bz\, dz$$

$$= e^{az}\left(\frac{\sin bz}{b}\right) - \int (ae^{az})\left(\frac{\sin bz}{b}\right) dz$$

$$= \frac{e^{az} \sin bz}{b} - \frac{a}{b}\left[e^{az}\left(\frac{-\cos bz}{b}\right) - \int (ae^{az})\left(\frac{-\cos bz}{b}\right) dz\right]$$

Therefore $\quad I\left[1 + \frac{a^2}{b^2}\right] = e^{az}\left[\frac{\sin bz}{b} + \frac{a\cos bz}{b^2}\right]$

Hence $\quad I = \frac{e^{az}}{a^2 + b^2}(a\cos bz + b\sin bz)$

Using this result with $a = j\frac{2\pi}{\lambda_0}\cos\theta,\ b = \frac{2\pi}{\lambda_0}$ and

$$a^2 + b^2 = \left(\frac{2\pi}{\lambda_0}\right)^2 (1 - \cos^2\theta) = \left(\frac{2\pi}{\lambda_0}\right)^2 \sin^2\theta$$

we get

$$[E_\theta]_{\substack{\text{dipole}\\ L=\frac{\lambda_0}{2}}} = \frac{60\pi\, I_{max}\, \sin\theta}{\lambda_0 r} \times \frac{1}{\left(\frac{2\pi}{\lambda_0}\right)^2 \sin^2\theta}$$

$$\times \left[e^{j\frac{2\pi}{\lambda_0} z\cos\theta}\left(j\frac{2\pi}{\lambda_0}\cos\theta\cos\frac{2\pi}{\lambda_0} z + \frac{2\pi}{\lambda_0}\sin\frac{2\pi}{\lambda_0} z\right)\right]_{-\frac{\lambda_0}{4}}^{+\frac{\lambda_0}{4}}$$

$$= \frac{60\pi\, I_{max}}{\lambda_0 r \left(\frac{2\pi}{\lambda_0}\right)^2 \sin\theta}\left[\left\{e^{j\frac{\pi}{2}\cos\theta}\left(0 + \frac{2\pi}{\lambda_0}(+1)\right)\right\}\right.$$

$$\left. - \left\{e^{-j\frac{\pi}{2}\cos\theta}\left(0 + \frac{2\pi}{\lambda_0}(-1)\right)\right\}\right]$$

or

$$[E_\theta]_{\substack{\text{dipole}\\ L=\frac{\lambda_0}{2}}} = \frac{60\, I_{max}}{r} \frac{\cos\left(\frac{\pi}{2}\cos\theta\right)}{\sin\theta} \text{ volt/metre} \qquad \text{...(15.17)}$$

By using Eqn. (1.36), hence

$$[H_\phi]_{\text{dipole}\atop L=\frac{\lambda_0}{2}} = \frac{1}{\eta_0}[E_\theta]_{\text{dipole}\atop L=\frac{\lambda_0}{2}}$$

$$= \frac{I_{max}}{2\pi r}\frac{\cos\left(\frac{\pi}{2}\cos\theta\right)}{\sin\theta} \text{ ampere/metre} \qquad ...(15.18)$$

Since in Eqns. (15.17) and (15.18), the multiplying factor that is not included is only the time-variation factor of Eqn. (15.7*a*) E_θ and H_ϕ are in same phase. Therefore we omit it when finding the radiated power, and write

Average power radiated in *r*-direction

$$= [P_r]_{\text{average}} = \frac{E_{\theta_{max}} H_{\theta_{max}}}{\sqrt{2}\sqrt{2}}$$

$$= \frac{15 I_{max}^2}{\pi r^2}\frac{\cos^2\left(\frac{\pi}{2}\cos\theta\right)}{\sin^2\theta} \text{ watt/m}^2 \qquad ...(15.19)$$

Hence for an end-fed quarter-wave monopole on the earth's surface, the power radiated through a hemisphere of radius *r* above the earth is found by the method used in deriving Eqn. (15.8). Thus putting $I_{rms}^2 = I_{max}^2/2$

$$[P_{rad}]_{\text{monopole}\atop H=\frac{\lambda_0}{4}} = \frac{15 I_{max}^2}{\pi r^2}\int_0^{\frac{\pi}{2}}\frac{\cos^2\left(\frac{\pi}{2}\cos\theta\right)}{\sin^2\theta}(2\pi\, r\sin\theta\, r\, d\theta)$$

$$= 60 I_{rms}^2 \int_0^{\frac{\pi}{2}}\frac{\cos^2\left(\frac{\pi}{2}\cos\theta\right)}{\sin\theta}d\theta$$

This integral cannot be evaluated exactly in any analytical form. Hence the solution can be found graphically by plotting the curve representing the function over the interval 0 to $\pi/2$ and then finding the area under the curve by Simpson's rule. It is found that

$$\int_0^{\frac{\pi}{2}}\frac{\cos^2\left(\frac{\pi}{2}\cos\theta\right)}{\sin\theta}d\theta = 0.60908 \qquad ...(15.20)$$

Thus $$[P_{rad}]_{\text{monopole}\atop H=\frac{\lambda_0}{4}} = 60\, I_{rms}^2 \times 0.60908$$

$$= 36.545\; I_{rms}^2 \text{ watt} \qquad ...(15.21)$$

This makes radiation resistance of

$$\text{Quarter-wave monopole} = 36.545 \text{ ohms} \qquad ...(15.22)$$

For the centre-fed half-wave dipole, the total power radiated through a sphere of radius r will be double of that given by Eqn. (15.21). Hence

$$\text{Radiation resistance of Half-wave dipole} = 73.09 \text{ ohm} \qquad ...(15.23)$$

15.6 ANTENNA GAIN AND EFFECTIVE LENGTH OF A HALF-WAVE DIPOLE

Equation (15.19) shows that at a point P, the power P_r radiated by the antenna is dependent on both the r and θ co-ordinates of P. For comparing the characteristics of different antennas, we are not interested in the actual magnitude of the power, but only in knowing what *fraction* of the *total* power radiated by the antenna goes in a particular direction θ. Noticing the factor $\frac{da}{r^2}$ in the power P_r radiated through the area da, we divide the expression by 4π to get $\frac{da}{4\pi\, r^2} = d\Omega$ the solid angle in the direction θ. If $d\Omega$ = unity (or if, as mentioned towards the end of Sec. 2.4, we divide the expression involving $\left(\frac{da}{4\pi r^2}\right)$ by $d\Omega$ to get a result which does not change when da and $d\Omega$ tend towards zero), we get P_{r-1} the radiation intensity in the direction θ, which can be defined as the power per unit solid angle radiated in the direction θ.

This leads us to think of another useful parameter for comparing antennas. Since the antenna does not radiate uniformally in all directions, we think of an imaginary isotropic antenna which radiates the same total power P_{rad} as the given antenna, but radiates it uniformally in all directions, so that its radiation intensity in all directions becomes

$$[P_{r-i}]_{\text{isotropic}} = P_{\text{rad}}/4\pi.$$

The ratio of the radiation intensity $[P_{r-i}]_{\text{antenna}}$ of the antenna *in a particular direction* θ to the radiation intensity of an isotropic radiator radiating *the same total* power is called the Directional Gain of an antenna. Thus

$$\text{Directional Gain, } G_{\text{dir}} = \frac{[P_{r-i}]_{\text{antenna}}}{[P_{r-i}]_{\text{isotropic}}}$$

$$= \frac{4\pi[P_{r-1}]_{\text{antenna}}}{[P_{\text{rad}}]_{\text{antenna}}} \qquad ...(15.24)$$

Since this is a power ratio, it can also be expressed in decibels as

$$[G_{\text{dir}}]_{\text{dB}} = 10 \log_{10} [G_{\text{dir}}] \qquad ...(15.25)$$

Usually this is computed for the direction in which the radiation intensity is a *maximum,* in which case it may be denoted as $[G_{\text{dir}}]_{\max}$.

For a half-wave dipole, it is easily verified that it radiates maximum power in the direction $\theta = 90°$, because this makes $\dfrac{\cos^2\left(\dfrac{\pi}{2}\cos\theta\right)}{\sin^2\theta}$ a maximum (equal to unity).

For a half-wave dipole

the total power radiated $= [P_{\text{rad}}]_{\text{antenna}}$

$$= 73.09\ I_{\text{rms}}^2$$

and from Eqn. (15.19)

$$[P_{r-i}]_{\max\,\theta=90°} = \frac{15\,I_{\max}^2}{\pi r^2}\left(\frac{da}{4\pi}\right)\left(\frac{4\pi r^2}{da}\right) = \frac{120\,I_{\text{rms}}^2}{4\pi}$$

Using this value in Eqn. (15.24), we get the maximum directional gain of a half-wave dipole as

$$[G_{\text{dir}}]_{\max\,\theta=90°} = \frac{4\pi\dfrac{120\,I_{\text{rms}}^2}{4\pi}}{73.09\ I_{\text{rms}}^2} = \frac{120}{73.09}$$

$$= 1.64 \quad \text{or} \quad 2.15\text{ dB} \qquad \text{...(15.26)}$$

Another useful parameter for antenna performance comparison and study is the Effective Length, L_{eff} of an antenna. It tells us how good or effective the antenna is as a radiator of power, or conversely, as a collector of energy from a passing electromagnetic wave when used as a receiving antenna. For a transmitting antenna, if $I(0)$ is the current at the feed point of the antenna (at $z = 0$) where the transmission line joins it, and if this *constant current* flows through the entire length L_{eff}, then the power radiated in a direction perpendicular to the antenna will be the same as that radiated by the actual antenna with a sinusoidal (or any other) current distribution along its length. Expressed mathematically, if L is the length of the transmitting antenna, and $H = L/2$,

$$I(0)\ L_{\text{eff}} = \int_{-H}^{+H} I\,dz$$

or

$$L_{\text{eff}} = \frac{\int_{-H}^{+H} I\,dz}{I(0)} \qquad \text{...(15.27)}$$

For the centre-fed dipole of Fig. 15.3, if we assume a sinusoidal distribution of currents as in Eqn. (15.14), then in Eqn. (15.27), for $z = 0$,

$$I(0) = I_{\max}\sin\beta H\,e^{j\omega t}$$

and

$$\int_{-H}^{+H} I\,dz = e^{j\omega t}\left[\int_{-H}^{0} I_{\max}\sin\beta(H+z)\,dz + \int_{0}^{+H} I_{\max}\sin\beta(H-z)\,dz\right]$$

$$= I_{max}\, e^{j\omega t}\left[\frac{-\cos\beta(H+z)}{\beta}\right]_{-H}^{0} + I_{max}\, e^{j\omega t}\left[\frac{-\cos\beta(H-z)}{-\beta}\right]_{0}^{+H}$$

$$= I_{max}\, e^{j\omega t}\left[\frac{-\cos\beta H+1}{\beta}\right] + I_{max}\, e^{j\omega t}\left[\frac{-1+\cos\beta H}{-\beta}\right]$$

$$= I_{max}\, e^{j\omega t}\, 2\left[\frac{1-\cos\beta H}{\beta}\right]$$

Hence with $\beta = \dfrac{2\pi}{\lambda_0}$, $H = \dfrac{L}{2}$

$$L_{eff} = \frac{2(1-\cos\beta H)}{\beta} \times \frac{1}{\sin\beta H} = \frac{\lambda_0}{\pi}\,\frac{\left(1-\cos\dfrac{\pi L}{\lambda_0}\right)}{\sin\dfrac{\pi L}{\lambda_0}} \qquad ...(15.28)$$

$$\left.\begin{array}{ll}\text{For a half-wave dipole, } L = \dfrac{\lambda_0}{2}, & \text{and } L_{eff} = \dfrac{\lambda_0}{\pi} \\ \text{For a quarter-wave monopole, } H = \dfrac{\lambda_0}{4}, & \text{and } L_{eff} = \dfrac{\lambda_0}{2\pi}\end{array}\right\} \qquad ...(15.29)$$

since the integration will be only from 0 to $+H$ in above derivation.

15.7 DIRECTIONAL CHARACTERISTICS OF ANTENNA

An antenna acts as an impedance matching device between the transmission line from the transmitter on the one hand and free space on the other. A study of the simplest Hertzian dipole and the half-wave dipole has shown that antennas do not radiate power equally in all directions. Instead of being a disadvantage, this directional characteristic of an antenna can be usefully employed. Some ingenious methods can be developed to enhance this variation in power radiation in different directions. The objective is to direct more power in some desired directions, and to reduce, or sometimes to prevent completely, power going in an undesired direction.

An isotropic antenna or radiator is one which radiates uniformally in all directions. Since a physical radiator must have a current flowing in a conductor in some direction, all actual physical radiators have directional property inherent in them. We may therefore be led to conclude that only a point source can be an isotropic radiator and thus this can only exist in our imagination, or perhaps serve as a useful mathematical abstract model simplifying the study of antennas.

Such is however not the case. Man is an earth bound creature, moving most of the time in a more or less horizontal plane. Medium frequency broadcasting stations always use a mast-radiator or a vertical tower, which is usually end-fed like a monopole. We are only interested primarily in the radiation from this vertical mast in a horizontal direction, because that is where the receiving set is going to be located. For this horizontal direction pattern, the vertical radiator can be regarded as a point source. Thus a study of isotropic radiators can be of practical use.

The directional characteristics of an antenna can be described by measurement of the electric field intensity E at every point on the surface of a large sphere (to get the radiated field component only) of a standard radius of one kilometre surrounding the antenna. Let E_{max} be the maximum value of E from amongst these readings. Then for every point P on the surface of the sphere, we draw a line from the origin (centre of sphere or location of antenna) towards the point P but of length E/E_{max} (to some scale), where E is the field intensity at P. When this is done for every point on the sphere of one kilometre radius, we shall get a three-dimensional surface, the radial lines to whose surface are proportional in length to the field intensity in that direction.

To depict the above three-dimensional surface on a two-dimensional paper, we take three sections of the surface by three mutually perpendicular planes passing through the origin. The sections will be in the form of closed curves, called directional or radiation patterns.

(*i*) Horizontal pattern is obtained by taking a section with a horizontal plane passing through the origin.

(*ii*) Vertical pattern is obtained by taking a section with a vertical plane passing through the origin.

Though there is only one horizontal plane passing through the origin, there can be an infinite number of vertical planes. Of the latter, only two vertical planes are selected – one passing through the axis of the antenna and the other at right angles to it and passing through the origin. These three patterns are called the Principal Plane Patterns.

The process of dividing each measured value by a constant quantity (in this case E_{max}) is called normalization and these radiation patterns are therefore called the normalized radiation patterns for $E_n = \dfrac{E}{E_{max}}$. The maximum value of E_n is of course unity.

15.8 RADIATION PATTERN OF DIPOLE ANTENNA

For a short dipole antenna, at a constant distance of one kilometre from O, the Eqn. (15.7*a*) can be written as $E_\theta = E_0 \sin\theta$ V/m. If we assume that the *z*-axis is vertical, then for the horizontal plane passing through O, $\theta = 90°$ [Fig. 15.5 (a)] and the horizontal radiation pattern simply becomes a circle of unit radius, because $E_\theta = E_0$ and hence $E_n = \dfrac{E_\theta}{E_0} = 1$ [Fig. 15.5 (b)]. The equation of the circle is $r = 1$. The vertical plane patterns are all alike, because E_θ is independent of the angle ϕ. Here we have to draw a line OA [Fig. 15.5 (c)] making an angle θ with OZ and of length $OA = E_n = \dfrac{E_\theta \sin\theta}{E_0} = \sin\theta$. Let the line AB drawn perpendicular to OA intersect x-axis in point B. Then $m\angle OBA = m\angle ZOA = \theta$. Hence $OA/OB = \sin\theta$ or $OB = OA/\sin\theta = \sin\theta/\sin\theta = 1$. Since OB is a constant and angle OAB is a right angle, for all values of θ, the point A will lie on a circle with OB as diameter. (Angle OAB in a semi-circle is always a right angle). For values of θ to the left of OZ in Fig. 15.5(c), we get a similar circle with diameter $OB' = 1$. These two equal circles touching each other at O, give what is called the figure-of-eight pattern. (This name is also used for two identical ovals, 180° apart, touching each other at O). The equation to the figure-of-eight pattern curve is $r = \sin\theta$.

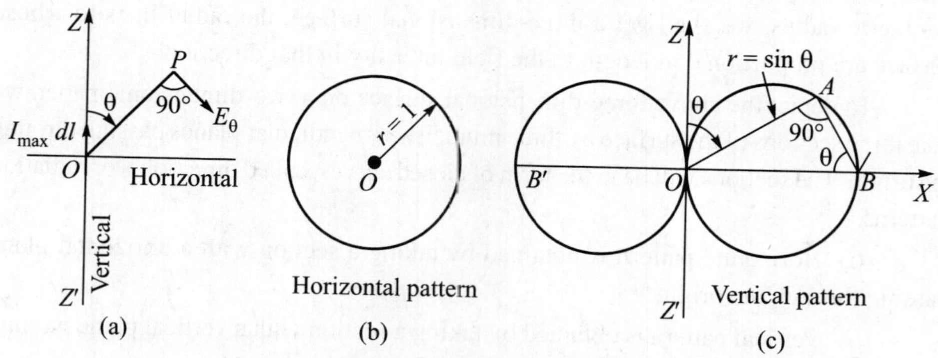

Fig. 15.5. Radiation pattern for dipole-antenna

For $\theta = 0$ or π, $E_\theta = 0$. For these values of θ, the line OA has zero length. Hence OZ and OZ' are called the null directions, and we say that the vertical radiation pattern has two nulls and two lobes, there being one lobe between two nulls (or one null between two lobes).

Since the vertical pattern is identical for all values of ϕ, the three-dimensional surface can be obtained by revolving the figure-of-eight about the vertical OZ axis, giving rise to a tube of diameter $OB = 1$. The section of this surface by a horizontal plane gives the circle, that was described by the points B and B', when the figure-of-eight pattern was revolved.

Finally, it is easy to visualize that if the axis of the dipole were horizontal, then

(i) the horizontal pattern would have been a figure-of-eight,

(ii) the vertical pattern for a plane passing through the axis of the antenna would also be a figure-of-eight, while

(iii) the pattern for a vertical plane perpendicular to the axis of the antenna and passing through O would have been a circle.

If the length of the vertical, centre-fed dipole antenna is increased, the vertical plane pattern undergoes a change, the circles of the figure-of-eight pattern change to ellipses, and for longer lengths of the antenna, smaller sized (called minor) lobes appear. This is shown in Fig. 15.6 for various lengths of the dipole. Since the selected lengths are multiples of $\lambda_0/2$ like $m\lambda_0/2$, m being an integer, these antennas are often described as resonant dipoles. In each figure, the assumed sinusoidal distribution of current in the antenna is shown by dotted lines. The general formula for all the patterns is

$$E_n = \frac{\cos\dfrac{m\pi}{2} - \cos\left(\dfrac{m\pi}{2}\cos\theta\right)}{\sin\theta} \quad ...(15.29a)$$

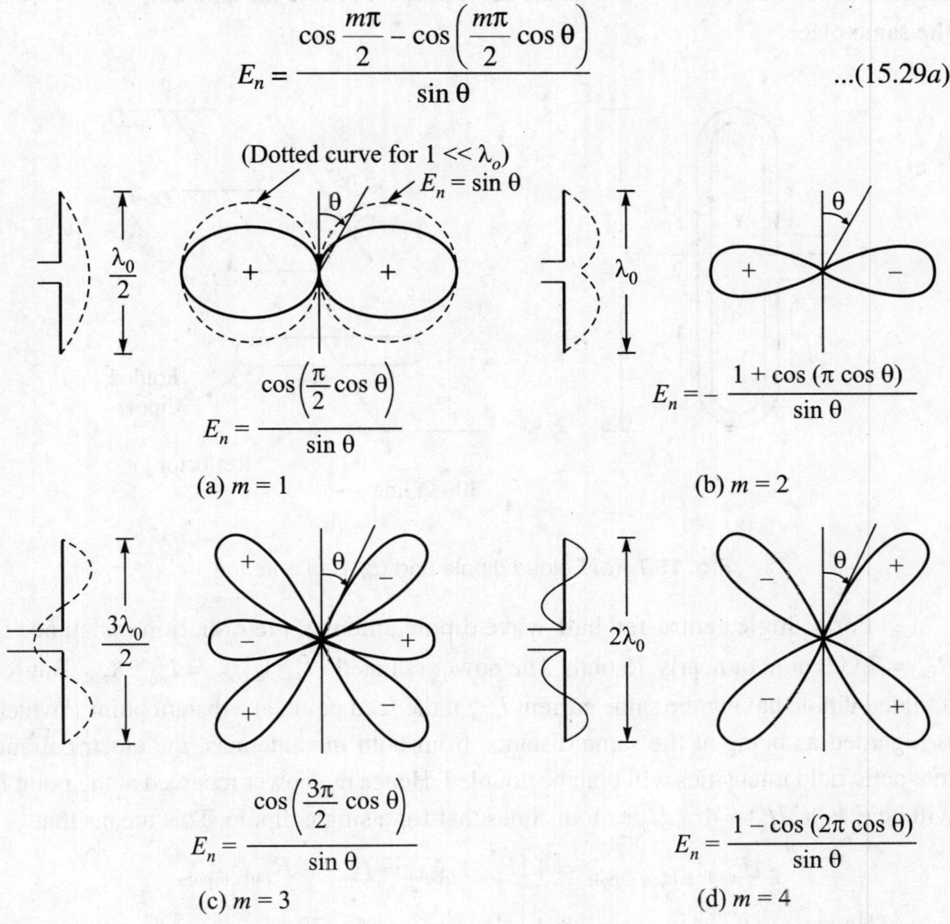

Fig. 15.6. Vertical radiation patterns for vertical, centre-fed resonant dipoles of different lengths

15.9 FOLDED DIPOLE AND YAGI ANTENNA

For certain applications like television, antennas were required which would have the ability to handle a signal of much larger bandwidth than radio broadcasting and which would also have a much larger impedance than the 73.09 ohm radiation resistance of a half-wave dipole. Of the many systems proposed, the one that became most popular because of its simplicity was the one developed by P.S. Carter of Radio Corporation of America (RCA) in 1939. He suggested the use of two half-wave dipoles instead of one, the two antennas being located at *almost the same place*. At the two ends, the antennas are joined together by a short conducting piece which may be straight, but is usually curved for ease of manufacture [Fig. 15.7(a)]. Since the ends of the antennas are held at the same potentials by these short-circuits at the two ends, they have identical potential differences between their ends, and hence the currents in the two antennas are equal in magnitude. The currents are also in the same phase, because the antennas are located at the same place.

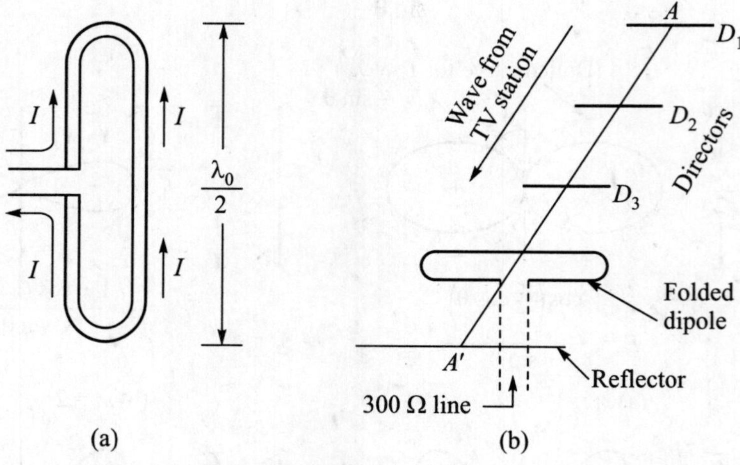

Fig. 15.7. (*a*) Folded dipole and (*b*) Yagi antenna

For a single centre-fed half-wave dipole antenna, the radiation resistance is R_{rad} = 73.09 ohm or nearly 75 ohm. The power radiated $[P_{rad}]_{dipole} = I_{rms}^2 R_{rad}$. But for a folded dipole having the same current I_{rms} at the feed point, at a distant point P which is regarded as being at the same distance from both the antennas, the electrical and magnetic field intensities will both be doubled. Hence the power received at the point P will be $(2E_\theta)(2H_\phi) = 4E_\theta H_\phi$ or four times that for a single dipole. This means that

$$[P_{rad}]_{\text{folded dipole}} = 4\,[P_{rad}]_{\text{dipole}} = I_{rms}^2\ (4R_{rad})_{\text{dipole}}$$

Hence $$[R_{rad}]_{\text{folded dipole}} = 4\,[R_{rad}]_{\text{dipole}} = 4 \times 75$$

$$= 300 \text{ ohm} \qquad \text{...(15.30)}$$

A simple PVC (polyvinyl chloride) twisted, flexible wire used in houses for temporary electrical wiring can be regarded as a two-wire transmission line having a characteristic impedance of 75 ohm. But the two-wire strip-line (with conductors parallel to each other) joining the TV-antenna to the TV-receiver has a characteristic impedance of 300 ohm. The twisted flexible can match the impedance of a half-wave dipole, but the two-wire strip line is needed to match the impedance of a folded dipole. The input impedance of a black-and-white TV receiver is 300 ohm, and hence the two-wire strip which matches it, can be directly connected to it. But a colour TV set has a 75-ohm coaxial cable input terminal. The two-wire line has both conductors balanced with respect to ground, but a coaxial cable is unbalanced. Therefore between the 300-ohm two-wire line and the 75-ohm coaxial cable, a BALUN (BALanced-to-UNbalanced 300/75 ohm) transformer has to be used.

Japanese engineers H. Yagi and S. Uda in 1926 succeeded in increasing the directivity of a receiving antenna by the use of "parasitic elements" with the antenna.

Figure 15.7 (*b*) shows such a Yagi antenna suitable for TV reception based on a folded dipole as the active element of the antenna. The metal rod *AA′* forming the axis of the antenna is aligned with the direction of the transmitting antenna. The order in which the radio wave from the transmitter reaches the different element is : The Directors D_1, D_2 and D_3, the folded dipole (active element), the reflector. The directors and reflectors are called "passive" elements, because they do not affect the radiation resistance of the folded dipole. However, these passive or parasitic elements alter the directional characteristic of the folded dipole. Only one reflector element is used, because the field behind the reflector becomes so weak, that additional reflectors will not materially alter the directional characteristic. The length of the reflector is greater than $\frac{\lambda_0}{2}$, of the folded dipole is $\frac{\lambda_0}{2}$ and the lengths of the directors are less than $\frac{\lambda_0}{2}$. The lengths of the directors decrease in the order D_3, D_2, D_1. As the mathematical analysis of the Yagi antenna becomes exceedingly complicated, the length of the directors and reflector and their spacing along the axis *AA′* is usually determined by trial and error.

The beam angle is defined as the angle between the half-power points. In other words, it is the angle between those points, where the electric field intensity is 0.707 E_0, where E_0 is the intensity in the direction *AA′* [Fig. 15.7 (*c*)]. The beam angle can be reduced from about 40° to about 10° by increasing the number of directors (3, 5, 7, 9, 11, 13).

It is obvious that since the lengths of folded dipole, reflector, and director are dependent on the wavelength λ_0, the same Yagi antenna cannot be used for receiving different TV channels as these have different wavelengths.

15.10 ANTENNA ARRAYS

In Sec. 15.8 we have seen that an isotropic antenna has no directional characteristic, but a dipole antenna gives lobes with nulls between them and is therefore more directional. Figure 15.6 shows that as the length of the antenna is increased, the width of the lobe decreases (or the angle between consecutive nulls decreases). But it is also obvious that the control over directivity obtained by changing the length of the antenna is very limited, with a single antenna.

It is found that greater flexibility and control over directivity becomes available if a number of isotropic radiators are used and the currents in them are of different amplitudes and phases. Such groups of radiators are called antenna arrays. The radiators may be arranged along a straight line to give a linear array, or along parallel straight lines in one plane to give a plane array in two-dimensions, or a three-dimensional array with many plane arrays grouped together.

An antenna array may consist of isotropic radiators or of identical antennas. In the latter case, all these similar antennas must be similarly oriented in space, because the antenna array uses the concept of interference of electromagnetic waves to produce cancellation or re-inforcement of the electric field intensity. The similar orientation of the antennas ensures that the electric field vectors for all of them are polarized in the same direction in space.

An array is said to be *linear* if all the antennas are spaced at equal distances (that is, d, the spacing between adjacent antennas, is a constant) along a straight line. The array will be called *uniform*, if currents of equal magnitude are fed to each antenna forming the array. The array will be called a *broadside array*, if the currents in all the antennas are in the same time phase. It will be called an *end-fire array*, if there is a progressive phase difference α in currents along the array equal (when measured in number of cycles) to the spacing d between two antennas (when this is measured in number of wavelengths). Stated mathematically, this means that if $d = n\,\lambda$, then $\alpha = n\,(2\pi)$.

It will be seen in the next section that the lobes obtained in the directional pattern of a broadside array are narrow, while those obtained with an end-fire array are wide or broad. One may therefore be inclined to think that the term broadside is a misnomer. The origin of the term broadside, is from naval usage. Cannons on a battleship are mounted along the long side of the ship (called the broadside). When all the guns on one side of the ship are discharged *simultaneously*, the ship is said to have fired a broadside. Similarly the antenna array in which all currents are *in the same time phase* is called a broadside array.

15.11 ARRAY OF TWO-ELEMENT ISOTROPIC RADIATORS

The simplest linear, uniform array has two identical isotropic radiators having currents of equal amplitude in the two radiators. It provides two variables for controlling the directivity of the array—the spacing d between the antennas, and measured in terms of wavelength, and the phase difference α between the currents, in the two antennas, measured in cycles or multiples of 2π radians. The standard nomenclature is illustrated in Fig. 15.8. The currents I_1 and I_2 in the two antennas are equal in magnitude and frequency, but the current in antenna 2 leads over the current in antenna 1 by an angle α radians and hence $I_2 = I_1 \angle + \alpha$. Since the radiation field alone is to be considered at the distant point P, we use the same concept as was employed in writing Eqn. (15.15) and in Fig. 15.8, we put $r_1 \approx r$ and $r_2 \approx r$ so far as amplitude of E are concerned, but use the relations $r_1 = r + \frac{d}{2}\cos\theta$ and $r_2 = r - \frac{d}{2}\cos\theta$ for phase considerations. Thus the positive term will produce a retardation and the negative term an advance of phase over what the phase would be if the source were at the origin O. Similarly, dividing the phase angle α between I_2 and I_1 in two equal parts, we write assuming that the electric field intensity at P would have been E_0 if the source were at the origin

$$\text{Field intensity due to antenna 1} = E_0\, e^{-j\frac{\alpha}{2}}\, e^{-j\frac{2\pi}{\lambda_0}\left(\frac{d}{2}\cos\theta\right)}$$

$$\text{Field intensity due to antenna 2} = E_0\, e^{+j\frac{\alpha}{2}}\, e^{+j\frac{2\pi}{\lambda_0}\left(\frac{d}{2}\cos\theta\right)}$$

The resultant field at P is the sum of these field intensities. Thus Resultant field intensity E

$$= 2E_0 \frac{e^{-j\left[\frac{\alpha}{2}+\frac{\pi d}{\lambda_0}\cos\theta\right]} + e^{-j\left[\frac{\alpha}{2}+\frac{\pi d}{\lambda_0}\cos\theta\right]}}{2}$$

$$= 2E_0 \cos\left[\frac{\alpha}{2}+\frac{\pi d}{\lambda_0}\cos\theta\right]$$

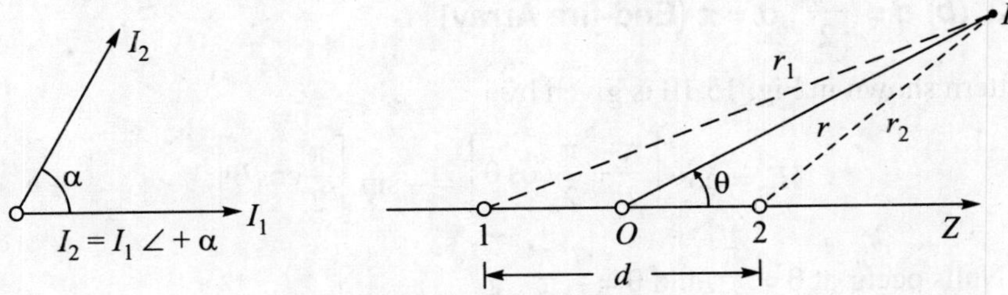

Fig. 15.8 Two-element array

Hence normalizing by dividing by the maximum value $2E_0$,

$$E_n = \cos\left[\frac{\alpha}{2}+\frac{\pi d}{\lambda_0}\cos\theta\right] \qquad ...(15.31)$$

Here d and α are the two variables which can be altered to obtain variety of directional patterns, some of which are described below. In each case, the three-dimensional surface is obtained by revolving the diagram about the axis of the array (z-axis). In the figures the short arrows below the antennas 1 and 2 indicate the phasors for I_1 and I_2.

15.11.1. (*a*) $d = \frac{\lambda_0}{2}$, $\alpha = 0$ (Broadside Array)

By Eqn. (15.31), the radiation pattern shown in Fig. 15.9 is

$$E_n = \cos\left[0+\frac{\pi}{\lambda_0}\frac{\lambda_0}{2}\cos\theta\right] = \cos\left[\frac{\pi}{2}\cos\theta\right]$$

Nulls occur at $\theta = 0$ and $\theta = \pi$.

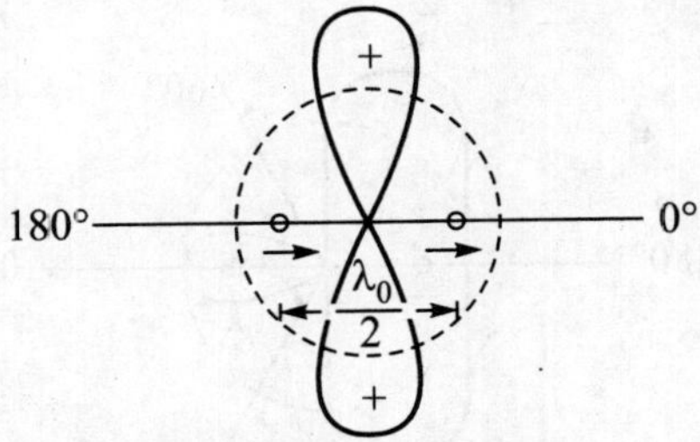

Fig. 15.9. Radiation pattern

Since $\theta = \dfrac{\pi}{2}$ and $\theta = -\dfrac{\pi}{2}$ give a positive value for E_n, the lobes are marked (+). These signs indicate the polarity of the electric field in the direction of the lobe. The dotted circle has the same area as the total area of all the lobes. It is seen that the lobes are narrow or they have a small beam angle. The lobes are normal to the array axis.

15.11.2. (*b*) $d = \dfrac{\lambda_0}{2}$, $\alpha = \pi$ (End-fire Array)

The pattern shown in Fig. 15.10 is given by

$$E_n = \cos\left[\frac{\pi}{2} + \frac{\pi}{2}\cos\theta\right] = -\sin\left[\frac{\pi}{2}\cos\theta\right]$$

Nulls occur at $\theta = \dfrac{\pi}{2}$ and $\theta = -\dfrac{\pi}{2}$.

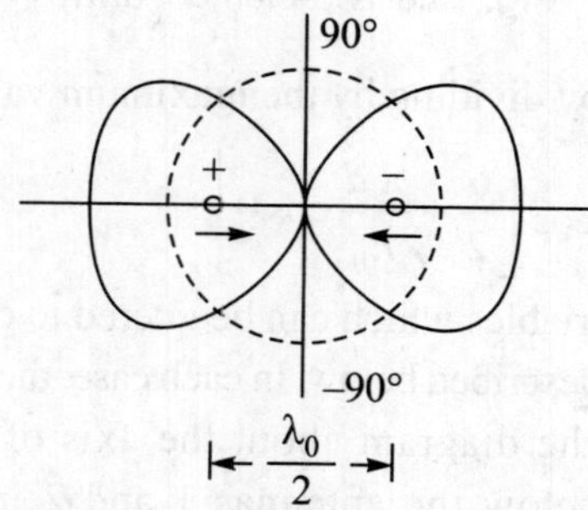

Fig. 15.10. Radiation pattern

Since $\theta = 0$ makes E_n negative, the right-hand lobe is marked (–). But $\theta = \pi$, makes left-hand lobe positive. The lobes are in the direction of the axis of the array and are broad (with larger beam angles).

15.11.3. (*c*) $d = \dfrac{\lambda_0}{2}$, $\alpha = \dfrac{\pi}{2}$

The radiation pattern shown in Fig. 15.11 is

$$E_n = \cos\left[\frac{\pi}{4} + \frac{\pi}{2}\cos\theta\right]$$

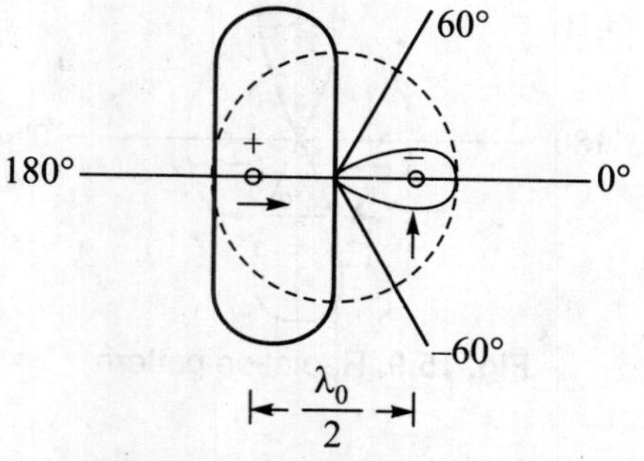

Fig. 15.11. Radiation pattern

Nulls occur where $\cos\theta = \frac{1}{2}$ or $\theta = \pm 60°$. The magnitude of E_n is the same for $\theta = 0$ and $\theta = \pi$.

15.11.4. (*d*) $d = \frac{\lambda_0}{4}$, $\alpha = \frac{\pi}{2}$: (End-fire Array)

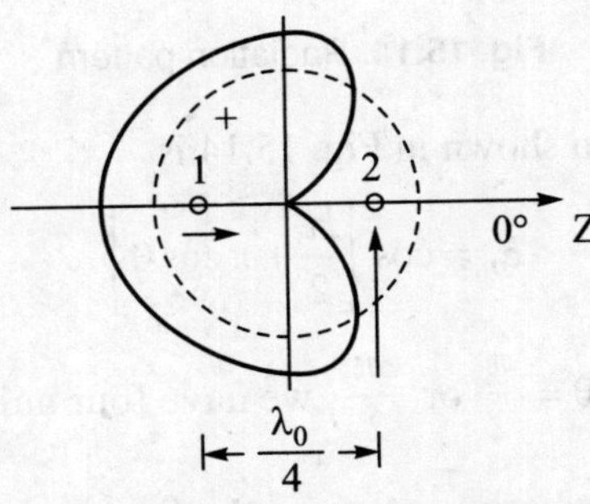

Fig. 15.12. Radiation pattern

The radiation pattern shown in Fig. 15.12 is

$$E_n = \cos\left[\frac{\pi}{4} + \frac{\pi}{4}\cos\theta\right]$$

There is only one null at $\theta = 0$.

For $\theta = \pi$, we get a maximum. The figure is called a cardiod, because it is heart-shaped.

15.11.5. (*e*) $d = \lambda_0$, $\alpha = 0$ (Broadside)

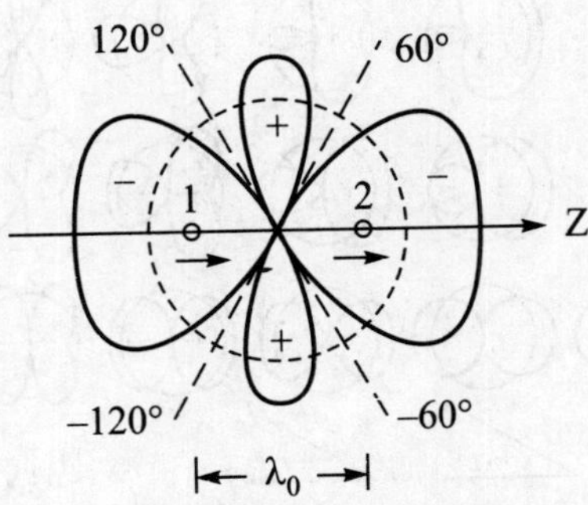

Fig. 15.13. Radiation pattern

The radiation pattern shown in Fig. 15.13 is

$$E_n = \cos\,[0 + \pi\cos\theta]$$

The four nulls are located at $\pm 60°$ and $\pm 120°$. The lobes corresponding to $\theta = 0$ and $\theta = \pi$ make E_n negative. Also they are broad. The lobes for $\theta = \pm 90°$ are narrow and positive.

15.11.6. (*f*) $d = \lambda_0, \alpha = \pi$

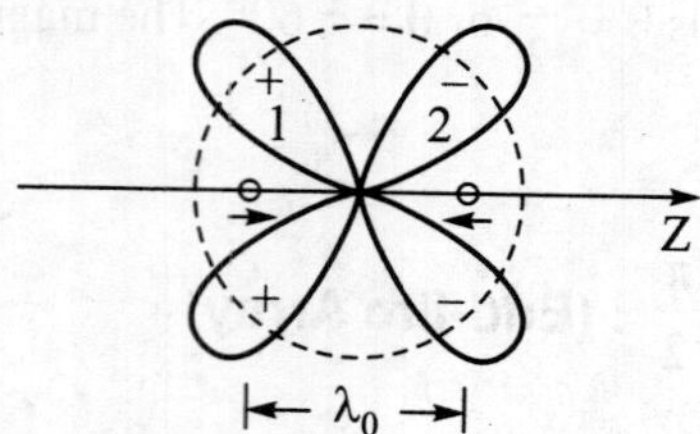

Fig. 15.14. Radiation pattern

The radiation pattern shown in Fig. 15.14 is

$$E_n = \cos\left[\frac{\pi}{2} + \pi \cos\theta\right]$$

Putting $\frac{\pi}{2} + \pi \cos\theta = \frac{\pi}{2}$ or $\frac{3\pi}{2}$, we have four nulls for $\theta = 0$, $\theta = \pm\frac{\pi}{2}$ and $\theta = \pi$. The maxima occur where $\frac{\pi}{2} + \pi\cos\theta = 0$ or π, that is where $\theta = \pm 60°$ and $\pm 120°$.

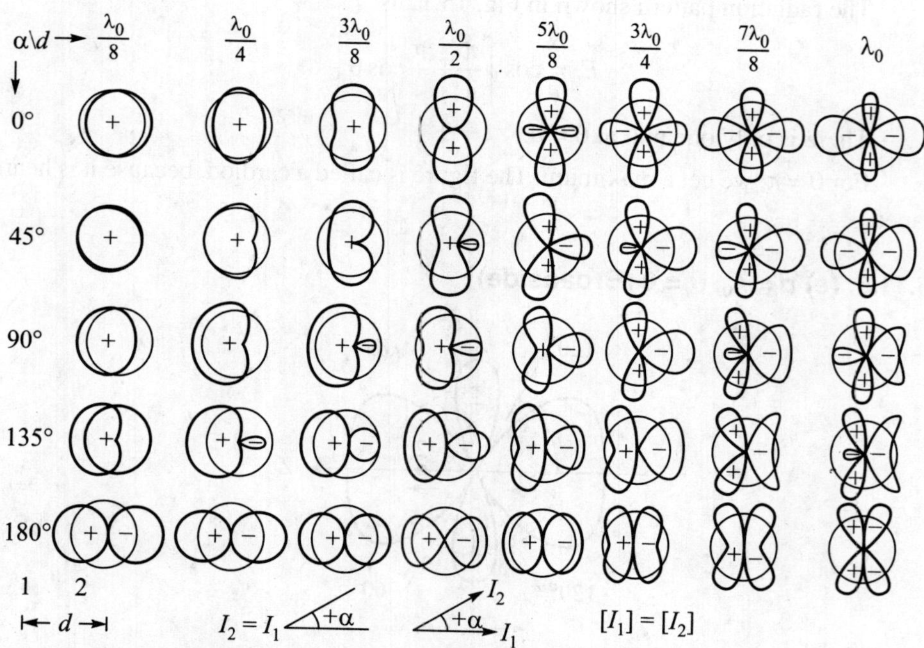

Fig. 15.15. Radiation pattern of two isotropic radiators with equal currents

For design purposes Fig. 15.15 will be found to be useful, because it shows the patterns for variations of d from 0 to λ_0 in steps of $\frac{\lambda_0}{8}$ and variations of α from 0 to 180° in steps of 45°.

15.12 LINEAR ARRAY OF ISOTROPIC RADIATORS

In a linear, uniform array of n isotropic radiators (Fig. 15.16), between adjacent antennas, there is a spacing of d metre and a leading phase angle of α radians for the current (so that $I_n = I_{n-1} \angle +\alpha$). If we take antenna 1 as the reference antenna giving a radiation electric field intensity E_0 volt/metre at a distant point P in the θ direction, then assuming that the length of the array is very small as compared to the distance to the point P, then the total phase lead in the electric field intensity of the next antenna 2 will be

$$\psi = \alpha + \frac{2\pi}{\lambda_0}(d \cos \theta) \qquad \text{...(15.32)}$$

This shows that the resultant radiated electric field intensity at P will be

$$E_{total} = | E_0 + E_0 e^{j\psi} + E_0 e^{j2\psi} + \ldots + E_0 e^{j(n-1)\psi} |$$

$$= E_0 \frac{|1 - e^{jn\psi}|}{|1 - e^{j\psi}|} \qquad \text{...(15.33)}$$

by the formula for the sum of a geometric series of n terms with common ratio $e^{j\psi}$. The modulus is written on the right-hand side, because we are interested only in the amplitude of the resultant field and not in its phase angle.

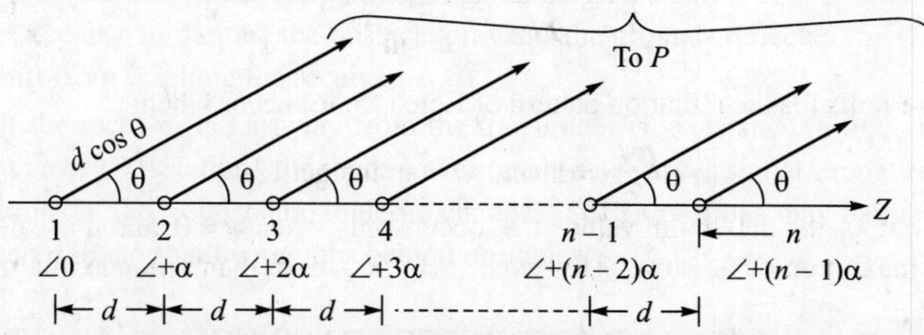

Fig. 15.16. Linear, uniform array of isotropic radiators

Now

$$|1 - e^{jn\psi}| = |1 - \cos n\psi - j \sin n\psi|$$

$$= \sqrt{(1 - \cos n\psi)^2 + \sin^2 n\psi}$$

$$= \sqrt{1 + (\cos^2 n\psi + \sin^2 n\psi) - 2 \cos n\psi}$$

$$= \sqrt{2[1 - \cos n\psi]} = \sqrt{2\left[1 - \left\{1 - 2\sin^2 \frac{n\psi}{2}\right\}\right]}$$

$$= 2 \sin \frac{n\psi}{2}$$

Similarly $\qquad \left|1 - e^{i\psi}\right| = 2 \sin \dfrac{\psi}{2}$

Hence Eqn. (15.33) becomes

$$E_{total} = E_0 \frac{\sin \dfrac{n\psi}{2}}{\sin \dfrac{\psi}{2}} \qquad \text{...(15.34)}$$

By taking logarithms and then differentiating with respect to ψ, it is found that $\left(\sin \frac{n\psi}{2}\right) \Big/ (\sin \psi/2)$ has a maximum value when $\psi = 0$. Since putting $\psi = 0$ makes the right-hand side of Eqn. (15.34) to be of the form, zero divided by zero, this maximum value is found by l'Hospital's rule as follows:

$$\lim_{\psi \to 0} \frac{\sin \frac{n\psi}{2}}{\sin \frac{\psi}{2}} = \lim_{\psi \to 0} \frac{\frac{\partial}{\partial \psi}\left(\sin \frac{n\psi}{2}\right)}{\frac{\partial}{\partial \psi}\left(\sin \frac{\psi}{2}\right)}$$

$$= \lim_{\psi \to 0} \frac{\frac{n}{2} \cos \frac{n\psi}{2}}{\frac{1}{2} \cos \frac{\psi}{2}} = n \qquad \text{...(15.35)}$$

Hence from Eqns. (15.34) and (15.35) we get the normalized pattern for linear, uniform array as

$$E_n = \frac{E_{\text{total}}}{nE_0} = \frac{\sin \frac{n\psi}{2}}{n \sin \frac{\psi}{2}} \qquad \text{...(15.36)}$$

The nulls for the radiation pattern of Eqn. (15.36) occur when

$$\frac{n\psi}{2} = \pm\, m\pi, \quad m = \text{an integer} \qquad \text{...(15.37)}$$

If $d < \lambda_0$, the maximum value of E_n occurs only when $\psi = 0$, and it is called the principal maximum of the array. However, if $d > \lambda_0$, more than one maxima may be obtained.

For a broadside array, $\alpha = 0$ in Eqn. (15.32). Since maximum occurs when $\psi = 0$, this makes $\cos\theta = 0$ or $\theta = \pm \frac{\pi}{2}$.

For an end-fire array, $\alpha = 2\pi d/\lambda_0$. This gives a maximum when

$$\psi = 0 = \frac{2\pi d}{\lambda_0} + \frac{2\pi d}{\lambda_0} \cos\theta$$

or

$$\theta = \pi.$$

Thus with end-fire array, the maximum occurs in the direction of the *most lagging phase*.

15.13 MULTIPLICATION OF PATTERNS

In the linear arrays, instead of the isotropic radiators, we could use identical non-isotropic radiators, provided they are all similarly oriented in space. In that case we would be required to replace E_0 in the above derivation of equation for the radiation pattern of the array E_{n1}, the radiation pattern of the antenna itself. If E_{n2} is the radiation pattern of the array consisting of only isotropic radiators, then the resultant pattern with isotropic radiators replaced by antennas having radiation pattern E_{n1} is

$$E_n = \underset{\text{(Pattern for antenna)}}{E_{n1}} \times \underset{\text{(Pattern for array)}}{E_{n2}} \qquad \text{...(15.38)}$$

Since multiplication by zero gives a zero, it follows that in the resultant pattern there will be a null corresponding to *each* null in patterns E_{n1} and E_{n2}. If N_1 and N_2 are the number of nulls in patterns E_{n1} and E_{n2} respectively, the number of nulls in E_n will be $(N_1 + N_2)$. This affords an excellent check on the correctness of the solution, provided care is taken to verify that the nulls of E_{n1} and E_{n2} do not occur for the same value of θ. This is because 0×0 will give only one null and not two. The number of such multiple nulls will have to be subtracted from $N_1 + N_2$, as in Secs (*g*) and (*i*) below.

This multiplication of patterns indicated by Eqn. (15.38) can be done analytically or graphically as shown below:

15.13.1. (*a*) Broadside Array with Dipoles Aligned $d = \frac{\lambda_0}{2}$, $\alpha = 0$

Notice that the distance *d* is the distance between the *centres* of the antennas. Figure 15.17 shows the steps.

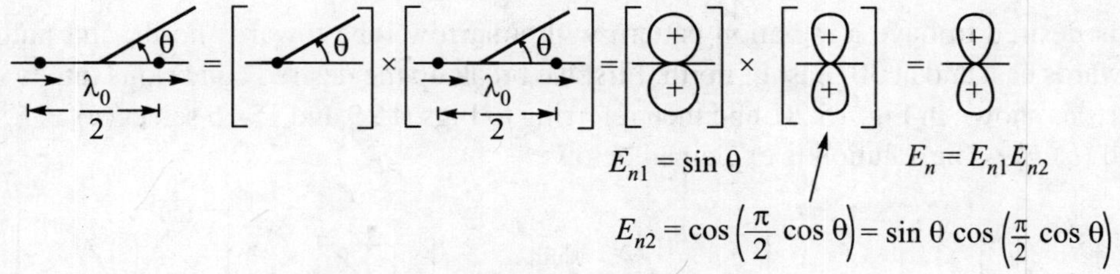

$E_{n1} = \sin\theta$ $\quad E_n = E_{n1}E_{n2}$

$$E_{n2} = \cos\left(\frac{\pi}{2}\cos\theta\right) = \sin\theta\cos\left(\frac{\pi}{2}\cos\theta\right)$$

Fig. 15.17

15.13.2. (*b*) Broadside Array with Dipoles Perpendicular to Axis of Array

$$d = \frac{\lambda_0}{2}, \alpha = 0$$

Notice that in Fig. 15.18, E_{n1} is written as cos θ, because the angle θ is here measured from the axis of the array (and not from the axis of the antenna as in Fig. 15.5).

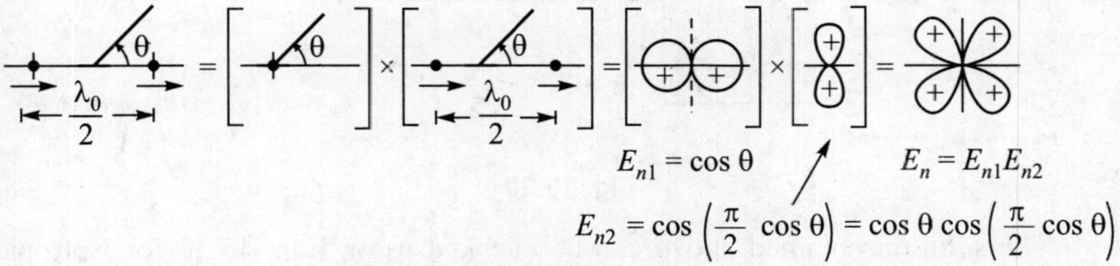

$E_{n1} = \cos\theta$ $\quad E_n = E_{n1}E_{n2}$

$$E_{n2} = \cos\left(\frac{\pi}{2}\cos\theta\right) = \cos\theta\cos\left(\frac{\pi}{2}\cos\theta\right)$$

Fig. 15.18

15.13.3. (*c*) End-fire Array with Dipoles Perpendicular to Axis of Array

$$d = \frac{\lambda_0}{2}, \alpha = \pi$$

The steps are shown in Fig. 15.19.

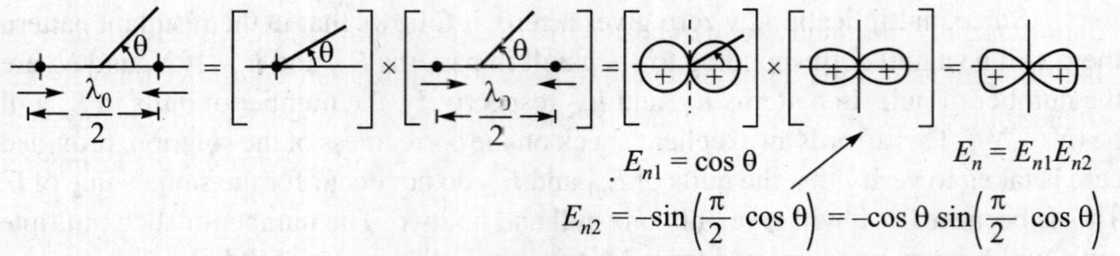
θ
λ0/2
+
$E_{n1} = \cos\theta$
$E_{n2} = -\sin\left(\frac{\pi}{2}\cos\theta\right) = -\cos\theta\,\sin\left(\frac{\pi}{2}\cos\theta\right)$
$E_n = E_{n1}E_{n2}$

Fig. 15.19

Since even a pair of isotropic radiators can give a wide variety of radiation patterns as in Fig. 15.15, it is obvious that we could use such a pair as E_{n1}, and make an array using two or more such pairs. Some examples will illustrate this concept in the next section.

15.14 FOUR AND EIGHT ELEMENT ARRAYS

15.14.1 (*a*)

It is desired to have a radiation pattern with a narrow lobe towards north, and nulls towards east and at 30° east of north. First we break up the desired conditions into two parts as shown in Fig. 15.20, and then referring to Figs. 15.9 and 15.15 select (*a*) as E_{n1} and (*b*) E_{n2}. The solution is explained below:

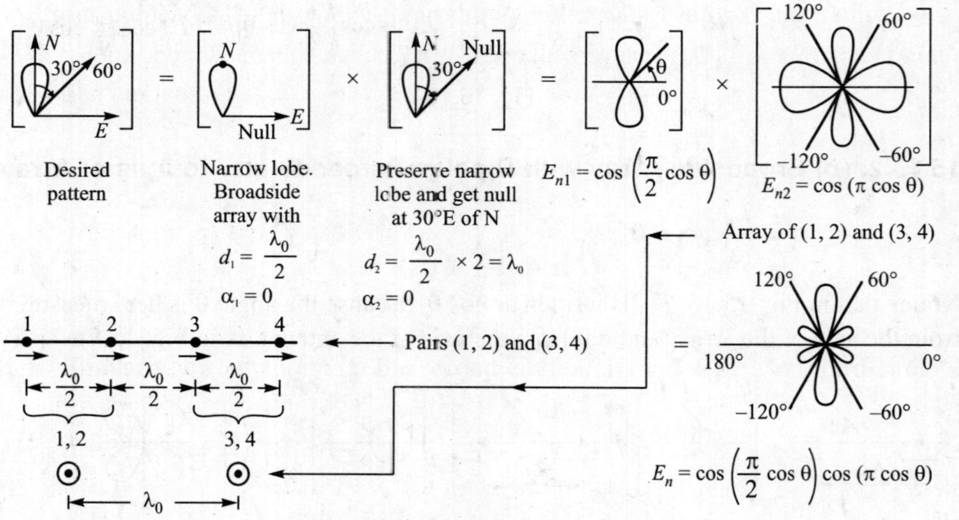
N
30°
60°
E
Desired pattern
Null
Narrow lobe. Broadside array with
$d_1 = \frac{\lambda_0}{2}$
$\alpha_1 = 0$
Preserve narrow lobe and get null at 30°E of N
$d_2 = \frac{\lambda_0}{2} \times 2 = \lambda_0$
$\alpha_2 = 0$
$E_{n1} = \cos\left(\frac{\pi}{2}\cos\theta\right)$
120°
-120°
-60°
$E_{n2} = \cos(\pi\cos\theta)$
Array of (1, 2) and (3, 4)
Pairs (1, 2) and (3, 4)
1
2
3
4
1, 2
3, 4
λ0
180°
0°
$E_n = \cos\left(\frac{\pi}{2}\cos\theta\right)\cos(\pi\cos\theta)$

Fig. 15.20

The solution obtained above can be checked using Eqn. 15.36 for isotropic radiators using the values $n = 4$ and

$$\psi = 0 + \frac{2\pi}{\lambda_0}\frac{\lambda_0}{2}\cos\theta = \pi\cos\theta.$$

Then
$$E_n = \frac{\sin\frac{4\psi}{2}}{4\sin\frac{\psi}{2}}$$

$$= \frac{\sin 2\psi}{4 \sin \frac{\psi}{2}} = \frac{2 \sin \psi \cos \psi}{4 \sin \frac{\psi}{2}} = \frac{4 \sin \frac{\psi}{2} \cos \frac{\psi}{2} \cos \psi}{4 \sin \frac{\psi}{2}}$$

$$= \cos \left(\frac{\pi}{2} \cos \theta \right) \cos (\pi \cos \theta)$$

which is the value obtained in Fig. 15.20.

15.14.2 (*b*) A Linear, Uniform Array of 8 Isotropic Radiators with $d = \frac{\lambda_0}{4}$, $\alpha = \pi$;

The step by step solution is obtained in Fig. 15.21 using Fig. 15.15.

The solution is checked using Eqn. 15.36 with $n = 8$ and

$$\psi = \pi + \frac{2\pi}{\lambda_0} \frac{\lambda_0}{4} \cos \theta = \pi + \frac{\pi}{2} \cos \theta$$

Then, $$E_n = \frac{\sin \frac{8\psi}{2}}{8 \sin \frac{\psi}{2}}$$

$$= \frac{8 \sin \frac{\psi}{2} \cos \frac{\psi}{2} \cos 2\psi}{8 \sin \frac{\psi}{2}} = \cos \frac{\psi}{2} \cos \psi \cos 2\psi$$

as found in Fig. 15.21.

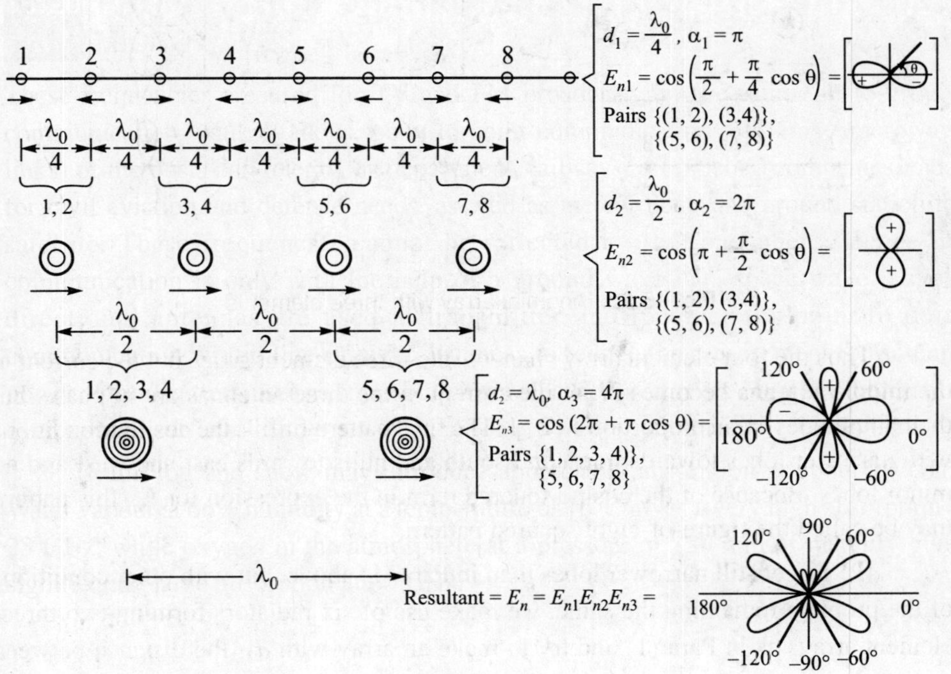

Fig. 15.21

15.15 BINOMIAL ARRAYS

In the examples of pattern multiplication it is observed that minor lobes appear in the resultant final pattern. In some applications this is not desired. Also it results in waste of power in undesired directions. This power would have been better utilised, if it were sent in the directions of the major lobes. Usually these minor lobes appear in E_{n2} and E_{n3}, etc. A perusal of Fig. 15.15 shows that minor lobes appear whenever the spacing between the isotropic radiators (d_2 or d_3) becomes greater than $\lambda_0/2$. Therefore the simplest way to eliminate the minor lobes is to prevent d_2 or d_3 from exceeding $\lambda_0/2$.

(1) It is desired to have a radiation pattern with narrow lobes towards the north and south and nulls towards east and west, but *with no minor lobes at all.* Since a narrow lobe is required, we start with a broadside array of isotropic radiators as in (*a*) with $d_1 = \dfrac{\lambda_0}{2}$ and $\alpha = 0$ as in Fig. 15.9. To get a narrower lobe than in Fig. 15.9, we use four such isotropic radiators with two pairs, each pair having $d_1 = \dfrac{\lambda_0}{2}$ and $\alpha_1 = 0$ as in Fig. 15.22. Now d_2 is the distance between the mid-points of the two pairs of radiators. To avoid minor lobes, d_2 must not exceed $\lambda_0/2$. This is only possible when antennas 2 and 3 coincide !!

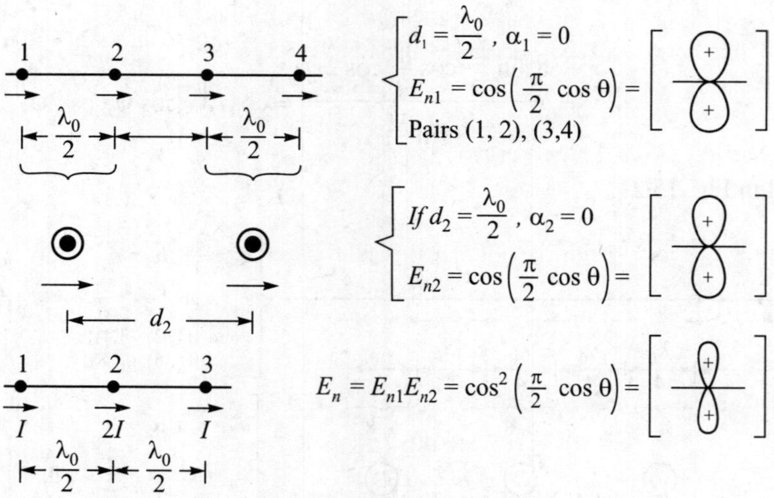

Fig. 15.22. Binomial array with three elements

Thus the four element array changes into three element array, but the current in the middle antenna becomes $2I$, or the current in the three antennas are in phase, but their amplitudes are in proportion 1 : 2 : 1. The final pattern fulfills the desired conditions with narrower lobes towards north and south and nulls towards east and west and no minor lobes. Because of the cosine squared term in the expression for E_n, this pattern may be called the figure-of-eight squared pattern.

(2) To get still narrower lobes than in Para (1) above, but with other conditions of the problem remaining the same, we make use of six radiators forming two three-element arrays as in Para (1) and try to make an array with d_3 (the distance between

mid-points of the three-element arrays) equal to $\lambda_0/2$ to avoid minor lobes (Fig. 15.23). This is possible only if antenna 2 and 4 coincide and also antenna 3 and 5 coincide. This reduces the six element array only to four elements, with currents in the two middle antennas being $3I$. Thus, the currents in the four antennas are in phase but their amplitudes are in the proportion 1 : 3 : 3 : 1. The cosine cubed term in the expression for E_n leads one to call it the figure-of-eight-cubed pattern.

By Binomial Theorem

$$(a + b)^2 = a^2 + 2ab + b^2$$

and
$$(a + b)^3 = a^3 + 3a^2b + 3ab^2 + b^3.$$

The coefficients (1, 2, 1) and (1, 3, 3, 1) in these give the proportions of currents in cases (1) and (2). Hence these arrays are called Binomial Arrays.

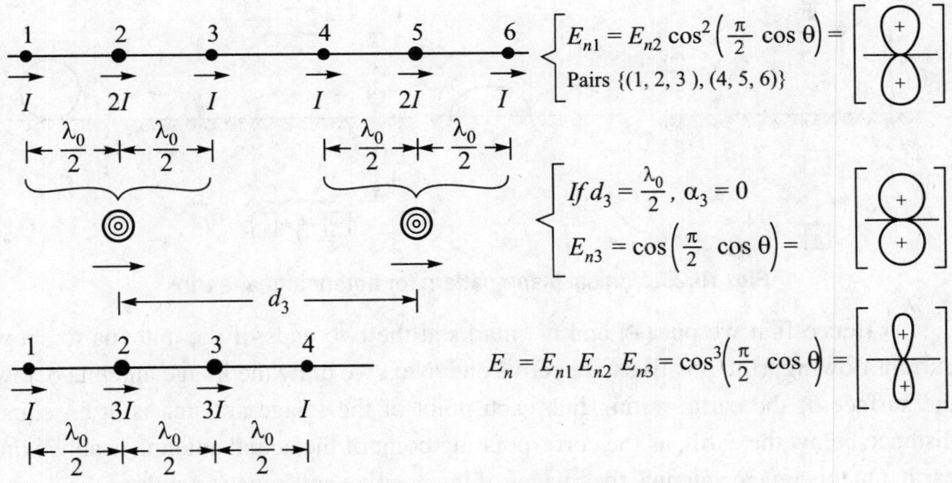

1 2 3 4 5 6

I $2I$ I I $2I$ I

$\frac{\lambda_0}{2}$ $\frac{\lambda_0}{2}$ $\frac{\lambda_0}{2}$ $\frac{\lambda_0}{2}$

$$E_{n1} = E_{n2} \cos^2\left(\frac{\pi}{2} \cos\theta\right) =$$

Pairs {(1, 2, 3), (4, 5, 6)}

d_3

$$\text{If } d_3 = \frac{\lambda_0}{2},\ \alpha_3 = 0$$

$$E_{n3} = \cos\left(\frac{\pi}{2} \cos\theta\right) =$$

1 2 3 4

I $3I$ $3I$ I

$\frac{\lambda_0}{2}$ $\frac{\lambda_0}{2}$ $\frac{\lambda_0}{2}$

$$E_n = E_{n1}\, E_{n2}\, E_{n3} = \cos^3\left(\frac{\pi}{2} \cos\theta\right) =$$

Fig. 15.23. Binomial array with four elements

15.16 EFFECT OF EARTH ON VERTICAL PATTERN

Most of the radio antennas used by mankind are located on the surface of the earth or not very far from the surface. At the frequencies used, the earth can be regarded as a very good (almost perfect) conductor. Hence the electric field produced or radiated by the antenna will cause currents to flow in the earth's surface, and these varying currents will modify the electromagnetic field above the surface. It is found that the principle of method of images can be applied in this case and we can locate suitable image antenna at equal distance below the earth's surface (assumed to be a plane), remove the conducting earth (mentally of course) and analyse the resultant array consisting of the actual antenna and its image, to get the field at any point (only) above the surface of the earth.

The directions of the currents in the image antennas are as shown in Fig. 15.24. Instead of trying to commit to memory a statement like "the currents in the actual and image antennas are in the same direction for vertical antennas, but in opposite directions for horizontal antennas", it may be easier to recollect that the method of images applies to stationary charges but not to currents.

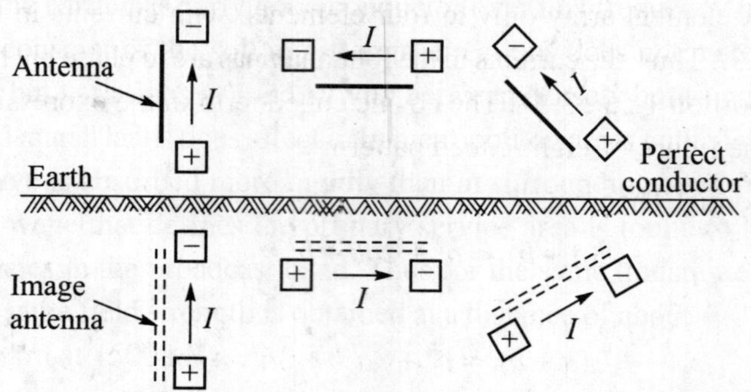

Fig. 15.24. Images of antennas in earth's surface

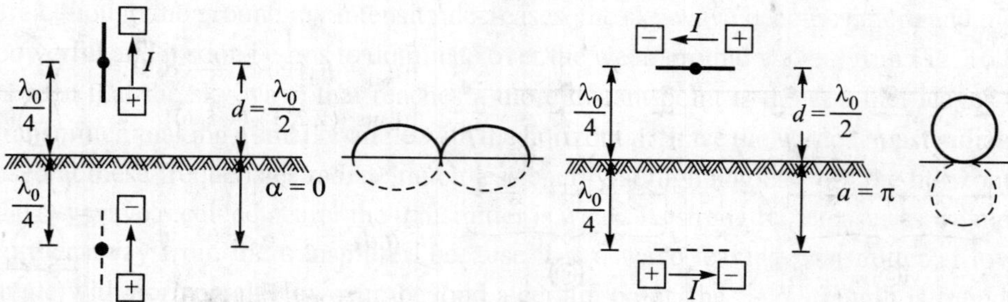

Fig. 15.25. Vertical plane pattern for antenna near earth

Hence first we put (+) and (–) marks at the two ends of the antenna to show current flowing from positive to negative end. Next we draw the image antenna below the surface of the earth, noting that each point of the image antenna is at an equal distance below the earth, as the corresponding point of the actual antenna is above the earth. On this image antenna, the images of the positive and negative ends of the actual antenna are marked with opposite polarity, as in Sec. 5.2. Then the current in the image antenna is from positive image change to the negative image charge. This procedure will automatically take care of the tricky situation with an antenna which is neither vertical nor horizontal but inclined. The vertical plane pattern can now be found for the array consisting of the actual and the image antenna, while ignoring the earth, and the pattern plotted only for points above the earth. This is shown for two cases of short dipoles in Fig. 15.25, which is based on (*g*) and (*i*).

15.17 THREE-DIMENSIONAL RECTANGULAR ARRAY OF IDENTICAL RADIATORS

Let a three-dimensional uniform array consist of identical radiators, whose radiation pattern is descibed as E_{n1}. Let

N_x, N_y, N_z be the number of such radiators along *x*, *y* and *z* axes,

d_x, d_y, d_z be the spacing between adjacent radiators respectively along *x*, *y* and *z* axes, measured as fractions of λ_0,

$\alpha_x, \alpha_y, \alpha_z$ be the phase advances between adjacent radiators along x, y and z axes respectively, measured as fractions of a cycle.

Then the radiation pattern for the array is

$$E_n = E_{n1}\left[\frac{\sin\{N_x\pi(\alpha_x + d_x\cos\phi\cos\theta)\}}{N_x\sin\{\pi(\alpha_x + d_x\cos\phi\cos\theta)\}}\right] \times \left[\frac{\sin\{N_y\pi(\alpha_y + d_y\sin\phi\cos\theta)\}}{N_y\sin\{\pi(\alpha_y + d_y\sin\phi\cos\theta)\}}\right] \times \left[\frac{\sin\{N_z\pi(\alpha_z + d_z\sin\theta)\}}{N_z\sin\{\pi(\alpha_z + d_z\sin\theta)\}}\right] \quad ...(15.39)$$

15.18 SUMMARY AND COMMENTS

The analysis in Sections 15.1 to 15.17 has shown that even such a simple device like a short Hertzian dipole gives an electromagnetic field, whose components can only be expressed in terms of rather complicated-looking equations. We simplified the analysis by concentrating our attention only on the radiation field term, which varies as $\frac{1}{r}$ and therefore which is the only term of importance at distances greater than a few wavelengths away from the antenna.

This led us to a very useful, though purely imaginary, concept of a radiation resistance of an antenna. We applied these concepts to the study of the simplest resonant antenna—an end-fed quarter wave monopole and a centre-fed half-wave dipole.

For the comparison of antennas, we developed the concept of the directional gain of an antenna and its effective length. The latter quantity is of importance, as a measure of the effectiveness of an antenna as a transmitter, or a receiver of electromagnetic waves in a direction perpendicular to its axis.

Finally a study of antennas arrays has shown that even a pair of isotropic radiators can be made to have many very interesting directional radiation patterns. By a large number of examples, the way in which various aspects of the radiation pattern can be manipulated has been illustrated.

EXERCISE 15

1. If $\overline{\nabla} \cdot \overline{A} = -\varepsilon\dfrac{\partial V}{\partial t}$, then prove that

$$\frac{1}{\varepsilon}\int \overline{\nabla}\times(\overline{\nabla}\times\overline{A})\,dt + \mu\frac{\partial\overline{A}}{\partial t} = -\overline{\nabla}V$$

where V is the electrostatic potential and $\overline{A}$ is the vector magnetic potential.

2. Putting $u = \cos\theta$, show that

$$\int_0^{\frac{\pi}{2}} \frac{\cos^2\left(\frac{\pi}{2}\cos\theta\right)}{\sin\theta}\,d\theta = \int_{-1}^{+1}\frac{1+\cos\pi u}{4(1+u)}\,du$$

3. (*a*) Putting $v = \pi(1 + u)$, show that

$$\int_{-1}^{+1} \frac{1 + \cos \pi u}{4(1 + u)} du = \frac{1}{4} \int_{0}^{2\pi} \frac{1 - \cos v}{v} dv$$

(*b*) Since $\cos v = 1 - \frac{v^2}{2!} + \frac{v^4}{4!} - \frac{v^6}{6!} + ...$, find the value of integral in (*a*) in the form of an infinite series. Taking terms upto v^{10}, find approximate value of the integral.

4. Assume that the Trichur station of All India Radio has two identical vertical mast radiators separated along the east-west line by a distance $d = 0.407\ \lambda_0$. If equal currents are fed in phase to the two mast radiators, determine the horizontal radiation pattern for the array and plot it. Comment on the need for this shape of the pattern.

[**Hint:** See a map of Kerala state, in which Trichur is situated].

5. Assuming that the earth is a perfect conductor, show that the electric field intensity at the surface of the earth for an end-fed quarter-wave monopole (Fig. 15.3*b*) is given by

$$E = \frac{3.816}{r} \sqrt{P_{\text{rad}}} \text{ volt/meter}$$

where r is the distance in kilometres and P_{rad} is the power in watts that is radiated by the monopole.

6. Following the method of analysis of Sec. 15.5, derive Eqn. (15.29*a*).

7. Using Eqn. (15.36), obtain the radiation pattern for an array of six isotropic radiators spaced $d = \frac{\lambda_0}{4}$ apart with $\alpha = \pi$. Sketch the pattern and indicate its nulls.

CHAPTER

16 PROPAGATION CHARACTERISTICS OF RADIO WAVES

16.1 THE ELECTROMAGNETIC WAVE SPECTRUM

Maxwell's remarkable 1864 discovery that time-varying electric and magnetic fields give rise to an electromagnetic wave travelling in space with the velocity of light, led to the identification of a whole family of these electromagnetic waves which differ in frequency (and wavelength), but all of which have the same velocity of propagation in space of approximately 3×10^8 m/s. Depending on their frequency (or wavelength), these waves can produce different effects in various materials and devices, and therefore different parts of the electromagnetic spectrum have been used for different purposes. Thus frequencies below 30 MHz are used for broadcasting and world-wide radio communication, while those between 30 MHz and 300 GHz (wavelength = 1 mm) are used for radar, point-to-point radio communication, and for baking in microwave oven. Waves of wavelength lying between 0.1 and 0.001 mm falling on our skin produce a sensation of heat (infra-red rays). Waves and wavelengths between 720 nanometers (nm) and 400 nm falling on our eyes give us a sensation of colours ranging from violet, indigo, blue, green, yellow and orange to red (visible spectrum). From 400 nm to 30 nm wavelength is the ultra-violet radiation. Soft to hard X-rays lie in the wavelength range from 1 nm to 0.01 nm, while gamma-rays range from 0.1 nm to 0.001 nm. Wavelengths below 10^{-5} nm (= 10 fm) are the cosmic rays arriving on the earth's surface from the entire universe.

These are the natural divisions of the electromagnetic spectrum as found in nature. They are depicted in Fig. 16.1. However, man has arbitrarily divided the radio-frequency part of this spectrum into frequency-bands, which have been given some standard names and corresponding letter symbols. These are shown in Tables 16.1 and 16.2.

16.2 TRANSMISSION PATHS FROM TRANSMITTER TO RECEIVER

When on 12 December 1901 Marconi succeeded in transmitting by wireless telegraphy the letter "S" in Morse Code (dit-dit-dit or ...) from Poldhu in Cornwall, England across some 1700 miles or the Atlantic Ocean to St. Johns in Newfoundland, he appeared to have achieved the impossible. The solutions of Maxwell's equations gave the theoretical basis and the laboratory work of Hertz, Sir Jagdish Chandra Bose and others had convinced everyone that like light, the electromagnetic radio waves followed the law of rectilinear propagation, that is, they travelled in a straight line from the transmitter to the receiver.

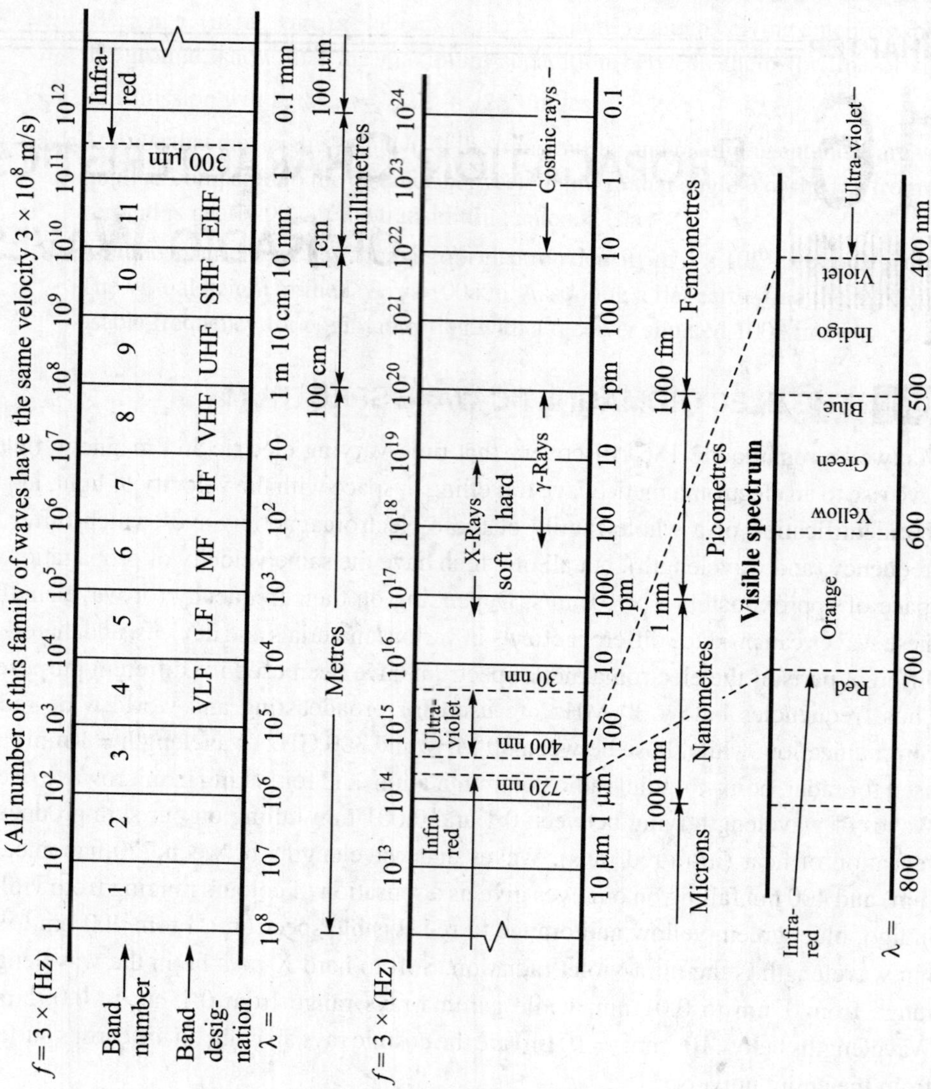

Fig. 16.1. Electromagnetic wave spectrum

Table 16.1 Nomenclature of frequency bands

Band number n	*Frequency range*		*Symbol and nomenclature*	*Uses*
4	3 to 30	kHz	VLF, Very Low Frequency	For maritime and naval use.
5	30 to 300	kHz	LF, Low Frequency	Maritime and aeronautical.
6	300 to 3000	kHz	MF, Medium Frequency	Distress signals and broadcasting.
7	3000 to 30000	kHz	HF, High Frequency	Long distance broadcasting.

(*contd...*)

8	30 to 300	MHz	VHF, Very High Frequency	Fixed & mobile AM & FM service, TV.
9	300 to 3000	MHz	UHF, Ultra-High Frequency	Fixed & mobile Am & FM
10	3 to 30	GHz	SHF, Super High Frequency (centimetre waves or microwaves)	Radar, satellite communication, microwave ovens.
11	30 to 300	GHz	EHF, Extremely-High Frequency (millimetre waves)	Research, Radio Astronomy

Thus, Band n extends from above 0.3×10^n Hz (excluding this frequency) to 3×10^n Hz (including this frequency).

This nomenclature was adopted in 1947 at Radio Convention held in Atlantic City in the USA.

Table 16.2 Standard letter designations for radar frequency bands (IEEE Standard 521–1976)

Letter designation	*Frequency band (GHz)*	*Letter designation*	*Frequency band (GHz)*
L	1 to 2	K	18 to 27
S	2 to 4	K_a	27 to 40
C	4 to 8	V	40 to 75
X	8 to 12	W	75 to 110
K_u	12 to 18	mm	110 to 300

An explanation was simultaneously offered by Oliver Heaviside in England and A.E. Kennelly in the USA, both of whom postulated the existence of an ionized conducting layer some 80 km or more above the earth. This layer acted like a mirror and reflected the radio wave back to earth, they suggested. A very crude simile will be to regard the ionized layer as a wire mesh. A large ball (= radio wave of long wavelength) striking it bounces back to earth, but a small ball (= radio wave of wavelength less than 10 m) will pass through the mesh, never to return. This explains why in an all-wave radio set, there is no band extending below 11 metres. This Kennelly-Heaviside layer is now called the ionosphere.

In this example, the path of the radio wave from the transmitter to the receiver lies entirely in the sky. Such a wave is therefore described as a sky wave. All long distance radio communication was dependent on this reflecting property of the ionosphere, till recent time. The space age dawned on 4th October, 1957 when the Russian satellite Sputnik began circling the earth. For sky wave communication with such a satellite situated on the other side of the ionosphere, the frequency had to be greater than 30 MHz to be able to penetrate the ionosphere (108 MHz, or 4 to 6 GHz with INSAT geo-synchronous satellites). The space probe Voyager 2 was launched from Cape Canaveral (Florida, USA) on 20 August 1977. It made its closest approach to Jupiter on 9 July 1979, Saturn in 1980, and Uranus in 1986. It passed Neptune in August 1989, and after sending us photographs of Neptune and its moon Triton, it has left the solar system, a never-to-return wanderer in space. This distance of 7×10^9 km represents the maximum distance from earth, where man has put a radio-transmitter and receiver and established two-way contact with it using space waves !! Though Pluto is the outermost planet of the solar system, in August 1989 it distance from the sun was less than that of Neptune, so that Neptune was the outermost planet at that time. This happened because the orbit of Pluto has a very high eccentricity, so that when it is closest to the sun, its orbit is within the orbit of Neptune.

Major Edwin H. Armstrong, who had in 1913 applied positive feedback in a triode amplifier and thus made an oscillator, was fighting in France against Germany during World War I. There in 1916 he conceived the idea of superheterodyne radio receiver, which was later patented. This made radio broadcasting commercially possible from 1920 onwards.

Broadcasters quickly discovered that the frequency band 535 to 1605 kHz was eminently suited for this purpose, because transmitting antennas in the form of vertical steel towers or masts of height that was stable could be easily fabricated by structural engineers. The electromagnetic wave radiating from a vertical mast uniformly in all directions in the horizontal plane travels out with its legs dragging along the ground or the surface of the earth. This causes the wavefront to tilt slightly forward, directing the signal energy towards the conducting ground, where it is slowly absorbed, thus producing attenuation of the wave as it progresses along the ground. Such a wave is called a surface wave, or a ground wave. Almost all the radio broadcasting in the world during daylight hours is done using ground waves.

With the development of air travel in this century, it became possible to have both the transmitter and the receiver at heights above the ground, but still within the troposphere—that is, within the first 15 km from the earth's surface, where the atmospheric temperature decreases with height. The receiver now gets a wave travelling in free space directly from the transmitter along a line-of-sight path, but it also receives another wave which is reflected from the ground or is diffracted around hills, very tall buildings, etc. Such a wave is called a space wave.

The third invention of Armstrong was in 1936 when he recommended the use of frequency modulation (FM) to get rid of noise or static in radio broadcasting. All television stations use FM for the audio or sound channel accompanying the amplitude modulated picture carrier.

For a short, vertical dipole antenna transmitting radio waves at an angle of, say, 10° to 40° above the horizon, this energy will travel like a wave in free space with electric field strength given by Eqn. (15.7*a*). This field strength will be decreasing inversely with distance from transmitter. For $\theta = 90°$ in Eqn. (15.7*a*) we get the amplitude of the electric field vector in the radiation field as

$$\text{Amplitude } E = \frac{60}{r}\left(\frac{\pi I_{max}\, dl}{\lambda}\right) = \frac{60\sqrt{2}}{r}\left(\frac{\pi I_{rms}\, dl}{\lambda}\right)$$

This result can be combined with Eqn. (15.8)

$$\text{Radiated power } P = 80\left(\frac{\pi I_{rms}\, dl}{\lambda}\right)^2$$

to give

$$E = \frac{60\sqrt{2}}{r}\sqrt{\frac{P}{80}} \text{ V/m} \qquad ...(16.1)$$

If the transmitter radiated power is $P = 1$ kW and the distance r from transmitter is 1 km = 1000 m,

$$E = \frac{60\sqrt{2}}{1000}\sqrt{\frac{1000}{80}} \text{ V/m}$$

$$= \frac{60\sqrt{2}\times 10}{2\sqrt{2}} \times \frac{1}{1000} \qquad ...(16.2)$$

or amplitude $E = 300$ mV/metre.

16.3 THE IONOSPHERE

The composition of the air by percentage of volume in the lower parts of the atmosphere is given in Table 16.3.

Table 16.3 Composition of Air (percentage by volume)

Gas		*Percentage*	*Gas*		*Percentage*
Nitrogen	(N_2)	78.094	Methane	(CH_4)	0.000 2
Oxygen	(O_2)	20.946	Krypton	(Kr)	0.000 114
Argon	(Ar)	0.934	Hydrogen	(H_2)	0.000 05
Neon	(Ne)	0.001 8	Nitrous Oxide	(N_2O)	0.000 05
Helium	(He)	0.000 524	Xenon	(Xe)	0.000 0087

Other gases present in variable concentrations are: water vapour 0 to 7%, carbon dioxide (CO_2) 0.01 to 0.1% (average 0.032%, ozone (O_3) 0 to 0.01%.

But at great heights, say above 90 km, the diffusion process dominates over the mixing by air currents. As a result the light gases hydrogen and helium are more abundant there than at ground level. Also some gases are found only below 100 km and not above. All these factors tend to give the ionosphere a layered appearance, rather than a homogeneous one, all the way from 50 to 250 km above the earth. In the region of the ionosphere, the atmospheric pressure is very low–of the same order as that in a high-vacuum triode tube !!

It was as early as 1839 that Gauss had tried to explain the measured variations in the earth's magnetic field by assuming that there exist large electric currents in an electrically conducting region in the upper part of the atmosphere. Interest in this region was revived by Heaviside and Kennelly. We now know that ionization of atoms and molecules of oxygen and to some extent of nitrous oxide by the ultraviolet rays of the sun is responsible for the presence of this region, where large electric currents can flow (also called the dynamo region or *E* layer, *E* originally signifying electrical). The production of ion-electron pairs per second at different heights by ultraviolet rays of different wavelengths (λ in nanometers) is shown in Fig, 16.2. Other agents like cosmic rays and meteors also produce these ion-electron pairs. Naturally the ions and electrons can recombine to form neutral atoms and molecules, but because of the extremely rarified atmosphere at these heights, this rate of recombination is far less than the rate of production. The result is a region that contains a large number of electrons, positive and negative ions.

However, in the shadow region of the earth, the absence of the ultraviolet radiation causes a marked decline in the rate of ionization and hence the charge density during the night (not on ground but at ionospheric height) is markedly lower than during day. It is only the electron density during day and night that is important for the study of radio wave propagation.

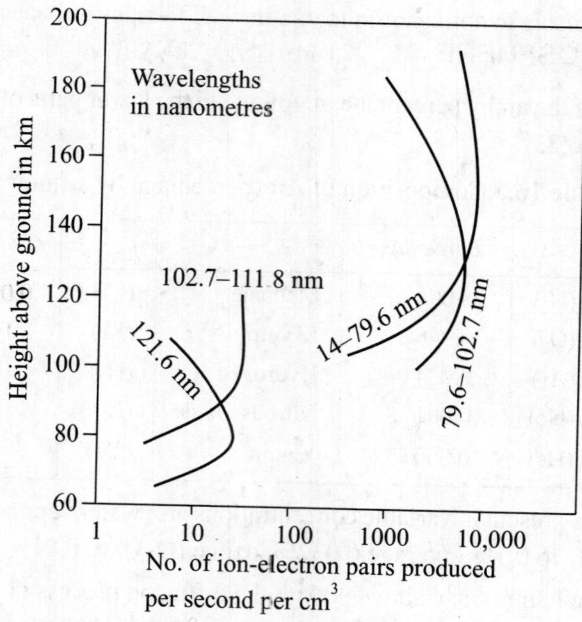

Fig. 16.2. Ionization rate in upper atmosphere

After the initial discovery of the so called *E* layer of the ionosphere, two more layers were discovered, one lying below and the other above the *E* layer. Naturally they came to be designated as the *D* layer and the *F* layer, respectively to maintain the alphabetical order. The F layer (formerly called the Appleton layer, since it was discovered by Sir. E. V. Appleton (1892-1965) in 1925) generally splits up during day time into F_1 (lower) and F_2 (upper) layers as seen in Fig. 16.3. The layers, their actual or true heights and other characteristics are described below:

D-layer extends from 50 to 90 km above the earth. It exists only during the day. It has a very high absorption rate for waves of frequencies in Band 6 (300 to 3000 kHz). This restricts the medium frequency radio broadcasts during day to ground wave only, and to a maximum distance of about 150 km. The D layer has only one electron/ion per 10^{11} neutral atoms and molecules.

E-layer occurs around 110-150 km above the ground. Its height remains more or less constant throughout day and night. The electron/ion concentration in it is about 1 per 10^8 neutral particles. This layer is useful for HF communication or broadcasting during daytime only up to a maximum distance of about 1500 km. At night, however, the MF broadcast range is increased beyond the day time limit of the ground wave, because of the reflection of the sky wave from the E-layer.

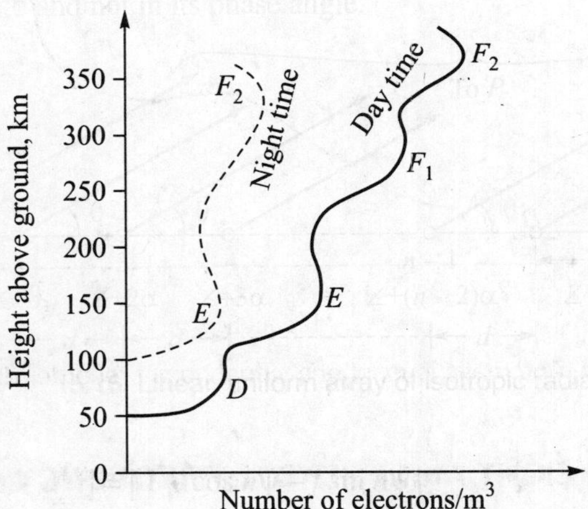

Fig. 16.3. Layers of ionosphere

F-layer can be divided into two layers. The F_1 layer at a height of about 220 km remains substantially at the same height throughout the day and night, while the F_2 layer has a height varying from 250 km to 350 km during day time. At night, however, the F_1 and F_2 layers coalesce to form a single layer designated as F_2 at about 250 km height. Usually a wave that penetrates the E-layer also penetrates the F_1 layer but with further attenuation. The ratio of ion/electron to neutral particles in the F-layer is $1 : 10^4$.

16.4 IONOSPHERIC INVESTIGATION

G. Breit and M. Tuve proposed in 1926 a radar-like echo method of sounding (= finding the depth of water under a ship) the ionosphere. Short pulses of radio waves of duration 0.1 millisecond were sent up, and this pulse and its echo displayed on a cathode-ray oscilloscope. Depending on the frequency of the electric power supply, 50 or 60 pulses are sent up per second. The time delay between the transmitted pulse and its echo from the ionosphere is then a measure of the to-and-fro path length. This gives the virtual height of the ionosphere, as described in the next section. Measurements have to be made with pulses of different frequencies to determine the heights of the different layers and to deduce their electron concentration.

Since the conditions in the ionosphere undergo drastic changes within a few minutes, it becomes necessary to make all the measurements over a range of frequencies (usually 1 to 10 MHz) in 15 minutes or less. The present day automatic measurement and recording of the virtual height of the ionosphere over range of frequencies is based on a method first used by Lal C. Verman. *et.* al.* in 1934. It gives a display of the type shown in Fig. 16.4.

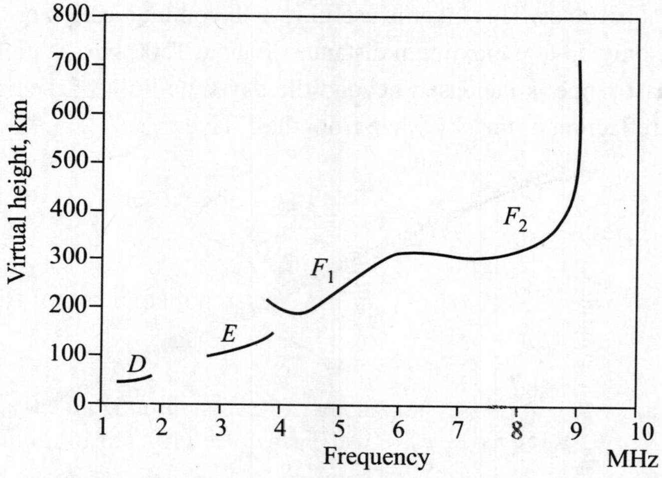

Fig. 16.4. Automatic ionospheric virtual height recorder display

16.5 VIRTUAL HEIGHT AND CRITICAL FREQUENCY

When a ray starting from a transmitter *T* reaches the receiver *R* via the ionosphere as in Fig. 16.5, a transfer of energy takes place from the electric field of the radio wave to the free electrons in the ionosphere. Some of these electrons collide with atoms and lose energy. This causes the group velocity of the wave in the ionosphere to be less than v_0, the velocity of the wave in the free space. There is a gradual turning of the wave, as it is propagated in the ionosphere along the path *CBD*. It is important to note that this reflection of the wave from the ionosphere is different from the sharp turning of a light beam from a plane mirror. If *TC* and *RD* produced intersect at *A* and *AN* is the perpendicular from *A* on the ground line *TR*, then *AN* is called the virtual height of the

*Lal C. Verman, S.T. Char and Aijaz Mohammed, *Continuous Recording of Retardation and Intensity of Echoes from the Ionosphere, Proc Inst. of Radio Engrs,* p. 906, July 1934. Lal C. Verman later became the Director of Indian Standards Institution. S.K. Mitra's book *The Upper Atmosphere* (1952) was a world classic on the subject of ionosphere.

ionosphere layer. In the actual time taken by the wave to travel along the path *TCBDR*, the wave could have travelled in free space with velocity v_0 along the path *TAR*.

The mechanism of this gradual turning of the wave can be understood in terms of the refractive index n and the dielectric constant ε_r of the ionosphere, which are related by the equation found in Chapter 11:

$$n = \sqrt{\varepsilon_r} \qquad \text{...(11.47)}$$

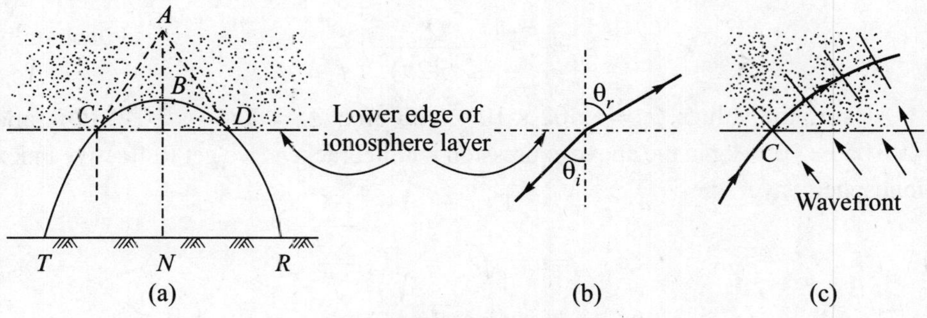

Fig. 16.5. Virtual height

Let N be the number of electrons per cubic metre in the layer of the ionosphere. Then under the influence of the sinusoidally varying electric vector component of the radio wave, the electrons will execute a motion with sinusoidally varying velocity along straight lines, so that $E = E_0\, e^{j\omega t}$, and velocity $v = v_0 e^{j\omega t}$. Denoting the charge and mass of an electron by Q_e and m, we can write

$$Q_e \overline{E} = m\frac{dv}{dt}$$

For the sinusoidal variation of E and v, the solution of this equation is

$$\overline{v} = \frac{Q_e \overline{E}}{j\omega m} \text{ m/s}$$

The current density resulting from this motion is

$$\overline{J} = NQ_e\, \overline{v} \text{ amp/m}^2$$

or

$$\overline{J} = \frac{NQ_e^2 \overline{E}}{j\omega m} = -j\omega\, \frac{NQ_e^2 \overline{E}}{\omega^2 m}$$

Actually the vibrating electrons lose a part of their kinetic energy when they collide with a gas molecule or atom, but such collisions are rather rare (say one per second). Such energy loss would give a small in-phase component of current, but since we have found that this is small, we can regard the current density $\overline{J}$ to be purely inductive, that is, we assume that it is lagging exactly 90° behind the voltage $\overline{E}$, as indicated by the coefficient $-j\,\omega$ in the above equation. This leads us to write Maxwell's curl equation for $\overline{H}$ as

$$\text{curl } \overline{H} = \overline{J} + \frac{\partial \overline{D}}{\partial t} = \overline{J} + \varepsilon_0 \frac{\partial \overline{E}}{\partial t} = \overline{J} + j\omega\varepsilon_0 \overline{E}$$

or using the above value of $\overline{J}$,

$$\text{curl } \overline{H} = j\omega \left[1 - \frac{NQ_e^2}{\varepsilon_0 \omega^2 m}\right] \varepsilon_0 \overline{E}$$

As in Eqn. (11.23), the right-hand side of the above equation should have been of the form $(\sigma + jw\varepsilon)\ \overline{E}$ or $jw\varepsilon \overline{E}$, since in this case we have put $\sigma \approx 0$. As $\varepsilon = \varepsilon_r\, \varepsilon_0$ and $\omega = 2\pi f$, it follows that

$$\varepsilon_r = 1 - \left[\frac{Q_e^2}{\varepsilon_0 m (2\pi)^2}\right] \frac{N}{f^2}$$

Using the values $Q_e = 1.602 \times 10^{-19}$ coloumb, $\varepsilon_0 = 8.854 \times 10^{-12}$ F/m, and $m = 9.1066 \times 10^{-31}$ kg in the above expression within brackets, we get Refractive index of ionosphere layer

$$= n = \sqrt{\varepsilon_r} = \sqrt{1 - \frac{81N}{f^2}} \qquad \text{...(16.3)}$$

where, N = Number of electrons per cubic metre, and

f = Frequency of the radio wave in Hz.

It is obvious that if $81N < f^2$, then the refractive index n is a real quantity and its value is less than unity. We know from optics that if a ray of light is going from a denser to a rarer medium (such as glass to air, for which $\mu_{ga} < 1$ then the ray bends away from the normal (Fig. 16.5*b*). This accounts for the bending of the path downwards from *C* to *B* in Fig. 16.5 as the electron density *N* increases from *C* onwards, as the ray penetrates more and more inside the layer. This increase of *N* with height as we go from the lower boundary of the layer towards the centre of the layer suggests that at some point we have a value of *N* such that $81\ N = f^2$. This makes the refractive index equal to zero. A higher value of *N* will make the refractive index an imaginary quantity, denoting that wave cannot be propagated there. The point corresponding to zero refractive index therefore marks the highest point up to which the wave can penetrate into the layer, and from where it, begins to come down towards the point *D*. Denoting by f_v the frequency of the wave which can reach a point *B* where the refractive index is zero, we write

$$f_v = \sqrt{81N} \text{ Hz} \qquad \text{...(16.4}a\text{)}$$

The point *B* corresponds to a point of total internal reflection of a ray from glass-to-air-surface, where the angle of refraction corresponding to the critical angle of incidence become equal to 90°. From *B* to *D* the mechanism is reversed and the ray is bent more and more towards the normal, as the value of *N* decreases and the refractive index increases towards the value unity at *D*.

If $N_{\max}$ is the maximum electron density in a layer of the ionosphere, then the frequency F_c at which this value will make the refractive index equal to zero will be given by

$$f_c = \sqrt{81 N_{max}} \text{ Hz} \quad ...(16.4b)$$

Then this critical frequency f_c is the highest frequency which, if incident on this layer of the ionosphere at any angle, will always be reflected back to the earth. A frequency higher than the critical frequency incident with $\phi_i = 0$ (vertical incidence) will never be able to reach the condition of zero refractive index in this layer of the ionosphere and will therefore penetrate this layer and proceed higher up.

Another way of looking at the above phenomenon is in terms of the phase velocity (see Sec. 14.14), which is related to the velocity v_0 in free space by the equation

$$\text{Phase velocity in layer of ionosphere} = \frac{v_0}{n} \quad ...(16.5)$$

Since $n < 1$, it follows that the phase velocity is greater in the ionosphere than in free space, in the same way as it is greater in a waveguide. An increase of N as the ray penetrates into the layer implies a decrease of the refractive index n, and hence a corresponding further increase in the phase velocity. This implies that the upper part of the wavefront (Fig. 16.5*c*) begins to move faster than the lower part, thus causing the path of the ray to turn. It is like a vehicle making a turn, in which the outer wheel turns faster than the inner wheel. In such a situation Snell's Law of Refraction is applicable and we have

$$n = \frac{\sin \theta_i}{\sin \theta_r} \quad ...(16.6)$$

At the point B, $\theta_r = 90°$ Hence, at that point

$$n_B = \sin \theta_i \quad ...(16.7)$$

It is conventional to call the point B as the point of reflection of the wave, even though, as we have seen above, the phenomenon involved is one of refraction.

Now we are in a position to state what would happen to a ray of frequency greater than the critical frequency incident at an angle on the layer. In such a case if Eqn. (16.7) can be satisfied, the ray will be reflected back to the earth; otherwise it will pass through the layer.

16.6 MAXIMUM USABLE FREQUENCY

Denoting by N_B the value of the electron density N at the point B, for which Eqn. (16.7) is true, we can rewrite Eqn. (16.7) as

$$\sqrt{1 - \frac{81 N_B}{f^2}} = \sin \theta_i$$

Squaring and transposing terms

$$\frac{81 N_B}{f^2} = 1 - \sin^2 \theta_i = \cos^2 \theta_i$$

or
$$N_B = \frac{f^2 \cos^2 \theta_i}{81} \qquad \text{...(16.8)}$$

Let D denote the distance TR from the transmitter to the receiver. A wave incident vertically and reflected from the ionosphere will return close to the transmitter with $D = 0$. But for the receiver situated at a distant point as in Fig. 16.5, the wave will have to be incident at an angle θ_i. If this wave is to be reflected from the ionosphere, Eqn. (16.8) must be satisfied. If N_{max} is the maximum value of the electron density in the layer, then replacing N_B by N_{max} in Eqn. (16.8), will enable us to find the maximum value of the frequency of the wave which will be able to reach the point R. Thus

$$81 N_{max} = f^2_{max} \cos^2 \theta_i$$

Combining this with Eqn. (16.4*b*), we get

$$f_c = \sqrt{81 N_{max}} = f_{max} \cos \theta_i$$

or Maximum Usable Frequency (MUF)

$$= f_{max} = f_c \sec \theta_i \qquad \text{...(16.9)}$$

Since sec θ_i is always greater than unity, it follows that MUF $> f_c$.

Also from the triangle TAN in Fig. 16.5*a*, with $m \angle TAN = \theta_i$.

$$\tan \theta_i = \frac{TN}{AN} = \frac{\frac{D}{2}}{\text{Virtual height}} \qquad \text{...(16.10)}$$

In practice, because of the earth's curvature must be taken into account when D is larger than say 100 km, both the transmitter-receiver line TR and also the ionosphere are parts of concentric circular arcs as shown in Fig. 16.6. Hence, for the F layer which is responsible for the major long distance transmission via the ionosphere, the largest possible value of θ_i is about 74°. This happens when the wave leaves the transmitter along the horizontal direction and also reaches the receiver in the horizontal direction (Fig. 16.6). Denoting by R_E (= 6350 km) the radius of the earth, we see that, if h is the height of layer above ground,

$$\sin \theta_{i\,max} = \frac{R_E}{R_E + h} \qquad \text{...(16.11)}$$

For this case,

$$(\text{MUF})_{max} = f_c \sec 74° = 3.628\, f_c$$

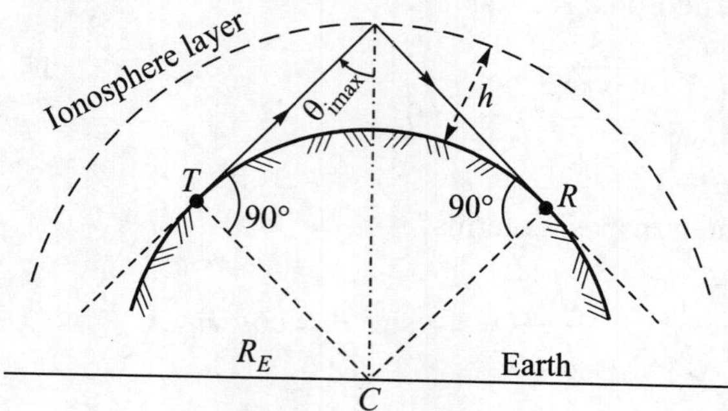

Fig. 16.6. Limiting case for MUF$_{max}$

16.7 EFFECT OF EARTH'S MAGNETIC FIELD

In the derivation of the refractive index of the ionosphere layer, we had taken into account only the effect of the electric vector of the radio waves on the electron. But we did not mention then that this electron is always situated in the magnetic field of the earth. The component of this magnetic field directed at right angles to the velocity of the vibrating electron will cause the electron to move along an arc instead of along a straight line (see Secs. 9.3 and 9.6).

Because of the earth's magnetic field, at frequencies of the order of 10 MHz or so, the electron moves along a narrow ellipse, whose major axis coincides with the electric vector of the radio wave (Fig. 16.7*a*). A part of this electrical path thus lies at right angles to the electric vector. The energy radiated by the moving electron (like a short dipole antenna) will thus give an electric vector perpendicular to the original electric vector of the radio wave. Thus an originally plane, linearly polarized wave will become an elliptically polarized wave.

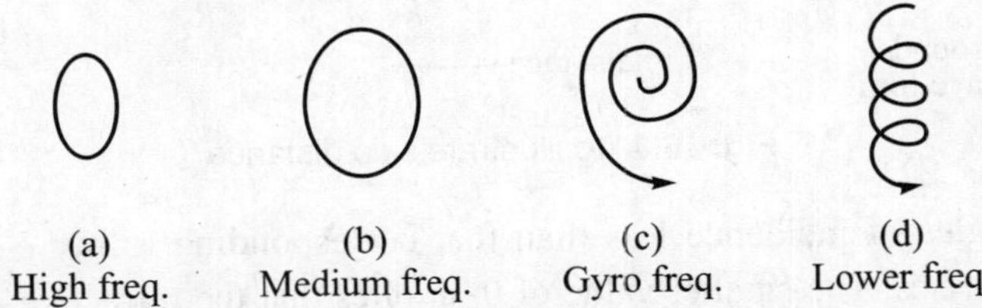

Fig. 15.7. Effect of earth's magnetic field on vibrating electrons

At frequencies lower than the above value, this effect will be more pronounced, because the average velocity of the electron is inversely proportional to the frequency of the radio wave. As a result the electron will move along a broader ellipse (Fig. 16.7*b*).

When the frequency is around 1400 kHz, an interesting phenomenon occurs. Now in the time that it takes for the electron to travel through 180° along its path, the electric field reverses. This synchronization between the change of direction of the velocity of the electron and the electric field is the basis used for accelerating charged particles in nuclear physics continuously, along paths of ever increasing radius in a device called cyclotron. Here the electron is said to undergo cyclotron-type resonance. The electron path now becomes a spiral instead of a closed loop (Fig. 16.7*c*). The frequency at which this occurs is called the gyro frequency. At frequencies below the gyro frequency, the path will be forming loops as shown in Fig. 16.7*d*.

The above elliptical polarization of the radio waves in the ionosphere can be compared to the production of ordinary and extra-ordinary rays in double-refracting crystal in optical studies.

16.8 SKIP DISTANCE

Consider a transmitter *T* sending out radio waves at different angles above the horizontal (Fig. 16.8). Let us study what happens to these rays when they reach the ionospheric layer at different angles. A radio wave *a* for which the angle of incidence θ_i is pretty large will get reflected almost from the lower edge of the layer and reach a very distant receiver at R_1. A ray *b* striking the ionosphere at a nearer point and therefore with a smaller value for θ_i, will penetrate a little into the layer and return to earth at a nearer point like R_2. This is the result of Eqn. (16.7), which has to be satisfied for the ray to be reflected. At a certain value of *N* corresponding to ray *c*, the ray is able to return to the nearest point R_3 from *T*.

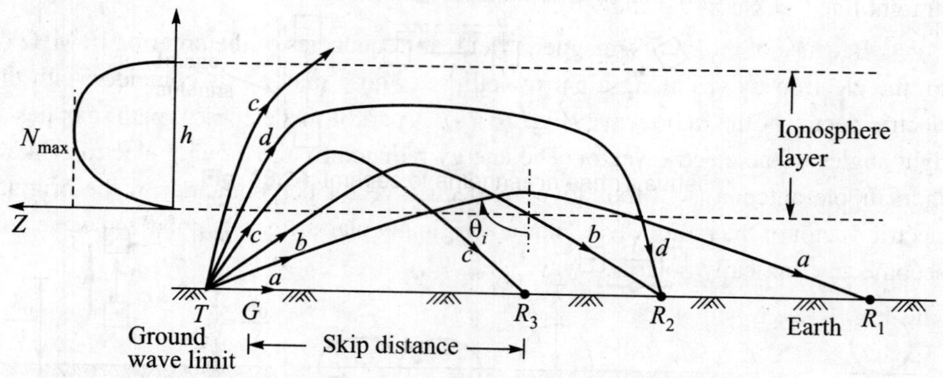

Fig. 16.8. To illustrate skip distance

For angles of incidence less than that corresponding to ray *c* an interesting complication arises. The smaller value of θ_i implies that the refractive index must be smaller, which means that the electron density must be higher. Now if we plot the variation of the electron density *N* with distance across the ionospheric layer as shown at the left of Fig. 16.8, we find that the resultant curve is having a broad flat top near N_{max}. For a ray like *d* with an angle of incidence less than that for ray *c*, the penetration brings the wave into this flat topped part of the curve. This slow rate of change of *N* near N_{max} causes the ray like *d* to travel a long distance almost parallel to the earth (curved earth if long distances are involved), before it is possible for Eqn. (16.7) to be satisfied and the ray reflected back to earth to reach a point *farther* than R_3.

It is clear from the above discussion that no sky wave can return to the earth at a point nearer than R_3. Since the ground wave will be completely absorbed within a short distance *TG* from the transmitter, it means that at this particular frequency, no signal can reach any points lying between *G* and R_3. This distance GR_3, which is approximately the same as TR_3, is referred to as the skip distance.

At still smaller angles of incidence like ray *e*, if N_{max} for the layer is insufficient to satisfy Eqn. (16.7), then the ray will fail to get reflected from this layer and will pass through it upwards.

From the foregoing, it is seen that the magnitude of the skip distance GR_3 (or TR_3) depends both on the frequency of the transmitter and the electron density variation in the ionosphere. Finally, one may wonder what would be the nature of the signal received at points beyond R_3, where, as shown in Fig. 16.8 for R_2, two rays along different paths reach at the same time. Of these two rays, the one designated as *d* in the figure (also called the Penderson ray) gets attenuated because of its longer distance of travel inside the layer of the ionosphere. Moreover, it does not come down at a single point like R_2 as suggested by the figure, but over an area around R_2. This distribution of energy of the wave over an area further reduces the intensity. The net result is that, the interfering effect of the Penderson ray is considerably less than what one would have expected, and effectively the signal along b is what is predominantly received by the receiver at R_2.

16.9 IONOSPHERIC BEHAVIOUR VARIATIONS

Because of the very mechanism of its coming into existence, the characteristics and hence the behaviour of the ionosphere is undergoing continuous changes. Some of these changes are of a regular cyclic type, and therefore, to a great extent, they are predictable. These are classified as the normal behaviour of the ionosphere. Thus the daily rising and setting of the sun (not on the ground level, but at the level of the ionospheric layer) produces a diurnal cycle of changes of the electron density and also the height of the layer, the density being higher during day time than at night. Superimposed on these variations is the seasonal change produced by the revolution of the earth around the sun, while its axis retains a fixed direction in space. This causes a cyclic change in the angle at which the sun's rays reach the earth and this therefore affects the rate of ionization.

While the variation in longitude does not appear to make any difference (that is, the same pattern of variation is obtained if, in each case, we take the *local* sunrise time as the reference point), a change of latitude results in marked variations, because of the change in the angle of incidence of the sun's rays. Thus there is greater ionization at lower latitudes. Curves giving the critical frequencies and virtual heights of the layers of the ionosphere at different local times are prepared for many places. When this is done over many years, it is found that there is a superimposed 11-year cyclic variation. Thus for the F_2-layer, the critical frequencies may vary from 5 to 11 MHz, tending to have a maximum value in those years in which the number of sun-spots are a maximum. This 11-year sun-spot cycle had reached maximum values in 1948, 1959, 1970 and 1981 and is expected to be maximum in 1992 and 2003. The observed values over many decades have shown the average period to be 11.11 years.

A few instances of abnormal behaviour of the ionosphere will now be considered. Sometimes in summer nights, sky waves of a frequency higher than the critical frequency of the E-layer are returned to the earth. This reflection occurs at a sporadic E-layer in the form of clouds of ions and electrons varying in size from one km to a few hundred km across and having a very high concentration of electrons/ions. The sporadic E-layer clouds have sharp boundaries. These clouds drift about in the E-layer, so that the signal is received after reflection from it only when the cloud is in a particular position. This unpredictability is described by the adjective, sporadic. The phenomenon can also occur at any time of the day and at any other season, but is more commonly observed during summer nights. Perhaps there tend to be more cases of sporadic E-layer formation when sun-spot cycle is at a maximum, but the correlation is not very good.

Dellinger in 1937 reported the sudden disappearance of the sky wave. He associated it with the sudden emission of large ionizing radiation from a solar flare. This produces an enormous increase in the value of the ionization below the E-layer. Radio waves passing up and then down through this layer are greatly attenuated, causing a radio fade-out called the Dellinger effect. The phenomenon may last from a few minutes to a few hours.

In Fig. 16.8 the sky wave is represented as a single ray coming down to the receiver. In reality, however, this is not a correct picture of the phenomenon. By its very nature, the ionospheric reflection is a statistical average involving a large number of electrons. As a result the electric field intensity received is the vector sum of the electric vectors of a very large number of waves coming from different groups of electrons. Since the phases of these vectors have random distribution which varies from moment to moment, the resultant electric field intensity undergoes variations from a maximum to a minimum value. This phenomenon is called fading. It occurs all the time. Another way in which fading can result is when two rays, following different paths through the ionosphere, arrive at the receiver. A third case of fading can result at medium wave frequencies, where the ground wave can travel fairly long distances and can therefore reach a receiver along with a sky wave reflected from the ionosphere. In each case it is the random variation of the phases of the two waves at the receiver which is responsible for fading.

To counteract the effect of this varying signal strength due to fading, we have to incorporate in all radio receivers an AGC (automatic gain control) circuit. This was formerly named AVC or automatic volume control circuit. In the presence of solar flares and magnetic storms, severe ionospheric storms can occur. In such cases fading can become pretty severe. If different frequencies fade at different rates, it is called selective fading. This can cause distortion, by modifying the sidebands of amplitude modulated signals.

Another phenomenon which also can produce fading, occurs when two rays reach the receiver at the same time after being reflected by different layers of the ionosphere such as E and F_2. Such fading can persist for a long time, until the constants of the layers change substantially.

All the year round a large number of meteors strike the upper atmosphere of the earth throughout day and night. It is only when they are of a somewhat larger size that they burn with a glow, which is visible at night and people call them shooting stars. These meteors also produce ionization, which contributes to the total ionization of ionosphere, particularly of the E-layer at night. Sometimes the earth in its passage around the sun passes through the gaseous tail of a comet, when such ionization assumes large magnitudes.

16.10 SKY WAVE PROPAGATION

Following the success of Marconi in 1901 to span the Atlantic Ocean with radio waves, long wave radio communication via the ionosphere developed rapidly. It is obvious from the geometry of Fig. 16.8 which is drawn for a flat earth, that for a single hop transmission via the ionosphere, the transmitter-receiver distance D will increase if the angle which the ray makes with the horizontal decreases. In the limit when these angles are zero and both transmitted and received rays are along the horizontal (for a curved earth) as shown in Fig. 16.6, the distance D will be a maximum. Further this distance will be greater for a ray reflected by a higher layer of the ionosphere (layer heights are: $D < E < F_1 < F_2$).

However, the sky wave reaching the ground via the ionosphere can get reflected from the earth as discussed in Chap. 12, because at the medium frequencies used for broadcasting, the earth can be regarded as a fairly good conductor. This ground-reflected ray can again be reflected by the ionosphere. There is no limit to the number of hops in such multi-hop transmission, but it must be remembered that every time a ray passes through a layer of the ionosphere (particularly the D-layer) while going up and coming down, it gets attenuated.

After 1920 when radio broadcasting became a successful commercial venture at medium frequencies using ground wave transmission, the radio amateurs were driven away from these frequencies below 3 MHz and assigned the useless frequencies above 3 MHz. It was a case similar to children being asked to leave the drawing room and to go to play in the backyard. Yet it was these radio amateurs who succeeded in sending signals at these frequencies from England to Australia towards the end of the third decade of this century. One of these amateurs, Gordon, came out to India to head the engineering department of the newly created All-India Radio in 1936. This discovery of the short waves led BBC (the British Broadcasting Corporation) to introduce in the 1930s the Empire Broadcasts, in the 31, 25, 19, 16 and 13 metre bands.

Those in India who listen to BBC late at night, particularly during winter often notice a peculiar phenomenon, as if an echo is being heard, lagging by about 140 m sec behind the main programme. This becomes possible because the signal from London to India travels on by multiple-hops over Australia, where it is *summer night.* The ionospheric conditions there are sometimes favourable for very long distance multi-hop transmission without much attenuation. As a result these signals travel completely round the world and again reach us a second time from the direction of London, to produce this echo effect. Since the radius of the earth is R_E = 6350 km and its circumference 39 880 km, a radio wave travelling at 300 000 km/s takes about one-seventh of a second to travel around the earth along the zig-zag path lying between the earth and the ionosphere. This is the basis for the 140 m sec delayed echo effect. On very rare occasions, as many as five such round-trip echoes have been reported, but one or two round-the-world signals are more common.

16.11 GREAT CIRCLE DISTANCES ON EARTH

In connection with a spherical earth, it is necessary to remember that the Euclidean geometrical ideas are no longer true here. Thus the shortest distance between two points on spherical earth is not a straight line but the arc of a circle on curved ground. Consider first two points *A* and *B* on a plane with *D* as the midpoint of the line *AB* (Fig. 16.9). If a circle is drawn with centre *D* and radius *DA* (or *DB*), it will be the longest arc of a circle joining *A* and *B*. If *NDN′* is the straight line which is the perpendicular bisector of the line *AB*, then with any point *C* on *NN′* as centre and radius *CA* (or *CB*), an arc can be drawn through *A* and *B*. Such an arc 2 will be shorter than arc 1 drawn first. If we place the restriction that *CA* cannot be greater than a certain constant value *R*, then the arc 2 will be the shortest if *CA* = *R*.

This is the concept used on the spherical surface of the earth. If *A* and *B* are on the earth's surface, straight line *AB* lies wholly inside the earth. Any plane passing through the straight line *AB* will intersect the earth's surface along a circle. Changing the position of the plane changes the size of the circle. The largest sized circle is obtained when the circle passes through the centre *C* of the earth, because then $CA = CB = R_E$, the radius of the earth. This circle which has its centre at *C* and whose radius is R_E is called the great circle, passing through the points *A* and *B*. It gives two arcs joining *A* and *B*: one is the shortest path from *A* to *B* *along the surface of the earth,* and the other arc is the longer great-circle path. All sky wave paths, including round the world signals, follow such great circle paths, and all rays from transmitter *T* to receiver *R* lie on the plane passing through the points *T, R* and *C,* the centre of the earth. Finally in connection with the round-the-world, 1/7 sec delay signals from, say, London mentioned in Sec. 16.10, it may be mentioned that, under favourable conditions, it is also possible to receive two signals from London, which have travelled in opposite directions, one along the shorter and the other along the longer, great circle paths. In this case the delay between the signals will be less than 1/7 second.

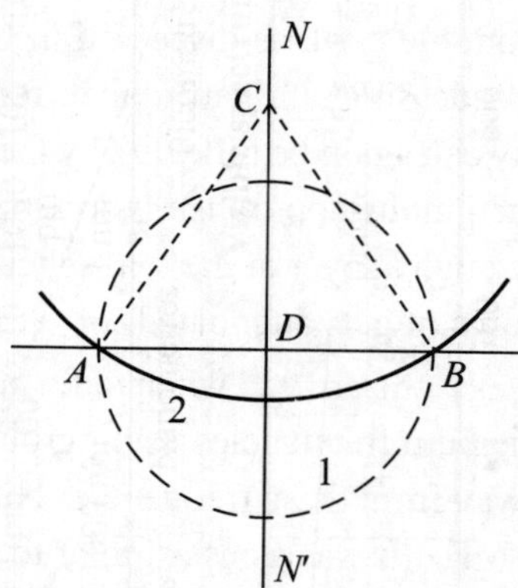

Fig. 16.9. Circular arcs joining points *A* and *B*

16.12 GROUND WAVE PROPAGATION

When the transmitting antenna is vertical and close to the ground like a tower or mast of length less than a wavelength, a ground or surface wave propagation results, in which the electric vector is everywhere vertical and terminated in positive or negative charges on the ground. Also the magnetic lines are everywhere horizontal and forming closed loops surrounding the electric field vectors. Thus the requirements of Sec. 14.1 are duly fulfilled and a side view of the wave looks the same as in the side view of

Fig. 14.5. For these frequencies, the earth appears to be a fairly good conductor, though not a nearly perfect conductor like a metal. This results in power loss in the ground and therefore the wave seems to drag its feet along the ground. This causes the electric vector to tilt slightly forward.

This short Hertzian dipole antenna radiates a field in the space above the ground in the form of a hemispherical wave spreading out from the antenna. The electric field is isotropic in the horizontal plane, but proportional to sin θ (θ being the angle with the vertical) as in Eqn. (15.7), or proportional to the cosine of the angle with the horizontal plane. By Eqn. (15.7), the amplitude of the electric field vector can be represented by E_0/D at the receiver, showing that it is inversely proportional to the distance D from the transmitter. The value of E_0 is found by an extension of Eqn. (16.2) to be $300\sqrt{P_{kw}}$ mV/metre at $D = 1$ km, where P_{kw} is the radiated power of the transmitter in kilowatts. But as the German physicist A. Sommerfeld pointed out in his paper in 1909, the actual field strength of the ground or surface wave is

$$E = A\frac{E_0}{D} \qquad ...(16.12)$$

where A is a factor whose magnitude depends in a complicated manner upon numerous factors like ground conductivity and permittivity, frequency, polarization (vertical and horizontal) of the wave and distance D. As a result it is always easier to obtain the values of A and E from curves drawn for different values of these parameters. If D is large and therefore involves the curvature of the earth, then the bending of the ray into the region below the optical horizon has to depend on the refracting property of the atmosphere, whose dielectric constant is slightly more than unity, as given in Table 7.2, or on the phenomenon of diffraction. For frequencies below 3 MHz, the earth acts like a purely resistive surface, but at frequencies higher than, say 10 MHz, it behaves more like a leaky capacitor (a capacitor shunted by a high resistance).

16.13 SPACE WAVE PROPAGATION

When the transmitting and receiving antennas are above the ground but situated anywhere in the troposphere within 15 km above the surface of the earth, then in general two waves can reach the receiver. One is a direct sky wave from the transmitter to the receiver and the other is a wave reflected by the ground as shown in Fig. 16.10. This combination is referred to as space wave propagation. Such a situation will arise for instance when the transmitting and receiving antennas are in two aeroplanes, or are situated at elevated positions above the general ground level.

It is obvious that for dealing with a wave reflected by the ground, the coefficient of reflection of the ground has to be taken into consideration as in Chap. 12. This reflection coefficient depends upon the conductivity and dielectric constant of the earth, the frequency and polarization of the wave, as also on the angle with the horizontal that the incident ray makes. Therefore, this complicated relationship is generally easier to depict in the form of charts.

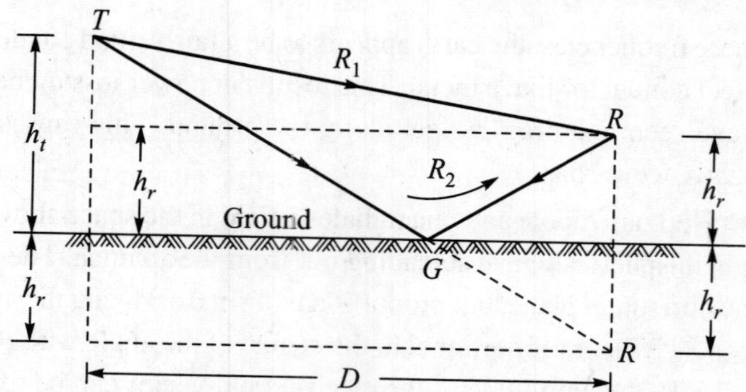

Fig. 16.10. Space wave propagation

But a very simplified analysis can be attempted here on the assumption that the distance D is so large as compared to the heights h_t and h_r of the transmitting and receiving antennas above the ground, that for all practical purposes, we can regard the ray incident on the ground at G to have the angle of incidence measured from the vertical as $\theta \approx 90°$. As in geometrical optics for reflection from a plane mirror, the point G is located by first finding the image point R' at a distance h_r below the ground, and then joining TR' to intersect the ground at G. Denote the direct distance TR by R_1 and the distance $TR' = TG + GR$ by R_2.

Since both the parts of R_2 are nearly parallel to the ground (horizontal, $\theta \approx 90°$) and $R_1 \approx R_2$, it is reasonable to assume that the two waves reaching the receiver R along the two paths have equal amplitudes, but they differ in phase due to the difference in lengths of the two paths. To this phase difference we have to add another 180°, because a phase reversal occurs on reflection at G, irrespective of whether the wave is horizontally or vertically polarized. This means that the reflection coefficient is taken as $\rho = 1 \angle 180°$.

From the geometry of Fig. 16.10, we write [using Eqn. (1.1)]

$$R_1 = \sqrt{D^2 + (h_t - h_r)^2} = D\left[1 + \left(\frac{h_t - h_r}{D}\right)^2\right]^{\frac{1}{2}} = D + \frac{(h_t - h_r)^2}{2D}$$

$$R_2 = \sqrt{D^2 + (h_t + h_r)^2} = D\left[1 + \left(\frac{h_t + h_r}{D}\right)^2\right]^{\frac{1}{2}} = D + \frac{(h_t + h_r)^2}{2D}$$

Hence, $$R_2 - R_1 = \frac{(h_t - h_r)^2 - (h_t - h_r)^2}{2D} = \frac{2h_t h_r}{D}$$

This path difference introduces a phase difference of

$$\frac{2\pi}{\lambda} \times \frac{2h_t h_r}{D}$$

radians. Introducing a sinusoidal time variation term, the two waves reaching the receiver R can be represented as:

$$\text{along } R_1 : E_0 \cos \omega t$$

$$\text{along } R_2 : - E_0 \cos \left(\omega t + \frac{4\pi h_t h_r}{\lambda D} \right).$$

the minus sign showing phase reversed at G.

Adding,

$$\text{resultant} = 2E_0 \sin \frac{1}{2} \left[\omega t + \left(\omega t + \frac{4\pi h_t h_r}{\lambda t} \right) \right] \times \sin \frac{1}{2} \left[\left(\omega t + \frac{4\pi h_t h_r}{\lambda D} \right) - \omega t \right]$$

$$= \left[2E_0 \sin \frac{2\pi h_t h_r}{\lambda D} \right] \sin \omega t \qquad \text{...(16.13)}$$

since, $\frac{2\pi h_t h_r}{\lambda D} \ll \omega t.$

Hence, the amplitude of the resultant shows that in expression for field strength at distance D, E_0 has to be replaced by the quantity in brackets in the above expression. Thus we get,

$$\text{Field strength at receiver } R = 2\frac{E_0}{D} \sin \frac{2\pi h_t h_r}{\lambda D}$$

$$= \left(\frac{4\pi h_t h_r}{\lambda} \right) \frac{E_0}{D^2} \qquad \text{...(16.14)}$$

using the relation $\sin x = x$ when x is very small.

In the absence of the ground-reflected wave, the field strength would have been $\frac{E_0}{D}$, that is varying inversely as D. But now with the ground-reflected wave present, Eqn. (16.14) shows that it varies inversely as the square of the distance D. In other words, the field strength now decreases more rapidly with distance. Also Eqn. (16.13) shows that for values of D which will make $\frac{2\pi h_t h_r}{\lambda D}$ = an integral multiple of π, the signal strength will be zero. We may have 1 to 10 such minima between the values 0 to $\pi/6$ for $2\pi h_t h_r/(\lambda D)$, the higher number occurring when the frequency is higher. In between these minima, maximas occur where the two signals reinforce each other and the resultant becomes greater than E_0/D.

16.14 SPACE WAVE PROPAGATION AFFECTED BY ATMOSPHERE

The density of the air in the troposphere within 15 km of the ground decreases rapidly with height. Moreover the air can hold varying amounts of water vapour in it. Since the dielectric constant of water is as high as 80, it follows that the dielectric constant of moist air can vary over a wide range. This means that the refractive index of air will be higher than one and it will be varying with height, because of changes of density and vapour content. To take into account this presence of moisture, we modify the refractive index μ of air to a new value defined as

Modified refractive index of air = μ_m

$$= \left(\mu - 1 + \frac{h}{R_E}\right) \times 10^6 \qquad ...(16.15)$$

where h is the height above the ground and R_E is the radius of the earth, both in metres.

In practice it has been found that most of the effects produced by the atmospheric refraction do not depend so much on the value of the modified refractive index μ_m, but on the rate at which it changes with height, that is, on $\frac{d\mu_m}{dh}$. At very great heights in the troposphere, the dielectric constant μ of the air does not vary with height. This gives a constant value of μ, but μ_m as defined by Eqn. (16.15) increases at the constant rate of 0.157 48 units per metre.

Measurements show that over *land,* near the earth's surface, with *typical* atmospheric conditions, the dielectric constant and hence the refractive index of the atmosphere *decreases* with height. In this *standard atmosphere,* therefore, the rate of increase of μ_m is less than the value given above, and it is only 0.118 11 units per metre as shown in Fig. 16.11(a). The decreasing value of μ with height in the standard

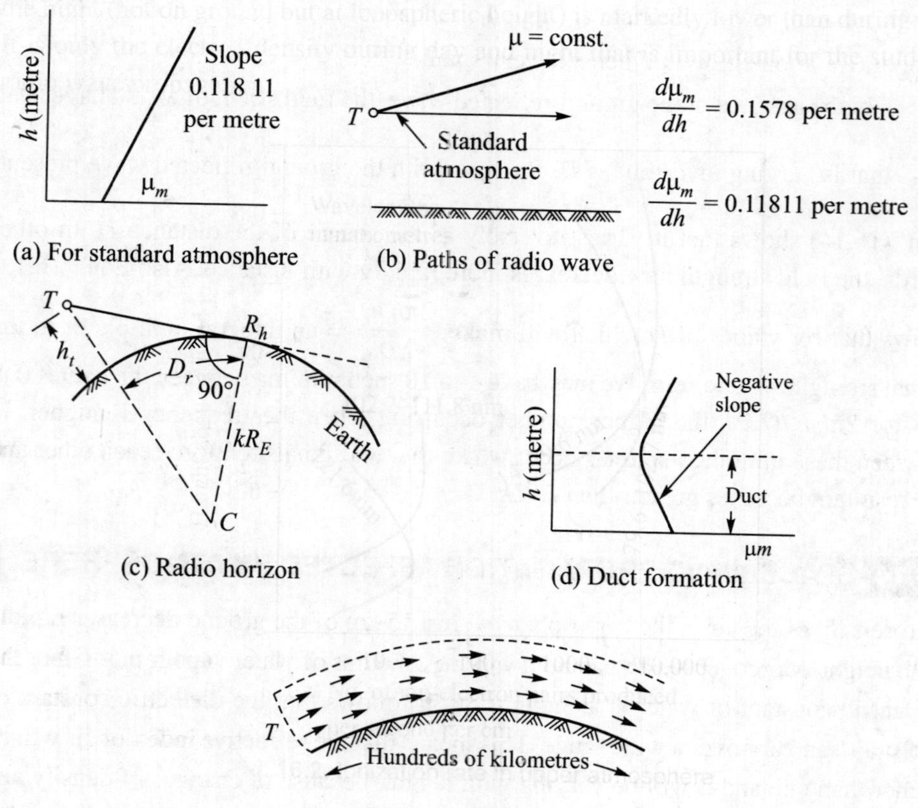

Fig. 16.11. Effect of variation of modified refractive index

atmosphere, is like a ray in optics travelling from an optically denser to a rarer medium and getting refracted away from the normal. Hence as shown in Fig. 16.11(b), the standard atmosphere causes the radio beam to bend towards the earth and thus be able to reach points that are farther away than the optical horizon.

It is inconvenient to work with a curved path for the radio beam. We search for some method by which we could replace this curved path from the transmitter to the receiver by a straight one. In Chap. 5, we had seen some methods of modifying co-ordinates in such a way that what comes out as a curve in one system of co-ordinates becomes a straight line in another system of co-ordinates. A similar transformation is used here. Figure 16.11(b) suggests that the desired result can be obtained if we use

Modified radius of the earth

$$= kR_E = \frac{0.157\,48}{0.118\,11} R_E$$

$$= \frac{4}{3} R_E = 1.33 R_E \qquad \text{...(16.16)}$$

The use of this modified radius is made in Fig. 16.11(c) to determine the distance from the transmitter T to the radio horizon R_h. Obviously this figure is not drawn to scale, because in reality, kR_E is very much larger than TR_h. Hence to a first approximation

$$\text{Distance to Radio Horizon} = D_r \approx TR_h = \sqrt{2(1.33R_E)h_t} \qquad \text{...(16.17)}$$

where h_t and R_E are both in metres, and also D_r is in metres.

However, a very neat and interesting modification of this equation is obtained if h_t is expressed in feet and D_r in miles. Then

$$\text{Distance to Radio Horizon} \approx D_r \approx \sqrt{2h_t} \text{ miles} \qquad \text{...(16.18)}$$

An interesting phenomenon is associated with exceedingly large moisture content in the air near the earth's surface, particularly over seas and oceans. This not only decreases $\frac{d\mu_m}{dh}$ from the value 0.157 48 to a lower value, but actually makes it negative as shown in Fig. 16.11(*d*). A duct is formed at the earth's surface, the top of the duct being at the point where $\frac{d\mu_m}{dh} = 0$. In the duct the radio beams bend so as to maintain a constant height above the ground as shown in Fig. 16.11(e). These radio waves can therefore travel to very long distances to many hundred kilometres without suffering the large attenuation to be expected at these distances.

Many years ago, the Pune TV station used to receive the Bombay TV signals on Channel 4 by sky wave at its transmitter situated on Simhgarh Fort at an altitude of 1280 metres, and then it would re-transmit the programme on Channel 5 of Pune station. Since Karachi, Bombay and Pune are on a straight line and Karachi TV station also operates on Channel 4, during the months of June-July with monsoon active all over the Arabian Sea, the above duct formation often used to occur, enabling Pune station to receive both Bombay and Karachi TV signals at the same time, causing mutual interference. This trouble is now eliminated as Bombay TV programme is being sent to Simhgarh by a microwave link.

16.15 PROBLEMS OF SPACE WAVE PROPAGATION

One of the problems encountered in space wave propagation such as TV reception, is the presence of obstacles like hills and tall buildings. These structures can reflect the radio waves from the transmitter on to the receiving antenna, thereby producing a slightly displaced (because of differences in path lengths) faint picture called ghosts, near the brighter main picture received directly by a sky wave. Another case arises when the hill or tall building is in the direct path from transmitter to receiver. Though the receiver is now located in the shadow of the obstacle, it is still possible to receive a satisfactory signal under certain conditions. To understand the phenomenon involved, let us first get the idea of Fresnel zones.

To explain the phenomenon of diffraction of light at a straight knife-edge, the French physicist Fresnel (1788- 1827) (the "*s*" in his name is not to be pronounced) considered first a ray *CR* along which light is travelling (Fig. 16.12). Then on the plane wavefront perpendicular to *CR,* we can draw circles with centre *C* and radii $\sqrt{d\lambda}$, $\sqrt{2d\lambda}$, $\sqrt{3d\lambda}$, etc. where distance $CR = d$. Then the distance RM_1 from *R* to any point on the first circle is

$$RM_1 = \sqrt{d^2 + \left(\sqrt{d\lambda}\right)^2} = (d^2 + d\lambda)^{\frac{1}{2}}$$

$$= d\left(1 + \frac{\lambda}{d}\right)^{\frac{1}{2}} = d\left(1 + \frac{\lambda}{2d}\right) = d + \frac{\lambda}{2}$$

Similarly, for any point on the second circle, $RM_2 = d + 2\frac{\lambda}{2}$, on third circle, $RM_3 = d + 3\frac{\lambda}{2}$, etc.

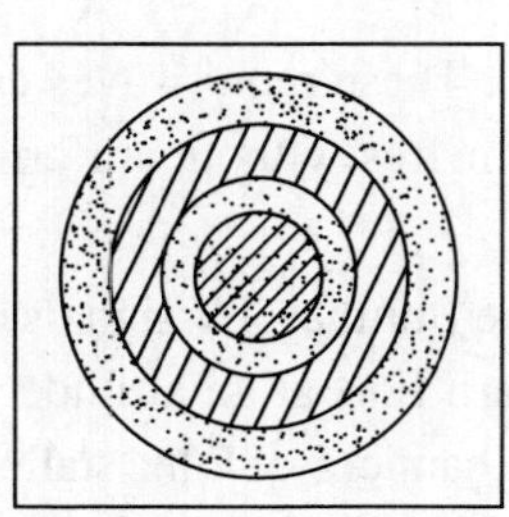

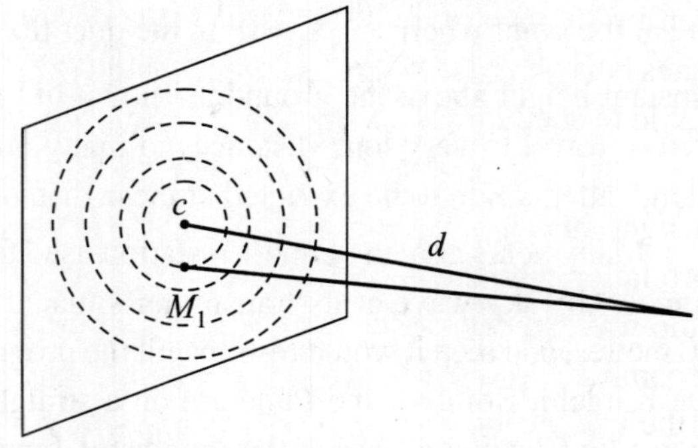

Fig. 16.12. Fresnel zones

The areas of the circles are π (radius)^2 or $\pi d\lambda$, $2\pi d\lambda$, $3\pi d\lambda$, etc. In other words, the area of the central circle (first Fresnel zone), is $\pi d\lambda$; the area of the surrounding ring (second Fresnel zone) between first and second circles is $(2\pi d\lambda - \pi d\lambda) = \pi d\lambda$; of the next surrounding ring (third Fresnel zone) is $(3\pi d\lambda - 2\pi d\lambda) = \pi d\lambda$, etc. Thus each Fresnel zone has the same area $\pi d\lambda$.

Because of the increase in path difference by $\frac{\lambda}{2}$ associated with each successive zone having same area $\pi d\lambda$, we can say that each successive Fresnel zone makes equal but opposite contributions to the electric field intensity at R.

To determine whether the earth is sufficiently removed from the radio wave path, to allow free-space propagation to be possible between T and R, is to have the First Fresnel zone to clear all obstacles in the path. For this the modified (or fictitious) earth radius 1.33 R_E is used.

In the shadow region behind an obstacle, the field strength can be determined by replacing the obstacle by a knife-edge and then determining ray in the shadow, where diffraction fringes result. Sometimes the signal strength is found to be larger in the shadow than the free-space field strength given by Eqn. (16.1). This obstacle gain is possible, because of the fact the obstacle prevents the ground- reflected ray with 180° phase shift from reaching the receiver.

If the receiver is far away from the transmitter (for instance in the secondary service area of a TV station), the space wave signals may vary in signal strength (fading) if conditions in the troposphere undergo changes. Such variations may be more often encountered in the shadow regions behind obstacles.

16.16 PROPAGATION CHARACTERISTICS OF RADIO WAVES 3 TO 100 kHz

These long waves (wavelength at 10 kHz is 30 km) are able to go as ground or surface waves to very long distances, such as 1000 km and as sky waves to many times longer distances, and still provide usable field strength. The sky wave attenuation is small, because it is reflected from the lower regions of the ionosphere, at heights of 60 to 90 km. Moreover, the very low frequency waves below 30 kHz frequency can be received even inside the sea by a submarine. It is because of these advantages that the United States Navy has a radio telegraphy transmitter in California working on 16 kHz, which is able to send messages to any US naval ship or submarine anywhere in the world.

There is hardly any fading of the signal at these frequencies. However, slow changes in field strength during night and day, over the seasons and also having a correlation with the 11-year sun spot cycle are always present. The field strength of the radio wave in this frequency range when propagated over the sea during daytime, can be computed using the empirical Austin-Cohen formula given in 1911 by these scientists of the US Bureau of Standards:

$$E = \frac{298}{D}\sqrt{P_{kw}\frac{D/R_E}{\sin\left(\frac{D}{R_E}\right)}}\,e^{-0.0015D/\sqrt{\lambda}} \qquad ...(16.19)$$

where D = Distance between transmitter and receiver in kilometers

P_{kw} = Transmitter radiated power in kilowatts

R_E = Radius of the earth (or D/R_E radians is the angle subtended by D at the centre of the earth

and λ = Wavelength of radio wave in kilometre $\left(= \frac{300}{\text{kHz}}\right)$.

The formula is valid with fair accuracy for distances between 600 and 10,000 km. The later theoretical considerations showed that the attenuation factor α which has the value 0.0015 in most cases [and is therefore used in Eqn. (16.19)], could vary from 0.0001 to 0.0050.

It is found that there is a difference in the field strength when the waves travel a distance in the north-south direction than when they travel the same distance in an east-west direction. Two factors obviously cause this difference. The first is the magnetic field of the earth, which is directed mostly north-south. The second is that the local sun-time is the same throughout a north-south path but not an east-west path. As a result magnetic storms which cause rapid and erratic variations in the magnetic field of the earth can affect radio propagation at these frequencies. Such effects may last as much as 100 hours.

16.17 PROPAGATION CHARACTERISTICS OF RADIO WAVES, 100 TO 535 kHz

For short, vertical antennas over ground of fairly good conductivity, the field strength of these waves can be obtained using modification of Eqn. (16.2) shown in Sec. 16.12. General characteristics of propagation of these waves is similar to those of longer wavelengths discussed in the previous section, except that in this case, the attenuation is much higher. As a result the ground wave can be received with sufficient signal strength only up to about 800 km, this distance being greater for the lower frequencies. At longer distances, one has to depend upon the sky wave, whose absorption is greater during day than at night. Besides the diurnal variation of the ionosphere, its seasonal and 11-year sun spot cycle variations also affect the sky wave propagation. Changes in the constants for the ground also materially affect the field strength at these frequencies, which is dependent to some extent on the latitude.

The frequency band 490–510 kHz assigned for mobile transmitters and particularly for distress signals (SOS or in Morse code, dit-dit-dit dah-dah-dah dit-dit-dit) on 500 kHz falls in this band.

16.18 PROPAGATION CHARACTERISTICS OF RADIO WAVES, 535 TO 1605 kHz

This band of frequencies was earmarked for commercial broadcasting in the 1920s as mentioned in Sec. 16.2. The choice of the lower end of the band at 535 kHz was dictated by the need to have sufficient separation from the 490-510 kHz band reserved for distress signals and mobile services. The higher limit was set by the following considerations.

In a radio set, for tuning a radio station, a variable capacitor C is used in parallel with a fixed inductance L (as per famous Marconi Patent No. 7777 of 1901). The capacitance variation is typically from 25 to 345 pF. When the capacitor is wired into a circuit, a constant wiring capacitance C_w = 15 pF (say), is added to the above values, giving

$$C_{min} = 25 + 15 = 40 \text{ pF and } C_{max} = 345 + 15 = 360 \text{ pF.}$$

Since resonance frequency of the tuned circuit is $f = \left(\frac{1}{2\pi\sqrt{L}}\right)\cdot\frac{1}{\sqrt{C}}$ it follows that the ratio of the limits of the frequency range of the tuned circuit will be

$$\frac{f_{max}}{f_{min}} = \sqrt{\frac{C_{max}}{C_{min}}} = \sqrt{\frac{360}{40}} = 3.$$

Hence, the upper limit of the frequency band is 535 × 3 = 1605 kHz.

All over the world the practice is to tell the frequency of a broadcasting station, because this is always a positive integer. Earlier this number was always divisible by 10, because the minimum separation between two radio stations was kept at 10 kHz. This gave a maximum of (1600-535)/10 = 107 different station frequencies. To accommodate more stations in the band, the current practice is to have a separation of 9 kHz. Hence the station frequencies are divisible by 9 (for instance, Bombay: 558 kHz, 1044 kHz; Pune: 792 kHz, 1602 kHz).

In this broadcast band frequency range, day time range is determined by the limit up to which the ground wave can be received with sufficient field strength, since daytime propagation via the sky wave is not possible, because of the complete absorption of these frequencies in the D-layer and the lower part of the E-layer during the day. This ground wave coverage region is called the primary service area of the broadcasting station. In this area the signal-to-noise ratio is sufficiently large to prevent interference by man-made static or atmospherics (see Sec. 16.22). The field strength has to be better than 5 mV/metre in heavily built-up residential districts in towns, but may be one- tenth this value in rural surroundings. The radius of the primary service area is typically 75 to 150 km. The region beyond this is called the secondary service area, in which the S/N ratio is not always adequate for noise-free reception, and where the sky wave also can reach the receiver.

Curves showing the field strength at different distances from the transmitter for the secondary service area often give the quasi-maximum value instead of the maximum value. This quasi-maximum value is that, below which the field strength never falls for 95% of the time.

Since the earth is a fairly good conductor over this frequency band, changes in the dielectric constant of the soil in different directions does not materially affect the field strength, but variations in conductivity between densely built-up areas in the city, and forests and marsh lands does affect it. In metropoliton areas with high-rise buildings, the ground wave is absorbed more rapidly than in surrounding open fields. The range of the ground wave that defines the primary service area is found to be greater at the lower frequencies in the broadcast band. Thus for the same transmitter power and the same soil, the same field strength is obtained at a distance of about 100 km at 550 kHz, as at about 60 km at 1500 kHz.

Reflections from the *E*-layer of the ionosphere at night gives a much larger secondary service area for the broadcast band stations. Beyond the primary service area, though the ground ray intensity decreases, the sky wave becomes more and more powerful and it soon begins to dominate over the weak ground wave. From Fig. 16.8 it is seen that the sky wave that reaches a more distant point is the one that leaves the transmitter, making a smaller angle with the horizontal. Since the vertical mast radiators used at these frequencies radiate much less energy at high angles from the horizontal, the sky wave received nearer the transmitter is weak. Its strength increases as we move further away from the transmitter, because it is a wave leaving transmitter at lower angle with horizontal. However, beyond a certain point, the field strength is found to decrease again with increasing distance. The reason is that this wave, though initially more powerful, travels a longer distance in the ionosphere. This reduces the reflection coefficient of the ionosphere to a low value like 0.25, so that only one-fourth the field strength is received (or 1/16 of the power). Moreover, the wave may have reached the receiver as a result of multi-hop transmission. The sky wave strength at a distant point increases rapidly after sunset as the *D*-layer disappears, but begins to decrease near sunrise time.

The sky waves are subject to fading because of changing ionosphere conditions. In the region where the intensities of the ground wave and the sky wave are nearly equal, this produces selective fading of the sidebands and this causes distortion in the output of the radio receiver. Because of this, if the distortion is unacceptable to the listener, then the primary service area will be found to be greater during daytime than at night. Increase in the transmitter strength does not improve the situation, because this increases the strengths of the ground wave and the sky wave in the same proportion.

16.19 PROPAGATION CHARACTERISTICS OF SHORT WAVES, 1.605 TO 30 MHz

At these frequencies the ground wave is useful only for such services as police radio, radio taxi service or ship-to-shore service in harbours. More useful coverage for broadcasting is almost exclusively obtained using sky waves for points beyond the skip distance zone around the transmitter.

For practical purposes, to avoid obstacles on the ground, the sky wave must leave the transmitter at an angle greater than about 3.5° above the horizontal. At this angle, the maximum skip distance is approximately 1700 km for the E-layer reflected ray and 3000 km for the F_2-layer reflected ray. For the case of short distances for which a flat earth may be assumed, the maximum usable frequency MUF is related to the skip distance Ds by a simplified formula

$$\text{MUF} = f_{\max} = f_c \sqrt{\left(\frac{D_s}{2h_i}\right)^2 + 1} \qquad \text{...(16.20)}$$

where f_c = The critical frequency of the layer

and h_i = The height of the layer of the ionosphere.

For broadcasts meant for listeners at very long distances like England to Australia or External Services of All India Radio, multi-hops transmission is used. In such cases, the ionospheric conditions at the top of each hop will always be different, the difference being greater if these points differ a lot in longitude, as their sun-times will be different.

Generally, the short wave services are designed for and operated with rays making angles between 10° to 25° to the horizontal. The frequency is generally chosen to be about 15% lower than MUF, so as to allow for variations in the conditions in the ionosphere. Usually different frequencies are used for daytime and night-time services. Thus higher frequencies in the 41 and 49 metre bands may be used during day and lower frequencies in 60, 75, and 90 metre bands may be used at night. The frequencies used during winter are lower than those used during summer.

16.20 PROPAGATION CHARACTERISTICS OF VHF, UHF AND MICROWAVES

These frequencies are used for TV and FM broadcasts, police radio, air to-ground communication, walkie-talkies, point-to-point communications systems (microwave links) of the posts and telegraphs department, railways, oil pipeline companies, radar for civil aviation and defence needs, as well as meteorology and ground station-to-satellite. These frequencies cannot be reflected by the ionosphere. Hence the communication is only with the help of a ground wave or a space wave. Highly directional antennas are used at transmitter and receiver for point-to-point communication within line-of-sight distances to reduce the requirement for antenna power at transmitter, just as a small bulb in an electric torch gives a bright illumination over a small surface.

Rain, fog and snow may introduce appreciable absorption at about 10 GHz. Water vapour at 66% humidity at a temperature of 16°C gives a very high absorption at 23 GHz, while oxygen in the atmosphere at a pressure of 150 mm of mercury gives significantly large absorption at 60 GHz.

16.21 TROPOSPHERIC SCATTER PROPAGATION

The atmosphere in the troposphere, particularly in the lower altitudes of 1 km to 3 km above ground cannot be considered to be homogeneous in the strict sense. There are numerous small regions where there are discontinuities in the dielectric constant, because of air masses at different temperatures and with different amounts of water vapour in them. These are the regions where clouds form and convection currents flow. These air masses have a reflection coefficient ρ for radio waves in the frequency bands above 30 MHz of the order of 0.0001 to 0.001. Though this value is extremely small, it is enough to provide a signal of sufficient strength at a distant point a few hundred kilometres away, by the vector addition of signals reflected (actually, scattered) by these numerous discontinuities. It must be remembered that at these low heights, there are hardly any electrons/ions present in large numbers, and therefore this mechanism is entirely different from the ionospheric reflection. VHF and UHF signals in the frequency range 40 MHz to 4 GHz can be used for such tropospheric scatter propagation at distances 50 km to 150 km beyond the radio horizon. It is found that vertical polarization provides a better communication link than horizontal polarization.

16.22 NOISE, ATMOSPHERICS AND MAN-MADE STATIC

The thermal agitation of atoms and electrons in a resistor produce a broad spectrum of electromagnetic waves, which produce a hissing sound in the output of a radio receiver. The rms value of this noise voltage is proportional to $\sqrt{TR}$, where T is the temperature in Kelvin of the resistor having a resistance of R ohms. The trouble with noise is that if it gets into the radio receiver along with a desired signal, then no amount of amplification in later stages can improve the signal-to-noise ratio, since both get amplified in the same proportion. The only remedy is to reduce the noise at the receiver input by use of smaller resistance in this part of the circuit and by operating it at lower temperature. It is for this reason that the Arvi Earth Station off Pune-Nashik road used by Videsh Sanchar Nigam for its satellite link, the input stage of the receiver is located close to the paraboloid dish antenna and operates at a temperature of 6 K by being immersed in a liquid helium bath.

Thunderstorms and lightning which are more numerous in the tropics, particularly within 10 degrees north and south of the equator, can be regarded as radio transmitters simultaneously emitting a large number of radio frequencies which travel to great distances and cause interference in radio reception, referred to as atmospherics. They produce all kinds of crackling and rumbling noises in receiver output. The only remedy to minimize their nuisance is to have a more powerful radio transmitter to give a strong signal.

In built-up areas in towns, operation of electric switches for lights and other domestic appliances as well as industrial machinery and electric traction, and sparks produced by spark plugs in petrol engines cause electromagnetic waves at all frequencies. They are referred to collectively as man-made static. These waves may travel to the set via the electric power mains or through space. The remedy for the first type is to install surge suppressors across the power line near the receiver. Since shielding will not only eliminate the noise coming through space, it will also eliminate the desired signal. Hence a better method is to use directional antennas.

A receiver in an aeroplane flying through a dust storm, rain or snow may experience a continuous hissing sound in its output, produced by the numerous electrically charged particles that strike its antenna in this situation and produce what is called precipitation static.

Since most of these noise sources can be regarded as amplitude modulated transmitters, there is no way of *completely* eliminating them, without at the same time completely eliminating the desired amplitude modulated broadcast signal. The only remedy is to use an entirely different type of signal, for instance, a frequency modulated signal as suggested by Armstrong. Very high quality with high fidelity and even stereophonic sound broadcasting is now possible with FM stations.

16.23 SUMMARY AND COMMENTS

In this chapter we have tried to consider the practical use of the knowledge gained in all the previous chapters. In the course of this study we came to know of the ionosphere and its various characteristics and how they affect the radio wave being propagated through it. The study of radio wave propagation involves a very large number of unknown quantities or parameters at different points along the path of the wave from the transmitter to the receiver. A very simple and elementary form of discussion has been attempted in this chapter only to make the reader acquainted with the various problems involved.

This elementary textbook will have provided very good grasp of the basic principles of electromagnetic field theory to enable the reader to embark on a study of a more advanced book on the subject.

EXERCISE 16

The following data may be used wherever necessary:

$$1 \text{ mile} = 5280 \text{ feet} = 1.609344 \text{ km}.$$

$$\text{Radius of Earth} = R_E = 6350 \text{ km}.$$

1. Show that if the power radiated by a transmitted is P_{kw} kW, then a result similar to Eqn. (16.2) will be that the electric field strength at a distance of 1 mile from the transmitter is $186.4\sqrt{P_{\text{kw}}}$ mV/metre.

2. If the refractive index μ of air is a constant in Eqn. (16.15), show that the modified refractive index μ_m will still vary with height, and that $\dfrac{d\mu_m}{dh} = 0.15748$ units/metre.

3. From Fig. 16.11(*c*), on the assumption that $h_t \ll R_E$, derive Eqn. (16.18).

4. If h_t and h_r (in fact) are the heights of the transmitting and receiving antennas above the ground, show that the maximum separation between them for line-of-sight transmission will be $D_{max} = \sqrt{2h_t} + \sqrt{2h_r}$ miles.
5. Assume that the heights h_t and h_r of the transmitting and receiving antennas are very small as compared to the distance between them, so that angle θ of the ray from the vertical is nearly 90°. If the transmitting antenna is a
6. An ionospheric layer has a maximum electron density of 5×10^{10} electrons/ metre^3. The virtual height of the layer is 100 km. Assuming a flat earth, find the maximum usable frequency for communicating with a receiver situated 100 km away.

Appendix
SI Units

After the French Revolution (1789), the French Academy of Sciences in 1793 sponsored the adoption of the system of units which came to be known as the Metric System, from the Greek word *metron* which means 'to measure'. The fundamental unit of length was called the metre and it was intended to be one ten-millionth of the quadrant (from north pole to equator) of the meridian passing through Paris. It is desirable to avoid the American spelling meter for the unit of length, and to use meter, only to denote an instrument for measuring any quantity.

Instead of the larger (or smaller) units being in multiples (or submultiples) of 2, in the metric system, the ratio of two units of a quantity is always a positive or negative integral power of 10. The purpose behind the adoption of this decimal system was the desire of the French Academy to serve all people, in all times.

The scientists began using the centimetre-gram-second (CGS) system of units, but during the next 100 years or so, it was discovered that workers in different fields like mechanics, thermodynamics, electricity, magnetism, acoustics, optics, illumination, etc. had developed their own secondary or derived units for many quantities. Conversion of such units used by one group into those used by another involved multiplying factors. Moreover, conversion of units between the electrical and the magnetic groups involved the velocity of propagation of electromagnetic waves in a vacuum. This was assumed to be 3×10^8 metres/sec., but more accurate determinations have shown the velocity to be less than this value (see Sec. 2.2).

The units centimetre and gram were too small for practical use in everyday life. In 1901 Professor Giovanni Giorgi of Rome University proposed the use of the metre and kilogram instead of the centimetre and the gram, because they were of more practical size, and their use eliminated many multiplying factors (powers of ten like 1 watt = 1 joule/sec. instead of 10^7 erg/sec). Another advantage was that, the metre and the kilogram were the primary prototype standards maintained at Sévres near Paris. Professor Georgi also recommended the adoption of Oliver Heaviside's recommendation to rationalize the units by putting a factor of 4π in the denominator of Coloumb's law [Eqn. (2.4)].

In 1954 CGPM (Conference Generale des Poids et Measures = General Conference on Weights and Measures) adopted a rationalized, coherent system of units based on six fundamental or base units. In 1960 the CGPM gave the name Systemé International d'Unites with the abbreviation SI to this system of units. These units do not involve any conversion factors based on the velocity of light. The second, the unit of time, which was originally related to the motion of the earth, is now derived from the atom, as accurate measurements have shown that the earth's motion varies slightly over long periods of time. Moreover, gravitational units (kgf = Kilogram-force) are avoided, because the value of gravitational acceleration varies from place to place.

The most important advantage of the SI units is that, the system is coherent. That is, if all the quantities in a formula are expressed in SI units, the answer will be in an SI unit *without any multiplying factor.*

BASE OR FUNDAMENTAL UNITS WITH THEIR SYMBOLS

1. *Length:* The meter (m) is the unit of length equal to 1 650 763.73 wavelengths in vacuum of unperturbed transition between the levels $2p_{10}$ and $5d_5$ of the krypton-86 atom (the orange-red line).
2. *Mass*: The kilogram (kg) is the unit having mass equal to the mass of the platinum-iridium prototype cylinder kept at the International Bureau of Weights and Measures at Sévres.
3. *Time:* The second(s) is the unit, equal to the duration of 9 192 631 770 periods of the frequency of radiation corresponding to the transition between the hyperfine levels $F = 4$, $m_f = 0$ and $F = 3$, $m_f = 0$ of the ground state $2s_{1/2}$ of the atom of caesium-133, unperturbed by external fields.
4. *Intensity of electric current:* The ampere (A) is the unit. It is that constant current which, if maintained in two straight, parallel conductors of infinite length, of negligible circular cross-section, and placed one metre apart in vacuum, would produce between these conductors a force equal to 2×10^{-7} newton (N) per meter of length.
5. *Thermodynamic temperature:* The kelvin (K) is the unit of thermodynamic temperature. It is the fraction $\frac{1}{273.16}$ of the thermodynamic temperature of the triple point of water.
6. *Luminous intensity:* The candela (cd) is the unit. It is the luminous intensity, in the perpendicular direction, of a surface of $\frac{1}{600\,000}$ square metre of a black body at the temperature of freezing platinum (= 2042 K) under a pressure of 101 325 N/metre^2.
7. A seventh base unit has been adopted after 1960 for the *Amount of substance.* It is called the Mole (unit abbreviation: mol). It is the amount of substance that contains as many elementary entities as there are atoms in 0.012 kg of carbon-12.

Secondary units

Quantity	*Unit*	*Abbreviation*
Plane angle	radian	rad
Solid angle	steradian	sr

Prefixes for metric multiples and submultiples

Multiplying factor	*Prefix*	*Pronunciation*	*Symbol (used as a prefix)*
10^{18}	exa	exa	E
10^{15}	peta	peta	P
10^{12}	tera	tera	T
10^{9}	giga	ji ga	G
10^{6}	mega	mega	M
10^{3}	kilo	kiil$\bar{o}$	k
10^{2}	hecto	hekt$\bar{o}$	h
10	deka	deka	da
10^{-1}	deci	desi	d
10^{-2}	centi	senti	c
10^{-3}	milli	milli	m
10^{-6}	micro	mi kr$\bar{o}$	μ
10^{-9}	nano	nan$\bar{o}$	n
10^{-12}	pico	pek$\bar{o}$	p
10^{-15}	femto	femt$\bar{o}$	f
10^{-18}	atto	att$\bar{o}$	a

Phonetic scheme for pronunciation : a h, d i p, f i g, j $\bar{o}$ ke, m an, m $\bar{e}$ te, m $\bar{i}$ te, r *e* ckl *e* ss, t *a* -t *a* , t *a* x.

Metric Mnemonics for remembering symbols in last column: *Ascending above dotted line* "DAvid Has Killed Mighty Goliath, The Philistine Enemy."

Descending below dotted line: "Don't Carry Malignant Microbes Near Patients Fighting Ailments".

NOTES

1. The symbols for units in SI units must be written exactly as has been given. No individual in the world has been given the authority to alter then as the fancy takes him. Thus write "kilogram" as kg (and not as Kg or KG or KGM or kgm)

and "kilohertz" as kHz (and not as KHz or KHZ) and "millilitre" as ml (and not as ML).

2. Do not add suffix "s" to make plurals to symbols. Thus write "five metres" as 5 m (and not as 5 ms, because ms will denote "metre" multiplied by "second").
3. When writing the name of a unit in full, capital letter should not be used in the beginning, even if the unit is named after a scientist. Thus write 3.5 newton (and not Newtons) or 40 watt (and not Watt).
4. The word "degree" or small circle should not be used with thermodynamic temperature. Thus triple point of water is 273.16 K (and not 273.16 degree kelvin or 273.16°K). For practical purposes, the temperature may be given in degree Celsius or °C. Thus the triple point of water can be written as + 0.01 °C and the melting point of ice as 0°C or 273.15 K.
5. The litre was called a "unit of capacity" and originally defined as the volume occupied by one kilogram of water at 4°C (temperature corresponding to maximum density). This volume is equal to 1.00 028 dm^3. However, in SI in 1964 this definition has been abandoned and now litre is only another name for one cubic decimetre. Hence one millilitre is now exactly equal to one cubic centimetre (1 ml = 1 cm^3).
6. Numbers involving large number of digits must be written by arranging the digits on either side of the decimal point, in groups of three, with space between groups, but no comma (as this may create confusion with a decimal point). Thus, the accurate value of the velocity of propagation of electromagnetic waves in a vacuum is 299 792 457.4 ± 1.1 m/s.

INDEX